T0215020

Symmetry and Physical Properties of Crystals

Grenoble Sciences

The aim of Grenoble Sciences is twofold:

- to produce works corresponding to a clearly defined project, without the constraints of trends nor curriculum,
- to ensure the utmost scientific and pedagogic quality of the selected works: each project is selected by Grenoble Sciences with the help of anonymous referees. In order to optimize the work, the authors interact for a year (on average) with the members of a reading committee, whose names figure in the front pages of the work, which is then co-published with the most suitable publishing partner.

Contact: Tel.: (33) 4 76 51 46 95
E-mail: grenoble.sciences@ujf-grenoble.fr
Website: https://grenoble-sciences.ujf-grenoble.fr

Scientific Director of Grenoble Sciences
Jean Bornarel, Emeritus Professor
at the Joseph Fourier University, Grenoble, France

Grenoble Sciences is a department of the Joseph Fourier University
supported by the **ministère de l'Enseignement supérieur
et de la Recherche** and the **région Rhône-Alpes**.

Symmetry and Physical Properties of Crystals is an improved version of the original book *Symétrie et propriétés physiques des cristaux* by Cécile Malgrange, Christian Ricolleau and Françoise Lefaucheux, EDP Sciences, Savoirs Actuels Series, 2011, ISBN 978 2 7598 0499 3.

The Reading Committee included the following members:

- **Ryszard Cach**, Professor, University of Wrocław (Wrocław, Poland)
- **Vladimir E. Dmitrienko**, Chief Researcher, A.V. Shubnikov Institute of Crystallography (Moscow, Russia)
- **José Goulon**, Emeritus CNRS Senior Researcher, former group leader at the European Synchrotron Radiation Facility (Grenoble, France)
- **Hiroo Hashizume**, Emeritus Professor, Tokyo Institute of Technology (Tokyo, Japan)
- **Jochen Schneider**, Professor, former DESY Research Director (Hamburg, Germany), former director of the Experimental Facilities Division for the LCLS (Stanford, USA)

Translation from original French version performed by Michel Schlenker; typesetting: Stéphanie Trine—special thanks to Konstantin Protassov and Alexei Voronine; figures: Sylvie Bordage, Alain Jeanne-Michaud; cover illustration: Alice Giraud, with sample of pyrite: mineral collection of Pierre and Marie Curie University, photograph Jean-Pierre Boisseau

Cécile Malgrange · Christian Ricolleau
Michel Schlenker

Symmetry and Physical Properties of Crystals

 Springer

Cécile Malgrange
Université Pierre et Marie Curie
Institut de Minéralogie, de Physique des
  Matériaux et de Cosmochimie
case 115, 4 place Jussieu
75252 Paris cedex 05
France

Michel Schlenker
Université Grenoble-Alpes
Grenoble-INP et Institut Néel du CNRS
B.P. 166
38042 Grenoble cedex
France

Christian Ricolleau
Université Paris Diderot-Paris 7
Laboratoire Matériaux et Phénomènes
  Quantiques
Bâtiment Condorcet, case 7021
75205 Paris cedex 13
France

Translated, revised and adapted from "Symétrie et Propriétés Physiques des Cristaux", Cécile Malgrange, Christian Ricolleau et Françoise Lefaucheux, EDP Sciences/CNRS Éditions 2011

ISBN 978-94-024-0177-6   ISBN 978-94-017-8993-6   (eBook)
DOI 10.1007/978-94-017-8993-6

Springer Dordrecht Heidelberg New York London

Printed on acid-free paper

Springer is part of Springer Science+Business Media (www.springer.com)

In memory of Hubert Curien (1924–2005), Professor at Pierre & Marie Curie University, Paris, and member of the French Academy of Sciences. A great administrator, Minister of Research in the French government, he shared his enthusiasm for crystallography with his students through the courses he gave throughout his career.

Foreword

Apart from amorphous bodies such as glasses, the arrangement of ions, atoms and molecules in solid matter is regular and presents properties of symmetry. Crystals present both translational and orientational properties of symmetry, which govern their physical properties. It is because of their triple periodicity that crystals grow with a regular habit: quartz prisms, diamond octahedra or garnet rhomb-dodecahedra, and that they present planar cleavage faces, those of the cube for rock-salt, and those of the rhombohedron for Iceland spar. It is this same periodicity that is at the origin of the band structure of solids and conditions the properties of materials used in microelectronics. It is thanks to the redundancy of the information provided by the diffraction in phase of X-rays, neutrons or electrons, by all the molecular groups repeated by the translations of a crystal that one can determine their atomic structure. If one wishes to find the structure of a complex molecule such as a protein or a virus, one should therefore crystallize it. This is done systematically during the elaboration of drugs by the pharmaceutical industry. The orientational symmetry of crystals is present in all their properties, first of all in their external shape. But it is often because a particular element of symmetry is absent that a property may exist. It is because hydrated ammonium tartrate has no centre of symmetry that it is optically active, which Louis Pasteur associated to its external shape, right or left, thus discovering chirality (1848). It is because of symmetry considerations that the Curie brothers, Jacques and Pierre, looked for, and found, piezoelectricity in 1880. If one cools barium titanate below 120°C, its structure is modified. Above, it is cubic holohedral, below, it is tetragonal, with as its only symmetry elements a four-fold axis and mirrors parallel to that axis. It is this symmetry which makes it possible for a spontaneous electric polarisation to appear, and for the crystal to become pyro- and ferroelectric. Generally speaking, the properties associated to structural phase changes depend on the relations between symmetry groups and their subgroups.

The description of the symmetry operations of crystals and of the resulting physical properties is the aim of the present book. These operations are part of a complex system which was not grasped at once. It was a long way from Aristotle's elements to the concept of space lattice. The ancients were always fascinated by the regularity of six-sided prisms of rock-crystal, as noted by

Pliny the Elder in his *Natural History* (77 A.D.). Kepler (1611), followed by Descartes (1637) and Bartholin (1661) tried to interpret the formation of six-cornered snowflakes by the agglomeration of six spherules round a seventh one. Kepler was the first to describe the primitive and face-centred cubic lattices and the hexagonal lattice, although he did not call them that way. Compact pile-ups were considered similarly by Hooke (1665) and Huygens (1678) to explain the shape of quartz and the cleavage rhombohedron of calcite, respectively. The constancy of interfacial angles was established in the seventeenth and the eighteenth centuries by Steno (1669), Carangeot (1780) and Romé de l'Isle (1783). At the end of the eighteenth century, Bergman (1773) for calcite and Haüy (1784) in a general way, showed that the external shape of crystal may be obtained by a regular and triply periodic pile-up of identical elemental parallelepipeds, now called unit cells, and called *molécules intégrantes* by Haüy. But Haüy confused the unit cell with its chemical content and did not accept the notion of isomorphism introduced by Mitscherlich (1819). His classification of crystals was based on their geometrical shape while that of the German School, led by Weiss (1817), was based on systems of symmetry axes. Haüy's main merit was nevertheless to have introduced the notion of three-dimensional periodicity which defines the crystalline medium and to have established the geometrical properties of lattice planes by means of the law of simple rational intercepts. The relations between symmetry and properties were established in steps all along the nineteenth century: the 7 crystal systems (Mohs, 1822), the notion of point lattice (Seeber, 1824), the 32 crystal classes (Frankenheim, 1826; Hessel, 1830), the notion of hemihedry (Delafosse, 1840), molecular chirality (Pasteur, 1848), the 14 Bravais lattices (1850), groups of motion including helicoidal motions, the 65 chiral groups (Sohncke 1865), and, finally, the 230 space groups (Fedorov, 1890; Schoenflies, 1891; Barlow, 1894).[1]

The description of physical properties in their geometrical form requires in general other mathematical tools, such as tensors, which were first introduced in elasticity by Voigt (1899) and Brillouin (1949). If one pulls a rod made of an isotropic material, its length increases and its cross-section decreases. A scalar is insufficient to describe the response of the material to an excitation and two constants are necessary to describe the resulting deformation. Similarly, the polarisation of an anisotropic dielectric can only be expressed by way of a tensor.

The book by Cécile Malgrange, Christian Ricolleau and Michel Schlenker is an extended translation of a first version in French by Cécile Malgrange, Christian Ricolleau and Françoise Lefaucheux. It comprises three parts of roughly equal importance: 1) a first part on crystal symmetry, symmetry operations, crystal lattices, point groups and space groups; 2) a second part on crystal anisotropy and tensors applied to elasticity, propagation of elastic waves in crystals and piezoelectricity; 3) a third one on the optical properties of crystals, polariza-

---

1. A. Authier, *Early Days of X-Ray Crystallography* (Oxford University Press, Oxford, 2013)

tion, optical activity and electro-optical effects. The various types of chemical bonds are also described in a small chapter between the first and the second parts. Most chapters include an annex on a particular topic related to one of the points developed in the chapter, aimed at exciting the curiosity of the reader, and these complements are a real added value to the book. Exercises are also included, the solutions of which are given at the end of the book. They constitute a very valuable pedagogical tool. The whole of the book is written in a very clear and rigorous way.

The publication of a new, up-to-date, textbook of crystallography and crystal physics will be very welcome at a time when crystallography courses are disappearing in most universities throughout the world, although these disciplines represent basic knowledge compulsory for anyone doing material science, Earth science, biochemistry, solid state physics or chemistry.

André Authier

Professor Emeritus, Pierre & Marie Curie University, Paris
Former President of the International Union of Crystallography
Member of the German Academy of Sciences

Preface

Crystallography constantly enjoys a new youth, thanks in particular to the spectacular progress in X-ray sources (synchrotron radiation from dedicated facilities and, very recently, free electron X-ray lasers), which make it possible to investigate an extremely broad range of materials. Its contribution is vital in many scientific and technological areas. It makes it possible, for example, to understand, and even to predict the properties of materials as diverse as high-T_c superconductors, high performance magnets, or biological materials such as proteins. The pharmaceutical industry invests large funds to set up data banks of structures, determined in the crystallized state, of molecules which could become medicines, in order to correlate their configuration to their healing action. Crystallography also is very important in nanoscience, since some materials, when prepared in the form of nano-objects, take on crystal structures which do not exist in the bulk and which necessarily affect their physical and chemical properties. Geophysicists obtain major information about the Earth mantle by investigating crystallized materials under extreme temperature and pressure conditions. In technology, we must mention structural wonders such as thin film based artificial crystals (multilayers and superlattices) of semiconductors or magnetic materials. These have for several years been present in the giant or tunnel magnetoresistance read heads of the hard drives in all our computers.

On the other hand, understanding the relation between the structure and the macroscopic properties of crystals opens the way to the use of physical properties with many applications. Thus piezoelectricity, *i.e.* the production of an electrical voltage under the effect of stress, is the basis of our quartz watches and plays a central role in cell phones. Electro-optical effects, the change in refractive indices under the effect of an electric field, entail a change in light propagation in materials.

This book is mostly a translation of *Symétrie et propriétés physiques des cristaux*, published by EDP Sciences, Orsay, France in 2011. It provides the fundamentals underpinning all these research and application areas.

In the first part, it describes in a clear and thorough way the periodicity of crystals and their symmetry, both at the microscopic level (space groups) and at the macroscopic level (point groups). It includes a presentation of the crystallography of two-dimensional systems such as surfaces and interfaces.

The second part is dedicated to various physical properties of crystals. They are described in the tensor formalism which is natural for them, emphasizing their link with crystal symmetry. We show in particular how a given property is allowed or forbidden by various crystal symmetries, and how the coefficients of the characteristic tensor are related for the various point groups. We deal not only with basic properties such as elasticity and birefringence, but also with properties which are less often dealt with, such as piezoelectricity, optical activity or the electro-optical effects.

Each chapter starts with a short presentation of its contents, and almost all chapters are complemented by exercises through which the notions introduced can be concretely applied, and thus mastered. For these exercises to be a real studying tool, the solutions are explicitly given as a final separate Chapter 20.

Many of the chapters also include a brief complement, intended to give a glimpse into an area related to that of the chapter and to whet the reader's scientific appetite, for example on how and what for crystals are grown, with what probes crystal structures can be determined, etc.

This book originates in courses we gave at Paris Diderot University and at Grenoble Institute of Technology, France. It is aimed, with no special pre-requisite, at students in crystallography, physics and engineering (especially materials science and electrical engineering) curricula, and at PhD students. It should also be a reference book for researchers and engineers involved, in their daily activity, in the properties of crystals.

The authors sincerely thank the five members of the Advisory Editorial Committee for their very valuable suggestions, the vast majority of which were followed. They are also very grateful to Professor André Authier for writing the Foreword, which actually is a very informative chapter on the history of crystallography.

The authors intend to correct the mistakes and typos which will, in spite of their efforts, have remained in this first edition. Errata sheets will be provided at

https://grenoble-sciences.ujf-grenoble.fr/ouvrage/symmetry-crystals

Comments from readers are very welcome. Please write to

symmetry-book@impmc.upmc.fr

Cécile Malgrange,
Christian Ricolleau,
Michel Schlenker.

Table of symbols

$\mathbf{a}, \mathbf{b}, \mathbf{c}$	Basis vectors of the unit cell in direct space
a, b, c	Lengths of the basis vectors of the unit cell in direct space
$\mathbf{a}^*, \mathbf{b}^*, \mathbf{c}^*$	Basis vectors of the reciprocal unit cell
a^*, b^*, c^*	Lengths of the basis vectors of the reciprocal unit cell
v_0	Volume of the direct unit cell
v_0^*	Volume of the reciprocal unit cell
$[uvw]$	Lattice row parallel to the vector $u\mathbf{a} + v\mathbf{b} + w\mathbf{c}$
$< uvw >$	Set of lattice rows equivalent to $[uvw]$ through all the symmetry operations of the crystal group
(hkl)	Family of lattice planes with Miller indices h, k and l
$(hkil)$	Family of lattice planes with Miller indices h, k, i and l for a hexagonal lattice
$\{hkl\}$	Set of lattice planes equivalent to (hkl) through the symmetry operations of the crystal group
$\{hkil\}$	Set of lattice planes equivalent to $(hkil)$ through the symmetry operations of the crystal group, for a hexagonal lattice
d_{hkl}	Distance between two neighboring (hkl) lattice planes, $i.e.$ the lattice spacing of family (hkl)
$\mathbf{e}_1, \mathbf{e}_2, \mathbf{e}_3$	Basis vectors of an orthonormal frame
$\{M_{ij}\}$	The matrix for which M_{ij} is the coefficient located on the i-th row and the j-th column
$\{a_{ij}\} = A$	Transition matrix for going over from an orthonormal frame to another one
$\mathbf{V}$	Polar vector
$\overset{\smile}{\mathbf{V}}$	Axial vector
$[T]$	Tensor
T_{ij}	Components, in a given orthonormal axis system, of a rank-2 tensor $[T]$. The associated matrix is noted T.
T_{ijk}	Components, in a given orthonormal axis system, of a rank-3 tensor $[T]$

e_{ij}	Components of the tensor of displacement gradients, or distorsion tensor, $[e]$
T	The set of the crystal lattice translations
G_s	Space group
G_p	Point group
$[S], S_{ij}$	Strain tensor, and its components
$[T], T_{ij}$	Stress tensor, and its components
S_α, T_α	Contracted notation (Voigt's notation) for the tensors $[S]$ and $[T]$. α ranges from 1 to 6, so that the components of these tensors form column vectors with 6 rows.
$[s], s_{ijkl}$	Compliance tensor, and its components
$[c], c_{ijkl}$	Stiffness tensor, and its components
Θ	Temperature
$\mathcal{S}$	Entropy
E	Electric field
D	Electric induction
P	Polarization
$[\varepsilon], \varepsilon_{ij}$	Relative permittivity (or dielectric constant) tensor, and its components
ε_0	Permittivity (or dielectric constant) of vacuum
$[\eta], \eta_{ij}$	Electric impermeability tensor, and its components
δ_{ij}	Kronecker symbol
δ_{ijk}	Levi-Civita, or permutation, tensor
$[G], G_{ij}$	Gyration tensor, and its components
$[d], d_{ijk}$	Piezoelectric tensor, and its components
$[r], r_{ijk}$	Linear electro-optic tensor, or Pockels tensor, and its components
$[z], z_{ijkl}$	Quadratic electro-optic tensor, or Kerr tensor, and its components
$[p], p_{ijkl}$	Elasto-optic tensor and its components
$[\pi], \pi_{ijkl}$	Piezo-optic tensor and its components

Contents

Chapter 1

Introduction

This chapter introduces the spatial periodicity and the macroscopic scale symmetry properties which characterize crystals.

1.1. Crystal order

Depending on temperature and pressure, all materials take on three different states: solid (S), liquid (L) and gas (G) (Fig. 1.1). However, when looking at matter from a microscopic point of view, there are only two really distinct states: the ordered state (grey area on Fig. 1.1) and the disordered state. It is clear that a gas at low pressure, consisting of a set of molecules separated by distances that are large compared to their size, is totally disordered. In a liquid, the molecules are in contact with one another, like marbles packed in a bag. A liquid is not completely disordered: there is some order, but only at short range.

Furthermore, it is possible to go continuously from the gas state to the liquid state. At a temperature Θ between the temperature Θ_T of the triple point and that Θ_C of the critical point, a gas undergoing compression goes from the gas state (A) to the liquid state (B) through a sharp transition, liquefaction or condensation. But it is possible to go from A to B around the critical point, for instance by going from A to D at constant pressure, then from D to F at constant temperature, and then from F to B at constant pressure. In contrast, it is impossible to go from the disordered area to the ordered area without crossing the line, shown as a thick line on the figure, that clearly separates the ordered and the disordered domains. This line consists of two parts: one corresponds to sublimation (solid–gas transition) and to condensation for the inverse transformation, the other to melting (solid–liquid transition) and to solidification for the inverse transformation.

The equilibrium diagram of Figure 1.1 applies to most solids and their transition to a disordered state, either liquid or gas. However, for some solids, the liquid–solid transition is not sharp but continuous: this is the case for amorphous or glassy solids. When the temperature of such a material in liquid form is lowered, its viscosity increases, which means that its flow is more and

© Springer Science+Business Media Dordrecht 2014
C. Malgrange et al., *Symmetry and Physical Properties of Crystals*,
DOI 10.1007/978-94-017-8993-6_1

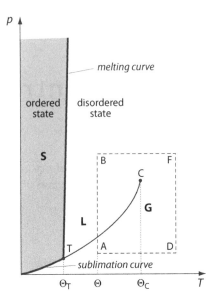

Figure 1.1: Equilibrium diagram for a pure substance, in a representation where pressure p is shown as a function of temperature T

more difficult. It forms a kind of paste, which ends up being solid but with a structure hardly more ordered than that of a liquid: it is a kind of set liquid – which corresponds to a metastable state.

This book deals only with crystalline solids, characterized by long-range order. The size of the ordered regions depends on the conditions in which crystallization occurs. If no special care is exercised when solidifying a pure liquid, microcrystals appear at various points of the liquid, grow and end up being contiguous. Each small crystal is perfectly ordered, but there is no reason for these crystals to have the same orientation. We get a polycrystal, *i.e.* a solid made up of small crystals (grains) with the same structure, but differently oriented. This is the most prevalent case for materials and in particular for metals (Fig. 1.2). However, some minerals occur in single crystal form, *i.e.* in a form where the solid is identically oriented over its entire volume. This is the case for the gems of jewelers, and for some beautiful minerals which can be found in private or public collections[1] (Fig. 1.3a). On the other hand, some crystals can be manufactured in the form of quite large single crystals. This is the case for silicon or quartz, which are produced as single crystals several tens of centimeters in size (Fig. 1.3b).

---

1.   All museums of natural history have a section on minerals. If you visit Paris (France), you will enjoy in particular the mineral collection of Pierre & Marie Curie University (UPMC) and of the Institut de Minéralogie et de Physique des Milieux Condensés (IMPMC), located at 4 place Jussieu, in the 5th district (arrondissement).

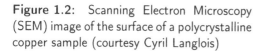

Figure 1.2: Scanning Electron Microscopy (SEM) image of the surface of a polycrystalline copper sample (courtesy Cyril Langlois)

Figure 1.3: (a) A sample of pyrite (iron sulfide, FeS_2), mineral collection of Pierre & Marie Curie University (Paris 6) (photograph by Jean-Pierre Boisseau); (b) two bars of single crystal synthetic quartz produced by the SICN company (photograph by Alain Jeanne-Michaud)

1.2. Order on a macroscopic scale

The existence of crystal order was sensed in the 17th century at the macroscopic scale, and recognized at the beginning of the 20th century at the microscopic scale. For instance, Niels Stensen, or Nicolaus Stenonius, or Steno, a Dane, noticed as early as 1669 that the angles between faces of natural crystals of quartz and some other minerals have a well defined, constant value, independent of their origin. He then spelled out the law of constancy of angles, which was to be generalized to all crystals by Romé de l'Isle in 1783. This constancy of angles also shows up on the small crystals which form naturally during solidification.

The mineralogist René Just Haüy showed in 1784 how this regularity in angles can be explained. He had noticed that, when he would cleave[2] a crystal of calcite, he would obtain smaller and smaller rhombohedra, all with the same

---

2. When a crystal of calcite is broken, the fracture surface features perfectly plane areas, with a restricted number of orientations. If one attempts, using a cutting tool and a small hammer, to cleave the crystal, *i.e.* to break it along one of these special areas, one obtains, provided the blade is properly oriented and the experimentalist has some experience, two fragments separated by a beautiful plane suface known as a cleavage surface.

angles. He then imagined that a final stage, *i.e.* a small elementary rhombohedron which can no more be cut, should eventually be reached. He deduced that the macroscopic crystal should be formed from the regular stacking of small rhombohedra, all identical to this elementary rhombohedron which he called the "integrating molecule". He then showed, and this is the strong point of his theory, that he could reconstruct all external forms of crystals by stacking identical elementary parallepiped-shaped blocks into successive layers which he shifted by a constant, integral number of units at each layer (Fig. 1.4). The parameters defining the basic parallelepiped were characteristic of the crystal. The notions of periodicity and of crystal unit cell, which we will define later, thus appear.

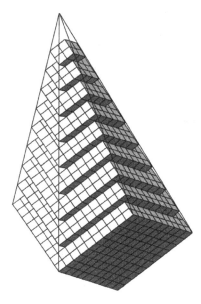

Figure 1.4: A stacking of layers of identical parallelepipeds, regularly shifted to reproduce the actual habit of the crystal

1.3. Order on a microscopic scale

After Haüy and until 1912 the concepts of symmetry and of internal periodicity in the crystal developed. However it was not evident, or at any rate not proved, that this internal order, *viz.* a triply periodic stacking, could exist in materials with any external shape.

It was in fact through the discovery of X-ray diffraction by Max von Laue in 1912, and thanks to the work of William and Lawrence Bragg, that the existence of a periodic stacking of atoms in solids, as predicted by Haüy, could be evidenced. They showed that X-rays are diffracted by crystals along special directions, forming a discrete set of spots which could be interpreted only through a triply periodic stacking of the atoms. Thus crystalline solids can be represented using a crystal motif,[3] comprising a finite number of atoms, periodically repeated along three non-coplanar directions with periods along these directions respectively a, b and c.

Let us discuss a few examples to illustrate this notion of periodicity. Figure 1.5 shows a frieze formed by repeating, with period a, a motif. Figure 1.6 shows a sample of wallpaper featuring periodicity in two directions (periods **a** and **b**).

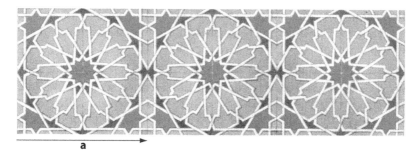

Figure 1.5: A decorative pattern from Arabic art, after J. Bourgoin, *Arabic Geometrical Pattern and Design* (Dover Publications, New York, 1973)

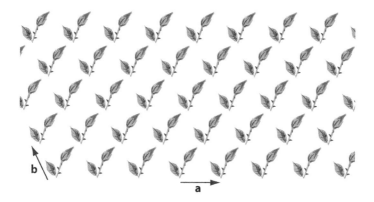

Figure 1.6: A sample of wallpaper

A crystal is thus formed from a motif consisting of a set of atoms. Repeating the motif through a translation **a** leads to a kind of linear crystal. If this linear crystal is submitted to a translation by a vector **b** (not collinear with **a**), then by 2**b**, 3**b**, etc., a plane crystal is obtained. To obtain the global crystal, the plane crystal is then thought to be submitted to a translation by a vector **c** not coplanar with plane (**a**, **b**), then by 2**c**, 3**c**, etc.

---

3. *Motif* is the French word for pattern. It is adopted in crystallographic English, with the present meaning.

1.4. Basic assumptions of geometrical crystallography

In order to describe crystal structures, we make three assumptions:

1. The crystal is assumed to be infinite. This assumption is reasonable, since the order of magnitude of the lengths a, b and c of the period vectors is a few angströms (1 Å $= 10^{-1}$ nm). A small crystal with dimensions of the order of a micrometer, for example, contains linear sequences several thousands of periods long.

2. The thermal vibrations of atoms about a perfectly defined average position is neglected. Our description assumes that the atoms are fixed at these average positions.

3. The crystal is assumed to be perfectly triply periodic, *i.e.* not to feature any defect. Crystal defects can then be described with respect to the perfect crystal. The study of crystal defects is a broad subject, outside the scope of this book. A brief introduction to one type of defect, the dislocation line, is given in Complement 13C.

1.5. Anisotropy of physical properties

The beautiful crystal faces that can be observed in natural minerals and in some synthetic crystals are related to the very strong anisotropy of the growth rates as a function of growth direction. Figure 1.7 explains how the only faces that remain after some growth has occurred are those perpendicular to the directions where the growth rate is much smaller than in other directions. We consider, at time $t = 0$, three face orientations denoted as 1, 2 and 3, with one of them (noted 2) growing at a velocity v_2 twice larger than the two others (faces 1 and 3). We see that at time t all faces are present, whereas, at time $2t$, face 2 has disappeared.

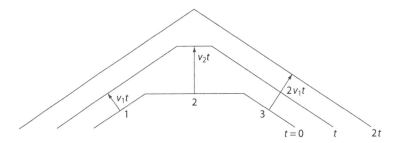

Figure 1.7: Growth at different rates of crystal faces. Face 2, with growth velocity v_2 twice larger than that, v_1, of faces 1 and 3, ends up disappearing.

The habit, *i.e.* the very regular special shape, of crystals thus shows that the growth velocity is much smaller along certain directions, finite in number, than in all other directions. This is one of the expressions of the anisotropy of the physical properties of crystals. In this case, anisotropy is discontinuous while, for most other physical properties, it is continuous, meaning that the value of a physical quantity such as the expansion of a material along a given direction, under the influence of a rise in temperature, varies continuously as direction is varied. This anisotropy of their physical properties is a characteristic of crystals. Describing it involves the use of tensors. This will be the topic of the second part of this book, starting with Chapter 9.

1.6. Remarks on the scope of this book : topics left out

The aim of this book is to describe the structure and properties of perfectly ordered crystalline solids. We have seen that we therefore leave aside glasses or amorphous solids, the structure of which is close to that of a liquid set at a given instant [1]. Neither shall we mention liquid crystals or mesophases [2]. They often consist of rod-shaped molecules, which assemble parallel to one another, forming plane layers. In a given layer, the molecules may be ordered only at short range, as in a liquid, but the existence of these parallel layers denotes partial order, whence the term crystal.

Neither will we deal with quasi-crystals, discovered in 1984 [3] (see also [4]). They can be obtained through ultra-fast quenching of metallic alloys (e.g. Al–Mn–Si or Al–Fe–Cu). They do not feature the triple periodicity of crystals. They do however, like crystals, feature orientation symmetry: all properties observed along a given direction are reproduced along the directions deduced from the former through the action of the symmetry elements in the point group of the quasi-crystal. Because there is no translation symmetry, these point groups are different from those that are consistent with the crystalline state.

Complement 1C. Crystal growth

This complement is no more than a glimpse into the fascinating art and science of crystal growth. The variety of techniques and of characterization methods it encompasses is exemplified by a recent handbook of crystal growth (see Further reading): it comprises over 1800 pages, with a short table of contents spanning 14 pages and a detailed table of contents of 30 pages.

1C.1. Natural growth of single crystals

The largest producer of single crystals is nature. Large (centimeter-range) single crystals of many naturally occurring compounds, usually called minerals, have long been valued for their beauty and have raised many questions that led to understanding the structure of matter. Their characteristic habits, with well-developed natural faces, occur when their growth process involves isotropic enough an environment as well as enough space. The typical process, hydrothermal growth, is based on the slow deposition of a solute as the temperature of the solution changes, and it is often associated with high pressures. Among the most popular examples of minerals are quartz SiO_2 (rock crystal), calcite $CaCO_3$ (Iceland spar), diamond C, pyrite FeS_2 (fool's gold, Fig. 1.3a). They may well, through their beauty, have kindled the reader's interest for crystals and symmetry. Other materials crystallize very easily, at room pressure and room temperature, from solutions: this is the case for sodium chloride NaCl (rocksalt), or copper sulfate $CuSO_4$. Crystallization is then a convenient way of purifying the material.

1C.2. Crystal growth for scientific purposes

Growing single crystals is an important support activity for basic research. A meaningful investigation of anisotropic macroscopic properties, for example magnetism, electrical or heat conduction, or the ferroelectricity (see Sect. 15.2) of materials, requires single crystals, where the directional properties are the same throughout the sample. The growth must then be followed by the precise determination of the orientation of the crystal, so that a sample with well-defined crystal directions can be cut out. The laboratory techniques cover a broad range of approaches, depending on the chemical properties, melting temperature, etc. of the material. Thus soluble materials can often be made to crystallize in a controlled way by slow evaporation of the solvent, while refractory materials will have to be molten in an induction furnace, often in the form of a levitating drop without contact with a crucible to avoid contamination, and cooled in a temperature gradient. The Bridgman method provides for cooling of a larger molten sample, in contact with a small crystal used as a seed, in a temperature gradient. Another broadly used technique is the Czochralski method, where the seed is dipped into the melt and slowly pulled out at the same time that it is rotated. In the flux-growth method, the material to be crystallized is molten together with another material which is not incorporated in the solid but lowers the melting temperature; with proper setting of the temperature evolution over days or weeks, good crystals can be obtained.

From another point of view, single crystals are necessary for many structure determinations using X-ray or neutron diffraction techniques. However, detailed examination of the validity of the usual data processing procedures ("kinematical approximation", if necessary with corrections) indicates that the crystals should not be big. Typical optimal sizes are tenths of a millimeter for X-rays, and a few millimeters for neutrons. Furthermore, nearly perfect crystals are not desirable for this kind of study, so that the requirements are, happily, close to what crystal synthesis produces. The preparation of single crystals of biological material, such as proteins, in their native hydrated form, is a crucial step toward the determination of their structure. The approach usually consists in trying many recipes at the same time, using boxes with tens of compartments which are filled with drops of the material with various values of the pH or other possibly relevant parameters. The growth of crystals of biomaterials was investigated in microgravity conditions in satellites, in order to eliminate the influence of convection. A similar quest is the motivation for crystal growth in gels.

1C.3. Industrial crystal growth

It may appear paradoxical to find that almost perfect crystals, the production of which is almost invariably fraught with difficulties, are required by industry rather than by basic research. Semiconductor device production is based on the preparation of many chips per single-crystal silicon substrate. They are cut from huge single crystal ingots, typically 300 mm in diameter in 2009 and a meter long, without any linear defect (dislocation line, see Complement 13C). This achievement in crystal growth technology was the result of large efforts toward a thorough understanding of crystal defects. These efforts were undertaken because it was vital to keep at a very low level the number of defective chips grown on a wafer. One side-effect of the mastery achieved in the preparation and processing of silicon crystals was that elaborately shaped parts like inkjet printer nozzles were made out of silicon single crystals.

Crystals with intermediate sizes are grown industrially for applications in optoelectronics (for example lithium niobate $LiNbO_3$ – see Sect. 19.2). Other materials routinely produced in the form of excellent crystals, but comparable neither in quantities nor in quality with silicon, are SiO_2 (quartz) and $AlPO_4$ (berlinite) for piezoelectric oscillators (see Chap. 15), and SiC (silicon carbide) for electronic devices.

Industrial single crystals of diamond are available in a wide range of crystal quality and crystal size, with use as different as oil drill rigs and optical elements (monochromators) for synchrotron radiation X-rays. Synthetic diamond does not reach into the jewelry range yet, but other gemstones (emerald, sapphire...) are routinely produced artificially as substitutes for their natural counterparts. Again in a very different area, turbine blades for jet planes are made of single crystals of an aluminum-base alloy in view of their mechanical strength at the high temperatures involved.

1C.4. Growth of single-crystal thin films

Films with very small thickness, down to the nanometre range, are important for nanotechnology and basic science, e.g. for the investigation of semiconductor quantum well or quantum dot devices, or of magnetic recording and readout heads for hard disk drives. Their manufacture involves a range of approaches, based on vapor-phase or liquid-phase epitaxy (the growth of single crystal films on single-crystal substrates, with a close relation between their orientations), or molecular-beam epitaxy (MBE). The latter is a highly sophisticated technique involving extreme vacuum and requiring stringent purity controls. Multilayers, consisting of several stacked thin films exhibiting structural coherence at the interfaces, are industrially manufactured by sputtering techniques. They make up the tunnel magnetoresistance (TMR) read heads[4] used in 2011 in the hard disk drive of all recent computers. Mastering epitaxic growth processes implies a very good knowledge of the surfaces on which the layers are to be grown. This makes the study of surface crystallography very important. The description of crystal surfaces is dealt with in Section 3.8 of this book.

Further reading

I.V. Markov, *Crystal Growth for Beginners* (World Scientific, Singapore, 1996)

G. Dhanaraj, K. Byrappa, V. Prasad, M. Dudley, *Springer Handbook of Crystal Growth* (Springer, 2010)

---

4. Wikipedia, *Tunnel magnetoresistance*, http://en.wikipedia.org/wiki/Tunnel_magneto-resistance (as of Nov. 7, 2013)

References

[1] S.R. Elliot, *Physics of Amorphous Materials*, 2nd edn. (Longman Scientific and Technical, Harlow, Essex, 1990)

[2] P.G. de Gennes and J. Prost, *The Physics of Liquid Crystals* (Oxford University Press, Oxford, 1993)

[3] D. Shechtman, I. Blech, D. Gratias, J.W. Cahn, Metallic phase with long-range orientational order and no translational symmetry, *Phys. Rev. Lett.* **53**, 1951–1953 (1984)

[4] C. Janot and J.M. Dubois, *Les quasicristaux, matière à paradoxes* (EDP Sciences, Les Ulis, 1998)

Chapter 2

Symmetry operations

*The aim of this chapter is to detail the notions of symmetry oper-
ations and symmetry elements, which are basic in the description
of crystals. The final section introduces some notions related to
symmetry groups.*

2.1. Isometries

An isometry is a geometrical transformation which transforms an object either
into an identical object (this is called a direct or proper isometry) or into its
image with respect to a mirror plane (it is then called an indirect or improper
isometry). A direct, or orientation preserving, isometry conserves the handedness (right or left) of a trihedron, whereas an indirect, or orientation reversing,
isometry reverses it. Successively applying two isometries, *i.e.* applying the
composition or "product" of these isometries obviously is an isometry. The
product of two direct isometries is a direct isometry since the transform is at
each stage identical to the object. The product of a direct and an indirect
isometry is an indirect isometry, and the product of two indirect isometries is
a direct one. More generally, the product of any number of isometries is indirect if it involves an odd number of indirect isometries, and a direct isometry
otherwise.

The isometries are: a rotation around a given axis, a translation (both of
these are direct isometries), symmetry with respect to a point, also called
inversion with respect to this point, symmetry with respect to a plane (both
being indirect isometries), and all operations resulting from the composition
(or product) of two or several of these isometries.

It is convenient to define a special indirect isometry which is very much used in
crystallography: rotoinversion (or rotation-inversion), the product of a rotation
around an axis and of an inversion with respect to a point located on this axis.

Two special rotoinversions should be mentioned here: rotoinversion by 2π,
which is nothing else than inversion, and rotoinversion with angle π, which is

© Springer Science+Business Media Dordrecht 2014
C. Malgrange et al., *Symmetry and Physical Properties of Crystals*,
DOI 10.1007/978-94-017-8993-6_2

equivalent to symmetry with respect to the plane perpendicular to the axis used for the rotation, and going through the point O used for the inversion (Fig. 2.1).

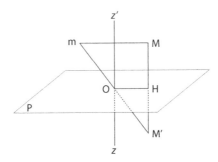

Figure 2.1: Rotoinversion by π around axis zz', the point used as a center of symmetry being O. M' is the transform of M through this rotoinversion because m is the transform of M through a rotation by π around axis zz' and M' is the transform of m through the symmetry with center O. M' is symmetric of M with respect to plane P, perpendicular to zz' and going through O.

We thus see that the set of isometries can also be defined as *the set of the rotations, rotoinversions, translations and their combinations,* since inversion and symmetry with respect to a plane are two special rotoinversions.

2.2. Symmetry operations. Symmetry elements

For some objects, an isometry (or several isometries) transform(s) the object into an object that is not only identical but superimposable on the object itself, and which is therefore undistinguishable from the latter. These isometries which leave the object invariant are the *symmetry operations* for this object.

2.2.1. *n*-fold rotations and axes

Consider as an example the two-handled jug shown on Figure 2.2a. A rotation by 180° around axis zz' changes the jug into a jug that superimposes upon itself: the rotation by 180° around zz' is a symmetry operation for this object. The product of this symmetry operation with itself is a rotation by 360° around zz', or the identity operation since, following this symmetry operation, the object is superimposed upon itself point for point. Axis zz' is said to be a 2-fold axis for the jug, or alternatively the jug is said to have a 2-fold axis as a symmetry element.

The regular square-based pyramid shown on Figure 2.2b has a 4-fold axis because the rotation by $2\pi/4 = \pi/2$ around zz' is a symmetry operation for this pyramid. The existence of a 4-fold axis implies that the object has 4 symmetry operations, *viz.* the rotation by $\pi/2$ around the axis, which can for example be noted as R, its product with itself R^2, which is the rotation by π, the product of R^2 by R, *i.e.* R^3, which is the rotation by $3\pi/2$, and operation R^4, equal to the rotation by 2π or identity operation, noted E.

In a general way, if an object has an n-fold rotation axis, it has n symmetry elements which are the rotations by $2\pi/n, 2(2\pi/n), \ldots, k(2\pi/n) \ldots$ and finally the rotation by $n(2\pi/n) = 2\pi$ which is the identity operation.

The cylinder of revolution shown on Figure 2.2c has an infinite-order axis, because a rotation by any angle, however small, is a symmetry operation.

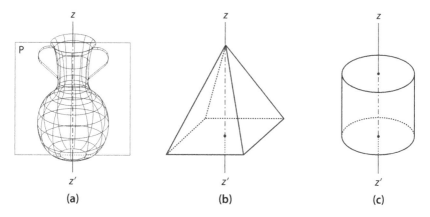

Figure 2.2: Objects featuring symmetry elements

We will see (Sect. 3.3.2) that crystals, because of their triple periodicity, can only have 2-, 3-, 4- or 6-fold axes.

2.2.2. n-fold rotoinversions and n-fold rotoinversion axes, denoted as $\bar{n}$-fold axes

If a rotoinversion by angle $2\pi/n$ (the product of a rotation by $2\pi/n$ around a given axis and an inversion with center O located on this axis) is a symmetry operation for an object, the latter is said to have an $\bar{n}$-fold rotoinversion axis. We have seen that the rotoinversion by 2π is identical with inversion. The existence of a $\bar{1}$ axis is therefore equivalent to the existence of a center of symmetry.

The existence of a $\bar{2}$-fold axis (for which the associated symmetry operation is the rotoinversion by π) is equivalent to the existence of a symmetry plane (or mirror) perpendicular to the axis and going through the center used for inversion (see Fig. 2.1). It should be noted, in contrast, that the axis defining the rotation is not a 2-fold axis for the object, just as the inversion center that is used in this rotoinversion is not a center of symmetry for the object. A rotoinversion is the product of a rotation and an inversion. It is therefore quite distinct from the rotation itself and the inversion itself.

We will see that crystals can only have as rotoinversion axes $\bar{1}$, $\bar{2}$, $\bar{3}$, $\bar{4}$ or $\bar{6}$ axes.

An example of an object having a $\bar{4}$-axis is shown on Figure 2.3a. It is a sphenohedron, a solid having four vertices A, B, C and D with the following properties: segments AB and CD have equal lengths and are perpendicular to each other. They are perpendicular to the axis zz' which intersects them in their middles, H and K respectively (Fig. 2.3b and 2.3c). The rotoinversion by $\pi/2$ around zz', with point O, the middle of HK, as the inversion center, transforms point A into C, point B into D, C into B and D into A. It thus transforms the sphenohedron into itself. Axis zz' is therefore a $\bar{4}$ axis for this solid. In contrast, it is quite clear that the sphenohedron has neither a 4-fold axis nor an inversion center.

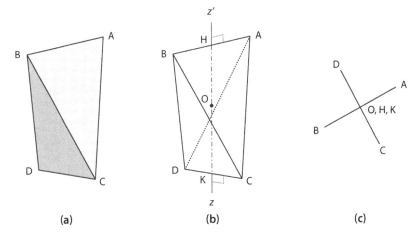

(a)   (b)   (c)

Figure 2.3: (a) A perspective view of a sphenohedron. (b) The sphenohedron ABCD. Axis zz' is a $\bar{4}$-fold axis for this object. (c) Projection of points A, B, C and D on a plane perpendicular to zz'.

2.2.3. Translations

If a translation by a vector **t** is a symmetry operation for an object, the product of n translations **t**, equal to the translation $n\mathbf{t}$, is also a symmetry operation. Since n can be arbitrarily large, *the object is necessarily infinite*, in contrast to the above cases of rotations and rotoinversions. Such an object is said to be periodic with period **t**. For example, Figures 2.4a and 2.4b, obtained from a simple pattern which is outlined by a thin dotted line (something like a music note for 2.4a and a set of two such notes for 2.4b), repeated to infinity by translation **t**, have translation **t** (and all the translations $n\mathbf{t}$) as symmetry operations. Figure 2.4b furthermore has a symmetry plane or mirror, perpendicular to the plane of the figure and going through the line uu'. If the infinite sequence of patterns of Figure 2.4a is submitted to a translation by a vector **t**$'$ (not collinear with **t**) and its multiples, the obtained figure (Fig. 2.5) has the translations $n\mathbf{t} + n'\mathbf{t}'$ (n and n' integers) as symmetry operations.

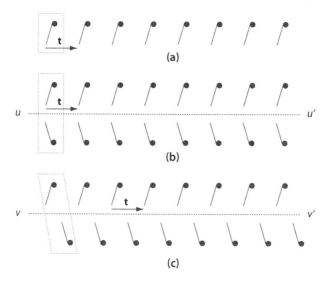

Figure 2.4: Figures which are periodic with period vector **t** (assuming they extend to infinity). The motif or pattern is marked with a thin dotted line on each figure.

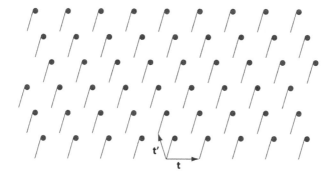

Figure 2.5: A plane figure with periodicity along two directions

2.2.4. Screw axes and glide mirrors

In Sections 2.2.1 and 2.2.2, we defined two types of symmetry elements : axes of order n and axes of order $\bar{n}$. In crystallography, further symmetry elements must be defined. They are associated to the symmetry operations which are the products (also called compositions) of rotations or rotoinversions and translations. These elements are on the one hand screw axes, the associated symmetry operations of which are the products of a rotation and a translation parallel to the rotation axis and, on the other hand, glide mirrors, the associated symmetry operations of which are the products of an operation of symmetry with respect to a plane (a mirror operation) and a translation parallel to this plane. Consider two examples: an object with a screw axis and an object with a glide mirror. Since the associated symmetry operations include a translation part, these objects are necessarily infinite.

Screw axis: consider a helical staircase, in which all steps have the same height, and such that when walking up by one step we rotate around the central axis by an angle equal to a submultiple of 2π, for example $30°$. The operation product of a rotation by $30°$ around the axis of the staircase and a vertical translation equal to the height of the steps is a symmetry operation for this staircase, under the proviso, of course, that we imagine it to have infinite height. This staircase has a screw axis coinciding with its axis.

Glide mirror: Figure 2.4c is formed from a motif (the pattern of Fig. 2.4b altered by translating one of the two notes by $\mathbf{t}/2$), repeated to infinity by translation $\mathbf{t}$. This periodic figure thus features translation $\mathbf{t}$ and its multiples as symmetry operations. It also features a less obvious symmetry operation: the product of the symmetry with respect to a plane perpendicular to the plane of the figure and going through line vv' and the translation by vector $\mathbf{t}/2$. The corresponding symmetry element is called a glide mirror, the glide being the translation $\mathbf{t}/2$.

2.2.5. Note

We have defined the following symmetry elements: n-fold axes, for which the associated symmetry operations are a rotation by $2\pi/n$ and its multiples, $\bar{n}$-fold axes, for which the associated symmetry operations are a rotoinversion by $2\pi/n$ and its multiples, mirrors for which the associated symmetry operation is symmetry with respect to the plane of the mirror, screw axes for which the associated symmetry operations are the product of a rotation and a translation and its multiples, and glide mirrors, for which the associated symmetry operations are the product of a symmetry with respect to the plane of the mirror and a translation and its multiples.

We see that the term "symmetry element" covers two notions: on the one hand a geometrical object, a line or a plane, and on the other hand the set of symmetry operations which are associated to it.

2.3. Introduction to symmetry groups

In this introduction, we restrict discussion to finite objects. This rules out symmetry operations that include a translation. The symmetry operations which a finite object can feature are thus the rotations and rotoinversions, the latter including symmetry with respect to a point and symmetry with respect to a plane (Sect. 2.1 and Fig. 2.1).

We show that the symmetry operations of a finite object form a group. Let us first recall the definition of a group.

A group is a set of elements among which one can define an internal composition law that is associative and includes a neutral element e such that, for any element a of the group, $ae = ea = a$. Each element of the group must have an inverse a^{-1} such that $aa^{-1} = a^{-1}a = e$.

For the symmetry operations featured by an object, the internal composition law is the product of two symmetry operations, *i.e.* the succession of these two operations. The neutral element is the identity operation, and the inverse operations of the elementary operations, such as rotation by an angle φ around an axis, symmetry with respect to a point O or symmetry with respect to a plane P are thus evident: they are respectively the rotation by an angle $-\varphi$ around the same axis, symmetry with respect to the same point O, and symmetry with respect to the same plane P. The symmetry operations of an object thus form a group.

Let us take as an example the symmetry operations of the jug of Figure 2.2a. We saw that the axis zz' is a 2-fold axis to which are associated the two following symmetry operations: rotation by an angle π around zz', which we will denote as A, and the operation A^2 or rotation by 2π around zz', which is the identity operation, noted as E. Furthermore, this jug has two other symmetry elements. The plane P going through zz' and parallel to the two handles is a symmetry plane or mirror. Let m_1 be the operation "symmetry with respect to plane P". It is clear that $m_1^2 = E$. In the same way, the plane P' going through zz' and perpendicular to plane P is a symmetry element. If m_2 is the operation "symmetry with respect to plane P'", then $m_2^2 = E$. We have thus defined four symmetry operations, *viz.* A, m_1, m_2 and E.

We can easily draw the multiplication table for these operations (Tab. 2.1).

Table 2.1: Multiplication table for the group of symmetry operations of the jug of Figure 2.2a

	E	A	m_1	m_2
E	E	A	m_1	m_2
A	A	E	m_2	m_1
m_1	m_1	m_2	E	A
m_2	m_2	m_1	A	E

The product of A and m_1 is noted m_1A, the rule being that the order of operations is read from right to left, so that operation A is here followed by operation m_1. This product m_1A is shown at the intersection of line m_1 and row A. It is easily checked that this product m_1A is equal to m_2. Consider any point M. The plane perpendicular to the 2-fold axis going through M is represented on Figure 2.6 where uu' and vv' are the traces, on this plane, of mirrors P and P' respectively.

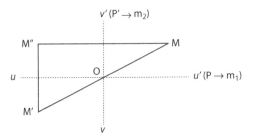

Figure 2.6: Showing the product of operation A (rotation by π around axis zz' perpendicular to the plane of the figure) and operation m_1 (symmetry with respect to the plane P, with trace uu' on the figure). The transform of M through A is M', the transform of M' through m_1 is M''. M'' is obtained from M through symmetry with respect to plane P', *i.e.* through operation m_2.

We notice that, on each row and each column of the table itself (surrounded by a bold line on Tab. 2.1), all the elements of the group appear, and each appears only once. This is a general result which is derived in Appendix 4A.

The generating elements of a group are the elements which are sufficient to define the group. The set of elements of the group is obtained by multiplying these elements with one another or with their inverses. In the example chosen, we see that A and m_1 are generating elements of the group, but alternative choices of generating elements are A and m_2 or m_1 and m_2.

A group is said to be commutative if, for any two elements a and b of the group, $ab = ba$. We see that the group considered above is commutative since the multiplication table is symmetric with respect to its diagonal.

2.4. Exercises

Exercise 2.1

We want to derive the following result: if an object has an n-fold axis and a mirror that goes through this axis, it has n mirrors which go through the axis and which enclose between one another angles π/n.

1. Consider two mirrors, M_1 and M_2, enclosing between them an angle α. We note zz' the intersection of the two mirrors. Let m_1 and m_2 be the symmetry operations with respect to plane M_1 and to plane M_2 respectively. Use a geometrical argument to show that the product m_2m_1 is equal to a rotation by 2α around zz', so that $m_2m_1 = \mathrm{Rot}_{zz'}(2\alpha)$.

2. Deduce from this result that $m_1 = m_2\mathrm{Rot}_{zz'}(2\alpha)$ and $m_2 = \mathrm{Rot}_{zz'}(2\alpha)m_1$.

3. Use the result of (2) to obtain the result we seek, *viz.* that, if an object has an n-fold axis and a mirror that goes through this axis, it has n mirrors deduced from the first one through rotations by π/n and its multiples around the axis.

4. Enumerate the eight symmetry operations of the square-based regular pyramid of Figure 2.2b.

5. Write down the multiplication table of the group consisting of these symmetry operations. Is it commutative?

Exercise 2.2

Determine the symmetry elements of the two molecules shown below:

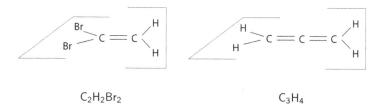

$$C_2H_2Br_2 \qquad\qquad\qquad C_3H_4$$

$C_2H_2Br_2$: the two H atoms and the carbon to which they are linked are coplanar. Their plane is perpendicular to the plane containing the other C atom and the two Br atoms which are linked to it.

C_3H_4: the three C atoms are aligned and equidistant. The plane containing the C atom at one end of the chain and the two H atoms which are linked to it is perpendicular to the plane containing the C atom at the other end of the chain and the two H atoms which are linked to it. The angle between the C–H bonds is identical in the two planes.

Exercise 2.3

Show that the proper symmetry elements of a cube are: three 4-fold axes, four 3-fold axes and six 2-fold axes.

Chapter 3

Crystal lattices

This chapter defines the crystal lattice, which directly expresses the triple periodicity of the crystal, and all the derived notions: reciprocal lattice, crystal systems and Bravais lattices. The final part of the chapter shows how this definition of the crystal lattice is extended for the study of the surface of crystals.

3.1. Direct lattice

We saw that crystals are characterized by their periodicity in three directions. They are defined by a crystal motif consisting of a well-defined set of atoms, which repeats through periodic translations in three non-coplanar directions. We will see that a lattice is associated to this triple periodicity, and that the crystal can thus be defined by its motif and its lattice, which is called direct (in contrast with the so-called reciprocal lattice which will be deduced from the direct lattice in Sect. 3.2).

3.1.1. Unit cell, row

Two-dimensional lattice

We start with examples of two-dimensional periodic systems because they are easier to represent and to understand, and are well suited for introducing many notions. Figure 3.1a is an example. It shows an elementary motif, a general triangle, periodically reproduced in two different directions with period vectors respectively $\mathbf{a}$ and $\mathbf{b}$, forming an infinite object. This infinite object features translation symmetry, meaning that any translation $\mathbf{t} = u\mathbf{a} + v\mathbf{b}$, where u and v are two integers, transforms the object into an object undistinguishable from itself. If we replace the triangle by a group of atoms, we obtain a two-dimensional crystal.

Now suppose we replace the elementary motif by a point located near the motif in some position, general (*i.e.* with no special requirement) but perfectly defined with respect to the motif (Fig. 3.1b). We obtain a regular lattice of points which deduce from one another through the set of translations $u\mathbf{a} + v\mathbf{b}$

© Springer Science+Business Media Dordrecht 2014
C. Malgrange et al., *Symmetry and Physical Properties of Crystals*,
DOI 10.1007/978-94-017-8993-6_3

(u and v being integers). This set of points (Fig. 3.1c) is called the crystal *lattice*, and each point is a lattice *node*. We can thus define a plane crystal lattice, with basis vectors **a** and **b**, as a set of points M, called nodes, defined by the relation

$$\mathbf{OM} = u\mathbf{a} + v\mathbf{b}$$

where O is any point, chosen as the origin node, and u and v are integers.

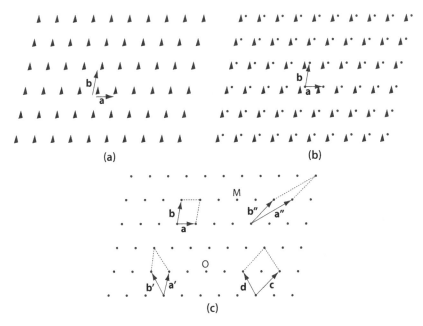

(a)   (b)

(c)

Figure 3.1: (a) A two-dimensional periodic figure; (b) the associated crystal lattice; (c) examples of primitive cells (**a**, **b**), (**a′**, **b′**), (**a″**, **b″**) and a multiple unit cell (**c**, **d**)

Any line going through two nodes is called a *row*. The parallelogram built on **a** and **b** is called the *primitive cell* for this lattice, and the vectors **a** and **b** are its basis vectors. We note that the whole crystal lattice is obtained by paving the whole plane with parallelograms identical to this one, and placing one node at each of the vertices. Each node then belongs to four cells, and counts for $1/4$ per cell. One cell then contains 4 times ($1/4$ node), *i.e.* one node per cell, and this is why such cells are called primitive cells. It is sometimes designated as a simple unit cell. Vectors **a** and **b** are not the only ones which can define a simple unit cell for a given lattice. Other possible choices are for example the vector pairs **a′** and **b′**, or **a″** and **b″** (Fig. 3.1c). In contrast, vectors **c** and **d** define a cell which also contains a node at its center. The cell is no more a primitive cell, it is termed a multiple cell. Alternative denominations are non-primitive or centered (or centred) cell. The whole crystal lattice is obtained by setting next to one another such cells, which have a node in common at each of their vertices and a node at their center. The cell spanning $\mathbf{c}$ and $\mathbf{d}$ is a double unit cell because it contains two nodes: the node at the center which belongs to only one cell, and the four vertex nodes, each belonging to four cells and counting for $1/4$, thus in total another node.[1] The number of nodes associated to one unit cell is called the multiplicity or order of the unit cell: thus the cell $(\mathbf{c}, \mathbf{d})$ has multiplicity 2. The area of a primitive cell defined by vectors $\mathbf{a}$ and $\mathbf{b}$ is equal to the norm of the cross product $\mathbf{a} \times \mathbf{b}$. The areas of all the other primitive cells (such as those defined by $\mathbf{a}'$ and $\mathbf{b}'$ or $\mathbf{a}''$ and $\mathbf{b}''$ for instance) are identical, each of them being equal to the area associated to one node. The area of a unit cell of order n is equal to n times the area of a primitive cell, since it is equal to the area associated to n nodes. The multiplicity of the cell defined by the vectors $\mathbf{f} = u\mathbf{a} + v\mathbf{b}$ and $\mathbf{g} = u'\mathbf{a} + v'\mathbf{b}$ (u, v, u', v' being integers) can also be determined by computing the cross product $\mathbf{f} \times \mathbf{g}$:

$$(u\mathbf{a} + v\mathbf{b}) \times (u'\mathbf{a} + v'\mathbf{b}) = (uv' - u'v)\,(\mathbf{a} \times \mathbf{b})$$

the norm of which is equal to $(uv' - u'v)\,\|\mathbf{a} \times \mathbf{b}\|$, thus an integer (possibly equal to 1) times the area of the primitive cell since u, v, u', v' are integers.

The crystal lattice is characterized by choosing, if possible, a primitive cell which directly reflects the symmetry of the crystal lattice. Suppose for example that the lattice consists of nodes at the vertices of rectangles positioned next to one another (Fig. 3.2a). The lattice is then defined by a rectangular primitive cell, $(\mathbf{a}, \mathbf{b})$, and not by a general parallelogram. Now suppose the primitive cell $(\mathbf{a}, \mathbf{b})$ of the lattice is a lozenge (Fig. 3.2b). The lattice features symmetry elements: mirrors perpendicular to the plane of the figure and parallel to the diagonals of the lozenges. These symmetry elements show up immediately if the choice is made of a unit cell with basis vectors $\mathbf{c}$ and $\mathbf{d}$, orthogonal to each other, with a node in its center. The lattice is thus defined by a double rectangular cell termed centered rectangular.

Three-dimensional lattice

The above notions are easily generalized to the case of three-dimensional crystals. The crystal motif is then reproduced through non-coplanar translations $\mathbf{a}$, $\mathbf{b}$, $\mathbf{c}$, and their multiples. Replacing the motif by a point located in some arbitrary but perfectly defined position with respect to the motif produces a regular lattice of points, deduced from one another by translations $u\mathbf{a}+v\mathbf{b}+w\mathbf{c}$ with u, v, w integers. The crystal lattice is then defined by choosing for the origin a point O and drawing all the points M that are the ends of vectors $\mathbf{OM} = u\mathbf{a} + v\mathbf{b} + w\mathbf{c}$ with u, v, w integers. The set of points M defines the crystal *lattice* and each of these points is called a lattice *node*. Point O is the origin node.

---

1. Another way of counting the nodes per unit cell is to imagine the cells to be slightly displaced, so that the lattice nodes no more coincide with the vertices. The nodes in each unit cell are then distinctly visible.

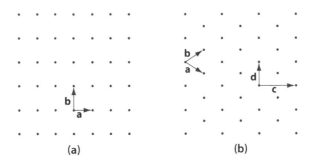

(a)   (b)

Figure 3.2: Rectangular plane lattices. The parallelograms built on **a** and **b** are primitive cells in both figures. However, in Figure (a), this parallelogram, actually a rectangle, describes the full symmetry of the lattice. In contrast, describing the symmetry in Figure (b) requires a multiple (centered) unit cell.

Any line going through two nodes M and M′ is called a *row*, and it contains an infinity of nodes (Fig. 3.3). The vector **MM′** joining nodes M and M′ has the form $u'\mathbf{a} + v'\mathbf{b} + w'\mathbf{c}$, where u', v' and w' are integers since **OM** and **OM′** are of this form. There is an infinity of rows parallel to **MM′**. The one that goes through the node at origin O goes through node N′ such that $\mathbf{ON'} = \mathbf{MM'} = u'\mathbf{a} + v'\mathbf{b} + w'\mathbf{c}$. Let u, v and w be integers proportional to u', v' and w' and mutually prime. The vector $\mathbf{ON} = u\mathbf{a} + v\mathbf{b} + w\mathbf{c}$ is the smallest lattice vector parallel to this direction. It defines the period of the nodes on this row, and more generally on the family of rows parallel to this one, which will be denoted as $[uvw]$.

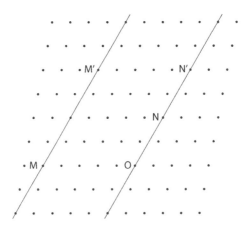

Figure 3.3: An example of a crystal row. The plane of the figure is defined by the nodes M and M′ and the node O taken as the origin.

The parallelepiped spanning the three basis vectors $\mathbf{a}$, $\mathbf{b}$ and $\mathbf{c}$ is the *primitive cell* of the crystal (Fig. 3.4a). The sign of the vectors $\mathbf{a}$, $\mathbf{b}$ and $\mathbf{c}$ is conventionally chosen so that the trihedron $\mathbf{a}$, $\mathbf{b}$, $\mathbf{c}$ be direct. The volume of the cell is equal to the mixed product $(\mathbf{a}, \mathbf{b}, \mathbf{c})$, equal to $\mathbf{a} \cdot (\mathbf{b} \times \mathbf{c})$.

We will see that the requirement for a unit cell to feature all the symmetry elements of the crystal lattice leads to defining multiple cells, just as in the case of two-dimensional lattices. Multiple cells can be base-centered if two opposite faces are centered, face-centered if all the faces are centered (Fig. 3.4b), or body-centered if there is a node at the center of the cell. This is why multiple cells are also called centered cells. The multiplicity (or order) of the unit cell is obtained by counting the number of nodes per unit cell, taking into account that a node at the center of the cell adds one node per cell, while a node at the center of a face, belonging to two distinct cells, counts for $1/2$.

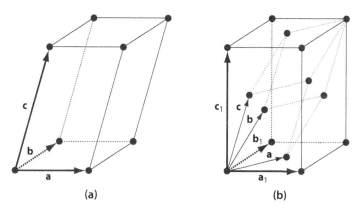

(a)   (b)

Figure 3.4: (a) Primitive cell of a crystal lattice; (b) multiple, face-centered unit cell with basis vectors $\mathbf{a_1}$, $\mathbf{b_1}$, $\mathbf{c_1}$, and the corresponding primitive cell with basis vectors $\mathbf{a}$, $\mathbf{b}$, $\mathbf{c}$

The multiplicity of a cell spanning the crystal lattice vectors $\mathbf{a_1}$, $\mathbf{b_1}$, and $\mathbf{c_1}$ such that

$$\mathbf{a_1} = u_1\mathbf{a} + v_1\mathbf{b} + w_1\mathbf{c},$$
$$\mathbf{b_1} = u_2\mathbf{a} + v_2\mathbf{b} + w_2\mathbf{c},$$
$$\mathbf{c_1} = u_3\mathbf{a} + v_3\mathbf{b} + w_3\mathbf{c},$$

(u_i, v_i, w_i being integers) can be determined from the volume of the unit cell, equal to the mixed product $(\mathbf{a_1}, \mathbf{b_1}, \mathbf{c_1})$. This can be written in matrix from:

$$\begin{pmatrix} \mathbf{a_1} \\ \mathbf{b_1} \\ \mathbf{c_1} \end{pmatrix} = U \begin{pmatrix} \mathbf{a} \\ \mathbf{b} \\ \mathbf{c} \end{pmatrix}$$

with

$$U = \begin{pmatrix} u_1 & v_1 & w_1 \\ u_2 & v_2 & w_2 \\ u_3 & v_3 & w_3 \end{pmatrix}$$

Hence $(\mathbf{a_1}, \mathbf{b_1}, \mathbf{c_1}) = \det U (\mathbf{a}, \mathbf{b}, \mathbf{c})$ where $(\mathbf{a}, \mathbf{b}, \mathbf{c})$ is the volume of the primitive cell. The multiplicity is thus given by $\det U$.

3.1.2. Lattice planes

Any plane that goes through three non-aligned nodes is called a *lattice plane*. It contains an infinity of nodes, which form a two-dimensional lattice. The set of lattice planes parallel to a given lattice plane describes the whole crystal lattice. In other words, the lattice can be described as the set of the lattice planes of a given family. This can be done in an infinity of ways, since any family of lattice planes can be chosen as the starting point. This is shown on Figure 3.5, based, for simplicity, on a lattice with a primitive cubic cell. The basis vectors $\mathbf{a}$, $\mathbf{b}$ and $\mathbf{c}$ are therefore mutually orthogonal and have the same length a. The figure represents the projections of the nodes of this lattice onto a plane perpendicular to one of the basis vectors, for example $\mathbf{c}$. The projected nodes form a square lattice of points, and we have represented the traces of a few families of lattice planes perpendicular to the plane of the figure.

Note, and this is valid whatever the crystal lattice, that the planes of a given family are equally spaced (equidistant) and that the node density in the planes is smaller the closer the planes are to one another.[2]

Now consider a general crystal lattice, with a primitive cell defined by vectors $\mathbf{a}$, $\mathbf{b}$ and $\mathbf{c}$ (forming a direct trihedron, as mentioned before). We choose an origin node O and axes Ox, Oy and Oz respectively parallel to $\mathbf{a}$, $\mathbf{b}$ and $\mathbf{c}$. We focus on the lattice plane parallel to vector $\mathbf{c}$ and containing two nodes M and N belonging to plane $(\mathbf{a}, \mathbf{b})$ (Fig. 3.6). The other lattice planes belonging to the same family intersect plane $(\mathbf{a}, \mathbf{b})$ along equidistant lines shown dotted on the figure. One of them goes through the origin O; we label it with number 0. The successive equidistant neighboring planes are numbered 1, 2, 3, ..., on the side towards which the vectors $\mathbf{a}$ and $\mathbf{b}$ point, and those located in the other half-space are numbered successively -1, -2, -3, etc. One of these planes necessarily goes through node A, the end of vector $\mathbf{a}$; let h be the integer that characterizes it. In the case shown in the figure, $h = 3$, so that it is the third plane. Plane number 1, the one closest to origin, intersects axis Ox at a point with coordinate a/h ($a/3$ on Fig. 3.6). In the same way, there is necessarily a plane of the same family, numbered k (2 on the figure), that goes through

---

2. This is easily verified by noting that the number n of nodes per unit volume has a given value for a given crystal lattice. On the other hand, on a lattice plane belonging to a given family, the number of nodes per unit area ρ is such that $\rho = nd$ where d is the distance between two successive lattice planes of this family.

node B, the end of vector **b**. Plane 1 intersects axis Oy at b/k ($b/2$ on the figure). Since we chose a family of planes parallel to the basis vector **c**, plane 1 intersects axis Oz at infinity.

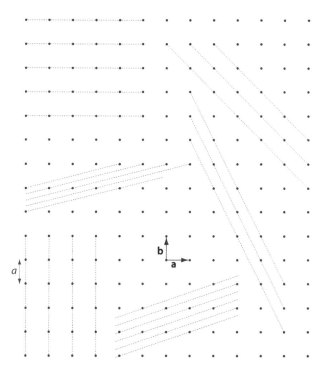

Figure 3.5: Projection of a simple cubic lattice on a plane perpendicular to the basis vector **c**. Various families of lattice planes perpendicular to the plane of the figure project along the dotted lines.

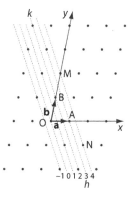

Figure 3.6: Traces on plane (**a**, **b**) of a family of lattice planes parallel to basis vector **c**

Now let us apply the same reasoning to a general family of lattice planes (Fig. 3.7). The reasoning is based on the fact that a family of lattice planes contains all the lattice nodes, and that all the planes of this family are equidistant. One of these planes, noted P_0, goes through the origin O of the lattice. Another one, called P, goes through the node located at A such that $\mathbf{OA} = \mathbf{a}$. There are, between these planes, a number n of equidistant intermediate planes, which divide segment $\mathbf{a}$ into an integral number of segments, each with length a/n. Let us call P_1 the one which is closest to P_0.

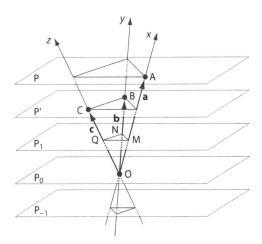

Figure 3.7: A family of lattice planes. P_0 is the plane belonging to the family that goes through the origin O of the crystal lattice. P_1 and P_{-1} are the two planes of this family that are closest to O.

This plane P_1 intersects segment $\mathbf{OA}$ at a point M with coordinates $(a/h, 0, 0)$ where h is a positive or negative integer depending on whether plane P_1 intersects axis Ox on the side of the positive or negative x, and $n = |h|$. Similarly, since one of the planes of the family necessarily goes through node B such that $\mathbf{OB} = \mathbf{b}$, plane P_1 intersects axis Oy at a point N with coordinates $(0, b/k, 0)$ where k is an integer. Finally a plane of the family goes through node C, such that $\mathbf{OC} = \mathbf{c}$, and plane P_1 intersects axis Oz at Q, with coordinates $(0, 0, c/l)$ where l is an integer. The numbers h, k, l are integers which can be positive or negative.

The equation of this plane P_1 in the coordinate system $Oxyz$ has the form $\alpha x + \beta y + \gamma z = \delta$. Writing that the coordinates of M, N and Q satisfy this equation, we obtain:

$$\alpha \frac{a}{h} = \delta, \qquad \beta \frac{b}{k} = \delta, \qquad \gamma \frac{c}{l} = \delta.$$

Replacing α, β and γ by their values as a function of δ, the equation of plane P_1 becomes

$$h \frac{x}{a} + k \frac{y}{b} + l \frac{z}{c} = 1.$$

Let us select as the length unit on each of the axes the length of the corresponding basis vector, *viz. a* on Ox, *b* on Oy and *c* on Oz, and *this is the choice we will use throughout this book*, this equation becomes

$$hx + ky + lz = 1. \tag{3.1}$$

The equation of the n-th plane of this family, on the same side as P$_1$ with respect to the origin, is

$$hx + ky + lz = n$$

while the n-th plane on the other side is

$$hx + ky + lz = -n.$$

The equation of the plane belonging to this family and going through the origin is

$$hx + ky + lz = 0. \tag{3.2}$$

The integers h, k and l are mutually prime. Let us show this by supposing they were not. Then we could write $h = mh'$, $k = mk'$ and $l = ml'$ where m, h', k' and l' are integers. Equation (3.1) would then become

$$h'x + k'y + l'z = \frac{1}{m}.$$

It must be satisfied by all the nodes of the plane, the coordinates of which are integers. h', k' and l' being integers, this is impossible if $m \neq 1$ since then $1/m$ would not be an integer.

This family of lattice planes is designated as (hkl) and we will remember that:

The notation (hkl), where h, k and l are mutually prime integers, designates a family of lattice planes such that the plane of the family closest to the node at origin intersects axis Ox at a/h, axis Oy at b/k and axis Oz at c/l. h, k and l are called the Miller indices of this family of lattice planes.

Note that the integers h, k and l are placed between parentheses while the integers defining a family of rows (Sect. 3.1.1) are noted between square brackets.

Note: the planes parallel to **a** have the designation $(0kl)$ since they intersect axis Ox at infinity. Similarly the planes parallel to **b** have the form $(h0l)$ and those parallel to **c** the form $(hk0)$.

Application: what condition must h, k, l and u, v, w satisfy for a plane (hkl) and a row $[uvw]$ to be parallel to each other?

Row $[uvw]$ is by definition parallel to the vector **OM** $= u\mathbf{a} + v\mathbf{b} + w\mathbf{c}$. The plane parallel to the row and going through origin therefore contains the point M, with coordinates u, v, w and, from (3.2), we must have $hu + kv + lw = 0$.

We note that:

A row $[uvw]$ is parallel to a plane (hkl) if and only if

$$hu + kv + lw = 0.$$

3.1.3. Wigner-Seitz cell

Solid-state physicists often use another way of subdividing direct space into identical cells containing one lattice node per cell. The center of these cells is chosen on one lattice node. The mediating planes of the segments joining the chosen node to the first nodes on the various rows which point out of this node are then drawn. The smallest volume thus created around the node is the Wigner-Seitz cell (Fig. 3.8). This cell is primitive because it is associated to a single lattice node, but it is not a parallelepiped. In Section 3.2, we will define another lattice of points, deduced from the direct lattice, which is called the reciprocal lattice. The same construction applied to this reciprocal lattice defines a cell called the first Brillouin zone.

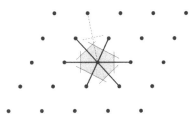

Figure 3.8: Wigner-Seitz cell of a plane lattice. The segment in dotted lines connects the origin to a node which does not take part in the construction of the cell because the mediator of this segment is outside the cell.

3.2. Reciprocal lattice

3.2.1. Introduction based on diffraction

The above considered lattice is defined in the space of the crystal, and this is why it is called the direct lattice. Another associated lattice is very much used: it is called the reciprocal lattice and plays a part in the phenomena of diffraction by crystals and more generally in all problems of wave propagation in crystals. It is the space of wave vectors, the dimensions of which are reciprocal lengths. It should be mentioned that the crystallographers' tradition is to write a plane wave w, with amplitude a and frequency ν (angular frequency $\omega = 2\pi\nu$) in the form $w = a \cos 2\pi(\mathbf{k} \cdot \mathbf{r} - \nu t)$, so that the norm of the wave-vector is equal to $1/\lambda$ where λ is the spatial period of the wave, *i.e.* the wavelength. This is in contrast with the tradition of optics and of most physicists, who write $w = a \cos(\mathbf{k} \cdot \mathbf{r} - \omega t)$, so that the norm of their $\mathbf{k}$ is equal to $2\pi/\lambda$. This is summarized in Table 3.1 below.

The triple periodicity in the position of atoms in a crystal leads to a plane wave of monochromatic X-rays incident on the crystal being diffracted into well-defined directions.

Table 3.1: Definition of the wave-vector **k** as used in various areas of physics

Crystallography	$w = a \cos 2\pi(\mathbf{k} \cdot \mathbf{r} - \nu t)$ $\|\mathbf{k}\| = 1/\lambda$
Optics, quantum mechanics, solid state physics	$w = a \cos(\mathbf{k} \cdot \mathbf{r} - \omega t)$ $\|\mathbf{k}\| = 2\pi/\lambda$

Consider a crystal whose primitive cell is defined by the vectors **a**, **b** and **c**. This crystal consists in a motif and its transforms by all the translations of the crystal lattice $\boldsymbol{\tau} = u\mathbf{a} + v\mathbf{b} + w\mathbf{c}$ (u, v, w integers). Let $\mathbf{k}_o$ be the wave-vector of the incident wave. This wave interacts with the electrons in the atoms of the crystal motif. Assume for simplicity that the motif consists of a single atom. This monoatomic motif scatters into all space a spherical wave with the same wavelength. For an observer located far enough from the crystal, this spherical wave can be considered as plane with a wave-vector $\mathbf{k}_d$. The waves scattered by two identical motifs located at P and Q, a vector $\boldsymbol{\tau} = u\mathbf{a} + v\mathbf{b} + w\mathbf{c}$ apart, have a path difference $\delta = \mathrm{PK} - \mathrm{HQ}$ such that:

$$\delta = (\mathbf{s}_d - \mathbf{s}_o) \cdot \boldsymbol{\tau}$$

where $\mathbf{s}_d$ and $\mathbf{s}_o$ are unit vectors along $\mathbf{k}_d$ and $\mathbf{k}_o$ respectively.

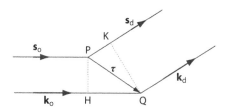

Figure 3.9: Scattering of a plane wave by two crystal motifs separated by a vector $\boldsymbol{\tau}$

All the plane waves scattered by all the motifs are in phase if $\delta = n\lambda$ (n an integer) for all $\boldsymbol{\tau}$, so that

$$(\mathbf{s}_d - \mathbf{s}_o) \cdot \boldsymbol{\tau} = n\lambda$$

or
$$(\mathbf{k}_d - \mathbf{k}_o) \cdot \boldsymbol{\tau} = n \qquad (3.3)$$

since $|\mathbf{k}_d| = |\mathbf{k}_o| = 1/\lambda$.

This is seen to involve calculating the dot product of a vector $\boldsymbol{\tau}$ of direct space, expressed in a non-orthonormal basis $\{\mathbf{a}, \mathbf{b}, \mathbf{c}\}$, and a vector of the space of wave-vectors. This dot product is easily written if the vector $\mathbf{k}_d - \mathbf{k}_o$ is expressed in a basis $\{\mathbf{a}^*, \mathbf{b}^*, \mathbf{c}^*\}$ such that the dot product of $\mathbf{a}^*$ and $\mathbf{a}$ is equal to 1 while the dot products of $\mathbf{a}^*$ and $\mathbf{b}$ or $\mathbf{c}$ are zero, with similar relations for $\mathbf{b}^*$ an $\mathbf{c}^*$:

$$
\begin{array}{lll}
\mathbf{a}^* \cdot \mathbf{a} = 1 & \mathbf{a}^* \cdot \mathbf{b} = 0 & \mathbf{a}^* \cdot \mathbf{c} = 0 \\
\mathbf{b}^* \cdot \mathbf{a} = 0 & \mathbf{b}^* \cdot \mathbf{b} = 1 & \mathbf{b}^* \cdot \mathbf{c} = 0 \\
\mathbf{c}^* \cdot \mathbf{a} = 0 & \mathbf{c}^* \cdot \mathbf{b} = 0 & \mathbf{c}^* \cdot \mathbf{c} = 1
\end{array}
\qquad (3.4)
$$

Let u^*, v^* and w^* be the components of the vector $\mathbf{k}_d - \mathbf{k}_o$ in the basis $\{\mathbf{a}^*, \mathbf{b}^*, \mathbf{c}^*\}$. The vector $\mathbf{k}_d - \mathbf{k}_o$ is expressed as

$$
\mathbf{k}_d - \mathbf{k}_o = u^*\mathbf{a}^* + v^*\mathbf{b}^* + w^*\mathbf{c}^*.
$$

Taking into account Equations (3.4), condition (3.3), with n an integer, takes the form

$$
u^*u + v^*v + w^*w = n
$$

which must be verified for all integral values of u, v and w. This requires that u^*, v^* and w^* be integers.

The reciprocal lattice of a crystal is defined as the set of points N obtained starting from an origin O and such that $\mathbf{ON} = h\mathbf{a}^* + k\mathbf{b}^* + l\mathbf{c}^*$ where h, k and l are integers. The points N are the nodes of the reciprocal lattice and O the origin node.

We just showed that the diffraction condition is equivalent to the condition that the components u^*, v^* and w^*, in the basis of the reciprocal lattice, of the scattering vector (or diffraction vector) $\mathbf{k}_d - \mathbf{k}_o$ be integers, *i.e.* that this scattering vector be a vector joining two nodes of the reciprocal lattice. The diffraction condition is thus very simply expressed thanks to the reciprocal lattice.

Furthermore, we see on Equations (3.4) that:

The reciprocal lattice of the reciprocal lattice is the direct lattice.

This results from the fact that the first column of Equations (3.4) defines $\mathbf{a}$ as a function of $\mathbf{a}^*$, $\mathbf{b}^*$ and $\mathbf{c}^*$ just like the first row defines $\mathbf{a}^*$ as a function of $\mathbf{a}$, $\mathbf{b}$ and $\mathbf{c}$, etc.

3.2.2. Alternative definition of the reciprocal lattice

The relations (3.4) show that $\mathbf{a}^*$ is perpendicular to $\mathbf{b}$ and $\mathbf{c}$, hence parallel to the cross product $\mathbf{b} \times \mathbf{c}$. We can thus write:

$$
\mathbf{a}^* = \alpha\,(\mathbf{b} \times \mathbf{c}).
$$

Furthermore, since $\mathbf{a} \cdot \mathbf{a}^* = 1$, we have:

$$
\mathbf{a} \cdot \mathbf{a}^* = \alpha\,\mathbf{a} \cdot (\mathbf{b} \times \mathbf{c}) = \alpha v_0 = 1,
$$

with $v_0 = (\mathbf{a}, \mathbf{b}, \mathbf{c})$ the volume of the crystal unit cell. Hence $\alpha = 1/v_0$ and therefore

$$
\mathbf{a}^* = (\mathbf{b} \times \mathbf{c})/v_0.
$$

In the same way, we obtain

$$\mathbf{b}^* = (\mathbf{c} \times \mathbf{a})/v_0,$$

and
$$\mathbf{c}^* = (\mathbf{a} \times \mathbf{b})/v_0. \tag{3.5}$$

Relations (3.5) directly define the reciprocal lattice with respect to the basis vectors of the direct lattice. This is the most usual definition of the basis vectors of the reciprocal lattice.

Further, since the direct lattice is the reciprocal lattice of the reciprocal lattice,

$$\mathbf{a} = (\mathbf{b}^* \times \mathbf{c}^*)/v_0^*,$$
$$\mathbf{b} = (\mathbf{c}^* \times \mathbf{a}^*)/v_0^*, \tag{3.6}$$
$$\mathbf{c} = (\mathbf{a}^* \times \mathbf{b}^*)/v_0^*,$$

with v_0^* the volume of the reciprocal unit cell, *i.e.* the volume of the parallelogram built on $\mathbf{a}^*$, $\mathbf{b}^*$ and $\mathbf{c}^*$.

3.2.3. Properties of the reciprocal lattice

Basic property

Consider the lattice plane P_1 of the family (hkl) that is closest to the origin O of the direct lattice. A point M of this plane has coordinates x, y, z in basis $\{\mathbf{a}, \mathbf{b}, \mathbf{c}\}$:

$$\mathbf{OM} = x\mathbf{a} + y\mathbf{b} + z\mathbf{c}.$$

From relation (3.1), x, y and z then satisfy the relation

$$hx + ky + lz = 1 \tag{3.7}$$

where we recall that the integers h, k and l are mutually prime.

Consider the vector $\mathbf{h}^*$ of the reciprocal lattice with components h, k and l

$$\mathbf{h}^* = h\mathbf{a}^* + k\mathbf{b}^* + l\mathbf{c}^*.$$

The properties of the reciprocal lattice encountered above show that

$$\mathbf{OM} \cdot \mathbf{h}^* = hx + ky + lz$$

and, taking into account (3.7),

$$\mathbf{OM} \cdot \mathbf{h}^* = 1.$$

Let H be the projection of M on the vector $\mathbf{h}^*$ (Fig. 3.10), with norm $\|\mathbf{h}^*\|$.

$$\mathbf{OM} \cdot \mathbf{h}^* = \mathrm{OH}\,\|\mathbf{h}^*\| = 1$$

and
$$\mathrm{OH} = 1/\|\mathbf{h}^*\| = \mathrm{constant}.$$

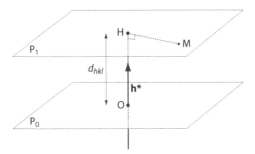

Figure 3.10: Two neighboring (hkl) planes: P_0 goes through the origin of the lattice and P_1 is its first neighbor.

We deduce that all points of plane P_1 project onto vector $\mathbf{h}^*$ at the same point H, which means that plane P_1 is perpendicular to the vector $\mathbf{h}^*$. On the other hand, OH is the distance of plane P_1 to the plane of the family (hkl) which goes through O. This is the distance d_{hkl} between two successive planes in the family (hkl). We thus obtain

$$d_{hkl} = \frac{1}{||\mathbf{h}^*||}. \tag{3.8}$$

Basic property: The planes (hkl) are perpendicular to the row $[hkl]^*$ of the reciprocal lattice and their distance d_{hkl} is equal to the reciprocal of the modulus of the vector $\mathbf{h}^* = h\mathbf{a}^* + k\mathbf{b}^* + l\mathbf{c}^*$.

It is important to recall that the vector $\mathbf{h}^*$ which is being discussed as a way of calculating the distance between two neighboring lattice planes of family (hkl) connects the origin of the reciprocal lattice[3] to the first reciprocal lattice node on the row $[hkl]^*$ that goes through the origin, since h, k, l are mutually prime.

X-ray diffraction can also be described in terms of selective mirror reflection of X-rays on the lattice planes. The condition for constructive interference is given by Bragg's law

$$2d_{hkl}\sin\theta = n\lambda$$

where λ is the wavelength, θ the complement of the angle of incidence on the lattice planes, and n the order of the reflection.

Bragg's law can also be written $2(d_{hkl}/n)\sin\theta = \lambda$ and the reflection of order n on the lattice planes (hkl) described by arbitrarily introducing reflecting planes parallel to the lattice planes (hkl), and n times closer to one another, their distance being d_{hkl}/n. This distance d_{hkl}/n between reflecting planes is equal to $1/(n||\mathbf{h}^*||)$, *i.e.* the reciprocal of the norm of a new vector $\mathbf{h}'^* = n\mathbf{h}^*$ of

---

3. For simplicity, we select the origin of the reciprocal lattice as coinciding with that of the direct lattice.

the reciprocal lattice connecting the origin of the reciprocal lattice to the n-th node on the row $[hkl]^*$. One then speaks of a reflection nh, nk, nl such that $2d_{nh,nk,nl} \sin \theta = \lambda$ with $d_{nh,nk,nl} = 1/\|\mathbf{h}'^*\|$ where $\mathbf{h}'^*$ is now the reciprocal lattice vector connecting the origin to the n-th node along row $[hkl]^*$. In this way each reciprocal lattice node is associated to a reflection, the first node in each row corresponding to the reflection of order 1 and the n-th node to the reflection of order n.

We note that everything concerning the reciprocal lattice is indicated by a star: the basis vectors $\mathbf{a}^*$, $\mathbf{b}^*$, $\mathbf{c}^*$, a row $[hkl]^*$ and a family of lattice planes $(hkl)^*$.

Since the lattice reciprocal to the reciprocal lattice is the direct lattice, we can similarly state that:

The planes $(hkl)^*$ of the reciprocal lattice are perpendicular to the row $[hkl]$ of the direct lattice.

Volume of the reciprocal unit cell

Let us substitute, in the expression (3.5) for $\mathbf{a}^*$, the vector $\mathbf{c}$ by its expression (3.6) as a function of the reciprocal lattice vectors. We obtain

$$\mathbf{a}^* = (\mathbf{b} \times (\mathbf{a}^* \times \mathbf{b}^*))/v_0 v_0^*.$$

Using the expression of the double cross product

$$\mathbf{A} \times (\mathbf{B} \times \mathbf{C}) = \mathbf{B}(\mathbf{A} \cdot \mathbf{C}) - \mathbf{C}(\mathbf{A} \cdot \mathbf{B}),$$

and relations (3.4), we obtain

$$\mathbf{a}^* = [\mathbf{a}^*(\mathbf{b} \cdot \mathbf{b}^*) - \mathbf{b}^*(\mathbf{b} \cdot \mathbf{a}^*)]/v_0 v_0^* = \mathbf{a}^*/v_0 v_0^*.$$

This shows that

$$v_0 v_0^* = 1. \tag{3.9}$$

Note: all of the above discussion refers to unit cells associated with a single lattice node (primitive cells). The reciprocal lattice of a multiple unit cell must be defined using a primitive cell. It is however possible to then describe the reciprocal lattice using a multiple reciprocal unit cell (see Sect. 3.7.3).

3.2.4. Crystallographic calculations

Calculation of lattice distances

The lattice distance d_{hkl} in a family of lattice planes (hkl) is determined by relation (3.8), which yields:

$$\frac{1}{d_{hkl}^2} = (h\mathbf{a}^* + k\mathbf{b}^* + l\mathbf{c}^*)^2,$$

where the basis vectors $\mathbf{a}^*$, $\mathbf{b}^*$ and $\mathbf{c}^*$ of the reciprocal lattice are determined as a function of the basis vectors $\mathbf{a}$, $\mathbf{b}$ and $\mathbf{c}$ of the direct lattice through relations (3.5).

We thus obtain

$$\frac{1}{d_{hkl}^2} = (h\mathbf{a}^* + k\mathbf{b}^* + l\mathbf{c}^*)^2$$

$$= h^2\mathbf{a}^{*2} + k^2\mathbf{b}^{*2} + l^2\mathbf{c}^{*2} + 2hk\,\mathbf{a}^* \cdot \mathbf{b}^* + 2kl\,\mathbf{b}^* \cdot \mathbf{c}^* + 2lh\,\mathbf{c}^* \cdot \mathbf{a}^*. \quad (3.10)$$

In Section 3.4.2, we will define the crystal systems, and derive the expressions for the values of d_{hkl} in each of them.

Search for the indices of a lattice plane parallel to two rows of the direct lattice

We are looking for the plane (hkl) which is parallel to rows $[u_1 v_1 w_1]$ and $[u_2 v_2 w_2]$.

First method:

We saw, at the end of Section 3.1.2, that the condition for row $[uvw]$ of the direct lattice to be parallel to plane (hkl) is that

$$hu + kv + lw = 0. \quad (3.11)$$

We apply this condition to both rows:

$$hu_1 + kv_1 + lw_1 = 0,$$
$$hu_2 + kv_2 + lw_2 = 0.$$

We deduce that

$$\frac{h}{v_1 w_2 - w_1 v_2} = \frac{k}{w_1 u_2 - u_1 w_2} = \frac{l}{u_1 v_2 - v_1 u_2}. \quad (3.12)$$

Second method:

Let $\mathbf{n}_1 = u_1\mathbf{a} + v_1\mathbf{b} + w_1\mathbf{c}$ and $\mathbf{n}_2 = u_2\mathbf{a} + v_2\mathbf{b} + w_2\mathbf{c}$ be vectors parallel to the two rows respectively.

The plane we are looking for, being parallel to both vectors, is perpendicular to the cross product of the vectors $\mathbf{n}_1$ and $\mathbf{n}_2$, which is expressed as

$$\mathbf{n}_1 \times \mathbf{n}_2 = (u_1\,\mathbf{a} + v_1\,\mathbf{b} + w_1\,\mathbf{c}) \times (u_2\,\mathbf{a} + v_2\,\mathbf{b} + w_2\,\mathbf{c})$$

$$= (v_1 w_2 - w_1 v_2)(\mathbf{b} \times \mathbf{c}) + (w_1 u_2 - u_1 w_2)(\mathbf{c} \times \mathbf{a})$$

$$+ (u_1 v_2 - v_1 u_2)(\mathbf{a} \times \mathbf{b}). \quad (3.13)$$

Taking into account Equations (3.5) which define the basis vectors of the reciprocal lattice, we obtain:

$$\mathbf{n}_1 \times \mathbf{n}_2 = v_0[(v_1 w_2 - w_1 v_2)\mathbf{a}^* + (w_1 u_2 - u_1 w_2)\mathbf{b}^*$$

$$+ (u_1 v_2 - v_1 u_2)\mathbf{c}^*]. \quad (3.14)$$

We see that

The cross product of two vectors of direct space lattice is parallel to a vector of the reciprocal lattice,

a result which will be used in the next paragraph.

Let us write the cross product (3.14) in the following form:

$$\mathbf{n}_1 \times \mathbf{n}_2 = v_0[h'\mathbf{a}^* + k'\mathbf{b}^* + l'\mathbf{c}^*]$$

where $h' = v_1 w_2 - w_1 v_2$, $k' = w_1 u_2 - u_1 w_2$ and $l' = u_1 v_2 - v_1 u_2$ are integers since u_i, v_i and w_i are integers. If h', k', l' are not mutually prime, call h, k, and l the mutually prime integers proportional to them ($h' = nh$, $k' = nk$, $l' = nl$). Thus $\mathbf{n}_1 \times \mathbf{n}_2$ is parallel to the vector $[hkl]^*$ of the reciprocal lattice, with h, k and l mutually prime. Such a vector is perpendicular to the plane (hkl) of the direct lattice, which is the plane we are looking for. We thus retrieve relations (3.12).

Search for the row [uvw] parallel to the intersection of planes $(h_1 k_1 l_1)$ and $(h_2 k_2 l_2)$

First method:

We apply condition (3.11) to the row $[uvw]$ and to the planes $(h_1 k_1 l_1)$ and $(h_2 k_2 l_2)$.

$$h_1 u + k_1 v + l_1 w = 0,$$
$$h_2 u + k_2 v + l_2 w = 0.$$

We deduce that

$$\frac{u}{k_1 l_2 - l_1 k_2} = \frac{v}{l_1 h_2 - h_1 l_2} = \frac{w}{h_1 k_2 - k_1 h_2}. \tag{3.15}$$

Second method:

The vectors $\mathbf{V}_1^* = h_1 \mathbf{a}^* + k_1 \mathbf{b}^* + l_1 \mathbf{c}^*$ and $\mathbf{V}_2^* = h_2 \mathbf{a}^* + k_2 \mathbf{b}^* + l_2 \mathbf{c}^*$ are perpendicular to the planes $(h_1 k_1 l_1)$ and $(h_2 k_2 l_2)$ respectively. The row we are looking for is therefore perpendicular to these vectors, hence parallel to the cross product $\mathbf{V}_1^* \times \mathbf{V}_2^*$. The latter is parallel to a vector of the space reciprocal to reciprocal space, *i.e.* a vector of direct space. Following the same approach as in (3.13) and (3.14), we obtain:

$$\mathbf{V}_1^* \times \mathbf{V}_2^* = v_0^*[(k_1 l_2 - l_1 k_2)\mathbf{a} + (l_1 h_2 - h_1 l_2)\mathbf{b} + (h_1 k_2 - k_1 h_2)\mathbf{c}]$$

and we retrieve the result (3.15).

3.3. Properties of crystal lattices

3.3.1. Centers of symmetry

All nodes of a crystal lattice are centers of symmetry for the lattice, as we can easily see, for example on the two-dimensional lattice of Figure 3.1c. The same applies to the centers of the unit cells. Therefore

All crystal lattices feature centers of symmetry.

3.3.2. n-fold and $\bar{n}$-fold axes consistent with the crystalline state

We will now show that:

A crystal lattice can only feature n-fold axes of order 2, 3, 4 or 6. The same applies to $\bar{n}$-fold axes, which can only be $\bar{1}$, $\bar{2}$, $\bar{3}$, $\bar{4}$ or $\bar{6}$ axes.

First derivation:

Consider a vector $\mathbf{V}$ of the crystal lattice. If the lattice features a symmetry element, one of the symmetry operations S associated to this element will transform $\mathbf{V}$ into a vector $\mathbf{V}' = S\mathbf{V}$ which is also a lattice vector.

$\mathbf{V}$ and $\mathbf{V}'$, being lattice vectors, have integral components with respect to the coordinate system defined by the basis vectors $\mathbf{a}$, $\mathbf{b}$ and $\mathbf{c}$ of the crystal unit cell. The coefficients in the matrix representing operation S in this basis are therefore integers, and its trace is also an integer.

If a lattice features an n-fold axis, let us call S the rotation by $\alpha = 2\pi/n$ around this axis, denoted A_n. The trace of the matrix representing an operator does not depend on the basis chosen (orthonormal or not) to represent this operator. We choose to represent this operator in an orthonormal frame such that Oz coincides with axis A_n hence through matrix S'

$$ S' = \begin{pmatrix} \cos\alpha & -\sin\alpha & 0 \\ \sin\alpha & \cos\alpha & 0 \\ 0 & 0 & 1 \end{pmatrix}. $$

The trace of S' is equal to $1 + 2\cos\alpha$: therefore $2\cos\alpha$ must be an integer. Furthermore, $|2\cos\alpha| \leq 2$ and the only possible values of $2\cos\alpha$ are therefore: -2, -1, 0, 1, 2.

- If $\cos\alpha = \pm 1$, $\alpha = 0$ or π. There is a 1-fold axis, *i.e.* no symmetry (if $\alpha = 0$), or a 2-fold axis (if $\alpha = \pi$).

- If $\cos\alpha = \pm 1/2$, $\alpha = \pm 2\pi/3$ or $\pm \pi/3$, and there is a 3-fold axis or a 6-fold axis.

- If $\cos\alpha = 0$, $\alpha = \pm\pi/2$ and there is a 4-fold axis.

The only possible values for n are therefore $n = 1$ (trivial), 2, 3, 4 and 6.

The derivation is the same for the $\bar{n}$ axes. Operation S is then the rotation by $\alpha = 2\pi/n$ followed by an inversion with respect to a center located on the rotation axis. The matrix representing this operation is then $-ES'$ where S' is the above matrix and E is the unit matrix. The trace of this matrix is $-(1 + 2\cos\alpha)$, which must be an integer; therefore $2\cos\alpha$ is an integer with modulus less or equal to 2. We thus retrieve the preceding results, hence the same consequences, *viz.* the possible existence of $\bar{1}, \bar{2}, \bar{3}, \bar{4}$ and $\bar{6}$ axes. A $\bar{1}$ axis is identical to a symmetry center, already mentioned in Section 3.3.1.

Second (geometrical) derivation

This derivation applies more conveniently to rotation axes. We use the result, independently derived in Section 3.3.4 below, that an n-fold axis is perpendicular to a family of lattice planes, hence to a set of lattice vectors.

Let $\mathbf{V} = \mathbf{AB}$ be an elementary vector of the crystal lattice (*i.e.* a vector connecting two neighboring nodes), perpendicular to the n-fold axis. Let us perform onto $\mathbf{AB}$ the symmetry operation S, a rotation by $\alpha = 2\pi/n$ around the n-fold axis that goes through node A (Fig. 3.11). The vector $\mathbf{AB}$ becomes $\mathbf{AB'}$, which therefore is a lattice vector. Operation S, performed $n-1$ times, is a symmetry operation. It is a rotation by $(n-1)\alpha = 2\pi - \alpha = -\alpha$ (modulo 2π). We recall that two angles are equal modulo 2π if their difference is an integer times 2π. Let us apply this rotation $-\alpha$ to vector $\mathbf{AB}$ using the axis located at B:[4] $\mathbf{BA}$ becomes $\mathbf{BA'}$, a lattice vector. It is easy to show that vector $\mathbf{B'A'}$ is parallel to $\mathbf{AB}$. Its length is equal to $\mathbf{AB}(1 - 2\cos\alpha)$ and it is a lattice vector. It is therefore equal to an integer times $\mathbf{AB}$. Thus $(1 - 2\cos\alpha)$ is an integer and $2\cos\alpha$ is an integer. We retrieve the same results as above.

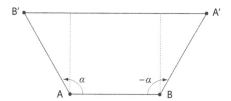

Figure 3.11: A and B are two neighboring nodes on a crystal lattice featuring n-fold axes perpendicular to the plane of the figure.

3.3.3. The direct lattice and the reciprocal lattice have the same symmetry elements

Let S be a symmetry operation (rotation or rotoinversion) featured by a given direct crystal lattice. It therefore transforms the direct lattice R into a lattice R′ with which it is superimposed, so that $SR = R' \equiv R$.

---

4. Translation symmetry implies the existence of an n-fold axis at B if there is one at A.

Let $\mathbf{a}'$, $\mathbf{b}'$ and $\mathbf{c}'$ be the transforms of $\mathbf{a}$, $\mathbf{b}$ and $\mathbf{c}$ under S. We can write

$$\mathbf{a}' = S\,\mathbf{a}, \qquad \mathbf{b}' = S\,\mathbf{b} \quad \text{and} \quad \mathbf{c}' = S\,\mathbf{c}.$$

The reciprocal lattice R^* of R can be obtained from $\mathbf{a}'$, $\mathbf{b}'$ and $\mathbf{c}'$ or from $\mathbf{a}$, $\mathbf{b}$ and $\mathbf{c}$ since R and R$'$ are identical.

We want to show that the symmetry operation S transforms R^* into itself. Let us define R^* through $\mathbf{a}'^*$, $\mathbf{b}'^*$ and $\mathbf{c}'^*$ such that:

$$\mathbf{a}'^* = (\mathbf{b}' \times \mathbf{c}')/v_0 = (S\mathbf{b} \times S\mathbf{c})/v_0 = \pm S(\mathbf{b} \times \mathbf{c})/v_0$$
$$= \pm S\mathbf{a}^* = S(\pm \mathbf{a}^*).$$

Similarly, $\mathbf{b}'^* = S(\pm \mathbf{b}^*)$ and $\mathbf{c}'^* = S(\pm \mathbf{c}^*)$.

The sign $+$ or $-$ depends on the nature (proper or improper) of S but R^* can be defined using $-\mathbf{a}^*$, $-\mathbf{b}^*$ and $-\mathbf{c}^*$ just as well as using $\mathbf{a}^*$, $\mathbf{b}^*$ and $\mathbf{c}^*$. We therefore see clearly that the lattice R^* is transformed by S into itself. S is thus a symmetry operation for the reciprocal lattice.

3.3.4. Geometrical relation between the symmetry axes and the crystal lattice

We now show that:

Any symmetry axis (rotation or rotoinversion axis) is parallel to a row of the direct lattice and perpendicular to a lattice plane.

Let S be the elementary symmetry operation associated to the n-fold axis (rotation by $2\pi/n$) or $\bar{n}$–fold axis (rotation by $2\pi/n$ followed by an inversion). It transforms a vector $\mathbf{V}$ parallel to the axis into an equal (rotation case) or opposite (rotoinversion case) vector.

$$S\mathbf{V} = \pm\mathbf{V} \qquad \text{or} \qquad (S \mp E)\mathbf{V} = \mathbf{0}$$

where E is the unit operator.

In the basis $\mathbf{a}$, $\mathbf{b}$, $\mathbf{c}$, the coefficients of the matrix S representing operation S are integers since S transforms a lattice vector into another lattice vector. The same applies to $S \pm E$. Solving the equation system $(S \pm E)\mathbf{V} = \mathbf{0}$, we find that the components u, v and w of vector $\mathbf{V}$ are such that $u/w = p/r$ and $v/w = q/r$ with p, q and r integers, whence $u/p = v/q = w/r$. Since p, q and r are integers, the symmetry axis is parallel to a row[5] $[p\,q\,r]$ of the lattice. The same derivation can be applied to the reciprocal lattice, which also features this symmetry element (Sect. 3.3.3). The symmetry axis is thus also parallel to a reciprocal lattice row, and thus perpendicular to a lattice plane of the direct lattice (see Sect. 3.2.3).

---

5.   The integers p, q and r are not necessarily mutually prime. If they are not, they should be replaced by numbers proportional to p, q and r and mutually prime, in order to define the row adequately.

3.3.5. The lattice is at least as symmetrical as the crystal

Pierre and Jacques Curie stated in 1894 a very far-reaching principle which is now called the Curie principle. It will be introduced in Chapter 9 (Sect. 9.5.2) and widely used in the second part of this book. One of the direct consequences of this principle can be stated as follows: the physical properties of a crystal are at least as symmetrical as the crystal. The crystal lattice describes one property of the crystal, *viz.* its triple periodicity. Therefore the crystal lattice must be at least as symmetric as the crystal. This property will be derived in Section 4.2.3 and used in Section 5.6 to determine the crystal lattice of a crystal as a function of the symmetry properties of the crystal itself.

3.4. Crystal systems

The results mentioned in this section are not derived but the enumeration of crystal systems introduced below is so natural that it can be performed before their precise definition is introduced (Sect. 5.6 and Chap. 6). Suffice it to say that a crystal system is characterized by the smallest unit cell, primitive or multiple (centered), which features all the symmetry elements of the corresponding lattice.

3.4.1. Two-dimensional crystal systems

Let us recall that the symmetry elements of a plane object are perpendicular to the plane of the object, and that a center of symmetry O is then equivalent to a 2-fold axis perpendicular to the plane and going through O.

There are four plane crystal systems, shown in Table 3.2.

Table 3.2: The plane crystal systems

System	Oblique	Rectangular	Square	Hexagonal
Unit cell shape	parallelogram	rectangle	square	lozenge with angle 120°
	(parallelogram diagram)	*(rectangle diagram)*	*(square diagram)*	*(lozenge diagram, 120°)*
Bravais lattices	p	p c (centered)	p	p

They are: the oblique (general primitive cell), rectangular (rectangular primitive cell), square (square primitive cell), and hexagonal (primitive cell in the shape of a lozenge with angle 120°) systems. We note that a lozenge-shaped primitive cell produces a rectangular centered lattice. The rectangular centered double cell evidences more explicitly the symmetry elements of the lattice, *viz.* the mirror symmetry planes parallel to the basis vectors of the centered rectangular cell. We will see (Sect. 3.7) that they are the primitive (p) and centered (c) Bravais lattices.

3.4.2. Three-dimensional crystal systems

There are seven three-dimensional crystal systems, defined and illustrated in Table 3.3. The basis vectors of the unit cell are always called $\mathbf{a}$, $\mathbf{b}$, $\mathbf{c}$ and the trihedron $\mathbf{a}$, $\mathbf{b}$, $\mathbf{c}$ is chosen as direct. The angles in the cell faces are denoted as $\alpha = (\mathbf{b}, \mathbf{c})$, $\beta = (\mathbf{c}, \mathbf{a})$ and $\gamma = (\mathbf{a}, \mathbf{b})$.

The seven systems are the following:

Triclinic: the unit cell is a general parallelepiped, defined by its three angles α, β, and γ (in Greek, *klinos* means angle) and the lengths a, b and c of the three basis vectors.

Monoclinic: the unit cell is a right prism with bases in the shape of a parallelogram, defined by just one angle (whence the name) since the two others are right angles, and the lengths a, b and c of the three basis vectors. The basis vector perpendicular to the two others was long selected as the vector $\mathbf{b}$, and the angle to be defined is then β. Nowadays a frequent choice, more consistent with the choices for the other systems, is to take it as vector $\mathbf{c}$, and the angle to be defined is then angle γ.

Orthorhombic (from the Greek *orthos* = right): the unit cell is a rectangle parallelepiped. All that is to be defined are the three lattice parameters a, b and c.

Tetragonal: the unit cell is a right prism with square base (a rectangle parallelepiped in which two opposite faces are square). One chooses to call $\mathbf{a}$ and $\mathbf{b}$ the basis vectors forming the square and $\mathbf{c}$ the basis vector which is perpendicular to them. Two lattice parameters must therefore be defined: $a = b$ and c.

Rhombohedral (or trigonal): the unit cell is a rhombohedron, *i.e.* a parallelepiped in which all the faces are equal (Fig. 3.14a). The only values to be defined are the lattice parameter $a = b = c$ and the angle $\alpha = \beta = \gamma$. We can visualize a rhombohedron as a cube which has been drawn out or compressed along one of its major diagonals. In a cube, the major diagonals are 3-fold axes. If a cube is drawn out or flattened along one of its major diagonals while retaining the equality of the edges, this diagonal remains a 3-fold axis while the others are no longer 3-fold axes. The rhombohedral lattice is described in more detail in Section 3.6.2.

Hexagonal: the unit cell is a special right prism. Its base is a lozenge with one angle equal to 120°. The basis vector perpendicular to the two others is $\mathbf{c}$. Two cell parameters must be defined: $a = b$ and c.

Cubic: the unit cell is a cube, and the only parameter to be defined is $a = b = c$.

Table 3.3: The seven three-dimensional crystal systems

Crystal system	Shape of unit cell	Parameters	Polyhedron	Bravais lattices
Triclinic		$a \neq b \neq c$ $\alpha \neq \beta \neq \gamma \neq \pi/2$	general parallelepiped	P
Monoclinic		$a \neq b \neq c$ $\alpha = \beta = \pi/2$ γ general	right prism with parallelogram-shaped base	P, A (or B)
		$a \neq b \neq c$ $\alpha = \gamma = \pi/2$ β general		P, C (or A)
Orthorhombic		$a \neq b \neq c$ $\alpha = \beta = \gamma = \pi/2$	right prism with rectangular base	P A (or B or C) I, F
Tetragonal		$a = b \neq c$ $\alpha = \beta = \gamma = \pi/2$	right prism with square base	P, I
Rhombohedral (Trigonal)		$a = b = c$ $\alpha = \beta = \gamma$	rhombohedron	R (P)
Hexagonal		$a = b \neq c$ $\alpha = \beta = \pi/2$ $\gamma = 120°$	right prism with 120° lozenge base	P
Cubic		$a = b = c$ $\alpha = \beta = \gamma = \pi/2$	cube	P, I, F

In each crystal system, the lattice distance d_{hkl} can be calculated as a function of the basis vectors $\mathbf{a}^*$, $\mathbf{b}^*$ and $\mathbf{c}^*$ of the reciprocal lattice using formula (3.10).

For all systems except for the triclinic and rhombohedral systems, this formula can be simplified, and d_{hkl} is easily expressed as a function of the parameters of the direct lattice, *i.e.* of the lengths a, b and c of the basis vectors and the angles α, β and γ between these vectors. Some of these calculations are the topics of exercises, and the result for the various crystal systems is given in Table 3.4.

For the triclinic and rhombohedral lattices, it is convenient to use a quantity called the metric tensor, which is introduced in Appendix 3A. The metric tensor is widely used in numerical calculations for crystallography.

Table 3.4: Lattice distances for the various crystal systems

Crystal system	d_{hkl}
Triclinic	$\sqrt{\dfrac{1 - \cos^2\alpha - \cos^2\beta - \cos^2\gamma + 2\cos\alpha\cos\beta\cos\gamma}{A}}$
Monoclinic	$\sin\beta \ / \ \sqrt{\dfrac{h^2}{a^2} + \dfrac{l^2}{c^2} + \left(\dfrac{k}{b}\right)^2 \sin^2\beta - 2\dfrac{hl}{ac}\cos\beta}$ if $\mathbf{b} /\!/$ 2-fold axis of the cell
	$\sin\gamma \ / \ \sqrt{\dfrac{h^2}{a^2} + \dfrac{k^2}{b^2} + \left(\dfrac{l}{c}\right)^2 \sin^2\gamma - 2\dfrac{hk}{ab}\cos\gamma}$ if $\mathbf{c} /\!/$ 2-fold axis of the cell
Orthorhombic	$1 / \sqrt{\dfrac{h^2}{a^2} + \dfrac{k^2}{b^2} + \dfrac{l^2}{c^2}}$
Tetragonal	$a / \sqrt{h^2 + k^2 + l^2\left(\dfrac{a}{c}\right)^2}$
Rhombohedral	$a\sqrt{\dfrac{1 + 2\cos^3\alpha - 3\cos^2\alpha}{(h^2+k^2+l^2)\sin^2\alpha + 2(hk+kl+lh)(\cos^2\alpha - \cos\alpha)}}$
Hexagonal	$a / \sqrt{\dfrac{4}{3}(h^2 + k^2 + hk) + l^2\left(\dfrac{a}{c}\right)^2}$
Cubic	$a / \sqrt{h^2 + k^2 + l^2}$

$$A = \frac{h^2}{a^2}\sin^2\alpha + \frac{k^2}{b^2}\sin^2\beta + \frac{l^2}{c^2}\sin^2\gamma + 2\frac{hk}{ab}(\cos\alpha\cos\beta - \cos\gamma)$$
$$+ 2\frac{hl}{ac}(\cos\alpha\cos\gamma - \cos\beta) + 2\frac{kl}{bc}(\cos\beta\cos\gamma - \cos\alpha)$$

3.5. Examples of reciprocal lattices

3.5.1. Monoclinic lattice

The unit cell is defined by the lengths a, b, c of the basis vectors and by the angle β between vectors $\mathbf{a}$ and $\mathbf{c}$ (if the choice is to denote as $\mathbf{b}$ the basis vector parallel to the 2-fold axis of the cell). The volume v_0 of the unit cell is equal to $abc\sin\beta$.

$$\mathbf{a}^* = \frac{\mathbf{b} \times \mathbf{c}}{v_0} \quad \text{and} \quad \| \mathbf{a}^* \| = \frac{1}{a\sin\beta} \; ;$$

$$\mathbf{b}^* = \frac{\mathbf{c} \times \mathbf{a}}{v_0} \quad \text{and} \quad \| \mathbf{b}^* \| = \frac{1}{b} \; ;$$

$$\mathbf{c}^* = \frac{\mathbf{a} \times \mathbf{b}}{v_0} \quad \text{and} \quad \| \mathbf{c}^* \| = \frac{1}{c\sin\beta}.$$

Vector $\mathbf{b}^*$ is parallel to vector $\mathbf{b}$. The vectors $\mathbf{a}^*$ and $\mathbf{c}^*$ are in the plane of vectors $\mathbf{a}$ and $\mathbf{c}$ and perpendicular to $\mathbf{c}$ and $\mathbf{a}$ respectively. The angle between $\mathbf{a}^*$ and $\mathbf{c}^*$ is equal to $\pi - \beta$ (Fig. 3.12).

The hexagonal lattice can be treated the same way (Exercise 3.4).

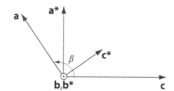

Figure 3.12: Basis vectors of a monoclinic lattice and of the corresponding reciprocal lattice projected on a plane perpendicular to vector $\mathbf{b}$ (parallel to the 2-fold axis). The vectors $\mathbf{b}$ and $\mathbf{b}^*$ are perpendicular to the plane of the figure and point upwards.

3.5.2. Orthorhombic, tetragonal and cubic lattices

In these three lattices, the basis vectors $\mathbf{a}$, $\mathbf{b}$ and $\mathbf{c}$ are mutually orthogonal: therefore the vectors $\mathbf{a}^*$, $\mathbf{b}^*$ and $\mathbf{c}^*$ are parallel to vectors $\mathbf{a}$, $\mathbf{b}$ and $\mathbf{c}$ respectively, and their lengths a^*, b^* and c^* are equal to $1/a$, $1/b$ and $1/c$ respectively.

In the tetragonal lattice, $a = b$ and $a^* = b^*$. In the cubic lattice where $a = b = c$, the three basis vectors of the reciprocal lattice have the same length $a^* = b^* = c^* = 1/a$.

3.6. Hexagonal lattice and rhombohedral lattice

3.6.1. Hexagonal lattice

The hexagonal lattice is defined from two basis vectors **a** and **b** with the same length, enclosing an angle of 120° (Fig. 3.13), and a vector **c** perpendicular to **a** and **b**. In the basal plane defined by the vectors **a** and **b**, we could just as well select the vector couples **b** and **d** (or **d** and **a**) where **d** is the vector deduced from **b** through a rotation by 120° around the 6-fold axis of the lattice (parallel to **c**). These three vectors **a**, **b** and **d** are related by:

$$\mathbf{a} + \mathbf{b} + \mathbf{d} = 0.$$

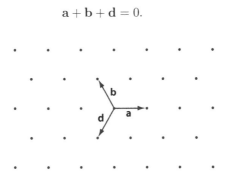

Figure 3.13: Hexagonal lattice; projection of the nodes onto a plane perpendicular to **c**

To evidence, in the lattice plane notation, the 3-fold symmetry of these vectors, a fourth index i is added between indices k and l, leading to the notation $(hkil)$ such that the lattice plane closest to origin intersects the vectors **a**, **b**, **d** and **c** at a/h, b/k, d/i and c/l. It can be shown (Exercise 3.5) that the relation

$$h + k + i = 0 \qquad \text{or} \qquad i = -(h + k)$$

applies.

This four-index notation makes it possible to easily determine the lattice planes equivalent through the symmetry operations of the crystal (Exercises 3.5 and 5.6). The notation $(hkil)$ is sometimes replaced by $(hk.l)$ to show that one deals with a hexagonal lattice. This four-index notation cannot apply to rows $[uvw]$, since it is then meaningless.

3.6.2. Rhombohedral lattice

Consider a rhombohedral lattice. The unit cell is a rhombohedron shown on Figure 3.14a.

Its basis vectors are $\mathbf{LP} = \mathbf{a}$, $\mathbf{LM} = \mathbf{b}$ and $\mathbf{LN} = \mathbf{c}$. The plane MNP is a (111) plane, noted as Π_1 on Figure 3.14b. We note as Π_0 the (111) plane containing node L, Π_2 the one containing nodes Q, R and S and Π_3 the one containing T.

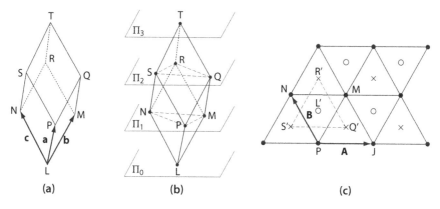

Figure 3.14: Rhombohedral lattice: (a) primitive cell; (b) showing the succession of (111) planes; (c) projection of the lattice onto plane Π_1, perpendicular to the 3-fold axis (empty circles for the nodes of Π_0 and Π_3, full circles for the nodes of Π_1, and crosses for the nodes of Π_2)

The nodes in each of these (111) planes form a hexagonal plane lattice since they are at the vertices of adjacent equilateral triangles. The nodes of planes Π_0 (containing L) and Π_3 (containing T) project onto Π_1 at L′, the center of triangle MNP, and on all equivalent points (empty circles in Fig. 3.14c), *i.e.* at the centers of every other of the equilateral triangles formed by the hexagonal plane lattice of Π_1. The nodes of plane Π_2 such as Q, R and S project at the centers of the other triangles (crosses, and for example Q′, R′ and S′, the projections of Q, R and S respectively). We immediately see on Figure 3.14c that the rhombohedral lattice can be described by a triple hexagonal unit cell, with basis vectors

$$\mathbf{A} = \mathbf{PJ} = \mathbf{NM} = \mathbf{NL} + \mathbf{LM} = \mathbf{b} - \mathbf{c},$$
$$\mathbf{B} = \mathbf{PN} = \mathbf{PL} + \mathbf{LN} = \mathbf{c} - \mathbf{a},$$
$$\mathbf{C} = \mathbf{LT} = \mathbf{LN} + \mathbf{NR} + \mathbf{RT} = \mathbf{a} + \mathbf{b} + \mathbf{c},$$

so that finally
$$\mathbf{A} = \mathbf{b} - \mathbf{c},$$
$$\mathbf{B} = \mathbf{c} - \mathbf{a},$$
$$\mathbf{C} = \mathbf{a} + \mathbf{b} + \mathbf{c}.$$

This unit cell is triple because it contains the nodes Q and T which project onto Π_1 at Q′ and L′ respectively.

We can add to the basis vectors $\mathbf{A}$ and $\mathbf{B}$ a vector $\mathbf{D}$ defined as in Section 3.6.1, *i.e.* such that

$$\mathbf{A} + \mathbf{B} + \mathbf{D} = 0$$

so that $\mathbf{D} = \mathbf{MP} = \mathbf{a} - \mathbf{b}$.

It is current practice to describe a rhombohedral lattice in this way, through its triple hexagonal unit cell, with basis vectors $\mathbf{A}$, $\mathbf{B}$ and $\mathbf{C}$, because it is much easier to handle.

Conversely, the basis vectors of the rhombohedral unit cell are obtained from those of the hexagonal unit cell through the relations

$$\mathbf{a} = \frac{\mathbf{C}}{3} + \frac{\mathbf{D} - \mathbf{B}}{3},$$
$$\mathbf{b} = \frac{\mathbf{C}}{3} + \frac{\mathbf{A} - \mathbf{D}}{3},$$
$$\mathbf{c} = \frac{\mathbf{C}}{3} + \frac{\mathbf{B} - \mathbf{A}}{3}.$$

3.7. Bravais lattices

3.7.1. Why they are needed

We are not going to systematically enumerate the Bravais lattices. This will be done in Chapter 6. Here we introduce them after showing why it is of interest to define them.

In many cases, the primitive cell considered as an individual object has symmetry lower than the symmetry of the lattice, while a multiple (or centered) cell can evidence the symmetry of the lattice. A first example is the case of a rhombohedral unit cell with angle equal to $60°$. When the whole lattice is represented, we find that the nodes are at the vertices of cubes and at the centers of the faces of these cubes (Fig. 3.15a). The derivation is simpler when starting from the face-centered cubic (fcc) lattice. It is then easy to see that the parallelepiped in which the basis vectors $\mathbf{A}$, $\mathbf{B}$, $\mathbf{C}$ connect a vertex P to the centers Q, R, S of the faces starting in P is a rhombohedron with apical angle $60°$, since the triangles PQR, PRS and PSQ are equilateral. Thus the primitive cell, taken individually, does not feature all the symmetry elements of the face-centered cubic lattice (in particular the 4-fold axes). This is the reason for choosing the quadruple face-centered cubic cell.

Thus, for three-dimensional lattices, several multiple cells are used to describe lattices featuring more symmetry elements than the primitive cell shows. Each multiple cell defines a lattice mode. If the cell that describes the lattice is a primitive cell, the lattice mode is said to be primitive, and it is noted as P. If the cell is base-centered (double cell), the lattice mode is A or B or C depending on whether the centered bases are respectively the faces $(\mathbf{b}, \mathbf{c})$ or $(\mathbf{c}, \mathbf{a})$ or $(\mathbf{a}, \mathbf{b})$. The generic designation for this type of centering is S (for side-face centered). If the cell is body-centered (double cell), the lattice mode is I (as in "inside"), and if the cell has all of its faces centered (quadruple cell), it is noted F (as in "face"). The reader will easily check that these multiple cells do define a crystal lattice. In contrast, a cell to which one would attempt to add

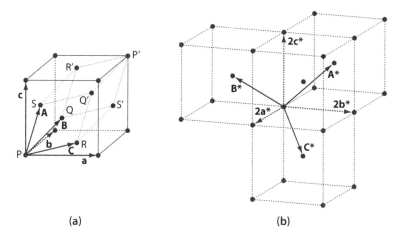

(a)   (b)

Figure 3.15: Face-centered cubic lattice (a) and its reciprocal lattice (b)

nodes at the centers of four pair-wise parallel faces does not define a crystal lattice (Exercise 3.6). To each crystal system there correspond one or several lattice modes. The set of all these lattice modes form the 14 Bravais lattices, distributed over the 7 crystal systems.

The rule for determining which Bravais lattice, hence which unit cell, should describe a given structure is to choose the cell
(a) featuring the symmetry of the lattice and
(b) with the smallest volume.
Two examples are very illuminating: the face-centered cubic lattice quoted above, and the body-centered cubic lattice, for which the primitive cell is a rhombohedron with angle $\alpha = 109.47°$ ($\cos\alpha = -1/3$) (Fig. 3.16).

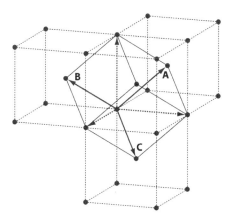

Figure 3.16: The body-centered cubic lattice. The primitive cell is defined by the vectors **A**, **B** and **C**.

One may ask why all of the lattice modes are not defined in each of the crystal systems. Actually, some multiple (centered) cells provide no improvement over the primitive cell or over some other multiple cell with lower multiplicity, or again the multiple cell does not feature the symmetry of the lattice any more. Consider for example the tetragonal lattice. There is no advantage in defining a tetragonal lattice in which the square base (**a**, **b**) is centered, since the unit cell built on the half-diagonals of the square, **a′** and **b′**, and the vector **c** is tetragonal too (Fig. 3.17). Furthermore, it can be shown that a face-centered tetragonal unit cell can be described by a centered cell twice as small (Exercise 3.10). A tetragonal unit cell with two opposite non-square faces centered is not a tetragonal cell, since the lattice built on this cell does not have a 4-fold axis any more. The only multiple tetragonal unit cell is the tetragonal body-centered cell, for which no primitive cell can represent the tetragonal symmetry of the lattice.

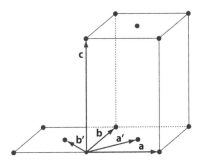

Figure 3.17: A tetragonal lattice with its square base centered does not need to be considered. The primitive cell, with basis vectors **a′**, **b′** and **c**, is tetragonal.

3.7.2. The fourteen Bravais lattices

The last column of Table 3.3 indicates for each crystal system the various lattice modes. There are all together 14 Bravais lattices (1 triclinic, 2 monoclinic, 4 orthorhombic, 2 tetragonal, 1 rhombohedral, 1 hexagonal, and 3 cubic). Two rows were set aside for the monoclinic system, in order to be able to state the denomination of the base-centered lattices. In this case, the bases which can be centered are the rectangular faces parallel to the 2-fold axis. If we choose to call **c** the basis vector parallel to this axis, then the lattice mode can be A or B-type. If we choose to call it **b**, then it can be C or A-type. The orthorhombic lattice modes can be primitive (P), body-centered (I), face-centered (F), or base-centered A, B or C (generically, side-face centered, *i.e.* S), since the three faces can be centered. The triclinic, rhombohedral and hexagonal lattices can only be primitive. The rhombohedral lattice mode is usually noted R rather than P. The reason will be given in Chapters 5 (Sect. 5.6) and 6 (Sect. 6.3.3). The tetragonal lattice modes can be primitive (P) or centered (I), and the cubic lattice modes can be primitive (P), body-centered (I) or face-centered (F).

It can be shown that an F tetragonal lattice mode can be described via an I lattice mode (Exercise 3.10) and an F or I monoclinic lattice mode via an A or C lattice mode (Exercise 3.11), in full agreement with their absence among the Bravais lattice modes.

3.7.3. Reciprocal lattices of the non-primitive lattices

We take as an example the face-centered cubic lattice, and search for its reciprocal lattice, the latter being defined by starting from a primitive cell of the direct lattice. Let $\mathbf{a}$, $\mathbf{b}$, $\mathbf{c}$ be the basis vectors of the multiple (quadruple) cubic unit cell, and $\mathbf{A}$, $\mathbf{B}$ and $\mathbf{C}$ the basis vectors of a primitive cell (Fig. 3.15a) such that

$$\mathbf{A} = \frac{1}{2}(\mathbf{b} + \mathbf{c}), \qquad \mathbf{B} = \frac{1}{2}(\mathbf{c} + \mathbf{a}), \qquad \mathbf{C} = \frac{1}{2}(\mathbf{a} + \mathbf{b}).$$

Let us denote the volume of this primitive cell by v_0, and let us determine the basis vectors $\mathbf{A}^*$, $\mathbf{B}^*$, $\mathbf{C}^*$ of the reciprocal lattice, starting with $\mathbf{A}^*$.

$$\mathbf{A}^* = \frac{1}{v_0}(\mathbf{B} \times \mathbf{C}) = \frac{1}{4v_0}[(\mathbf{c} + \mathbf{a}) \times (\mathbf{a} + \mathbf{b})]$$

$$= \frac{1}{4v_0}[(\mathbf{c} \times \mathbf{a}) + (\mathbf{c} \times \mathbf{b}) + (\mathbf{a} \times \mathbf{b})]. \tag{3.16}$$

Let $\mathbf{a}^*$, $\mathbf{b}^*$, $\mathbf{c}^*$ be the basis vectors of the lattice reciprocal to the direct lattice for which the primitive cell would be the cube built on $\mathbf{a}$, $\mathbf{b}$ and $\mathbf{c}$. The volume $(\mathbf{a}, \mathbf{b}, \mathbf{c})$ of this cubic unit cell of the direct lattice is $4v_0$, so that

$$\mathbf{a}^* = \frac{1}{4v_0}(\mathbf{b} \times \mathbf{c}),$$

$$\mathbf{b}^* = \frac{1}{4v_0}(\mathbf{c} \times \mathbf{a}),$$

$$\mathbf{c}^* = \frac{1}{4v_0}(\mathbf{a} \times \mathbf{b}).$$

Equation (3.16) thus yields

$$\mathbf{A}^* = \mathbf{b}^* - \mathbf{a}^* + \mathbf{c}^*.$$

Similarly, we obtain

$$\mathbf{B}^* = \mathbf{c}^* - \mathbf{b}^* + \mathbf{a}^*,$$
$$\mathbf{C}^* = \mathbf{a}^* - \mathbf{c}^* + \mathbf{b}^*.$$

If we build the reciprocal lattice starting from this primitive cell $(\mathbf{A}^*, \mathbf{B}^*, \mathbf{C}^*)$, we see that the lattice is body-centered cubic, the edges of the cube being the vectors $2\mathbf{a}^*$, $2\mathbf{b}^*$, $2\mathbf{c}^*$ (Fig. 3.15b).

Note: the volume V^* of this body-centered cubic reciprocal unit cell is 8 times the volume v^* of the unit cell built on $\mathbf{a}^*$, $\mathbf{b}^*$ and $\mathbf{c}^*$. Since v_0 is the volume of the primitive cell (the rhombohedron with angle $60°$), the volume of the unit cell built on $\mathbf{a}$, $\mathbf{b}$, $\mathbf{c}$ is $4v_0$ and the volume of its reciprocal cell (see Sect. 3.2.3) is $v^* = 1/(4v_0)$. We thus obtain $V^* = 8v^* = 2/v_0$. Since this reciprocal cell is body-centered, the volume of the primitive reciprocal unit cell is half this volume, $i.e.$ $1/v_0$ as expected.

The above derivation, which showed that the reciprocal lattice of a face-centered cubic lattice is a body-centered cubic lattice, was made on the basis of vector relations, without using the fact that the unit cell is cubic. The same derivation would apply for the orthorhombic face-centered lattice, and this would yield a body-centered reciprocal orthorhombic lattice. The orthorhombic reciprocal cell would have basis vectors $2\mathbf{a}^*$, $2\mathbf{b}^*$, $2\mathbf{c}^*$ if we again call the basis vectors of the direct orthorhombic cell $\mathbf{a}$, $\mathbf{b}$ and $\mathbf{c}$ and those of the corresponding reciprocal lattice $\mathbf{a}^*$, $\mathbf{b}^*$ and $\mathbf{c}^*$.

The same approach shows (Exercise 3.13) that the reciprocal lattice of a body-centered cubic lattice with basis vectors $\mathbf{a}$, $\mathbf{b}$ and $\mathbf{c}$ is a face-centered cubic lattice with basis vectors $2\mathbf{a}^*$, $2\mathbf{b}^*$ and $2\mathbf{c}^*$.

In the same way (Exercise 3.14), the reciprocal lattice of a base-centered lattice is shown to be a base-centered lattice.

3.8. Surface crystal lattice

This section is related to a special area of crystallography: the study of surfaces. It is mandatory for whoever has any interest in investigating crystal surfaces. This research area developed considerably over the last years due to the importance of two-dimensional surfaces and interfaces in many phenomena such as catalysis, corrosion, the manufacturing of metal-semiconductor junctions, etc.

To understand the mechanisms involved in these processes, the first step consists in investigating plane surfaces with well-defined crystal orientation and with low Miller indices, in order to determine their structure. The atoms located near the surface lose, due to the cut, some of their bonds. They must rearrange, together with their neighbors, in order to find a new equilibrium position. This rearrangement can extend over several atomic layers near the surface. One then defines on the one hand the "bulk" of the crystal, with its three-dimensional periodic structure, and on the other hand the "surface", which usually consists of several atomic layers, displaced with respect to their position in the bulk, and possibly with a different stoichiometry. We also note that the study of surfaces is related not only to clean crystal surfaces, but also to the way foreign atoms, called adatoms, deposit on the surface.

The "surface", thus defined as the set of atomic layers that are modified by the cut and possibly by the deposited adatoms, retains a two-dimensional periodicity. It is thus possible to define a surface crystal lattice. Determining this lattice is the first step in an investigation of surfaces.

3.8.1. Cut surface

Imagine we cut a crystal parallel to a given plane (hkl), and suppose the cut does not alter the atom positions. We can then determine a surface unit cell, defined as the section of the crystal lattice in the bulk by a plane (hkl). This surface unit cell depends on the surface orientation. Consider the example of a primitive cubic lattice, with basis vectors $\mathbf{a}$, $\mathbf{b}$ and $\mathbf{c}$. For an (001) surface, the surface cell is square and has basis vectors $\mathbf{a_1} = \mathbf{a}$ and $\mathbf{b_1} = \mathbf{b}$, while a (101) surface produces a rectangular surface cell with basis vectors $\mathbf{a_1} = \mathbf{a} - \mathbf{c}$ and $\mathbf{b_1} = \mathbf{b}$ (Fig. 3.18a).

Now consider a (101) surface in a body-centered lattice, then in a face-centered lattice. The surface unit cells are different. If the bulk lattice is body-centered, the (101) surface unit cell is centered rectangular with basis vectors $\mathbf{a_1} = \mathbf{a} - \mathbf{c}$ and $\mathbf{b_1} = \mathbf{b}$ (Fig. 3.18b). In the case of the face-centered bulk unit cell, the surface unit cell is simple rectangular, with basis vectors $\mathbf{a_1} = 1/2(\mathbf{a} - \mathbf{c})$ and $\mathbf{b_1} = \mathbf{b}$ (Fig. 3.18c). We thus determine easily a surface unit cell called the bulk-derived surface unit cell. We call $\mathbf{a_1}$ and $\mathbf{b_1}$ the basis vectors of this bulk-derived surface unit cell.

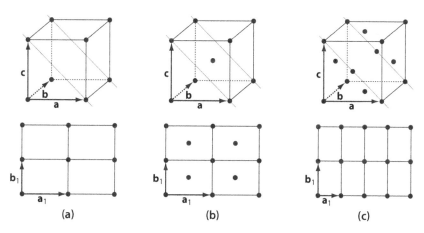

Figure 3.18: Bulk-derived (101) surface unit cells in the (a) primitive cubic, (b) body-centered cubic and (c) face-centered cubic lattices

3.8.2. Real surface

The existence of the surface alters the atom positions in the surface layer. If the displacements affect only the direction perpendicular to the surface without altering the stoichiometry, the bulk-derived unit cell is retained at the surface. This case is referred to as relaxation of the atomic positions. However, surface atoms usually rearrange, thus altering the lattice unit cell at the surface. The surface is said to reconstruct. A simple case of rearrangement or reconstruction is that of the (101) surface of gold. Gold has in the bulk a face-centered cubic lattice. The crystal motif consists of one atom only, so that Figure 3.18c can represent the atom positions in a unit cell. Figure 3.19a shows the atom positions which would prevail on the (101) surface if there were no reconstruction. The full circles correspond to the atoms in the first layer, and the empty circles to the second layer. It so happens that the surface reconstructs so that the atoms of every other row parallel to the [10$\bar{1}$] row in the first layer disappear (Fig. 3.19b). This is described by saying that there is a missing row. The reconstructed surface unit cell, *i.e.* the unit cell of the real surface, then has as its basis vectors $\mathbf{a}_2 = \mathbf{a}_1 = \frac{1}{2}(\mathbf{a} - \mathbf{c})$ and $\mathbf{b}_2 = 2\mathbf{b}_1 = 2\mathbf{b}$.

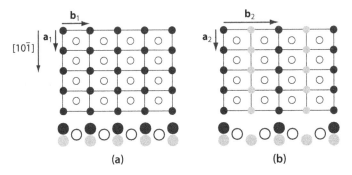

(a)   (b)

Figure 3.19: Reconstruction of the (101) surface of gold. The full circles represent the positions of atoms in the first layer, the empty circles those of the second layer and the gray ones those of the third layer. The upper part of each figure shows a top view; the lower part shows a cross-section. (a) The state of the surface if no reconstruction occurred. (b) Reconstructed surface, with its missing row parallel to [10$\bar{1}$].

3.8.3. Notations

The basis vectors $(\mathbf{a}_2, \mathbf{b}_2)$ of the real surface unit cell are generally expressed as a function of the basis vectors $(\mathbf{a}_1, \mathbf{b}_1)$ of the bulk-derived unit cell through the relations:

$$\mathbf{a}_2 = m_{11}\mathbf{a}_1 + m_{12}\mathbf{b}_1, \qquad\qquad \mathbf{b}_2 = m_{21}\mathbf{a}_1 + m_{22}\mathbf{b}_1$$

or
$$\begin{pmatrix} \mathbf{a}_2 \\ \mathbf{b}_2 \end{pmatrix} = M \begin{pmatrix} \mathbf{a}_1 \\ \mathbf{b}_1 \end{pmatrix}.$$

Matrix M defines the surface lattice completely. In the case of the (101) surface of gold which we just discussed, M is especially simple:

$$\begin{pmatrix} 1 & 0 \\ 0 & 2 \end{pmatrix}.$$

Another notation, due to Elizabeth A. Wood, is frequently used although it is not as general since it applies only to cases where:

1. The angle between vectors $\mathbf{a}_2$ and $\mathbf{b}_2$ is the same as that between $\mathbf{a}_1$ and $\mathbf{b}_1$.

2. The lengths of $\mathbf{a}_2$ and $\mathbf{b}_2$ are expressed as a function of those of $\mathbf{a}_1$ and $\mathbf{b}_1$ by relations: $\|\mathbf{a}_2\| = p \|\mathbf{a}_1\|$ and $\|\mathbf{b}_2\| = q \|\mathbf{b}_1\|$.

Noting φ the angle by which the surface lattice possibly has turned with respect to the bulk-derived lattice, the lattice for the (hkl) surface of a material with formula Y is indicated as:

$$Y(hkl)(p \times q)R\varphi.$$

The simple case of the (101) surface of gold is thus described as Au (101) (1×2). We note that the rotation part is omitted if the lattice did not turn. In this notation, the bulk-derived cell is often designated as the (1×1) unit cell.

Two examples of surface reconstructions where Wood's notation can be used are shown on Figures 3.20 and 3.21. Figure 3.20 shows the reconstruction of a (001) surface of tungsten (W), cooled to below room temperature. This crystal is body-centered cubic with edge a, and the motif contains only one W atom. Figure 3.20a shows the atoms of this surface, as seen from above, if there were no reconstruction. The atoms in the first layer are the empty circles, and those of the second layer are gray circles. The bulk-derived surface cell is a square with edges $\mathbf{a}$ and $\mathbf{b}$, so that $\mathbf{a}_1 = \mathbf{a}$ and $\mathbf{b}_1 = \mathbf{b}$. Due to the reconstruction, the first layer atoms undergo a displacement along the arrows, thus creating, in a $[1\bar{1}0]$ direction, a zigzag-like pattern, where the angle is no more 90° (Fig. 3.20b). The surface unit cell is a square (full lines on Fig. 3.20b), rotated by 45° with respect to the bulk-derived cell, and its edge is $a\sqrt{2}$. It is described as W (001) $(\sqrt{2} \times \sqrt{2})$ R45°. It is sometimes described using the square face-centered cell (dotted lines) which is not rotated with respect to the bulk-derived cell, and it is then noted as c(2×2) with c standing for "centered".

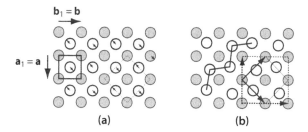

Figure 3.20: Reconstruction of the W (001) surface. The empty circles represent the atoms of the first layer, the gray circles those of the 2nd layer. (a) Bulk-derived cell. The arrows indicate the displacements which the atoms undergo during reconstruction. (b) Reconstructed surface and its primitive (full lines) and centered (dotted lines) unit cells.

The second example deals with a surface covered with adatoms, the (111) surface of silicon (face-centered cubic lattice), covered with 1/3 of a mono-layer of silver. The bulk-derived unit cell of a (111) surface of a fcc lattice is a hexagonal unit cell with basis vectors $\mathbf{a}_1 = \frac{1}{2} [10\bar{1}]$ and $\mathbf{b}_1 = \frac{1}{2} [\bar{1}10]$

represented on Figure 3.21a, where the atoms in the first layer are full circles and those of the second layer are empty circles. The Ag atoms deposit above every third atom of the second layer, forming a regular lattice (large circles in thick lines in Fig. 3.21b). The surface unit cell is hexagonal, turned by 30° with respect to the bulk-derived cell, and the edge length is $\sqrt{3}$ times larger than the edge of the bulk-derived hexagonal unit cell. It is described as Ag/Si (111) ($\sqrt{3} \times \sqrt{3}$) R30°.

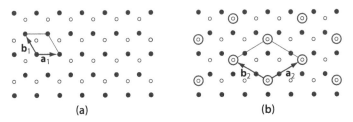

(a)   (b)

Figure 3.21: Reconstruction of a (111) surface of silicon covered with a third of a monolayer of silver. (a) Clean bulk-derived unit cell. The full circles represent the first-layer atoms and the empty circles the second-layer atoms. (b) Surface covered with Ag atoms, represented by large thick-line circles and located above second-layer atoms.

3.8.4. Reciprocal lattice

Interpreting (X-ray or electron) diffraction by surfaces involves defining a surface reciprocal lattice. It is therefore necessary to associate to the basis vectors of the surface unit cell a vector which is not in the surface plane. The appropriate vector is the smallest vector of the bulk crystal lattice perpendicular to the surface, and it is noted as $\mathbf{n}$.

$\mathbf{a}_2$ and $\mathbf{b}_2$ being the basis vectors of the unit cell of an (hkl) surface, the reciprocal lattice of the unit cell $(\mathbf{a}_2, \mathbf{b}_2, \mathbf{n})$ is then defined by the vectors $\mathbf{a}_2^*$, $\mathbf{b}_2^*$, $\mathbf{n}^*$ such that

$$\mathbf{a}_2^* = \frac{\mathbf{b}_2 \times \mathbf{n}}{V},$$

$$\mathbf{b}_2^* = \frac{\mathbf{n} \times \mathbf{a}_2}{V},$$

$$\mathbf{n}^* = \frac{\mathbf{a}_2 \times \mathbf{b}_2}{V}.$$

V, the volume built on vectors $\mathbf{a}_2$, $\mathbf{b}_2$ and $\mathbf{n}$, is such that $V = \mathbf{n} \cdot (\mathbf{a}_2 \times \mathbf{b}_2)$. We note that the vector $\mathbf{n}^*$ is perpendicular to the surface and its norm is equal to the reciprocal of the norm of vector $\mathbf{n}$.

For the reconstructed surfaces discussed above, the relevant vectors $\mathbf{n}$ are:
surface Au (101) (1 × 2) : $\mathbf{n} = \frac{1}{2}[101]$,
surface W (001) ($\sqrt{2} \times \sqrt{2}$) R45°: $\mathbf{n} = [001]$,
surface Ag/Si (111) ($\sqrt{3} \times \sqrt{3}$) R30°: $\mathbf{n} = [111]$.

3.9. Exercises

Exercise 3.1

1. Determine the indices (hkl) of the plane that goes through the three nodes (1, 0, 0), (0, 1, 0) and (0, 0, 3).

2. Show that the three rows [211], [120] and [302] are coplanar. What are the indices of their common plane?

3. Determine the row common to the planes (311) and ($1\bar{1}0$).

Exercise 3.2

Titanium oxide TiO_2 has a tetragonal lattice with parameters $a = 4.59$ Å and $c = 2.96$ Å.

1. What is the angle between faces (101) and ($10\bar{1}$)?

2. Show that, in this system, row $[hk0]$ is perpendicular to plane $(hk0)$, but that the same does not apply to row $[0kl]$ and plane $(0kl)$.

3. Calculate the lattice distance of planes (321).

Exercise 3.3

1. Consider a plane crystallographic lattice with basis vectors **a** and **b**. We choose an axis system Oxy parallel to **a** and **b**, with unit lengths a and b respectively. Consider the row D that goes through points M and N, with coordinates respectively (2, 0) and (0, 3). Determine:

   (a) the equation of row D.

   (b) the number of rows parallel to D located between D and the origin O of the lattice.

   (c) the equations of the rows belonging to this family that go through the point with coordinates (1, 0) and (0, 1) respectively.

2. Consider the three-dimensional crystallographic lattice with basis vectors **a**, **b** as given in (1) and **c**, where **c** is a general lattice vector which does not belong to plane (**a**, **b**).

   (a) Determine the values of h, k and l for the family (hkl) of lattice planes parallel to vector **MN** and **c**.

   (b) Determine the values of h, k and l for the family (hkl) of lattice planes parallel to the plane that goes through **MN** and through the point P with crystallographic coordinates (0, 0, 1).

Exercise 3.4

Hexagonal lattice: we recall that the basis vectors **a**, **b** and **c** of the unit cell are such that $a = b \neq c$ and that the angles are $\alpha = \beta = 90°$ and $\gamma = 120°$.

1. Determine the reciprocal lattice of the hexagonal lattice, and give its orientation with respect to the direct lattice.

2. Calculate the length of the vector $[110]^*$ of the reciprocal lattice. Deduce the value of the lattice distance between (110) planes. Retrieve this result through a geometrical argument without using the reciprocal lattice.

3. Calculate the angle between planes (120) and (100).

Exercise 3.5

Consider a hexagonal lattice. The plane $(hkil)$ closest to the origin of the lattice intersects plane (**a**, **b**) along a line which intersects vector **a** at point M such that $\mathbf{OM} = \mathbf{a}/h$, vector **b** at point N such that $\mathbf{ON} = \mathbf{b}/k$ and vector **d** at point Q such that $\mathbf{OQ} = \mathbf{d}/i$. Vector **d** is deduced from **b** through a rotation by 120° around **c**. Draw a figure in plane (**a**, **b**) and note that $\mathbf{a} + \mathbf{b} + \mathbf{d} = \mathbf{0}$.

1. Express the area of triangles OMN, OMQ and OQN as a function of the cross product of two of the basis vectors **a**, **b** or **d**.

2. By writing that the area of OMN is the sum of the areas of OMQ and OQN and by substituting for **d** its value as a function of **a** and **b**, show that you obtain the relation $h + k + i = 0$.

3. Consider the $(1\,2\,\bar{3}\,1)$ lattice plane family. Determine the lattice plane family which deduces from it through a rotation by $2\pi/3$, then the family obtained through a rotation by $4\pi/3$. In the same way, determine the families of planes obtained through symmetry with respect to plane (**a**, **c**) and with respect to plane (**b**, **c**) respectively.

Exercise 3.6

Show that the set of points consisting of the nodes of a crystal lattice and the points at the centers of four pairwise parallel faces of the unit cell, do not form a crystal lattice. In other words, a crystal lattice cannot have two distinct faces that are centered. Note that crystal periodicity implies that every face deduced through a lattice translation from one that is centered must also be centered.

Exercise 3.7

Consider the crystal structures, with cubic lattice, shown on Figures 8.3 (CsCl, NaCl, ZnS) and 8.6 (diamond). Determine their Bravais lattice and their motif.

Same question for the structures of Figure 8.4: cubic CaF_2, tetragonal TiO_2 and cubic $CaTiO_3$.

Exercise 3.8

Consider the structure of graphite shown on Figure 8.11.

1. Draw a projection onto the plane of the hexagons, indicating their heights, of the atom positions. Take as a unit twice the distance between two neighboring carbon planes.

2. Determine the unit cell, its parameters using the data shown on Figure 8.11, and the number of atoms per unit cell.

Exercise 3.9

The structure of a compound consisting of nickel and gadolinium can be represented as follows. Decompose space into a set of identical right prisms, with regular hexagons as their bases, and contacting one another. Each apex and each center of the lateral faces of the prism is occupied by a nickel atom. The Gd atoms are located at the centers of the hexagonal bases of the prism. In the basal plane, the shortest Ni–Gd distance is 2.83 Å and the height of the prisms is 3.97 Å.

1. Determine the chemical formula of this compound.

2. Draw a projection of the atom positions onto the basal plane, stating the heights, and deduce the lattice and the motif.

3. With the lattice now defined, give the Miller indices of the prism faces.

4. Calculate the angle between planes $(1\bar{1}0)$ and (110).

Exercise 3.10

Show that a tetragonal F (face-centered) lattice mode can be described by a tetragonal body-centered unit cell (I type lattice mode) with half the volume. It will be useful to make a projection of the lattice nodes onto a plane perpendicular to the basis vector $\mathbf{c}$, mentioning the heights.

Exercise 3.11

Show that a monoclinic F lattice mode can be described by a monoclinic I lattice mode with a unit cell half the volume, and that this I lattice mode can be described by a base-centered monoclinic unit cell with the same volume. It will be useful to draw a projection of the lattice nodes onto the plane normal to the 2-fold axis of the initial monoclinic lattice, specifying the heights.

Exercise 3.12

Draw a projection onto the plane (111) of the nodes of a cubic F lattice, indicating the heights. Show that this yields a figure similar to Figure 3.14c. Use the fact that direction [111], parallel to a major diagonal of the cube, is a 3-fold axis for the (111) planes which are perpendicular to it.

Exercise 3.13

Consider a body-centered cubic lattice with basis vectors $\mathbf{a}$, $\mathbf{b}$ and $\mathbf{c}$. Let $\mathbf{A}$, $\mathbf{B}$ and $\mathbf{C}$ be the basis vectors of the primitive rhombohedral unit cell:

$$\mathbf{A} = \frac{1}{2}(\mathbf{b} + \mathbf{c} - \mathbf{a}), \qquad \mathbf{B} = \frac{1}{2}(\mathbf{c} + \mathbf{a} - \mathbf{b}), \qquad \mathbf{C} = \frac{1}{2}(\mathbf{a} + \mathbf{b} - \mathbf{c}).$$

Call $\mathbf{a}^*$, $\mathbf{b}^*$ and $\mathbf{c}^*$ the basis vectors of the reciprocal unit cell of a primitive cubic unit cell with basis vectors $\mathbf{a}$, $\mathbf{b}$ and $\mathbf{c}$.

1. Determine the basis vectors $\mathbf{A}^*$, $\mathbf{B}^*$ and $\mathbf{C}^*$ of the reciprocal lattice as a function of $\mathbf{A}$, $\mathbf{B}$ and $\mathbf{C}$, then as a function of $\mathbf{a}^*$, $\mathbf{b}^*$ and $\mathbf{c}^*$.

2. Show that the reciprocal lattice of the body-centered cubic unit cell with basis vectors $\mathbf{A}^*$, $\mathbf{B}^*$ and $\mathbf{C}^*$ is a face-centered cubic lattice with basis vectors $2\mathbf{a}^*$, $2\mathbf{b}^*$ and $2\mathbf{c}^*$.

Exercise 3.14

Consider any base-centered lattice, for example A-type, with basis vectors $\mathbf{a}$, $\mathbf{b}$ and $\mathbf{c}$. Choose as basis vectors of the primitive unit cell the vectors

$$\mathbf{A} = \mathbf{a}, \qquad \mathbf{B} = \frac{1}{2}(\mathbf{b} - \mathbf{c}), \qquad \mathbf{C} = \frac{1}{2}(\mathbf{b} + \mathbf{c}).$$

Let $\mathbf{a}^*$, $\mathbf{b}^*$ and $\mathbf{c}^*$ be the basis vectors of the reciprocal unit cell of a primitive lattice, with basis vectors $\mathbf{a}$, $\mathbf{b}$ and $\mathbf{c}$.

Show that the reciprocal lattice of the A-type base-centered lattice is a base-centered A-type lattice with basis vectors $\mathbf{a}^*$, $2\mathbf{b}^*$ and $2\mathbf{c}^*$.

Appendix 3A. The metric tensor

The metric tensor is particularly useful for numerical calculations.

3A.1. Definition

The term tensor will be explicitly defined and discussed in Chapter 9, but the formal definition is not needed here. The term metric means it deals with measure, here the lengths and angles between the basis vectors of the crystal unit cell.

The components of the metric tensor are defined by the dot products of the basis vectors of the unit cell, which we note as $\mathbf{a}_1$, $\mathbf{a}_2$ and $\mathbf{a}_3$, *i.e.* the $\mathbf{a}_i$ with i ranging from 1 to 3, so that

$$g_{ij} = \mathbf{a}_i \cdot \mathbf{a}_j.$$

The metric tensor is also defined in reciprocal space:

$$g_{ij}^* = \mathbf{a}_i^* \cdot \mathbf{a}_j^*.$$

3A.2. Volume of the unit cell

The volume V of the direct unit cell is:

$$V = \mathbf{a}_1 \cdot (\mathbf{a}_2 \times \mathbf{a}_3) = (\mathbf{a}_1, \mathbf{a}_2, \mathbf{a}_3).$$

We express the basis vectors of the unit cell in an orthonormal basis $\{\mathbf{e}_1, \mathbf{e}_2, \mathbf{e}_3\}$:

$$\mathbf{a}_1 = \sum_{j=1}^{3} B_{1j}\mathbf{e}_j$$

which can be written using Einstein's convention, defined in Section 9.3.1. Einstein's convention consists in suppressing the summation sign (which becomes implicit) in a product of factors when the subscript over which the summation is to be performed is repeated.

Thus we can write

$$\mathbf{a}_1 = B_{1j}\mathbf{e}_j.$$

Since the subscript j appears twice, it is clear that summation must be performed over j. In a similar way, we write for each of the basis vectors $\mathbf{a}_i$:
$\mathbf{a}_i = B_{ij}\mathbf{e}_j.$

It is easy to check that

$$(\mathbf{a}_1, \mathbf{a}_2, \mathbf{a}_3) = \det B \, (\mathbf{e}_1, \mathbf{e}_2, \mathbf{e}_3)$$

where $\det B$ is the determinant of the matrix $\{B_{ij}\}$ or B matrix.

Since the basis $\{\mathbf{e}_1, \mathbf{e}_2, \mathbf{e}_3\}$ is orthonormal, its volume is equal to 1, hence

$$V = \det B, \tag{3A.1}$$

$$g_{ij} = \mathbf{a}_i \cdot \mathbf{a}_j = B_{ik}\mathbf{e}_k \cdot B_{jl}\mathbf{e}_l = B_{ik}B_{jl}\mathbf{e}_k \cdot \mathbf{e}_l$$
$$= B_{ik}B_{jl}\delta_{kl} = B_{ik}B_{jk}. \tag{3A.2}$$

B_{jk} can be substituted by B_{kj}^T where B^T is the transposed matrix of B.

We obtain $g_{ij} = B_{ik}B_{kj}^T$, so that $g = BB^T$ if we call g the matrix $\{g_{ij}\}$, and

$$\det g = \det B \det B^T = (\det B)^2 \tag{3A.3}$$

since the determinants of a matrix and of its transpose are equal.

Comparing (3A.1) and (3A.3), we see that

$$V^2 = \det g. \tag{3A.4}$$

It is on the basis of this formula that, in Section 3A.5 below, the volume of the triclinic unit cell will be easily calculated.

3A.3. Product of the matrices associated to the direct and reciprocal metric tensors

The reciprocal metric tensor is defined by the coefficients $g_{ij}^* = \mathbf{a}_i^* \cdot \mathbf{a}_j^*$, where the $\mathbf{a}_i^*$ ($i = 1$ to 3) are the basis vectors of the reciprocal unit cell.

Let us express the vectors $\mathbf{a}_i^*$ in the direct basis $\{\mathbf{a}_1, \mathbf{a}_2, \mathbf{a}_3\}$:

$$\mathbf{a}_i^* = M_{ik}\mathbf{a}_k$$
and $$g_{ij}^* = \mathbf{a}_i^* \cdot \mathbf{a}_j^* = M_{ik}\mathbf{a}_k \cdot \mathbf{a}_j^* = M_{ik}\,\delta_{kj} = M_{ij}$$

since the reciprocal lattice vectors are defined by relations (3.4), which can be written:

$$\mathbf{a}_i^* \cdot \mathbf{a}_k = \delta_{ik}$$

where δ_{ik} is Kronecker's symbol, equal to 1 if $i = k$ and equal to 0 otherwise.

Using

$$\mathbf{a}_i^* \cdot \mathbf{a}_k = M_{ij}\mathbf{a}_j \cdot \mathbf{a}_k = M_{ij}g_{jk} = g_{ij}^*g_{jk},$$

we obtain

$$g_{ij}^*g_{jk} = \delta_{ik}.$$

This relation shows that the product of the matrices g and g^* is equal to the unit matrix E

$$gg^* = E. \tag{3A.5}$$

The calculation of the reciprocal metric tensor is therefore performed by inverting the matrix associated to the direct metric tensor. Taking the determinant of both sides of Equation (3A.5), we obtain:

$$\det g \det g^* = 1.$$

It can be shown that $\det g^* = (V^*)^2$, as was done for the volume of the direct unit cell, and we retrieve the fact that the volume of the reciprocal unit cell is equal to the reciprocal of the volume of the direct unit cell.

3A.4. Calculation of lattice distances

The lattice distance of a family of planes (hkl) is given by relation (3.8):

$$d_{hkl} = \frac{1}{\|\mathbf{h}^*\|} = \frac{1}{\|h_i \mathbf{a}_i^*\|}.$$

We can write $\|h_i \mathbf{a}_i^*\|^2 = (h_i \mathbf{a}_i^*) \cdot (h_j \mathbf{a}_j^*) = h_i h_j g_{ij}^*$ and

$$d_{hkl} = \frac{1}{\sqrt{h_i h_j g_{ij}^*}} \tag{3A.6}$$

which is easily calculated knowing the reciprocal metric tensor which is calculated by inverting matrix g (Eq. (3A.5)).

3A.5. Applications

The metric tensor can be written, with the usual notations for the parameters of the crystal lattice:

$$g = \begin{pmatrix} a^2 & ab\cos\gamma & ac\cos\beta \\ ab\cos\gamma & b^2 & bc\cos\alpha \\ ac\cos\beta & bc\cos\alpha & c^2 \end{pmatrix}.$$

The determinant of this matrix is given by

$$\det g = a^2 b^2 c^2 (1 - \cos^2\alpha - \cos^2\beta - \cos^2\gamma + 2\cos\alpha\cos\beta\cos\gamma)$$

whence we deduce through (3A.4) the volume V of the unit cell:

$$V = abc\sqrt{1 - \cos^2\alpha - \cos^2\beta - \cos^2\gamma + 2\cos\alpha\cos\beta\cos\gamma}.$$

Inverting matrix g leads to matrix g^*, so that

$$g^* = \frac{1}{V^2} \begin{pmatrix} b^2 c^2 \sin^2\alpha & abc^2 K_{\alpha\beta\gamma} & ab^2 c K_{\gamma\alpha\beta} \\ abc^2 K_{\alpha\beta\gamma} & a^2 c^2 \sin^2\beta & a^2 bc K_{\beta\gamma\alpha} \\ ab^2 c K_{\gamma\alpha\beta} & a^2 bc K_{\beta\gamma\alpha} & a^2 b^2 \sin^2\gamma \end{pmatrix}$$

where we use the temporary notation $\cos\alpha\cos\beta - \cos\gamma = K_{\alpha\beta\gamma}$, with two other coefficients $K_{\beta\gamma\alpha}$ and $K_{\gamma\alpha\beta}$ being obtained through a circular permutation of the subscripts.

Since g^* is a symmetrical matrix, the general expression (3A.6) of d_{hkl} is given by

$$d_{hkl} = \frac{1}{\sqrt{g_{11}^* h^2 + g_{22}^* k^2 + g_{33}^* l^2 + 2g_{12}^* hk + 2g_{13}^* hl + 2g_{23}^* kl}}$$

and, in the most general case of the triclinic unit cell, we obtain:

$$d_{hkl} = \sqrt{\frac{1 - \cos^2 \alpha - \cos^2 \beta - \cos^2 \gamma + 2\cos \alpha \cos \beta \cos \gamma}{A}}$$

with

$$A = \frac{h^2}{a^2} \sin^2 \alpha + \frac{k^2}{b^2} \sin^2 \beta + \frac{l^2}{c^2} \sin^2 \gamma + 2\frac{hk}{ab}(\cos \alpha \cos \beta - \cos \gamma)$$
$$+ 2\frac{hl}{ac}(\cos \alpha \cos \gamma - \cos \beta) + 2\frac{kl}{bc}(\cos \beta \cos \gamma - \cos \alpha)$$

Chapter 4

Relationship between space groups and point groups

After showing that the symmetry operations of a crystal form a group, we show how to go over from this group to the space group characterizing the crystal structure on the microscopic scale, and then to the point group, which governs its macroscopic physical properties. The rigorous proof is preceded by an example which allows the reader to grasp these notions intuitively.

4.1. Introduction

The most general symmetry operation which any object, and in particular a crystal, can have can be noted [S(O), $\mathbf{t}$] where S represents a rotation or a rotoinversion (defined in Sect. 2.1) whose axis goes through point O, and $\mathbf{t}$ a translation. In the case of rotoinversion, O is also the associated symmetry center. Note that point O, which defines the position of the axis, has no reason not to change from one operation to another. We will show in the next section that the symmetry operations [S(O), $\mathbf{t}$] of the crystal make up a group. Among all these symmetry operations, there is the set of lattice translations T, and it is interesting to consider the symmetry operations [S(O), $\mathbf{t}$] modulo the lattice translations. The expression "[S(O), $\mathbf{t}$] modulo the lattice translations" means that we do not distinguish operations [S(O), $\mathbf{t}$] and [S(O), $\mathbf{t+n}$] if $\mathbf{n}$ is any lattice vector ($\mathbf{n}$ belonging to T). It will be shown that these operations form a group, called the position group or space group of the crystal.

Consider the direction of a straight line in the crystal, parallel to a vector $\mathbf{D}$, and submit it to any one of these symmetry operations. It becomes another straight line, parallel to vector $\mathbf{D}'$ such that

$$\mathbf{D}' = [S(O),\ \mathbf{t}]\,\mathbf{D} = S(O)\,\mathbf{D} = S\,\mathbf{D} \tag{4.1}$$

because $\mathbf{D}$ is unaltered by any translation $\mathbf{t}$. Reference to point O was suppressed because the result $\mathbf{D}'$ does not depend on the position of the rotation axis, nor on the point on the axis which is used for inversion when the operation

© Springer Science+Business Media Dordrecht 2014
C. Malgrange et al., *Symmetry and Physical Properties of Crystals*,
DOI 10.1007/978-94-017-8993-6_4

is a rotoinversion. The only relevant data are the direction of the axis and the rotation angle.

Since $\mathbf{D'}$ is deduced from $\mathbf{D}$ through a symmetry operation of the crystal, the macroscopic properties measured along directions $\mathbf{D}$ and $\mathbf{D'}$ will be the same (for example, the elongation of a bar parallel to $\mathbf{D}$ or to $\mathbf{D'}$ under the influence of a given increase in temperature). We thus understand that the symmetry of the macroscopic properties of the crystal is governed by the operations S solely defined by the direction of a rotation (or rotoinversion) axis and the rotation (or rotoinversion) angle. We will show in Section 4.3 that these operations S comprise a group, called the orientation group of the crystal or the point group. The latter term reminds us that, for this group, all the symmetry elements can be made to go through a given point, since the position of the symmetry element does not matter. The point group is thus obtained from the space group of the crystal by ignoring all translations (or, equivalently, by replacing all translations with the zero translation), and by having all the rotation or rotoinversion axes go through one given point, which will also be the center of symmetry used if there is a rotoinversion.

In order to illustrate the notions of space group and point group, consider the plane periodic figures of Figure 4.1, which we assume to extend to infinity. They can be considered as sorts of two-dimensional crystals.

Figure 4.1a (the crystal lattice of which is shown in Fig. 4.1b) features two families consisting of an infinity each of mirrors, parallel to one another and perpendicular to the plane of the figure. The mirrors of one of the families are denoted as M_i and the symmetry operation associated to mirror M_i is designated as m_i (symmetry with respect to plane M_i). The mirrors of the other family, perpendicular to the mirrors M_i, are denoted as M'_j and the associated symmetry operations as m'_j. The symmetry operations of this plane crystal have the form $[m_i, u\mathbf{a}+v\mathbf{b}]$ and $[m'_i, u'\mathbf{a}+v'\mathbf{b}]$,[1] where u, v, u' and v' are integers and $\mathbf{a}$ and $\mathbf{b}$ are the basis vectors of the lattice. We must add to these operations the products of these operations. These product operations are rotations by $180°$, denoted R_k, around axes perpendicular to the plane lattice, located at the intersections of the mirrors, followed by lattice translations, hence $[R_k, u''\mathbf{a} + v''\mathbf{b}]$. We will show (Exercise 4.1 and Sect. 7.2.2) that the product of a mirror operation m (symmetry with respect to a mirror plane M) by a translation $\mathbf{t}$ perpendicular to the mirror is a pure mirror operation where the new mirror is deduced from M by a translation $\mathbf{t}/2$. We thus obtain:

$$[m_1, \mathbf{b}] = [m_2, \mathbf{0}]; \quad [m_1, 2\mathbf{b}] = [m_3, \mathbf{0}]; \quad [m_1, -\mathbf{b}] = [m_0, \mathbf{0}]; \quad \text{etc.} \quad (4.2)$$

---

1. Here we simplified the notation. A mirror is equivalent to a $\bar{2}$ axis normal to the mirror for which the reference symmetry center is located at the point where the axis intersects the mirror. The notation consistent with $[S(O), \mathbf{t}]$ would, for operation $(m_i, u\mathbf{a}+v\mathbf{b})$ (see Sect. 4.2.1), be $[\bar{2}(O_i), u\mathbf{a} + v\mathbf{b}]$, with point O_i located at the intersection of the $\bar{2}$ axis with the plane M_i.

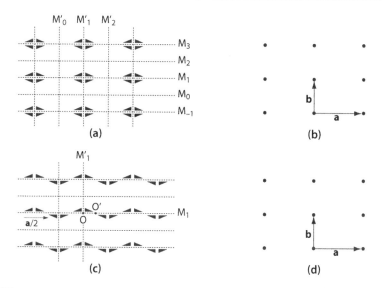

Figure 4.1: Examples of plane "crystals" (a) and (c) and their crystal lattices (b) and (d)

The mirror operations $[m_1, v\mathbf{b}]$ with v an integer thus describe the set of operations $[m_i, \mathbf{0}]$.

The set of operations $[m_i, u\mathbf{a}+v\mathbf{b}]$ can thus be replaced by the set $[m_1, u\mathbf{a}+v\mathbf{b}]$ where m_1 is the operation associated to any one particular mirror chosen as the reference (here M_1), and u and v any integers. This set of symmetry operations, modulo the lattice translations T, can be represented by operation $[m_1, \mathbf{0}]$. The same reasoning applied to operations $[m'_j, u'\mathbf{a} + v'\mathbf{b}]$ modulo T shows that $[m'_j, \mathbf{0}]$ can represent the set of operations $[m'_j, u'\mathbf{a}+v'\mathbf{b}]$ modulo T and that $[R_1, \mathbf{0}]$ can represent the set of operations $[R_k, u''\mathbf{a} + v''\mathbf{b}]$ modulo T. The symmetry operations of this crystal, modulo the lattice translations T, are thus identity, a mirror operation m_1, a mirror operation m'_1, and operation R_1, a rotation by π around the intersection of the mirrors M_1 and M'_1. This is a set G_s such that:

$$G_s = [E, \mathbf{0}] + [m_1, \mathbf{0}] + [m'_1, \mathbf{0}] + [R_1, \mathbf{0}] \quad \text{modulo T}.$$

These symmetry operations make up the space group. The point group G_0, obtained by deleting all translations and having all the symmetry elements go through the same point, is thus:

$$G_0 = [E, \mathbf{0}] + [m_1, \mathbf{0}] + [m'_1, \mathbf{0}] + [R_1, \mathbf{0}].$$

Now consider the "crystal" of Figure 4.1c, whose lattice, shown on Figure 4.1d, is identical to that of Figure 4.1a. It also features two families of mirrors, but the mirrors of family M_i are glide mirrors (Sect. 2.2.4), the symmetry operation of which is the product of operation mirror m_i and of translation $\mathbf{a}/2$.

The symmetry operations of this crystal have the form $[m_1, (u + \frac{1}{2})\mathbf{a} + v\mathbf{b}]$, $[m'_1, u'\mathbf{a} + v'\mathbf{b}]$ and their products. The mirror operations of the space group are now $[m_1, \mathbf{a}/2]$ and $[m'_1, \mathbf{0}]$ modulo the lattice translations. The space group is thus:

$$G'_s = [E, \mathbf{0}] + [m_1, \mathbf{a}/2] + [m'_1, \mathbf{0}] + [R', \mathbf{0}] \quad \text{modulo T.}$$

Operation R' is a rotation by $180°$ around the axis intersecting the plane of the figure at point O' deduced from O, the intersection of the traces of M_1 and M'_1, through translation $\mathbf{a}/4$ (Exercise 4.2).

The point group G_0 is obtained from G'_s by suppressing all translations and having all the symmetry elements go through one single point. We obtain:

$$G_0 = [E, \mathbf{0}] + [m_1, \mathbf{0}] + [m'_1, \mathbf{0}] + [R_1, \mathbf{0}].$$

The two "crystals" we are discussing have different space groups but the same point group G_0.

Further in this chapter, we will show rigorously how to go over from the group of symmetry operations which the crystal features, to its space group, and from the space group to the point group. This will require the use of the quotient group notion. It is defined in Appendix 4A, a refresher on some properties of groups which may be useful reading.

However, the reader may be content, on a first quick reading, with the approach just given. It provides a feeling for the point group and space group notions and their relationship. The reader can then go over directly to the next chapter, Chapter 5, which lists in a logical way the various point groups which crystals can feature.

4.2. Symmetry operations of the crystal

4.2.1. Change in origin

A symmetry operation $[S(O), \mathbf{t}]$ transforms point M_1 into point M_2 such that

$$\mathbf{OM_2} = S(\mathbf{OM_1}) + \mathbf{t}.$$

Let us show that we can change origin, *i.e.* find S', O' and $\mathbf{t}'$ such that

$$[S'(O'), \mathbf{t}'] = [S(O), \mathbf{t}].$$

Then $$\mathbf{O'M_2} = S'(\mathbf{O'M_1}) + \mathbf{t}' \qquad (4.3)$$

for any M_1.

Let us explicit both sides of Equation (4.3).

$$\mathbf{O'M_2} = \mathbf{O'O} + \mathbf{OM_2} = \mathbf{O'O} + S(\mathbf{OM_1}) + \mathbf{t}. \qquad (4.4)$$

$$S'(\mathbf{O'M_1}) + \mathbf{t}' = S'(\mathbf{O'O} + \mathbf{OM_1}) + \mathbf{t}' = S'(\mathbf{O'O}) + S'(\mathbf{OM_1}) + \mathbf{t}'. \qquad (4.5)$$

Since the right hand sides of Equations (4.4) and (4.5) are equal, we deduce that

$$S(\mathbf{OM_1}) + \mathbf{t} = S'(\mathbf{OM_1}) + S'(\mathbf{O'O}) - \mathbf{O'O} + \mathbf{t'}$$

which must be true for any point M_1. This implies that:

$$S' = S, \tag{4.6}$$
$$\mathbf{t'} = \mathbf{t} + (S - E)(\mathbf{OO'}) \tag{4.7}$$

where E is the identity operation.

We can therefore replace all the symmetry operations $[S(O), \mathbf{t}]$ featured by the crystal with operations $[S(O'), \mathbf{t'}]$ where O' is now common to all the symmetry operations, provided each translation $\mathbf{t}$ is replaced by translation $\mathbf{t'}$ given by Equation (4.7). We can thus choose a reference point common to all the symmetry operations, hence simplify the notation by not mentioning this reference point. This will be done throughout the remainder of this chapter.

We denote as $(S, \mathbf{t})$ *the symmetry operations featured by a crystal, with the understanding that a point* O *common to all the symmetry operations of the crystal has been chosen.*

We now show that these operations form a group, the symmetry group of the crystal.

4.2.2. The operations (S, t) form a group

The product of two symmetry operations of a crystal is of course also a symmetry operation. Let us look for the expression of the product of two of these operations, *viz.* $(S_1, \mathbf{t_1})$ which changes M into M_1 and $(S_2, \mathbf{t_2})$ which transforms M_1 into M_2.

$$\mathbf{OM_1} = S_1(\mathbf{OM}) + \mathbf{t_1}$$
$$\mathbf{OM_2} = S_2(\mathbf{OM_1}) + \mathbf{t_2} = S_2 S_1(\mathbf{OM}) + S_2 \mathbf{t_1} + \mathbf{t_2}$$

whence
$$(S_2, \mathbf{t_2})(S_1, \mathbf{t_1}) = (S_2 S_1, S_2 \mathbf{t_1} + \mathbf{t_2}). \tag{4.8}$$

Search for the neutral element: by using relation (4.8), we easily check that the neutral element is $(E, \mathbf{0})$ where E is the identity operation. The proof is

$$(E, \mathbf{0})(S, \mathbf{t}) = (S, \mathbf{t})(E, \mathbf{0}) = (S, \mathbf{t}).$$

Search for the inverse element: we want to determine $(S, \mathbf{t})^{-1} = (S', \mathbf{t'})$ such that

$$(S', \mathbf{t'})(S, \mathbf{t}) = (E, \mathbf{0}).$$

From (4.8), $(S', \mathbf{t'})(S, \mathbf{t}) = (S'S, S'\mathbf{t} + \mathbf{t'})$ hence $S'S = E$ and $S'\mathbf{t} + \mathbf{t'} = 0$.

Therefore
$$S' = S^{-1},$$
$$\mathbf{t'} = -S^{-1}\mathbf{t}. \tag{4.9}$$

S^{-1} is the inverse operation of S. Consider an example: if S is a rotation by $2\pi/3$ around a given axis (which goes through the point O common to all the symmetry operations), S^{-1} is a rotation by $-2\pi/3$ around the same axis. The translation $\mathbf{t}'$ is the opposite of the vector deduced from $\mathbf{t}$ through a rotation by $-2\pi/3$ around the axis (Fig. 4.2).

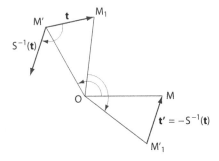

Figure 4.2: Illustrating the inverse operation of (S, t), where S is the rotation by $2\pi/3$ around an axis perpendicular to the plane of the figure and going through O. M_1 is the transform of M under this operation. The rotation by $-2\pi/3$ around the same axis transforms M_1 into M'_1, and translation $\mathbf{t}' = - S^{-1} \mathbf{t}$ brings M'_1 back to M.

4.2.3. The lattice translations form an invariant subgroup of the group of symmetry operations of the crystal

To prove this result and that of the next section, we use a few definitions and properties of invariant subgroups and of quotient groups given in Appendix 4A.

The lattice translations are the symmetry operations $(E, \mathbf{n})$ where E is the identity operation and $\mathbf{n} = u\mathbf{a} + v\mathbf{b} + w\mathbf{c}$, $\mathbf{a}$, $\mathbf{b}$ and $\mathbf{c}$ are the basis vectors of the unit cell and u, v, w are integers. We show that the set of operations $(E, \mathbf{n})$ form a group, which is a subgroup of the symmetry operations $(S, \mathbf{t})$ of the crystal. The product of two lattice translations is a lattice translation, and the neutral element $(E, \mathbf{0})$ is part of this set.

Let us show that this subgroup is invariant (Appendix 4A), *i.e.* that, for any S, we have:

$$
\begin{aligned}
(S, \mathbf{t})(E, \mathbf{n})(S, \mathbf{t})^{-1} &= (E, \mathbf{n}'). \\
(S, \mathbf{t})(E, \mathbf{n})(S, \mathbf{t})^{-1} &= (S, \mathbf{t})(E, \mathbf{n})(S^{-1}, -S^{-1}\mathbf{t}) \\
&= (S, \mathbf{t})(S^{-1}, -S^{-1}\mathbf{t} + \mathbf{n}) \\
&= (E, -SS^{-1}\mathbf{t} + S\mathbf{n} + \mathbf{t}) \\
&= (E, S\mathbf{n}).
\end{aligned}
$$

$(E, S\mathbf{n})$ is a pure translation. It is a symmetry operation of the crystal since it was obtained as the product of symmetry operations of the crystal. Since its rotation part is identity, it is necessarily a crystal lattice translation.

Conclusion

1. *The set of lattice translations is an invariant subgroup of the group of the symmetry operations of the crystal.*

2. *We showed that* **Sn** *is a crystal lattice vector. This result shows that the operations* S *are symmetry operations of the lattice. We will see at the end of Section 4.3 that the reciprocal proposition is not necessarily true.*

4.3. Space groups and point groups

The subgroup $(E, \mathbf{n}) = T$ of the lattice translations being invariant, we can define a quotient group G/T starting from the partition of G by T (Appendix 4A). The partition of G into cosets products of T by symmetry operations of the crystal which are not lattice translations can be expressed as

$$G = T + T(S_1, \mathbf{t}_1) + T(S_2, \mathbf{t}_2) + T(S_3, \mathbf{t}_3) + \ldots + T(S_m, \mathbf{t}_m).$$

This partition is performed in the following way: we isolate within G a first packet of operations, here the lattice translations T. Among the remaining operations, we choose an operation $(S_1, \mathbf{t}_1)$. We perform the product of this operation by all the lattice translations, forming the second packet $T(S_1, \mathbf{t}_1)$, and so forth until all the elements of G have been used. We here operate with the smallest possible vectors (null or smaller than a lattice vector) by subtracting from the initial translation vector $\mathbf{t}_j$ as large a lattice vector $\mathbf{n}_i$ as possible. This lattice vector $\mathbf{n}_i$ may then be included in T. This is possible since

$$(S_i, \mathbf{t}_j) = (S_i, \mathbf{t}_i + \mathbf{n}_i) = (E, \mathbf{n}_i)(S_i, \mathbf{t}_i)$$

so that
$$T(S_i, \mathbf{t}_j) = T(S_i, \mathbf{t}_i).$$

We will determine further (Sect. 7.2) the symmetry operations $(S_i, \mathbf{t}_i)$, where $\mathbf{t}_i$ is a non-zero vector smaller than a crystal lattice vector, which may exist in a crystal. We will see that they are the symmetry operations associated to screw axes or to glide mirrors (Sect. 7.2.1 and 7.2.2).

The quotient group is defined as $G_s = G/T$, the group whose elements are the cosets $T(S_i, \mathbf{t}_i)$. If we consider the symmetry operations modulo T, we can replace coset $T(S_i, \mathbf{t}_i)$ by $(S_i, \mathbf{t}_i)$. We can then write

$$G_s = (E, \mathbf{0}) + (S_1, \mathbf{t}_1) + (S_2, \mathbf{t}_2) + (S_3, \mathbf{t}_3) + \ldots + (S_m, \mathbf{t}_m) \quad \text{modulo}^2 \ T.$$

In the introduction to this chapter, we defined two operations equivalent modulo T as two operations for which the difference in the translation parts

---

2. Adding "modulo T" is mandatory in order for the above operations to form a group because it can happen that the product of two of the operations $(S_i, \mathbf{t}_i)$ is equal to $(S_j, \mathbf{t}_j + \mathbf{n})$ while, as mentioned above, the $\mathbf{t}_i$ are smaller than lattice translations.

is equal to a lattice translation $\mathbf{n}$. This is indeed the case here since $(\mathrm{E},\mathbf{n})(\mathrm{S},\mathbf{t}) = (\mathrm{S},\mathbf{t}+\mathbf{n})$.

G_s is called the space group, or position group, of the crystal.

Now consider the operations $(\mathrm{E},\mathbf{0})$, $(\mathrm{S}_1,\mathbf{0})$, $(\mathrm{S}_2,\mathbf{0})$, etc. which deduce from the elements of G_s by canceling all the translations, *i.e.* by canceling the "modulo T" and replacing the $\mathbf{t}_i$ by $\mathbf{0}$. Using the composition law (4.8), it is easy to show that these operations form a group G_0, isomorphous to G_s, which can be written

$$\mathrm{G}_0 = (\mathrm{E},\mathbf{0}) + (\mathrm{S}_1,\mathbf{0}) + (\mathrm{S}_2,\mathbf{0}) + (\mathrm{S}_3,\mathbf{0}) + \ldots + (\mathrm{S}_m,\mathbf{0}).$$

This group is the point group (or orientation group) of the crystal. It is obtained by considering all the symmetry operations of the crystal, defined, as we remember, with a common origin (Sect. 4.2.1), and by canceling all translations. We saw, at the end of Section 4.2.3, that the operations $(\mathrm{S}_i,\mathbf{0})$ are symmetry operations of the lattice. The reciprocal proposition is not necessarily true (and indeed it is often wrong), *i.e.* all symmetry operations of the lattice are not necessarily symmetry operations of the point group. We note that:

The symmetry operations of the point group of the crystal form a subgroup of the symmetry operations of the crystal lattice.

A simple example, chosen from among the plane lattices, is given on Figure 4.3a. The crystal features a family of pure plane mirrors M_i, and does not feature any mirror perpendicular to the M_i. The point group of the crystal is

$$\mathrm{G}_0 = (\mathrm{E},\mathbf{0}) + (\mathrm{m},\mathbf{0})$$

where m is the symmetry operation associated to one of the mirrors M_i.

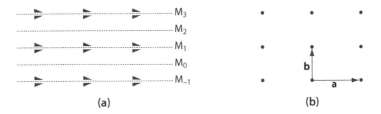

(a)   (b)

Figure 4.3: (a) An example of a plane "crystal", with its crystal lattice in (b)

The crystal lattice (Fig. 4.3b), for which the unit cell is a rectangle, features a family of mirrors M'_i perpendicular to the M_i. Let us denote with m′ the symmetry operation associated to one of the mirrors M'_i, and with A the product of operations m and m′, *i.e.* the rotation by π around their intersection. The point group of the crystal lattice is

$$\mathrm{G}'_0 = (\mathrm{E},\mathbf{0}) + (\mathrm{m},\mathbf{0}) + (\mathrm{m}',\mathbf{0}) + (\mathrm{A},\mathbf{0})$$

and indeed G_0 is a subgroup of G_0'.

As was discussed at the beginning of this chapter, the point group of the crystal is very important because it governs the symmetry of the macroscopic properties of crystals. Furthermore, the construction of space groups is easy to explain when starting from the point groups. We therefore start by investigating the point groups or orientation groups (Chap. 5) and then, in Chapter 7, we show how the space groups or position groups can be built starting from the point groups.

For a given point group, it is useful to know the directions that are equivalent to a given direction through the symmetry operations which are elements of the group. These equivalent directions are represented using the stereographic projection, defined in the next chapter.

4.4. Exercises

Exercise 4.1

Show that the product of a symmetry with respect to a plane P and a translation by a vector $\mathbf{V}$ perpendicular to plane P is a symmetry with respect to the plane P' deduced from P by a translation $\mathbf{V}/2$. Consider a general point M, its image M_1 by symmetry with respect to P, and then the point M_2, image of M_1 through translation $\mathbf{V}$. Draw the figure in the plane Π perpendicular to P and going through M.

Exercise 4.2

1. Show that the product of a symmetry with respect to a plane P_1 and a symmetry with respect to a plane P_2 perpendicular to P_1 is a rotation by π around zz', the intersection of planes P_1 and P_2.

2. Now replace the second symmetry operation of (1) by the operation product of the symmetry with respect to P_2 and a translation $\mathbf{V}$ perpendicular to zz' and parallel to plane P_2. Show that this final operation is a rotation by π around an axis deduced from zz' through a translation $\mathbf{V}/2$.

Complement 4C. Probes used
for crystal structure determination

4C.1. Possible probes and criteria for choice

Crystal structure research mainly involves the determination of the nature of
the atoms comprising the material and their arrangement in space. This process is invariably based on coherent scattering, *i.e.* diffraction. It therefore
requires probes that interact with atoms, with a wavelength smaller than the
distances between atoms, *viz.* 1 Å $= 1 \times 10^{-10}$ m $= 0.1$ nm. The wavelength
is given by $\lambda = h/p$ where h is Planck's constant $h = 6.63 \times 10^{-34}$ J s and
p the probe momentum. There are three possibilities: X-rays, neutrons and
electrons. Electrons can be effectively focused through magnetic lenses, giving
the possibility of either imaging or diffraction modes in the electron microscope. Neutrons and X-rays can be focused, either by reflection from curved
mirrors or crystals, or through diffractive lenses (zone plates, or Fresnel lenses),
or through the use of refractive lenses consisting of stacks of small spheres or
paraboloids, or holes in materials. Much better results are obtained in the
case of spatially coherent beams of synchrotron X-rays[3]. These three probes
interact with different features of matter. X-rays interact with the electrons
within materials, stronger scattering being associated with atoms with more
electrons, *i.e.* with heavier atoms. Neutrons are sensitive to the nature of the
nuclei (number of protons, *i.e.* chemical nature, but also number of neutrons,
hence isotope) and also to the magnetic fields within matter. Electrons sense
the electrostatic potential due to both the nuclei and the electrons. In structural work, the fact that there is no systematic relation between the strength of
nuclear scattering (the nuclear scattering length) of thermal neutrons and the
atomic number of the scatterer opens up possibilities for elegant approaches
complementary to X-ray diffraction. Electrons are strongly absorbed by materials, and can go through thin samples only (typically less than 1 micrometer),
while X-rays, and even more neutrons, can go through thick samples. These
probes are also very different in their ease of production and in the intensity
of the beams that can be obtained. The probe with by far the widest use is
X-rays.

4C.2. X-rays

X-rays are electromagnetic waves. They have the proper range of wavelength
λ, around 1 Å, when their photon energy $E = hc/\lambda$ where c is the velocity of light in vacuum, is of the order of 10 keV. They are produced, like
all electromagnetic waves, when light electric charges (in effect electrons) are
accelerated. The standard sources for X-rays are laboratory X-ray tubes, in
which electrons are accelerated by a voltage V on the order of 10^4 V and hit
a metal target, the anode. X-rays are emitted both in the form of "character-

---

3.   Christian David, http://www.psi.ch/lmn/x-ray-optics

istic" radiation, *i.e.* fairly fine spectral lines, and in the form of a continuous spectrum, *Bremsstrahlung*.[4] The characteristic lines correspond to transitions of atomic electrons between discrete atomic energy levels as the electron vacancies induced by the incoming electrons heal, and their energies or wavelengths depend on the target material. X-rays are also produced, with much higher intensity than in laboratory X-ray tubes, as part of the synchrotron radiation emitted when the trajectories of ultra-relativistic electrons or positrons (energy E much larger than their rest mass energy $m_e c^2 = 0.511$ MeV) are curved by a magnetic field. Synchrotron radiation covers a huge spectral range, from the infrared, through the visible and ultra-violet range, to high-energy X-rays if the energy of the electrons or positrons is high enough, and it is the hard X-ray part (typically 10 to 100 keV energy, or about 0.1 to 0.01 nm wavelength) that is put to use for structural work. Monochromatic beams (beams with a given energy or wavelength) must be prepared through monochromators. In practice, synchrotron radiation sources are storage rings into which the particles are fed after being accelerated to energies of several GeV. Initially considered by high energy physicists as a nuisance, synchrotron radiation is now the sole motivation for the construction of "dedicated" machines. Beyond intensity, synchrotron radiation offers, among the features useful in structure determination, the possibility of tuning the wavelength, and in particular of aiming for selected values where the response of one given type of atoms is enhanced by resonance.

For most structure investigations, laboratory X-rays, obtained from X-ray generators which are affordable for normal size laboratories, are satisfactory; access is then easy for the researchers. In some special cases, higher intensity or other requirements make it desirable to use hard X-rays from synchrotron radiation sources. These large facilities, with high construction and operation cost, are mostly available in developed countries. For example, in 2011, APS (Advanced Photon Source) in the USA, Spring8 in Japan and ESRF in Europe, Diamond the British source, Petra III the new German source, Swiss Light Source (SLS), Soleil the French one, all are third-generation, *i.e.* highest-performance sources. They are accessible to researchers on the basis of applications for beam-time which are ranked by committees, a procedure which takes time – and does not always lead to acceptance. Thus their use is effectively reserved for work, much of which is outside structural crystallography, which cannot be performed with laboratory X-rays. Novel large facilities for the production of X-rays are appearing: free electron lasers or XFEL (X-ray Free Electron Lasers). They produce pulses of coherent X-rays with extremely large intensity and a duration of less than 100 femtoseconds. This makes it possible to envisage watching atomic movements in solids, understanding chemical reac-

---

4. *Bremsstrahlung* (with a capital *B*) is the German word for "radiation from braking". It is associated with the decrease in velocity of electrons through interaction with the target material. Energy conservation immediately yields a useful feature: the energy of a photon cannot be more than that of the electrons, thus the short-wavelength limit of the spectrum is such that $hc/\lambda_{\min} = eV$.

tion processes on the molecular scale, etc. Two such sources are operational in 2011: LCLS (Linac Coherent Light Source) at Stanford, CA, USA since 2009, while SACLA (Spring-8 Angstrom Compact free electron Laser) in Harima, Japan produced its first beams in 2011. Others are being built: the European source XFEL in Hamburg, Germany and SwissFEL in Villigen, Switzerland, PAL (Pohang Accelerator Laboratory) in Korea.

4C.3. Neutrons

Neutrons are particles with rest mass $m_n = 1.67 \times 10^{-27}$ kg, with no electric charge but with a magnetic moment $\mu_n = -0.97 \times 10^{-26}$ J T^{-1} with the minus sign indicating that the magnetic moment is in the direction opposite to the spin. Neutron beams are produced either from nuclear reactors, using fission, or from spallation sources in which high energy proton beams interact with a target, usually in a pulsed mode[5]. Examples of research reactors are the multi-national Institut Laue-Langevin (ILL), and Orphée at the Laboratoire Léon Brillouin in Saclay, near Paris. ISIS, in England, or the Spallation Neutron Source (SNS) in Oak Ridge, TN, USA, are spallation sources. In either case, the initial energy of the neutrons, in the MeV range, is much too high for structural investigations, and they have to be submitted to a decrease in their kinetic energy, a process called moderation. They have De Broglie wavelength in the right range, around 1 Å, if their kinetic energy $E_c = p^2/2m_n = h^2/2m_n\lambda^2$, where p is their momentum, is a few tens of meV. A different way of expressing this condition is that the kinetic energy then corresponds to a thermal energy $k_B T$ where $k_B = 1.38 \times 10^{-34}$ J K^{-1} is Boltzmann's constant and T a temperature in the range around room temperature, *i.e.* 300 K. This description is meaningful because preparing neutron beams in the right range of energy is actually performed by letting the neutrons, which cannot be tailored by electric fields, collide with a moderator material, such as heavy water D_2O, at this temperature. They are usually called "thermal neutrons", while neutrons with higher energy, obtained through moderation in a hot object, are "hot neutrons", and those with lower energy are "cold" or "ultra-cold" neutrons. As in the case of synchrotron radiation, single-energy beams must be extracted through the use of monochromators. The intensity of neutron beams is smaller than that of X-ray beams from laboratory generators, and very much smaller than X-ray beams from synchrotron radiation. Furthermore, neutrons are only available at large facilities, which implies the same difficulties in access as for synchrotron radiations. This leads to their use being far less wide-spread than X-rays, and restricted to investigations where their special features are a bonus.

---

5. Recommended reading includes two courses by Roger Pynn:
    http://www.ncnr.nist.gov/summerschool/ss09/pdf/Lecture_1_Theory.pdf,
    and http://link.springer.com/book/10.1007/978-0-387-09416-8/page/1.

4C.4. Electrons

Electrons are particles with electric charge $-e = -1.6 \times 10^{-19}$ C, with rest mass $m_e = 0.9 \times 10^{-30}$ kg. Their magnetic moment, associated to the intrinsic angular momentum or spin, has a direction opposite to the spin. Its absolute value is 1 Bohr magneton $(1\mu_B) = 0.93 \times 10^{-23}$ J T^{-1}, much larger than the neutron's but whose influence is hidden by the effect of the charge. The relation between the kinetic energy eU, obtained through acceleration under voltage U, and the De Broglie wavelength of an electron[6] is:

$$\lambda = \frac{h}{\sqrt{2m_0 eU}} \frac{1}{\sqrt{1 + \frac{eU}{2m_0 c^2}}}.$$

Electrons with wavelength in the range of 10^{-10} m are unable to penetrate materials, and they can provide information from a very thin region near the surface (Low Energy Electron Diffraction, LEED). Higher energy electrons (100 keV or more) are used for the investigation of thin films (less than 1 micrometer) in electron microscopes. These allow switching between the imaging and diffraction modes, *i.e.* between the observation of direct and reciprocal space, at the press of a button. They are complementary to X-rays for some crystal structure investigations, in particular for materials where only tiny crystalline regions can be obtained.

Further reading

J. Als-Nielsen, D. McMorrow, *Elements of Modern X-ray Physics*, 2nd edn. (Wiley, 2011)

R. Scherm, B. Fåk, Neutrons, in *Neutron and Synchrotron Radiation for Condensed Matter Studies*, vol. 1: *Theory, Instruments and Methods*, ed. by J. Baruchel *et al.* (Les Editions de Physique and Springer-Verlag, 1993)

R. Pynn, *Introduction to Neutron Scattering*. A clear and comprehensive exposition is available on http://knocknick.files.wordpress.com/2008/04/neutrons-a-primer-by-rogen-pynn.pdf. The "rogen" in the address is a misprint for "roger"; in case it gets corrected at some point, the revised address could be tried in case of difficulty.

---

6. The general expression for the De Broglie wavelength of a particle is the relativistic expression $\lambda = h/p = hc/pc$, with $pc = \sqrt{E^2 - m_0^2 c^4}$. It is well introduced in http://hyperphysics.phy-astr.gsu.edu/hbase/quantum/debrog2.html#c2. The limiting cases for zero rest mass (ultrarelativistic case, e.g. photons) and for kinetic energy small with respect to the rest mass energy (non-relativistic case, e.g. thermal neutrons) are easily retrieved.

Appendix 4A. General features of groups

Definition

A group is a set G of elements g_1, $g_2 \ldots$ featuring an internal composition law, here denoted as $\bullet$, so that f $= g_i \bullet g_k$, and f is an element of G. This composition law

- is associative: $g_i \bullet (g_k \bullet g_l) = (g_i \bullet g_k) \bullet g_l$, so that the parentheses can be omitted in this product

- has a neutral element e such that $g_i \bullet e = e \bullet g_i = g_i$

- is such that each element g_i has an inverse g_i^{-1} such that $g_i \bullet g_i^{-1} = g_i^{-1} \bullet g_i = e$.

In what follows, we will omit the sign $\bullet$.

Order of the group

If G is an infinite set of elements, we say that the group is infinite. Otherwise, we say that it is finite, and the order of the group is the number of its elements.

Generating elements

These are elements of the group such that all the elements of the group can be obtained as products of powers of these elements with one another or with their inverses.

Rearrangement theorem

The multiplication table of group is a dual-entry table (Sect. 2.3). All elements of the group appear on each row and each column of the table, and no element appears twice on a given row or column. This means that if $x \neq y$, then $xz \neq yz$.

Proof: if $xz = yz$, then $xzz^{-1} = yzz^{-1}$ and $x = y$, which is contrary to the assumption.

Note that each row or column of the multiplication table contains all the elements of the group, but ordered differently from the preceding row or column. This is why this is referred as rearrangement theorem.

Subgroup

A subgroup H is a part of the elements of group G that forms a group with the same composition law as group G itself. We will show that, if group G is finite, the order n_G of G is a multiple of the order, n_H, of H. The ratio $i = n_G/n_H$ is called the subgroup index.

Invariant subgroup

A subgroup H (with elements denoted as h_i) of G is invariant if, for any element s of G and h_i of H, we have $s^{-1}h_i s = h_j$ or $h_i s = s h_j$.

Partition of a group G modulo a subgroup H

This is a classification of elements of group G. We start out by considering the elements of H and withdrawing them from G. Then there remains in G at least one element x. We form all the products xh_i. This set is denoted as xH and we remove it. We show that xH is disjoint from H. Assume in contrast that $xh_i = h_j$. Right-multiplying both sides of this equation by h_i^{-1}, we obtain $x = h_j h_i^{-1}$, which is an element of H since H is a group. But this is contrary to the starting assumption (x is not in H).

If there still remains in G an element y, we form the subset yH, we withdraw it from what remains of G, and so forth till there is nothing left. We can write

$$G = H + xH + yH + \ldots$$

We show that xH and yH are disjoint by noting that, if we had $yh_i = xh_j$, then $y = xh_j h_i^{-1} = xh_k$ and this is contrary to the assumption (y is not in xH).

Each subset xH or yH is called a coset. Since each coset of this partition has the same number of elements as H, we deduce that, if H and G are finite, the order of G is a multiple of the order of H.

The above partition was performed by left-multiplying the elements of H with x, y, etc. It could have been made by choosing right-multiplication. Both partitions have no reason to be identical. In contrast, they are identical if H is invariant, because then $xh_i = h_j x$ and in this case the set xH is identical to the set Hx.

Quotient group

Consider the partition of G by an invariant subgroup H. We show that the cosets xH, yH, etc. form a group which is called the quotient group of G by H and is denoted by G/H. The product of two elements xH and yH of this group is defined as the set of elements obtained through the product of one of the elements of xH by one of the elements of yH, $i.e.$

$$
\begin{aligned}
xh_i(yh_k) &= x(h_i y)h_k \quad \text{because of the associativity in G,} \\
&= x(yh_p)h_k \quad \text{since H is invariant,} \\
&= xy(h_p h_k) \quad \text{through associativity in G,} \\
&= xyh_m \quad \text{because H is a group.}
\end{aligned}
$$

We deduce that the product of xH by yH is xyH, $i.e.$ $(xH)(yH) = (xy)H$. The product xy cannot be within H since x and y are outside H, and thus $xy = z$ with $z \notin H$ and $z \neq x$ or y; otherwise x or y would be the neutral element of G which is also the neutral element of H, and thus within H.

Thus $(xH)(yH) = zH$. Associativity is obvious since associativity applies within G. The neutral element is H, since $h_i(y h_n) = h_i(h_p y) = (h_i h_p)y = h_v y = yh_w$ and thus $H(yH) = yH$.

The inverse of xH is $x^{-1}H$ since $xh_i x^{-1}h_n = xx^{-1}h_p h_n = h_q$.

Abstract groups

A group can be defined by the name of its elements and their multiplication table without reference to concrete examples. It is then referred to as an abstract group. Here are two examples:

1. Cyclic groups

   They are groups denoted as C_n which have n elements such that, if one is denoted as A, the others are A^2, A^3,..., $A^n = E$ where E is the neutral element.

   For example, group C_3 consists of an element A, of element A^2 and of the neutral element $E = A^3$. Its multiplication table is given below

	A	A^2	E
A	A^2	E	A
A^2	E	A	A^2
E	A	A^2	E

   As we will see, a concrete realization of this group is the group of rotations by $2\pi/3$ around an axis, *i.e.* in terms of point groups, group 3.

2. Dihedral groups

   A dihedral group is defined using two generating elements A and B which satisfy relations $B^2 = E$, $A^n = E$ and $BA = A^{n-1}B$. It includes $2n$ elements and is denoted as D_n.

   We will see that a concrete realization of group D_3 is for example point group 32, for which A is the rotation by $2\pi/3$ about a given axis and B is the rotation by π around an axis perpendicular to the above.

Chapter 5

Point groups

The point groups are enumerated in a rigorous way. On first reading, the reader can be content with their geometrical representation using the stereographic projection (Sect. 5.2) and with their denomination and classification by crystal system (Sect. 5.6).

5.1. Introduction

The point group of a crystal governs the symmetry of its physical properties. We recall that the point group consists of the symmetry operations of the crystal with all the translations removed. All the symmetry elements are assumed to go through a given point O. Further, we saw (Sect. 2.1) that the symmetry elements which do not contain a translation are either rotations or rotoinversions. The rotations are operations which conserve the handedness of a trihedron (proper operations), while the rotoinversions, which do not conserve the handedness of a trihedron, are said to be improper operations. We showed (Sect. 4.3) that the symmetry operations of the point group form a subset of the point group of its crystal lattice. This property also results from Curie's principle (Sect. 3.3.5) because the crystal lattice is a property of the crystal and therefore it must be at least as symmetric as the crystal. This is a basic result, because it restricts the number of proper crystallographic point groups (groups which contain only proper operations, *i.e.* rotations). Since the only symmetry axes which a crystal lattice can feature are the 2-, 3-, 4- or 6-fold axes, the proper symmetry operations of crystals are the symmetry operations associated to 2-, 3-, 4- or 6-fold axes. We will first enumerate the proper crystallographic point groups (Sect. 5.4), after showing that the improper groups always contain a proper subgroup with order half their own (Sect. 5.3). This property will be used in Section 5.5 to enumerate the improper groups starting from the proper groups.

Sections 5.3, 5.4 and 5.5 show how to logically enumerate the 32 crystallographic point groups. In a first approach, the reader may accept that there are 32 point groups, and get acquainted with them straight away in Section 5.6, where they are grouped into crystal systems. In contrast, reading

© Springer Science+Business Media Dordrecht 2014
C. Malgrange et al., *Symmetry and Physical Properties of Crystals*,
DOI 10.1007/978-94-017-8993-6_5

Section 5.2, which introduces the stereographic projection used to represent the point groups, is mandatory.

The chapter closes with a short discussion of isotropy groups in Section 5.9. They are not crystal point groups, but they enter the investigation of the physical properties of crystalline materials.

5.2. Stereographic projection

5.2.1. Definition

It is very useful to represent, for a given point group, the directions equivalent to a given general direction, *i.e.* the directions obtained from this direction by applying all the symmetry operations which make up this group. For a general direction, *i.e.* a direction which is not parallel to a symmetry axis (n-fold or $\bar{n}$-fold axis), there is one direction for each symmetry operation, and therefore there are as many equivalent directions as there are symmetry operations in the group. A precise two-dimensional representation of these equivalent directions requires a projection of three-dimensional space onto the plane of the drawing. The projection used by crystallographers is the stereographic projection, defined below.

Let O be the intersection of the various symmetry elements (rotation or rotoinversion axes).[1] A given general direction is represented by a half-line Ov starting from this point O. This half-line intersects a sphere with center O and with any radius R at a point P which characterizes the direction (Fig. 5.1a). Representing the half-lines is tantamount to projecting the points P of the sphere onto a plane. We choose a projection plane Π which goes through the center of the sphere. This plane Π intersects the sphere along a circle C. We draw the diameter NS perpendicular to plane Π; N is the north pole, S the south pole, Π is the equatorial plane and circle C the equatorial circle. Let Ov be a direction in the northern hemisphere. Consider the meridian plane defined by axis NS and point P (Fig. 5.1b). The line SP of this plane intersects the equatorial projection plane Π at n, which is the required projection. The triangles NPS and nOS are similar, and SN/Sn = SP/SO. We obtain $\overline{\text{Sn}}.\overline{\text{SP}} = \overline{\text{SN}}.\overline{\text{SO}} = 2R^2$. Point n is therefore deduced from P through an inversion with pole S and power $2R^2$.

Recall the definition of the inversion transformation. An inversion with pole O and power k transforms a general point M into an image point M$'$ such that O, M and M$'$ are aligned and $\overline{\text{OM}}.\overline{\text{OM}'} = k$. Inversion conserves angles and transforms a circle that does not go through pole O into a circle.

---

1. For rotoinversion axes, this is also the position of the center of symmetry used for the rotoinversion operations.

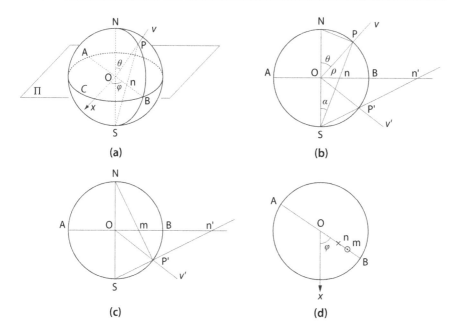

Figure 5.1: Principle of the stereographic projection. (a) The general direction Ov intersects the sphere at P. (b) n and n′ are respectively the transforms of P and P′ in inversions with pole S, the south pole. (c) m is the transform of P′ through inversion with pole N, the north pole. (d) n and m are the stereographic projections of the directions Ov and Ov' respectively.

Point P is defined by its spherical coordinates R, θ and φ (axis Ox is a general direction in plane Π), and its projection n by its polar coordinates ρ and φ in plane Π. The length ρ = On is easily calculated when we notice that $(\mathbf{SP}, \mathbf{SN}) = \alpha = \theta/2$. Hence On = $R\tan\alpha$ = $R\tan(\theta/2)$. It is important to note that the azimuthal coordinate φ is conserved by the stereographic projection.

Now consider a direction Ov' in the southern hemisphere, and, in order not to clutter the text with more figures than necessary, assume it is in the same meridian plane as the above direction Ov (Fig. 5.1b). It intersects the sphere at P′. Its projection n′, defined as above, is outside the equatorial disk. The closer P′ comes to the south pole, the further n′ moves away from O. This would lead, in order to represent all the directions of space, to using an infinite surface! To avoid this difficulty, we use, for the directions of the southern hemisphere, an inversion with pole N, and with the same power $2R^2$ (Fig. 5.1c). The projection m of direction Ov' is now located inside the equatorial disk. The stereographic projection is thus defined as an inversion with pole S (the south pole) for the directions located in the northern hemisphere, and an inversion with pole N (the north pole) for the directions in the southern hemisphere. Thus the same point will be the projection of two directions, symmetric to each other with respect to the projection plane Π. To distinguish these projections, one uses two different symbols, for example crosses for the projections of the directions in the northern hemisphere, and circles for those of the southern hemisphere (Fig. 5.1d). The equatorial disk thus contains the stereographic projections of all the directions of space.

5.2.2. Examples

Consider the point group 3, *i.e.* the group whose elements are the rotations by $2\pi/3$, $4\pi/3$ and 2π (identity) around a given 3-fold axis denoted as A_3. We choose as projection plane for the equivalent directions a plane perpendicular to this axis A_3. The projection of the axis itself is point O, the common projection of the directions ON and OS. Consider a general direction Ov, the stereographic projection of which is n, with polar coordinates ($\rho =$ On, φ) (Fig. 5.1a). Rotation around axis NS causes rotation of the meridian half-plane containing Ov and conserves angle θ, hence ρ. The projections of the three directions equivalent to Ov through the three symmetry operations of the group are n$'$ (ρ, $\varphi + 2\pi/3$) for the rotation by $2\pi/3$, n$''$ (ρ, $\varphi + 4\pi/3$) for the rotation by $4\pi/3$ and n itself for the rotation by 2π or identity (Fig. 5.2). We understand, using this simple example, that the number of directions equivalent to a given general direction is equal to the number of elements in the symmetry group, or order of the group. Of course this is no more true if the direction is special, as would be the direction of the axis itself, which is invariant through the symmetry operations, and the stereographic projection of which is O.

Figure 5.2: Stereographic projection of a general direction (projection at n) and its equivalent directions through the symmetry operations associated to an axis A_3

The point symmetry elements featured by a crystal with a given crystal group appear directly on the stereographic projection of the directions equivalent to a general direction through the symmetry operations of the group. In what follows, we will simplify our description by simply referring to the "stereographic projection of the equivalent directions", the remainder of the sentence being implied.

A few examples are illustrated on Figure 5.3. The derivation of the results is either evident, or given in Appendix 5A.1. The thick lines represent the stereographic projections of planes, while the thin lines are just aids for the drawing.

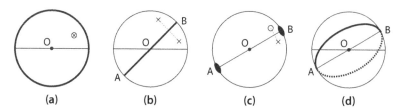

(a)	(b)	(c)	(d)

Figure 5.3: Stereographic projections of various symmetry elements and of the directions equivalent through the symmetry operations to which they are associated: (a) mirror in the projection plane; (b) mirror perpendicular to the projection plane; (c) 2-fold axis in the projection plane; (d) general mirror intersecting the projection plane along AB

– *Mirror parallel to the projection plane*: it coincides with plane Π. Its stereographic projection is the equatorial circle. Two directions which are equivalent through the mirror operation project on the same point with different symbols, since one of the directions is in the northern hemisphere and the other in the southern hemisphere (Fig. 5.3a and Appendix 5A.1.2).

– *Mirror perpendicular to the projection plane*: it goes through axis NS and its projection is the line where the mirror intersects the plane of projection, *i.e.* AB on Figure 5.3b. The equivalent directions are in the same hemisphere and their stereographic projections are two points with the same nature (cross or circle), symmetric with respect to the line representing the projection of the mirror (Appendix 5A.1.3).

– *n-fold axis, noted* A_n *or* n, *perpendicular to the projection plane*: the projection of the axis consists in two points coinciding with the center of the equatorial circle. It is represented by a single symbol, characteristic of its order n. The conventional symbols are given in Table 5.1.

The equivalent directions, all located in the same hemisphere, project on points which deduce from one another through rotations by $2\pi/n$ around the axis, and which are represented by the same sign (Fig. 5.4).

Table 5.1: Graphical representation of the symmetry axes

Axis type	Graphical representation	Axis type	Graphical representation
2	⬬	$\bar{1}$	○
3	▲	$\bar{3}$	◬
4	◆	$\bar{4}$	◈
6	⬢	$\bar{6}$	⬡

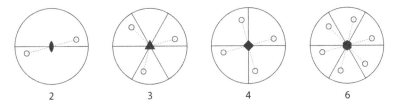

2   3   4   6

Figure 5.4: Stereographic projections of directions equivalent through the symmetry operations associated to 2-, 3-, 4- and 6-fold axes

– *2-fold axis parallel to the projection plane*: the projection of the axis consists in two points A and B, the intersections of the axis and the equatorial circle (Fig. 5.3c). These points are displayed using the symbol characteristic of the 2-fold axes. The equivalent directions are in different hemispheres, and their projections are symmetric with respect to line AB (Appendix 5A.1.4).

– *Mirror with general orientation*: such a mirror intersects the projection plane along a line AB. The stereographic projection of the mirror consists of two circular arcs going through A and B and symmetric with respect to AB (Fig. 5.3d and Appendix 5A.2.1).

5.2.3. Application to rotoinversion axes or $\bar{n}$-fold axes

The definition of $\bar{n}$-fold axes was given in Section 2.2.2, but their properties can become clearer using the new tool we now have. Figure 5.5 shows the stereographic projections of equivalent directions for these various rotoinversion axes. The projection plane is perpendicular to the $\bar{n}$ axis, and goes through the inversion center used for the rotoinversion.

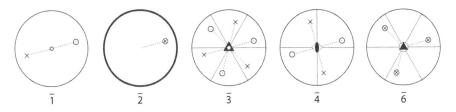

$\bar{1}$ $\bar{2}$ $\bar{3}$ $\bar{4}$ $\bar{6}$

Figure 5.5: Stereographic projections of directions equivalent through the symmetry operations associated to $\bar{1}$, $\bar{2}$, $\bar{3}$, $\bar{4}$ and $\bar{6}$-fold axes

We note that:

– a $\bar{3}$-fold axis is equivalent to the set consisting of a 3-fold axis (coinciding with the $\bar{3}$-fold axis) and a center of symmetry;

- a $\bar{6}$-fold axis is equivalent to the set consisting of a 3-fold axis (coinciding with the $\bar{6}$-fold axis) and a mirror perpendicular to this axis and going through the inversion center used for the rotoinversion;

- a $\bar{4}$-fold axis has no equivalent.

We retrieve the fact that a $\bar{1}$-fold axis is equivalent to a center of symmetry and a $\bar{2}$-fold axis to a mirror perpendicular to the axis and going through the center of symmetry used for the rotoinversion.

We can note that the order of the $\bar{n}$ groups is equal to n, except for the group $\bar{3}$ for which the order is 6.

5.2.4. Family of equivalent directions

We saw in Chapter 3 that a crystal row is noted $[uvw]$ if the vector $\mathbf{V}$ joining two neighboring nodes on the row can be expressed as:

$$\mathbf{V} = u\mathbf{a} + v\mathbf{b} + w\mathbf{c},$$

where $\mathbf{a}$, $\mathbf{b}$ and $\mathbf{c}$ are the basis vectors of the crystal unit cell. Since the physical properties of a crystal are identical along two directions equivalent through the symmetry operations of its point group, it is often necessary to consider not only a direction parallel to a row $[uvw]$, but the set of directions equivalent through the symmetry operations of the point group. The set of these equivalent directions is noted $< uvw >$. In the same way, the set of lattice planes equivalent to a lattice plane (hkl) through the symmetry operations of the point group is noted $\{hkl\}$. However, we must note that $[uvw]$ represents a row of nodes and is not oriented. In the same way, (hkl) is a lattice plane and its normal is not oriented. When directions equivalent through the symmetry operations of the point group are considered, it is necessary, for the non-centrosymmetric groups, *i.e.* those which do not contain inversion, to orient the direction. For example, in a crystal with the cubic point group $\bar{4}$3m, $<111>$ is different from $< \bar{1}\bar{1}\bar{1} >$ and, in the same way, $\{111\}$ is different from $\{\bar{1}\bar{1}\bar{1}\}$ (Exercise 5.6).

The name given to $\{hkl\}$ is a "crystal form". The reason is that the growth rate of a crystal is the same for all the equivalent (hkl) planes. If the external shape or form of the crystal (its "habit") features an (hkl) orientation, then the whole set of faces equivalent to (hkl) are encountered, whence the name "crystal form" for $\{hkl\}$ (assuming of course that special external interactions, which could hamper the growth of some faces, are not present).

5.3. Improper groups

5.3.1. Preliminary remark

For a given group, we will note S_i its proper symmetry operations (rotations) and $\bar{S}_j$ its improper symmetry operations (rotoinversions). We recall that the product of two proper operations or of two improper operations is a proper operation, and that the product of a proper operation and an improper operation is an improper operation.

By definition, a proper group contains only proper operations, and an improper group contains one or several improper operations. However an improper group contains at least one proper operation: the neutral element. We can therefore write any group G as the sum of a set of proper elements, among which the neutral element E or identity, and of a set of improper elements which is an empty set when G is a proper group:

$$G = \{E, S_1, S_2, \ldots\} + \{\overline{S}_1, \overline{S}_2, \ldots\}.$$

5.3.2. Properties of improper groups

The proper elements of an improper group form an invariant subgroup with order half that of the improper group.

1. They form a group because the product of two proper elements is a proper element, the neutral element is a proper element and the inverse of a proper element is a proper element.

2. This subgroup is invariant. This requires showing that, for any element s of the group, $s^{-1}S_i s = S_j$ (Appendix 4A). This is evident if s is a proper element. If s is improper and noted as $\overline{S}_k$, its inverse $\overline{S}_j$ is also improper, and we obtain a proper element for the product $\overline{S}_k S_i \overline{S}_j$.

3. This subgroup is of order half that of the group. We will show that an improper group contains as many proper elements as improper elements.

Let n_p be the number of proper elements and n_i the number of improper elements of the group. Consider a given improper element $\overline{S}_m$, and multiply it successively by

i)   all the n_p proper elements:

$$\overline{S}_m S_k = \overline{S}_l.$$

We obtain n_p improper elements, all different from one another (see the rearrangement theorem derived in Appendix 4A). Therefore $n_p \leq n_i$.

ii)   all the n_i improper elements:

$$\overline{S}_m \overline{S}_k = S_l.$$

We obtain n_i different proper elements, and $n_i \leq n_p$.

The conclusions of (i) and (ii) imply that $n_i = n_p$.

We deduce from (1), (2) and (3) that an improper group always contains a proper subgroup with order half its own and invariant.[2]

---

2.   It can be shown that a subgroup of order half is always invariant. Therefore we could have just shown that the subgroup is of order half.

We now enumerate first the proper groups, then the improper groups starting from the proper groups.

5.4. Enumeration of the proper point groups

5.4.1. Preamble

We recall that the elements of a proper point group are rotations by $2\pi/n$ and their multiples around a given axis, with the condition (Sect. 3.3.2):

$$n = 1, 2, 3, 4, \text{or } 6 \quad \text{(condition 1)}.$$

The simplest groups are those which contain only the symmetry operations resulting from the presence of a single symmetry axis, an n-fold axis, often noted as A_n. They will be enumerated in Section 5.4.2. We then consider the groups which contain symmetry operations associated with the existence of two axes, A_n and $A_{n'}$, where n and n' satisfy condition 1. The combination of two of these operations, i.e. the product of a rotation by $2\pi/n$ (or one of its multiples) around axis A_n and a rotation by $2\pi/n'$ (or one of its multiples) around axis $A_{n'}$ must also satisfy condition 1, i.e. be a rotation by $2\pi/n''$ (or one of its multiples) around an axis, with n'' also fulfilling condition 1. This restriction on n'' leads to the following results:

1. For a general value of n, n' is equal to 2 and the angle between the two axes is equal to 90°. The corresponding groups are described in Section 5.4.3.

2. For $n = 2$ or 4, n' can also be equal to 3 and the angle ω between the 3-fold axis and the 2- or 4-fold axis is then such that $\cos^2 \omega = 1/3$, i.e. the angle between a side and a major diagonal of a cube. The corresponding groups (cubic groups) are described in Section 5.4.4.

The approach used to derive these results is the following. Let ω be the angle between axes A_n and $A_{n'}$ and O their intersection point (Fig. 5.6). We choose a system of axes $Oxyz$ such that axis Oz be parallel to axis A_n and axis Ox be perpendicular to the plane defined by A_n and $A_{n'}$. The matrix S_n representing the rotation around axis A_n is easily expressed in the axis system $Oxyz$. The matrix $S_{n'}$ representing the rotation around axis $A_{n'}$ is easily expressed in a system of axes $Oxy'z'$ where Oz' is parallel to $A_{n'}$, and it is noted $S'_{n'}$. To obtain the matrix $S_{n'}$ for the rotation around $A_{n'}$ in the axis system $Oxyz$, we use the transition matrix R for passage from the referential $Oxy'z'$ to the referential $Oxyz$. We know that $S_{n'} = R^{-1}S'_{n'}R$. The product $S_nS_{n'}$ is a matrix for which the trace must be an integer in order to satisfy the condition $n'' = 1, 2, 3, 4$ or 6 (Sect. 3.3.2). The calculation is simple but tedious, and will not be presented here.

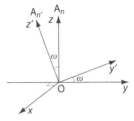

Figure 5.6: Systems of orthonormal axes used to obtain the product of a rotation around an axis A_n and a rotation around another axis $A_{n'}$. Axis Ox is perpendicular to the plane containing A_n and $A_{n'}$.

5.4.2. Groups containing only the symmetry operations associated to an axis A_n (cyclic groups)

Note: for simplicity, the title of this section could be replaced by "groups featuring only an axis A_n", and this is what we will often do in what follows even though this formulation is not very accurate. The elements of the group are the symmetry operations associated to the symmetry element, for example the rotations by $\pi/2$, π, $3\pi/2$ and 2π for a group "featuring an A_4 axis".

There are five such groups, corresponding to $n = 1, 2, 3, 4$ and 6, and they are noted n in the Hermann-Mauguin notation which we use in this book.

Group 1: no symmetry operation in this group, which contains only the identity element.

Group 2: existence of an axis A_2. The symmetry operations are the rotation by π around this axis and the rotation by 2π, *i.e.* identity.

Group 3: existence of an axis A_3. The symmetry operations are the rotations by $2\pi/3$, $4\pi/3$ and 2π (identity) around the axis.

Group 4: existence of an axis A_4. The symmetry operations are the rotations by $2\pi/4 = \pi/2$, π, $3\pi/2$ and 2π (identity) around the axis.

Group 6: existence of an axis A_6. The symmetry operations are the rotations by $2\pi/6 = \pi/3$, $2\pi/3$, π, $4\pi/3$, $5\pi/3$ around the axis, and identity.

Figure 5.4 shows the stereographic projection of the equivalent directions for each of these groups. The projection plane is perpendicular to axis A_n. The number of equivalent directions, equal to the order of the group (Sect. 5.2.2), is equal to n. The corresponding abstract groups are the cyclic groups C_n (Appendix 4A).

5.4.3. Groups containing the symmetry operations associated to an axis A_n and to a perpendicular axis A_2, or dihedral groups

An elementary geometrical argument shows that, if there exist a symmetry axis A_n and an axis A_2 which is perpendicular to it and intersects it at point O, then there exist n axes A_2 going through O, orthogonal to axis A_n, and with an angle π/n between them (Exercise 5.1). This can also be shown using the stereographic projection (Exercise 5.2). These groups are called $n2$ if n is odd (group 32) and $n22$ if n is even (groups 222, 422, 622). The second "2" in the latter cases may seem redundant. Its presence will be justified when we describe the space groups. The corresponding abstract groups are the dihedral groups D_n (Appendix 4A).

Figure 5.7 shows, for each of these groups, the stereographic projection of the equivalent directions and of the symmetry elements.

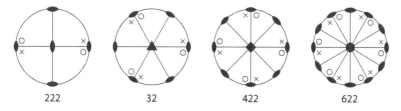

222　　　　　32　　　　　422　　　　　622

Figure 5.7: Dihedral point groups. They feature one A_n axis and n 2-fold axes which are perpendicular to A_n.

5.4.4. Cubic proper groups

These are groups for which the generating elements are the symmetry operations associated to a 2-fold (or 4-fold) axis and a 3-fold axis, with the angle between them equal to that between one side of a cube and a major diagonal of this cube. By working out the product of the various symmetry operations associated to the 2-fold (or 4-fold) axis and the 3-fold axis, it can be shown that there are then three 2-fold (or 4-fold) axes parallel to the three sides of a cube, and four 3-fold axes, parallel to the major diagonals of the cube (Exercise 5.3). The two groups thus defined are called 23 and 432. The notation for the cubic groups consists of two or three symbols. The first one concerns the symmetry axes parallel to the sides of the cube, the second one the 3-fold axes parallel to the major diagonals of the cube, and the third, if it exists, indicates the presence of axes parallel to the diagonals of the cube faces. We will see later that, in the cubic groups featuring mirrors, the same rule applies when considering the directions perpendicular to the mirrors. Thus the notation 23 indicates the presence of 2-fold axes parallel to the sides of a cube and of 3-fold axes parallel to the major diagonals of the cube. There is no third symbol because there are no symmetry axes parallel to the diagonals of the cube faces. The group 432 involves three 4-fold axes parallel to the sides of the cube, four 3-fold axes parallel to the cube diagonals, and six 2-fold axes parallel to the diagonals of the six cube faces.

Figure 5.8 shows, for each of these groups, the stereographic projection of the equivalent directions and the symmetry elements.

23   432

Figure 5.8: Proper cubic groups

5.5. Enumeration of the improper point groups

We saw that any improper group G contains a proper (and invariant) subgroup of order half. Let us note it as G_p. The group G can thus be decomposed into cosets starting from G_p:

$$G = G_p + \bar{A}G_p = \{S_1, S_2, \ldots, S_p\} + \{\bar{A}S_1, \bar{A}S_2, \ldots, \bar{A}S_p\}$$

where $\bar{A}$ is any improper element of G. If G contains inversion, a partition into these two cosets is evident since inversion itself can be chosen as the element $\bar{A}$. This is discussed in the next section. We will see in Section 5.5.2 how to obtain those improper groups that do not contain inversion.

5.5.1. Improper groups containing inversion

These are obtained by considering successively all the proper groups previously defined, and adding to them one generating element, inversion I (and consequently all the elements that are the product of an element of G_p and of inversion):

$$G = \{G_p\} + I\{G_p\}.$$

The stereographic projection of the directions equivalent through the symmetry operations of the group is a good tool to determine all the symmetry operations of the groups G thus obtained.

The first, almost trivial, example, is that of group 1, the only element of which is identity, to which one adds inversion I. The group contains two elements, the neutral element or identity E, and inversion I. It is noted as $\bar{1}$, since a $\bar{1}$ axis is equivalent to a center of symmetry (Fig. 5.5).

Groups deduced from the cyclic groups

Consider the proper group 2 and the improper group to which it is associated by adding inversion. We draw the stereographic projection of the equivalent directions onto a plane perpendicular to the 2-fold axis (Fig. 5.9). Consider an initial direction, and suppose it is in the northern hemisphere. Its stereographic projection is a cross at point 1. The direction equivalent through a rotation by π around the axis has for its stereographic projection a cross at point 2. Let us add a center of symmetry. The projection of the direction equivalent to 1 through inversion is a circle (since it is a direction in the southern hemisphere) at point 3 coinciding with point 2, and that of the direction equivalent to 2 through inversion is a circle at point 4 coinciding with 1. We therefore immediately see there is a mirror perpendicular to the 2-fold axis.

Figure 5.9: Point group obtained by adding inversion to group 2

The group we are discussing has 4 elements, corresponding to the four stereographic projections: identity E associated to the projection 1 of the initial direction, rotation by π around the axis, which we note as A, associated with projection 2 (2 is the transform of 1 through A), inversion I associated to 3 (3 is the transform of 1 through I), and the mirror operation m perpendicular to the axis, associated to 4 (4 is the transform of 1 through m = IA).

The generating elements of the group are A and I (the fourth element is then m = AI) or A and m (and the fourth element is I = Am). The notation for this group refers to the second choice. It is noted as $2/m$, the position of m as the denominator to 2 meaning that the mirror is perpendicular to the 2-fold axis.

Figure 5.10 shows the stereographic projections of the improper groups thus determined by adding inversion to the cyclic groups of Figure 5.4, and their denominations. They are noted $2/m$, $\bar{3}$, $4/m$ and $6/m$. We recall that the existence of a $\bar{3}$-fold axis is equivalent to the presence of a 3-fold axis and of a center of symmetry.

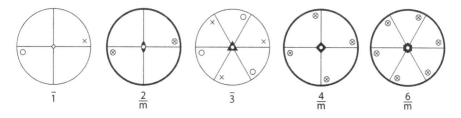

Figure 5.10: Point groups obtained by adding inversion to groups 1, 2, 3, 4 and 6

Groups deduced from the dihedral groups

Consider group 222 and the stereographic projection of the 4 directions equivalent through the symmetry operations of the group onto a plane perpendicular to one of the 2-fold axes (Fig. 5.11a). Let us add inversion to these symmetry operations. We obtain, on the stereographic projection, four further directions, the directions opposite to the above (Fig. 5.11b).

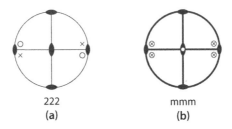

222
(a)
         mmm
(b)

Figure 5.11: Group 222 and group mmm, deduced from 222 by adding inversion

This projection reveals three mirrors perpendicular to the 2-fold axes, and mutually perpendicular. The group is noted mmm because the three mutually orthogonal mirrors can be chosen as generating elements. The presence of the 2-fold axes then results from the product of the mirror operations taken pairwise. In the same way, if inversion is added to group 422, we obtain a 4-fold axis, four mirrors going through the 4-fold axis and enclosing an angle of $\pi/4$ between each other, and a mirror perpendicular to the 4-fold axis. The generating elements of the group are the 4-fold axis, a mirror going through the axis, and a mirror perpendicular to the axis. Its denomination is 4/m mm. As above, the m as the denominator of the fraction indicates that the mirror is perpendicular to the 4-fold axis. The final m is redundant, but its presence is justified when space groups are considered (Chap. 7).

Figure 5.12 shows the stereographic projections of the equivalent directions in the improper groups thus determined, and their denominations ($\bar{3}$m, 4/m mm, 6/m mm).

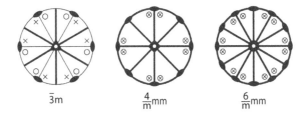

$\bar{3}$m $\frac{4}{m}$mm $\frac{6}{m}$mm

Figure 5.12: Groups obtained by adding inversion to the dihedral groups 32, 422 and 622. The white circles at the centers of the symbols for the 3-, 4- and 6-fold axes indicate the existence of a center of symmetry.

Groups deduced from the proper cubic groups

Adding inversion to group 23 gives group m$\bar{3}$ (Fig. 5.13, to be compared with Fig. 5.8). The first sign (m) indicates that there are mirrors perpendicular to the sides of the cube, and the second sign indicates that the axes parallel to the major diagonals are $\bar{3}$ axes. There is no third sign because there are no 2-fold axes parallel to the face diagonals, nor mirrors perpendicular to these diagonals.

m$\bar{3}$ m$\bar{3}$m

Figure 5.13: Groups obtained by adding inversion to the proper cubic groups

Adding inversion to group 432 leads to group m$\bar{3}$m which features all the symmetry elements of a cube. The reader will check that the symmetry elements of a cube all go through the center of the cube and are the following: a center of symmetry, three 4-fold axes parallel to the cube sides and the mirrors perpendicular to them, four 3-fold axes coinciding with the major diagonals of the cube, six 2-fold axes parallel to the face diagonals of the cube, and the six mirrors which are respectively perpendicular to them, called diagonal mirrors. The denomination of the group also states that there are three mirrors perpendicular to the sides of the cube (first m), four 3-fold axes parallel to the major diagonals of the cube and an inversion center ($\bar{3}$), and six mirrors perpendicular to the face diagonals of the cube (m in third position). The 2-fold axes result from the combination of two orthogonal mirrors and lie at their intersection. The 4-fold axes result from the joint presence of mirrors parallel to the cube faces and of diagonal mirrors.

5.5.2. Improper groups which do not feature inversion

We saw in Section 5.3.2 that any improper group G contains a proper invariant subgroup with order half. Let us denote it as G_p. Group G can therefore be split into two cosets starting from G_p:

$$G = G_p + \bar{A}G_p = \{S_1, S_2, \ldots, S_p\} + \{\bar{A}S_1, \bar{A}S_2, \ldots, \bar{A}S_p\}$$

where $\bar{A}$ is an improper element which we assume not to be inversion. Consider the proper element $A = I\bar{A}$ obtained by multiplying $\bar{A}$ and inversion I. This may look arbitrary, but we will soon see why it is of interest. Let us show, through a reduction toward the absurd, that A is not part of G or, more precisely, of G_p, since A is a proper element. If A belonged to G_p, then its inverse A^{-1} would also be in G_p, and its product with $\bar{A}$, which belongs to G, *i.e.* $\bar{A}A^{-1}$ would be part of G. But $\bar{A}A^{-1} = IAA^{-1} = I$ and, by assumption, I is not in G. Therefore A is not an element of G_p.

Consider the set

$$G' = G_p + AG_p = \{S_1, S_2, \ldots, S_p\} + \{AS_1, AS_2, \ldots, AS_p\}.$$

The elements AS_k do not belong to G_p. To show this, assume AS_k is part of G_p. Since S_k^{-1} is in G_p, the product $(AS_k)S_k^{-1} = A(S_kS_k^{-1}) = A$ would also be in G_p, which is contrary to the above result. All the AS_k are distinct, since if $AS_k = AS_l$, then left-multiplying with A^{-1} would yield $S_k = S_l$, which is contrary to the assumption. We now show that G and G' are isomorphous, with the following correspondence: the elements of G_p correspond with one another, and $\bar{A}S_p$ is associated to AS_p. Compare the multiplication tables of G' and G (Fig. 5.14), which we divided into four quadrants. The top left quadrants are identical. So are the bottom right quadrants, since $(\bar{A}S_j)(\bar{A}S_k) = (IAS_j)(IAS_k) = (AS_j)(AS_k)$ because of the commutativity of I with all the elements and because $I^2 = E$. The top right quadrant contains, for G', the products $S_j(AS_k)$ and for G, the products $S_j(\bar{A}S_k) = S_j(IAS_k) = IS_j(AS_k)$. The bottom left quadrants show for G' the products $(AS_j)S_k$ and for G the products $(\bar{A}S_j)S_k = (IAS_j)S_k = I(AS_j)S_k$. There is thus a one-to-one correspondence between the products in G' and G.

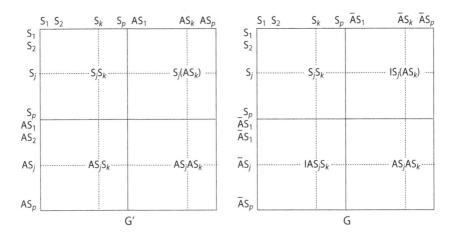

Figure 5.14: Multiplication table of the isomorphous groups G' and G

The elements of the set G' thus form a proper group of order $2p$, isomorphous to G. To each group G'_{2p} containing a proper subgroup with order half there will correspond an improper group G as searched for. To obtain G starting from G'_{2p} and its subgroup G_p, we proceed in the following way: we separate in G'_{2p} the subgroup G_p and its complementary subgroup K, consisting of the operations $K_1, K_2, \ldots, K_p$ equivalent to the operations AS_i:

$$G'_{2p} = G_p + \{K_i\}.$$

The operations $\bar{A}S_i$ required to obtain G are the operations $(IA)S_i = I(AS_i) = IK_i$. The required group G is:

$$G = G_p + \{IK_j\}.$$

We start with the simplest example: that of group 2 (consisting of identity E and of the rotation A by π around the axis, *i.e.* $\{E, A\}$). This has as its subgroup with order half the group 1, which has as its sole element E.

$G' = E + A$ and $G = E + IA = \{E, m\}$ where $m = IA$ is the symmetry operation with respect to a plane perpendicular to the 2-fold axis. This 2-fold axis was used as an intermediate device in the reasoning, but it is not a symmetry element of the group we are investigating. The improper group we are searching for has for its symmetry elements a mirror, and this group is denoted as m. In more complicated cases, the definition of G starting from G' is made much easier by using the projection of the directions equivalent through the elements of G'_{2p} and of G_p. We take as an example group 422 and its subgroup with order half, 222. Figure 5.15a is the stereographic projection of the directions equivalent through the symmetry operations of group 422 (G'_{2p}). We choose an initial projection direction, and denote it as 1. Figure 5.15b shows the directions equivalent to 1 through the symmetry operations of group 222 (G_p). Figure 5.15c shows the equivalent directions for the required group G. The additional directions $5', 6', 7', 8'$ are the reverse of directions 5, 6, 7 and 8, the complementary of directions 1, 2, 3 and 4 in obtaining 422 starting from 222. We see, on the projection of Figure 5.15c, that this group contains a $\bar{4}$-fold axis, two 2-fold axes perpendicular to this axis $\bar{4}$, and two mirrors parallel to axis $\bar{4}$ and enclosing angles of 45° to the 2-fold axes. This group is noted as $\bar{4}2m$.

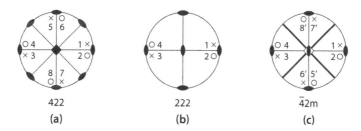

422   222 $\bar{4}2m$

(a)   (b)   (c)

Figure 5.15: Construction of group $\bar{4}2m$ starting from group 422 (a) and its subgroup with order half, 222 (b)

Figure 5.16 shows the stereographic projection of the equivalent directions in these various improper groups that do not contain inversion. We note that

1. Any group $2n$ (abstract group C_{2n}) admits group n (abstract group C_n) as a subgroup with order half $(2 \to 1)$, $(4 \to 2)$, $(6 \to 3)$ and these associations lead to groups m, $\bar{4}$ and $\bar{6}$ respectively (Fig. 5.16a).

2. Groups $n22$ admit groups n as subgroups of order half, with the associations $(222 \to 2)$, $(32 \to 3)$, $(422 \to 4)$, $(622 \to 6)$, which lead to the

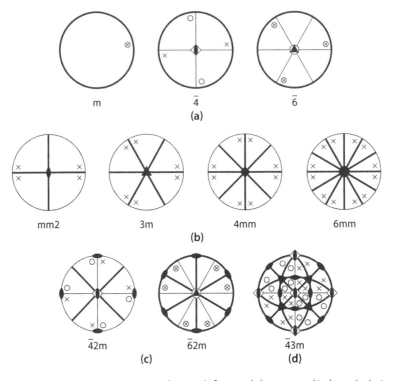

Figure 5.16: Point groups obtained from: (a) groups $(2n)$ and their subgroups n; (b) groups $n22$ and their subgroups n; (c) groups $2p22$ and their subgroups $p22$; (d) group 432 and its subgroup 23

groups mm2, 3m, 4mm and 6mm respectively, called pyramidal groups because they have the symmetry of regular pyramids with an n-fold axis and mirrors going through the axis (Fig. 5.16b).

3. Groups $n22$ with n even, *i.e.* $n = 2p$, also admit the subgroups with half order $p22$, with the associations $(422{\to}222)$ and $(622{\to}32)$, which lead to groups $\bar{4}2m$ and $\bar{6}2m$ respectively (Fig. 5.16c).

4. The cubic group 432 admits group 23 as a subgroup with order half, which leads to the group $\bar{4}3m$ (Fig. 5.16d). This is the symmetry group of a regular tetrahedron.

We thus defined the 32 crystallographic point groups, which are

-   11 proper groups,
-   11 improper groups containing inversion,
-   10 improper groups that do not contain inversion.

5.6. Classification of the point groups

The set of crystals having a given point group form a crystal class. All classes with the same lattice symmetry form a crystal system. Figures 5.17a and b show the classification of point groups (hence of crystal classes) by crystal system. This classification will be fully justified in Chapter 6. The qualitative presentation made here is an easy approach, sufficient on first reading.

We saw that there are seven crystal systems, defined on the basis of the symmetry of the crystal lattice. We also saw that the crystal lattice is at least as symmetric as the crystal or, to put it differently, that the point group of the crystal is a subgroup of the point group of the crystal lattice or the group itself (Sect. 4.3). Simple examples of crystal point groups which are subgroups of the point group of the crystal lattice are the 21 point groups which do not contain inversion, while the point groups of all crystal lattices do contain inversion.

We thus consider the seven point groups that have the symmetry of the seven crystal systems. Each of these groups is called a holohedral group. Each holohedral group is taken as the head of a row on which we arrange its respective subgroups. It is placed at the right-hand end of the row. On a given row, the number of elements of each group (the order of the group) grows from left to right, ending with the holohedral point group, the group of the crystal lattice. Figures 5.17 show, for all groups, the stereographic projection of their symmetry elements (Fig. 5.17a) and the directions equivalent to a given starting direction (Fig. 5.17b). The projection plane is perpendicular to the axis with the highest symmetry, except for groups 23 and m$\bar{3}$, where the projection is performed, as for all cubic groups, on a plane perpendicular to one side of the cube. The starting direction is the same for all groups, so that the subgroups of the groups show up easily.

Consider for example the crystals with an orthorhombic lattice. An orthorhombic lattice (unit cell in the shape of a rectangle parallelepiped) has for its symmetry elements 2-fold axes parallel to the sides of the parallelepiped, mirrors perpendicular to these axes, and a center of symmetry. The corresponding point group is group mmm (the third group starting from the left on the third row). The notation indicates the existence of three mutually orthogonal mirrors, the symmetry operations of which are the generating elements of the group.

We showed the existence of crystal point groups 222 (three mutually orthogonal 2-fold axes) and mm2 (two mutually orthogonal mirrors, leading to the presence of a 2-fold axis at their intersection), which are subgroups of group mmm. It follows that a crystal with orthorhombic lattice can have for its point group 222, mm2 or mmm.

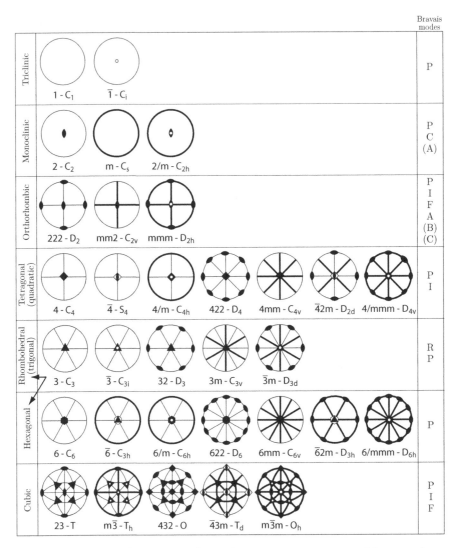

Figure 5.17a: The 32 crystallographic point groups arranged by crystal system: stereographic projection of the symmetry elements. The rightmost column indicates the Bravais modes consistent with each of the systems.

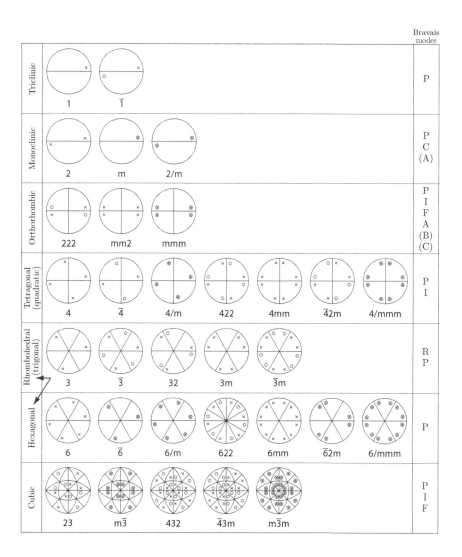

Figure 5.17b: The 32 crystallographic point groups arranged by crystal system: stereographic projection of the equivalent directions. The rightmost column indicates the Bravais modes consistent with each of the systems.

The arrangement of entries on the four first rows is such that the crystal lattices are ranked in order of growing symmetry (triclinic, monoclinic, orthorhombic, tetragonal), easy to identify since the point group of a crystal lattice is a subgroup of the point group of the crystal lattice on the next row. Group $\bar{1}$ of the triclinic lattice is a subgroup of group $2/m$ of the monoclinic lattice, which itself is a subgroup of group mmm (orthorhombic lattice), while the latter is a subgroup of $4/m$ mm (tetragonal lattice).

We note that, to determine the row corresponding to a given point group, we look for the holohedral group with the lowest symmetry of which it is a subgroup: for example point group mm2 is a subgroup of the holohedral group mmm of the orthorhombic lattice. It is also a subgroup of point group $4/m$ mm of the tetragonal lattice, but a crystal with point group mm2 has an orthorhombic, not a tetragonal, crystal lattice. It may occur, under specific temperature or pressure conditions, that a crystal with point group mm2 happens to have two lattice parameters equal, for example $a = b$. Such a crystal is then accidentally tetragonal.

The rhombohedral and hexagonal lattices appear on the following rows. The point group of the rhombohedral lattice is a subgroup of that of the hexagonal lattice. We make an important note: we will show in Chapter 6 that the crystals with point group 3, $\bar{3}$, 32, 3m and $\bar{3}$m, all shown on the 5th row, can have either a rhombohedral crystal lattice (noted R) or a hexagonal crystal lattice (noted P). This is indicated by the two arrows pointing from this row to the rhombohedral and hexagonal lattices, as well as by the P associated to R in the column showing the Bravais modes.

The cubic groups are shown on the last row.

If the point group of the crystal is a subgroup of the holohedral group, the group is said to be merohedral. The table below shows the type of merohedry as a function of the ratio between the order of the holohedral group n_H, *viz.* the order of the symmetry group of the crystal lattice, and the order of the merohedral group n_M of the crystal.

n_H/n_M	Designation of the group
1	holohedral
2	hemihedral
4	tetartohedral
8	ogdohedral

We recall the rules, already stated in Sections 5.4.2 to 5.4.4, of the Hermann-Mauguin notation. A mirror parallel to an n-fold (or $\bar{n}$-fold) axis is noted through an m next to the n (or $\bar{n}$), and a mirror perpendicular to an n-fold axis by an m as a denominator. For example group 3m involves a 3-fold axis and three mirrors parallel to this 3-fold axis. Group $4/m$ has a 4-fold axis and a mirror perpendicular to it. For the cubic groups, the first symbol shows the order of the axes parallel to the cube axes (2 or 4) or the existence of mirrors perpendicular to these axes (m). The second symbol shows the existence of 3-fold axes parallel to the major diagonals of the cube, or $\bar{3}$ if there is also an inversion center. The third symbol, if it exists, shows the existence of 2-fold axes parallel to the diagonals of the cube faces or of mirrors perpendicular to these directions.

Figure 5.17a also shows, next to this nomenclature which is the usual one for crystallographers, the Schönflies notation, frequently used in group theory, and summarized below:

For all groups except for the cubic groups and groups $\bar{4}$ and m:

– find the axis with the highest order, call it n, and assume it is vertical,

– associate to this group its cyclic (C_n) or dihedral (D_n) subgroup with the highest order, and start the notation with C_n or D_n respectively.

This notation is complemented, if necessary, with

– a subscript i if there is also a center of inversion

– one of the following subcripts if there are mirrors: h if the mirror is horizontal, v if it is vertical and there is no horizontal 2-fold axis, or if there is a horizontal 2-fold axis in the plane of the mirror, d if the mirrors are vertical with 2-fold axes in the bisecting planes of the mirrors. Example: group 4mm, with subgroup C_4, has a vertical mirror as an extra generating element. It is noted C_{4v}.

The other notations are the following: notation S_n refers to the presence of a rotoinversion axis $\bar{n}$, but it is used only for group S_4 ($\bar{4}$), because group $\bar{3}$ is noted as C_{3i} (a $\bar{3}$-fold axis being equivalent to a 3-fold axis and a center of symmetry), and group $\bar{6}$ is noted C_{3h} (a $\bar{6}$-fold axis being equivalent to a 3-fold axis and a mirror perpendicular to it). The notation C_s indicates group m (the s is from the German word *Spiegel* = mirror). The notations T (tetrahedron) and O (octahedron) are used for the cubic groups. T is the group of the rotations of the tetrahedron (group 23). It is easy to check, with the help of Figure 5.17b, that adding a horizontal mirror to group 23 yields group $m\bar{3}$, therefore noted T_h, while adding a mirror in the diagonal planes gives group $\bar{4}3m$, noted T_d. Group O is the group of the rotations of the cube (group 432), which becomes O_h ($m\bar{3}m$) if a horizontal mirror is added.

5.7. Laue classes

The investigation of X-ray diffraction by crystals shows that the intensity of the beams diffracted by a given crystal is the same whether the reflection[3] occurs off one side or the other of the lattice planes. This result, known as Friedel's law, only breaks down when the wavelength of the radiation is very near an absorption edge, *i.e.* when the energy of an X-ray photon is equal or nearly equal to that of an electronic energy level of one of the atoms in the crystal. Away from the absorption edges, diffraction thus introduces a center of symmetry even if the crystal structure has none. This is why the 32 symmetry groups are arranged into 11 sets such that the effects of the crystal symmetry on diffraction are the same within each set. Each of the 11 centrosymmetric crystallographic groups defines one of these sets, called Laue classes. For any of the 21 remaining groups, just multiplying the group with (E, I) produces the centrosymmetric group which defines the Laue class to which it belongs. We thus obtain Table 5.2.

Table 5.2: The 11 Laue classes. Each class is surrounded by a thick line.

1	$\bar{1}$					
2	m	$\dfrac{2}{m}$				
222	mm2	mmm				
4	$\bar{4}$	$\dfrac{4}{m}$	422	4mm	$\bar{4}2m$	$\dfrac{4}{m}mm$
3	$\bar{3}$	32	3m	$\bar{3}m$		
6	$\bar{6}$	$\dfrac{6}{m}$	622	6mm	$\bar{6}2m$	$\dfrac{6}{m}mm$
23	$m\bar{3}$	432	$\bar{4}3m$	$m\bar{3}m$		

---

3. Crystal diffraction can be described, as in Section 3.2.1, through the condition that the scattering vector be a reciprocal lattice vector $\mathbf{h}^* = h\mathbf{a}^* + k\mathbf{b}^* + l\mathbf{c}^*$ where $\mathbf{a}^*$, $\mathbf{b}^*$, $\mathbf{c}^*$ are the basis vectors of the reciprocal lattice. It is often convenient to describe it as mirror reflection off the lattice planes (hkl), but selective reflection because it only occurs when the wavelength satisfies Bragg's law. This is why crystal diffraction is often referred to as Bragg reflection.

5.8. Plane point groups

It is useful, both for later discussions on lattices, and for surface crystallog-raphy, to determine the plane crystallographic groups. Their determination can be performed in the same way as for 3-dimensional groups, by looking for the proper groups, then the improper groups. The operation of symmetry with respect to a given point O in the plane is identical to a rotation by 180° around an axis perpendicular to the plane that goes through O. It is no more an improper operation. The basic improper operation is now the operation of symmetry with respect to a plane perpendicular to the plane of the object, or mirror operation, noted m.

All the plane symmetry elements are perpendicular to the reference plane. The only proper plane crystallographic groups are 1, 2, 3, 4, 6. The improper groups are obtained by multiplying these proper groups with group (E, m). We then obtain groups m, 2m or mm,[4] 3m, 4m, 6m. We note that these 10 plane crystallographic point groups are identical to 10 of the 3-dimensional point groups.

5.9. Isotropy point groups

The investigation of the physical properties of crystals will benefit from a gener-alization of the discussion in this chapter to infinite-order non-crystallographic point groups. These are groups featuring an axis of revolution, which is none other than an infinite-order axis, or isotropy axis. It will be performed accord-ing to the same rules. The result is given below. For each of these groups, we give an example of an object with the symmetry elements of the group (Fig. 5.18).

Proper groups

(a) There is a single isotropy axis, and no other symmetry operation. The group has for its symbol ∞. It corresponds to the cyclic groups (Sect. 5.4.2). This is the symmetry of a half cone rotating around its axis.

(b) There is a single isotropy axis and an infinity of 2-fold axes perpendicular to it. This group is noted $\infty 2$, and it corresponds to the dihedral groups of Section 5.4.3. This is the symmetry of two half-cones with the same apical angle and with the same axis, but opposed at the apex and rotating in opposite directions.

---

4. This group contains the symmetry operations associated to a 2-fold axis and a mirror going through this axis. The existence of these two symmetry elements entails the existence of another mirror going through the axis and perpendicular to the first one. This is why this group is usually called mm.

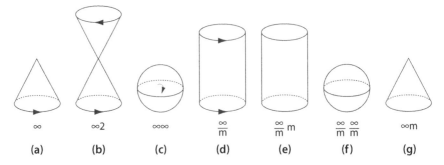

$$\infty \qquad \infty 2 \qquad \infty\infty \qquad \frac{\infty}{m} \qquad \frac{\infty}{m}\,m \qquad \frac{\infty}{m}\,\frac{\infty}{m} \qquad \infty m$$

(a)   (b)   (c)   (d)   (e)   (f)   (g)

Figure 5.18: Examples of objects having the symmetry of each of the 7 isotropy groups

(c) There is an infinity of isotropy axes. This group, noted $\infty\,\infty$, is the limiting case of groups involving several axes of order larger than 2 (corresponding to the cubic proper groups of Sect. 5.4.4). This is the symmetry of a sphere containing a liquid featuring rotatory power, hence not superimposable with its mirror image.

Improper groups containing inversion

We multiply by (E, I) the groups obtained in (a), (b) and (c). Since these groups contain at least one isotropy axis (which therefore includes a 2-fold axis), the presence of the center of symmetry is equivalent to the existence of a mirror perpendicular to the axis. We obtain:

(d) one isotropy axis and a perpendicular mirror, *i.e.* the group ∞/m (corresponding to groups $2/m$, $4/m$, $6/m$). This is the symmetry of a cylinder rotating around its axis, and also that of an axial vector (Sect. 10.6.1).

(e) one isotropy axis, an infinity of mirrors going through this axis, and a mirror perpendicular to the axis (and a center of symmetry). This group is called ∞/mm (it can be associated to groups $4/mmm$ and $6/mmm$). This is the symmetry of a stationary cylinder.

(f) an infinity of isotropy axes, an infinity of mirrors (and a center). This group is noted $\infty/m\,\infty/m$. This is the symmetry of a sphere.

Improper group which does not contain inversion

(g) Group $\infty 2$ admits group ∞ as its half order subgroup. The process described for the crystallographic groups now reduces to replacing the infinity of 2-fold axes by an infinity of mirrors going through the isotropy axis. This group is called ∞m. It generalizes the pyramidal groups. This is the symmetry of a stationary half-cone, and also that of a polar vector (Fig. 9.2).

These groups can be classified into two systems: the cylindrical system, which contains a single axis of revolution (a, b, d, e and g) and the spherical system which contains an infinity of axes of revolution (c and f).

These infinite-order groups will be used in the second part of the book to characterize the symmetry of some physical quantities. For example an electric field has for its symmetry group ∞m, a magnetic field ∞/m and a uniaxial stress (which will be defined in Sect. 11.5.1) ∞/mm.

5.10. Exercises

Exercise 5.1

Consider two axes $x'x$ and $y'y$, intersecting at point O, and enclosing an angle θ. Let A be the operation consisting of a rotation by π around $x'x$ and A′ the operation rotation by π around $y'y$.

1. Show that the product A′A is equal to a rotation by 2θ around an axis that goes through O and is perpendicular to the plane Π formed by $x'x$ and $y'y$. Denote this operation as $R_\perp(2\theta)$.

2. Deduce that $A' = R_\perp(2\theta)A$ and $A = A'R_\perp(2\theta)$, so that, if an object features an n-fold axis and a 2-fold axis perpendicular to it, then it also features n 2-fold axes with angles π/n between them.

Exercise 5.2

Consider the point group 32, *i.e.* a group comprising the symmetry operations associated to a 3-fold axis and to a 2-fold axis perpendicular to it.

Draw the stereographic projection, onto a plane perpendicular to the 3-fold axis, of the directions equivalent to a general direction through these symmetry operations. Deduce that this group includes three 2-fold axes perpendicular to the 3-fold axis, with angles of $2\pi/3$ between them. This is an easy way of retrieving the result of the preceding exercise in the special case where $n = 3$. It can be easily generalized to any value of n.

Exercise 5.3

1. Show that the existence of two intersecting and perpendicular 4-fold axes leads to the existence of four 3-fold axes parallel to the major diagonals of a cube with sides parallel to the 4-fold axes. The derivation can be made either using the rotation matrices representing the symmetry operations, or by considering a cube ABCDEFGH and submitting it to the successive operations.

2. Show that the reciprocal proposition is not true. The existence of four 3-fold axes along the major diagonals of a cube does not entail the existence of three 4-fold axes parallel to the sides of the cube. It only entails the existence of 2-fold axes parallel to the cube sides.

Exercise 5.4

Consider Figure 5A.5b showing the stereographic projection of some of the symmetry elements of a cube. If the projection plane is perpendicular to the basis vector $\mathbf{c}$ of a cubic lattice, and if b is the projection of [100], the circular arcs b'Fb and b'F'b are the projections of the diagonal plane $(01\bar{1})$.

Show that the centers of the circles to which these arcs belong are points a and a' respectively.

Note: it may be helpful to note that $\tan \frac{\pi}{8} = \sqrt{2} - 1$, a value which can be retrieved using $\tan 2\theta = \frac{2 \tan \theta}{1 - \tan^2 \theta}$ and $\tan \frac{\pi}{4} = 1$.

Exercise 5.5

Consider the various point groups consistent with a hexagonal lattice, and rank them according to their merohedry. Figure 5.17b can be useful.

Exercise 5.6

A form $\{hkl\}$ of a crystal comprises the set of faces in this crystal which are equivalent to face (hkl).

1. Give the Miller indices of the various planes in the form $\{31\bar{4}0\}$ for crystals with point group (a) 6mm, then (b) $\bar{6}2m$.

2. Same question for form $\{111\}$ in crystals belonging to the various cubic groups.

Appendix 5A. Complements on the stereographic projection

5A.1. Stereographic projection of the transform of a given direction through the symmetry operations associated to various symmetry elements

The notations are as follows. We call Ov the starting direction. It intersects the sphere at P, the stereographic projection of which is called n. Direction Ov', the transform of Ov through the considered symmetry operation, cuts the sphere at P', and its stereographic projection is called n'. The meridian half-planes containing OP and OP' intersect the projection plane along OQ and OQ' respectively (Fig. 5A.1). Projections n and n' are then located on OQ and OQ' respectively.

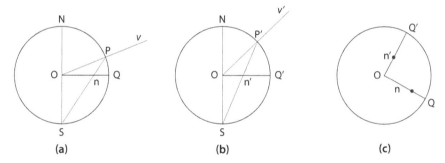

Figure 5A.1: Illustrating the notations used in this appendix: (a) meridian plane of the initial direction Ov; (b) meridian plane of direction Ov', the transform of Ov through the symmetry operation; (c) stereographic projections n and n' of directions Ov and Ov'

5A.1.1. n-fold axis perpendicular to the projection plane

The projection of the axis consists of two points coinciding with the origin O. The meridian planes containing Ov and Ov' enclose an angle $2\pi/n$. Ov and Ov' are at the same angle θ to NS and On $=$ On'. Projection n' is the transform of n through a rotation by $2\pi/n$ around an axis perpendicular to the projection plane and going through O (Fig. 5.2 and 5.4).

5A.1.2. Mirror coinciding with the equatorial plane

The stereographic projection of the mirror is the equatorial circle. The meridian plane containing Ov transforms into itself in the mirror operation, and in this plane Ov and Ov' are symmetric with respect to the trace of the equatorial plane (Fig. 5A.2a). The projections coincide but they are represented by different signs since Ov and Ov' are in two different hemispheres (Fig. 5A.2b).

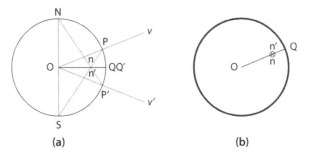

Figure 5A.2: (a) Meridian plane containing a general direction Ov and its transform Ov' through symmetry with respect to a mirror coinciding with the equatorial plane; (b) stereographic projections of the mirror (thick line) and of directions Ov and Ov'

5A.1.3. Mirror going through axis NS

The projection of the mirror is the line AB, intersection of the mirror and the equatorial plane (Fig. 5.3b). The meridian plane containing OP$'$ is symmetric of the one containing OP with respect to the mirror plane. In each of these planes, OP and OP$'$ enclose the same angle θ with axis NS, so that On $=$ On$'$. Their stereographic projections n and n$'$ are therefore symmetric with respect to the stereographic projection of the mirror.

5A.1.4. 2-fold axis in the equatorial plane

The projection of the axis consists of two opposite points on the equatorial circle, A and B (Fig. 5A.3c). Direction Ov intersects the sphere at point P, which projects on the equatorial plane at H. The meridian plane containing Ov is at an angle φ with AB. The transform of this meridian plane through a rotation by π around AB is a meridian plane enclosing the same angle φ with AB. The transforms of P and H are P$'$ and H$'$ respectively, and it is easy to show, using Figure 5A.3a, that $(\mathbf{OH}, \mathbf{OP}) = (\mathbf{OH'}, \mathbf{OP'}) = \beta$, P being in the northern hemisphere and P$'$ in the southern hemisphere. The stereographic projections n and n$'$ are therefore at the same distance from O on OQ and OQ$'$ (Fig. 5A.3b), both of which enclose with AB the angle φ. They are represented with different symbols (Fig. 5A.3c).

5A.2. Stereographic projections of the symmetry elements of a cube

5A.2.1. Preamble

We look for the stereographic projection of a plane going through the center O of the sphere and which does not go through axis NS. Its intersection with the sphere is a major circle Γ which intersects the equatorial circle at a and a$'$ (Fig. 5A.4a).

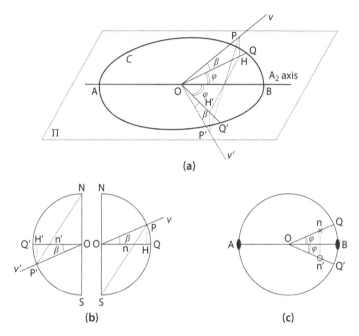

(a)

(b)   (c)

Figure 5A.3: Construction of the stereographic projection of the transform of a general direction Ov through rotation by π around an axis A_2
located in the equatorial plane: (a) perspective view; (b) meridian planes
containing Ov and Ov'; (c) stereographic projection

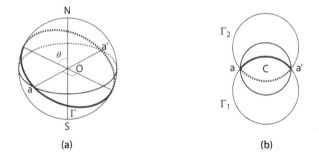

(a)   (b)

Figure 5A.4: A plane which does not go through axis NS. (a) Perspective
view. (b) The stereographic projection of the plane consists of two arcs
in full and dotted thick line, going through a and a'.

The stereographic projection of a circle being a circle, the projection of Γ starting from the south pole is a circle Γ_1 going through a and a' (Fig. 5A.4b). This circle has two parts: arc aa' (thicker line), located in the equatorial disk C, which corresponds to the projection of the half major circle located in the northern hemisphere. The second half of major circle Γ, located in the southern hemisphere, must be transformed through an inversion with the north pole as pole. The transform of Γ through the north pole inversion is circle Γ_2, symmetric of Γ_1 with respect to aa', with only the circular arc contained in the equatorial disk retained. It is represented in dotted line to indicate that it is the image of an arc in the southern hemisphere.

5A.2.2. Cubic groups

The stereographic projection is performed onto a plane parallel to one cube face, for example face HJKL of the cube shown on Figure 5A.5a.

1. Stereographic projection of one of the A_3 axes of the cube

    Select for example the A_3 axis coinciding with the major diagonal HK'. This diagonal is at the intersection of two diagonal planes of the cube, HJK'L' and HLK'J'. The stereographic projection of plane HJK'L' consists of two arcs intersecting along diameter aa' parallel to HJ. That of plane HLK'J' consists of two arcs intersecting along bb' parallel to side HL (Fig. 5A.5b). The stereographic projections of the axis are the intersections of the stereographic projections of the two diagonal planes, $i.e.$ point F is in the northern hemisphere and point F' in the southern hemisphere. They are represented by their characteristic symbol (an equilateral triangle).

    Axis A_3 is contained in a third diagonal plane of the cube, HKK'H'. This plane goes through axis NS. Its projection coincides with its intersection with the projection disk, $i.e.$ the segment cc' at 45° to aa' and bb'. It goes through F and F'.

2. Stereographic projection of all the symmetry elements of the cube, $i.e.$ of the point group m$\bar{3}$m (Fig. 5A.5c)

    –   The three A_4 axes represented by their characteristic symbols.

    –   The mirrors perpendicular to these A_4 axes, $i.e.$ the mirror parallel to the projection plane (and to face HJKL), the projection of which is the equatorial circle itself, and the mirrors parallel to axis NS, the projections of which are the segments aa' and bb' (mirrors parallel to HJJ'H' and JKK'J' respectively).

    –   The diagonal mirrors. There are six of them. We investigated three of them, which intersect along an A_3 axis. To each of these three mirrors we can associate another diagonal mirror which is perpendicular to it and intersects it along an A_4 axis. For example, mirror LJJ'L' is associated to mirror HKK'H'. These two mirrors go through axis NS and their stereographic projections are cc' and dd'. Mirror H'J'KL is associated to HJK'L'. They are symmetric of each other with respect to the projection plane. Their stereographic projections are identical (arcs aFa' and aF'a'), but with the characteristic symbols reversed. The arc in full line of one is superimposed on the arc in dotted lines of the other, and vice-versa. Finally the full lines cover the dotted lines. The same applies to the projections of the diagonal mirror HLK'J' and H'L'KJ, which consist of the arcs bFb' and bF'b'.

- Six A_2 axes perpendicular to the diagonal mirrors, and therefore located in the associated diagonal mirror. They connect, within these mirrors, the middles of two opposite sides. For example MP, located in the diagonal mirror HJK'L', is the A_2 axis perpendicular to mirror H'J'KL. Its stereographic projection consists of two points G and G' (Fig. 5A.5b). The axis A_2 located in the mirror H'J'KL projects on the same points but with different symbols. On the general figure, we show these two axes (as well as the four others) using the representative symbols of 2-fold axes.

- A center of symmetry. Its stereographic projection is at the center of the projection disk. Furthermore, the presence of a center of symmetry changes the A_3 axes into $\bar{3}$-fold axes.

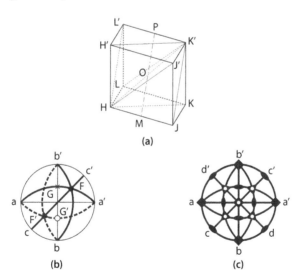

(a)

(b) (c)

Figure 5A.5: (a) Perspective view of a cube; (b) stereographic projection of some of its symmetry elements and (c) of all its symmetry elements

Chapter 6

Bravais lattices

This chapter shows how the Bravais lattices, already introduced in a qualitative way in Section 3.7, are enumerated. It can be skipped on first reading.

6.1. Introduction

Bravais lattices were introduced in Section 3.7. We showed that, in some cases, a multiple (or centered) cell evidences symmetry elements of the lattice which fail to appear in any primitive cell. We listed the 14 Bravais lattices (Sect. 3.7.2 and Tab. 3.3), without deriving how they are enumerated. The aim of the present chapter is to show how they are obtained logically. It can therefore be skipped on first reading. The derivation rests on the result obtained in Section 4.3, namely that the symmetry operations of the point group of the crystal form a subgroup of the symmetry operations of the crystal lattice.

The aim is therefore to find out what type of crystal lattice is consistent with each of the 32 point groups. However, since all crystal lattices feature centers of symmetry (on the nodes, at the centers of the unit cells and in the middle of the vectors connecting neighboring nodes), the point symmetry group of a crystal lattice necessarily contains inversion. It is therefore sufficient to know the lattice types consistent with each of the 11 centrosymmetric point groups, each of which is characteristic of its Laue class, defined in Section 5.7. The lattices of the 21 other groups are obtained by considering, for a given group, the group representative of its Laue class, *i.e.* the supergroup obtained by multiplying it with the group (E, I). This supergroup is one of the 11 centrosymmetric groups for which we determined the compatible lattice.

In practice, we will use this remark only in the course of the derivation, because it is simpler to start out by determining which lattice is consistent with a low-symmetry point group. We will see that we reach a lattice for which the point group is much more symmetric than the starting group, and therefore necessarily includes a set of subgroups. Before that, it is necessary and illuminating to enumerate the plane lattices. The results obtained are no surprise; they were already given without derivation in Section 3.4.1.

© Springer Science+Business Media Dordrecht 2014
C. Malgrange et al., *Symmetry and Physical Properties of Crystals*,
DOI 10.1007/978-94-017-8993-6_6

6.2. Plane lattices

We recall that the 10 plane point groups are the groups 1, 2, 3, 4, 6, m, mm,[1] 3m, 4m, 6m (Sect. 5.8). We will investigate successively the plane lattices compatible with the low-symmetry point groups, *viz.* groups 1, 3, 4 and m. The results obtained make it possible to determine the plane lattices consistent with each of the 10 groups.

6.2.1. Group 1

The point group of any (general) plane lattice features a center of symmetry, *i.e.*, in terms of plane groups for which the symmetry elements are normal to the plane, a 2-fold axis. The unit cell is a general parallelogram, a result valid for groups 1 and 2.

6.2.2. Group 3

Let **a** be the smallest of all lattice translations. Through a rotation by $2\pi/3$ around the 3-fold axis, vector **a** becomes vector **b** (Fig. 6.1a). Vectors **a** and **b** define a primitive cell, since there cannot be a node within this parallelogram, or else **a** would not be the smallest lattice vector. Only one type of plane lattice is consistent with a 3-fold axis. Its primitive cell is a parallelogram with apical angles $2\pi/3$ and $\pi/3$, and the lattice is called hexagonal. The symmetry of this lattice is 6m and the point groups consistent with this lattice are thus 3, 3m, 6, 6m (Fig. 6.1b).

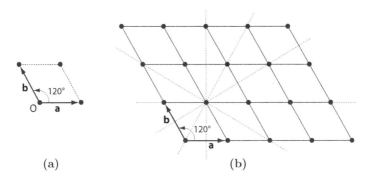

(a)   (b)

Figure 6.1: Plane lattice with point group 3: (a) primitive cell; (b) lattice, evidencing six concurrent mirrors, the trace of which is indicated by dotted lines

---

1. We saw that this notation is equivalent to the notation 2m.

6.2.3. Group 4

The same argument (rotation by $2\pi/4 = \pi/2$ of the smallest lattice vector $\mathbf{a}$, leading to vector $\mathbf{b}$ defining, together with $\mathbf{a}$, a square unit cell within which there can be no extra node) leads to a square primitive cell. The plane lattice, with square primitive cell, has point symmetry 4m and is consistent with the point groups 4 and 4m.

6.2.4. Group m

Let $\mathbf{a}'$ be a translation of the plane lattice that is neither parallel nor perpendicular to the trace uu' of the mirror on the plane of the lattice, and $\mathbf{b}'$ the transform of $\mathbf{a}'$ in the mirror operation m (Fig. 6.2a). Translations $\mathbf{a}' + \mathbf{b}'$ and $\mathbf{a}' - \mathbf{b}'$ define orthogonal rows, respectively parallel and perpendicular to the mirror. Let $\mathbf{a}$ and $\mathbf{b}$ be the smallest lattice vectors along these directions (Fig. 6.2b). They define a rectangular unit cell for which one of the apices is noted O. Consider a lattice vector $\mathbf{n}$ starting from O and for which the end M is inside the rectangle defined by the vectors $\mathbf{a}$ and $\mathbf{b}$ with origin O. Let $\mathbf{n}' = \mathbf{OM}'$ be its image through operation m. Draw the vector $\mathbf{n} + \mathbf{n}'$, parallel to $\mathbf{b}$. By construction, it is smaller than $2\mathbf{b}$ and can only be a multiple of $\mathbf{b}$, i.e. $\mathbf{0}$ or $\mathbf{b}$. In the same way, $\mathbf{n} - \mathbf{n}'$ can only be $\mathbf{0}$ or $\mathbf{a}$.

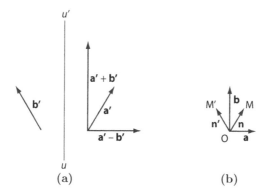

(a)   (b)

Figure 6.2: Plane lattice with point group m. uu' is the trace of one of the mirrors.

Consider the various possible cases:

1. $\mathbf{n} + \mathbf{n}' = \mathbf{0}$ and $\mathbf{n} - \mathbf{n}' = \mathbf{0}$, i.e. $\mathbf{n} = \mathbf{n}' = \mathbf{0}$.
   The rectangular unit cell with basis vectors $\mathbf{a}$ and $\mathbf{b}$ is primitive.

2. $\mathbf{n} + \mathbf{n}' = \mathbf{b}$ and $\mathbf{n} - \mathbf{n}' = \mathbf{a}$, i.e. $\mathbf{n} = (\mathbf{a} + \mathbf{b})/2$.
   The rectangular unit cell is double. It is a centered unit cell.

3. $\mathbf{n} + \mathbf{n}' = \mathbf{b}$ and $\mathbf{n} - \mathbf{n}' = \mathbf{0}$, i.e. $\mathbf{n} = \mathbf{b}/2$ or
   $\mathbf{n} + \mathbf{n}' = \mathbf{0}$ and $\mathbf{n} - \mathbf{n}' = \mathbf{a}$, i.e. $\mathbf{n} = \mathbf{a}/2$.

These two solutions must be rejected because they contradict the starting assumption (**a** and **b** the smallest lattice vectors along these directions).

These two plane lattices, with the rectangular unit cell respectively primitive and centered, have point symmetry mm, and the compatible crystal groups are m and mm.

6.2.5. Conclusion

The survey of plane lattices is finished because, by searching for the lattices associated with the lowest symmetries, we have determined lattices with higher symmetry which finally include all the plane point groups. The results are gathered in Table 6.1.

Table 6.1: Point groups consistent with the various plane lattices

Plane lattice	Compatible point groups
Oblique	1, 2
Hexagonal	3, 3m, 6, 6m
Square	4, 4m
Rectangular Primitive (p) Centered (c)	m, mm

6.3. 3-dimensional lattices

6.3.1. Group 1

This group features no symmetry element. It is therefore consistent with a general parallelepiped-shaped primitive cell. The lattice is called triclinic. As for all lattices, it involves centers of symmetry and its point group is thus $\bar{1}$. It is therefore consistent with crystals with point groups 1 and $\bar{1}$.

6.3.2. Group 2

We showed (Sect. 3.3.4) that any rotation axis is parallel to a row of the direct lattice and perpendicular to a lattice plane. Consider row zz' parallel to the 2-fold axis, and going through a given lattice node O. Let P_0 be the lattice plane which is perpendicular to this row and goes through O (Fig. 6.3). In plane P_0, the primitive cell is a general parallelogram. Let **a** and **b** be two vectors defining this unit cell.

Let P_1 be one of the two lattice planes parallel to P_0 and closest to P_0. The lattice nodes in this lattice plane project onto P_0 at the rate of one node per plane unit cell in P_0. Let F′ be the projection of one of these nodes, located in

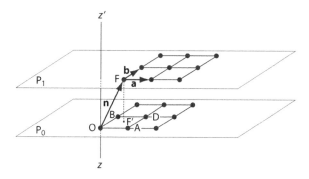

Figure 6.3: Three-dimensional lattice for a crystal with point group 2. zz' is a 2-fold axis. P_0 and P_1 are two consecutive lattice planes. Vector $\mathbf{n} = \mathbf{OF}$ connects node O in P_0 to a node of plane P_1 such that its projection is inside the plane unit cell OADB.

one of the unit cells containing lattice node O, and let F be the node associated to it in P_1. F' can also be located on the edges OA and OB of this unit cell, but A and B are excluded: if F' were at A (or B) we would consider $\mathbf{OF} - \mathbf{a}$ (or $\mathbf{OF} - \mathbf{b}$) and F' would then be at O. Let us note vector $\mathbf{OF}$ as $\mathbf{n}$. The three vectors $\mathbf{a}$, $\mathbf{b}$ and $\mathbf{n}$ define a primitive cell.

Let $\mathbf{n}'$ be the lattice vector resulting from the transformation of $\mathbf{n}$ through a rotation by π around the 2-fold axis (Fig. 6.4a). The component $\mathbf{n}_{//}$ of $\mathbf{n}$, parallel to the axis, is conserved in the rotation, and the component $\mathbf{n}_\perp$ is transformed into $-\mathbf{n}_\perp$. Also, $\mathbf{n} - \mathbf{n}'$ is a lattice vector. We obtain:

$$\mathbf{n} = \mathbf{n}_{//} + \mathbf{n}_\perp, \qquad \mathbf{n}' = \mathbf{n}_{//} - \mathbf{n}_\perp,$$
$$\mathbf{n} - \mathbf{n}' = 2\mathbf{n}_\perp = p\mathbf{a} + q\mathbf{b} \qquad (p \text{ and } q \text{ integers})$$

Hence

$$\mathbf{n}_\perp = \frac{p}{2}\mathbf{a} + \frac{q}{2}\mathbf{b}. \tag{6.1}$$

Taking into account the assumption we made about F and F', $p/2$ and $q/2$ are less than 1. Two cases must be considered:

1. $\mathbf{n}_\perp = \mathbf{0}$ $(p = q = 0)$, so that $\mathbf{n}$ is perpendicular to P_0.
    The unit cell is a right prism with general parallelogram base. It is a primitive cell and the lattice is called monoclinic primitive.

2. $\mathbf{n}_\perp = \mathbf{a}/2$ or $\mathbf{n}_\perp = \mathbf{b}/2$ or $\mathbf{n}_\perp = (\mathbf{a} + \mathbf{b})/2$.
    These three cases are equivalent, since nothing basically distinguishes lattice vectors $\mathbf{a}$, $\mathbf{b}$ and $\mathbf{a} + \mathbf{b}$, any one of which can be chosen as a basis vector for a plane primitive cell in P_0.

Assume that $\mathbf{n}_\perp = \mathbf{a}/2$ (Fig. 6.4b). Plane P_2, parallel to P_0 and P_1 and located immediately above P_1, deduces from P_0 through translation $2\mathbf{n}$, *i.e.*

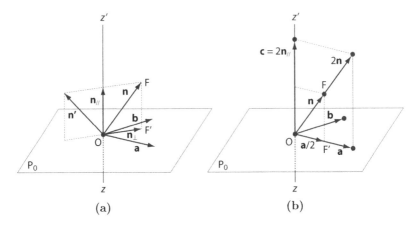

Figure 6.4: Search for the lattice of a crystal with point group 2. (a) Vector **n** transforms into **n'** by rotation around the 2-fold axis zz'. (b) The situation when the component $\mathbf{n}_\perp$ of **n** perpendicular to zz' is equal to $\mathbf{a}/2$.

$2\mathbf{n} = 2\mathbf{n}_{//} + \mathbf{a}$, which is equal to $2\mathbf{n}_{//}$ modulo the lattice translations. The vectors **a**, **b** and $\mathbf{c} = 2\mathbf{n}_{//}$ define a monoclinic unit cell. This unit cell contains node F at the center of the rectangular face $(\mathbf{a}, \mathbf{c} = 2\mathbf{n}_{//})$. It is a double unit cell, called base-centered monoclinic. The centered faces are parallel to the 2-fold axis. If, as we just did, we call **c** the basis vector parallel to the 2-fold axis, the lattice is called monoclinic B if the centered faces are $(\mathbf{a}, \mathbf{c})$, or monoclinic A if the centered faces are $(\mathbf{b}, \mathbf{c})$. If, as is still frequently the case, the basis vector parallel to the 2-fold axis is called **b**, the lattice is C or A depending on whether faces $(\mathbf{a}, \mathbf{b})$ or $(\mathbf{b}, \mathbf{c})$ are centered.

The point group of a monoclinic lattice is $2/m$. It is consistent with subgroups 2 and m and with group $2/m$.

6.3.3. Group 3

In the lattice planes perpendicular to the 3-fold axis, the lattice is hexagonal. It is defined by vectors **a** and **b**, with the same length and enclosing an angle $2\pi/3$. The same argument as above leads to considering a lattice plane P_0 perpendicular to the 3-fold axis, and its nearest neighbor P_1, which is parallel to it. Let **n** be the lattice vector connecting lattice node O in plane P_0 to lattice node F located in P_1 and which projects onto P_0 inside a primitive plane unit cell containing O, or one of its sides OA and OB, excluding A and B. Let **n'** be the transform of **n** through the operation rotation by $2\pi/3$ around the 3-fold axis going through O. The component $\mathbf{n}_{//}$ of **n** parallel to the axis is conserved in the rotation.

$$\mathbf{n} = \mathbf{n}_{//} + \mathbf{n}_\perp, \qquad \mathbf{n'} = \mathbf{n}_{//} + \mathbf{n}'_\perp,$$

where $\mathbf{n}_\perp$ and $\mathbf{n}'_\perp$ are respectively the components of **n** and **n'** perpendicular to the 3-fold axis (Fig. 6.5).

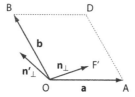

Figure 6.5: Primitive cell in plane P_0 perpendicular to the 3-fold axis

Taking into account the assumption made about F', the end of $\mathbf{n}_\perp$ is located inside the parallelogram OADB (or on OA or AB, excluding A and B) and thus its norm is smaller than **a**. We have

$$||\mathbf{n}_\perp|| < a.$$

Since $\| \mathbf{n}_\perp - \mathbf{n}'_\perp \| = \sqrt{3} \| \mathbf{n}_\perp \|$, we obtain $\| \mathbf{n}_\perp - \mathbf{n}'_\perp \| < a\sqrt{3}$.

As $\mathbf{n} - \mathbf{n}'$ is a lattice vector, $\mathbf{n}_\perp - \mathbf{n}'_\perp$, which is equal to it, is also a lattice vector, the norm of which is less than $a\sqrt{3}$. This norm can therefore have only two values, 0 or a.

1. $\| \mathbf{n}_\perp - \mathbf{n}'_\perp \| = 0$, so that $\mathbf{n}_\perp = \mathbf{n}'_\perp = \mathbf{0}$. Vector $\mathbf{n}$ is perpendicular to P_0 and the unit cell is a right prism whose base is an isosceles parallelogram with apical angle $2\pi/3$. The unit cell is primitive, and the lattice is called hexagonal primitive. The point group of such a unit cell is $6/mmm$, of which 3 is a subgroup (of index 8). The hexagonal lattice is therefore consistent also with groups 3, $\bar{3}$, 32, 3m, $\bar{3}$m, 6, $\bar{6}$, $6/m$, 622, 6mm, $\bar{6}$2m, subgroups of $6/mmm$. This result was shown on Figures 5.17a and 5.17b.

2. $\| \mathbf{n}_\perp - \mathbf{n}'_\perp \| = a$, and then $\mathbf{n}_\perp - \mathbf{n}'_\perp$ is a lattice vector in P_0, with length a. It is therefore equal to **a** or **b** or $\mathbf{a} + \mathbf{b}$. Assume it is equal to **a** (Fig. 6.6a). The end F' of $\mathbf{n}_\perp$ is then at the center M of the equilateral triangle OAD, and the nodes of the plane lattice P_1 project at M and at all the points deduced from M through lattice translations **a** and **b**, thus, for example, at N and P. The lattice of plane P_2, immediately above P_1, deduces from the lattice of plane P_0 through translation $2\mathbf{n} = 2\mathbf{n}_{/\!/} + 2\mathbf{n}_\perp$, and its nodes project onto P_0 at N, such that $\mathbf{ON} = 2\mathbf{OM}$, and at all the points deduced from N through the lattice translations, for example at Q, R and S.

The lattice in plane P_3 immediately above P_2 is deduced from that of P_0 through translation $3\mathbf{n} = 3\mathbf{n}_{/\!/} + 3\mathbf{n}_\perp$. But $3\mathbf{n}_\perp = \mathbf{OK}$, a lattice vector equal to $2\mathbf{a} + \mathbf{b}$. The nodes of plane P_3 project on the nodes of P_0. Comparing Figures 6.6a and 3.14c, we see that the unit cell is a rhombohedron (Fig. 6.6b), *i.e.* a parallelepiped for which the basis vectors $\mathbf{a}'$, $\mathbf{b}'$ and $\mathbf{c}'$ have the same length, and for which the major diagonal LT is a 3-fold axis. All its faces are lozenges, equal to one another. The point group of a rhombohedral unit cell is $\bar{3}$m. Such a lattice is therefore consistent with the point group $\bar{3}$m and its subgroups 3, $\bar{3}$, 32, 3m.

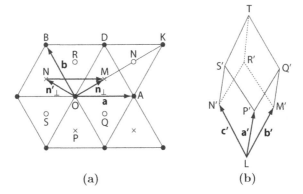

Figure 6.6: Rhombohedral lattice: (a) projection onto plane P_0 of the lattice nodes; (b) rhombohedral unit cell. P′, M′, N′ are in plane P_1 and project onto P_0 at P, M and N (crosses); Q′, R′, S′ are in plane P_2 and project onto P_0 at Q, R and S (empty circles); T is in plane P_3 and projects at O (full circle).

We thus obtained the two following results:

1. The crystals with point group 3, $\bar{3}$, 32, 3m and $\bar{3}$m are consistent with a primitive rhombohedral crystal lattice, noted R, and also with a primitive hexagonal lattice, noted P. Some crystals with one of these point groups have a rhombohedral lattice while others have a hexagonal lattice. For example, crystals of calcite $CaCO_3$, with point group $\bar{3}$m, have a rhombohedral lattice while crystals of $Ca(OH)_2$, with the same point group, have a hexagonal lattice.

2. The hexagonal lattice is consistent with groups 6, $\bar{6}$, 6/m, 622, 6mm, $\bar{6}$2m and 6/mmm. It will therefore not be necessary to examine group 6.

6.3.4. Group 4

Plane P_0, perpendicular to the 4-fold axis, is a lattice with a square primitive cell, with basis vectors **a** and **b** with the same norm a. The same argument as for groups 2 and 3 leads to considering the vector **n** connecting lattice node O in P_0 to node F in lattice plane P_1 (parallel to P_0 and nearest neighbor to it) which projects onto P_0 at point F′ (Fig. 6.7a), located within the plane unit cell OADB (or on OA and OB, excluding A and B). Let **n**′ be the transform of **n** through rotation by $\pi/2$ around the 4-fold axis. The component $\mathbf{n}_{/\!/}$ of **n** parallel to the axis is conserved in the rotation. We have

$$\mathbf{n} = \mathbf{n}_{/\!/} + \mathbf{n}_{\perp}, \qquad \mathbf{n}' = \mathbf{n}_{/\!/} + \mathbf{n}'_{\perp}$$

and

$$\|\mathbf{n}_{\perp}\| = \|\mathbf{n}'_{\perp}\|.$$

Also, the assumption made on F′ leads to $\|\mathbf{n}_{\perp}\| < a\sqrt{2}$.

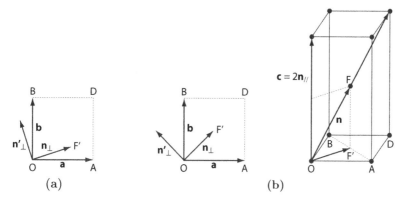

Figure 6.7: Search for the lattice of a crystal with point group 4. (a) Plane P_0, perpendicular to the 4-fold axis: $\mathbf{n}'_\perp$ is the transform of $\mathbf{n}_\perp$ through a rotation by $\pi/2$ around the 4-fold axis. (b) Case where $\mathbf{n}_\perp - \mathbf{n}'_\perp = \mathbf{a}$. The end of vector $\mathbf{n} = \mathbf{OF}$ is at the center of the unit cell.

Thus $$\mathbf{n} - \mathbf{n}' = \mathbf{n}_\perp - \mathbf{n}'_\perp$$
and $$\|\mathbf{n}_\perp - \mathbf{n}'_\perp\| = \|\mathbf{n}_\perp\|\sqrt{2}.$$

Therefore $\mathbf{n} - \mathbf{n}'$, and $\mathbf{n}_\perp - \mathbf{n}'_\perp$ which is equal to it, are lattice vectors parallel to P_0 with norm smaller than $2a$. Two cases should be distinguished:

1. $\mathbf{n}_\perp - \mathbf{n}'_\perp = \mathbf{0}$, so that $\mathbf{n}_\perp = \mathbf{n}'_\perp = \mathbf{0}$ and $\mathbf{n} = \mathbf{n}' = \mathbf{n}_{/\!/}$.
   The unit cell is primitive. It is a rectangle parallelepiped with square base. The unit cell is called tetragonal primitive.

2. $\mathbf{n}_\perp - \mathbf{n}'_\perp = \mathbf{a}$ so that $\mathbf{n}_\perp = (\mathbf{a} + \mathbf{b})/2$.
   There is a lattice node which projects at the center of the square plane unit cell OADB (Fig. 6.7b). The plane P_2, nearest neighbor to P_1, deduces from P_0 through translation $2\mathbf{n} = 2\mathbf{n}_{/\!/} + \mathbf{a} + \mathbf{b}$. The nodes of P_2 therefore project on the nodes of P_0. The lattice has a tetragonal double unit cell, with a node at the center of the unit cell.

This is a tetragonal body-centered lattice, noted I.

The point group of a tetragonal unit cell is $4/m\,mm$. This unit cell is consistent with point groups 4, $\bar{4}$, $4/m$, 422, 4mm, $\bar{4}2m$ and $4/m\,mm$. A tetragonal unit cell can therefore be primitive (P) or body-centered (I).

It is then unnecessary to deal with group 6, since it occurred in a natural way when considering group 3 (Sect. 6.3.3).

The above discussions led to the determination of the lattices consistent with 24 of the 32 point groups. We now search for the lattice consistent with the least symmetric of the remaining point groups, *viz.* 222.

6.3.5. Group 222

This group features three 2-fold axes, perpendicular to one another. Each of these axes being parallel to a lattice row (Sect. 3.3.4), the unit cell, be it primitive or centered, is therefore a rectangle parallelepiped with basis vectors $\mathbf{a}$, $\mathbf{b}$, and $\mathbf{c}$, orthogonal to one another. The lattice is called orthorhombic. Consider the 2-fold axis parallel to vector $\mathbf{c}$. The lattice plane perpendicular to $\mathbf{c}$ has a rectangular unit cell which, as we saw in Table 6.1, can be either primitive or centered. We will therefore successively apply the result of Section 6.3.2 to each of these unit cells.

The plane unit cell (a, b) is primitive (plane lattice p)

The approach of Section 6.3.2 leads to the result given in Equation (6.1):

$$\mathbf{n}_\perp = \frac{p}{2}\mathbf{a} + \frac{q}{2}\mathbf{b}.$$

Three cases must be considered:

1. $p = q = 0$, so that $\mathbf{n}_\perp = \mathbf{0}$.
    The unit cell is primitive.

2. $p = 1$, $q = 0$ or $p = 0$, $q = 1$, so that $\mathbf{n}_\perp = \mathbf{a}/2$ or $\mathbf{b}/2$.
    The lattice is base-centered B or A. The unit cell is double.

3. $p = 1$, $q = 1$, so that $\mathbf{n}_\perp = (\mathbf{a} + \mathbf{b})/2$.
    The lattice is body-centered, *i.e.* I. The unit cell is double.

The plane unit cell (a, b) is centered (plane lattice c)

We can choose as basis vectors of a primitive cell of the 2-dimensional lattice $\mathbf{a}$ and $(\mathbf{a} + \mathbf{b})/2$.

We then obtain

$$\mathbf{n}_\perp = \frac{p}{2}\mathbf{a} + \frac{q}{4}(\mathbf{a} + \mathbf{b})$$

with p and q integers. The cases to be considered are:

1. $p = 0$, $q = 0$ so that $\mathbf{n}_\perp = \mathbf{0}$.
    The lattice is base-centered, with the centered base being $(\mathbf{a},\mathbf{b})$, and the lattice is C-type.

2. $p = 1$, $q = 0$, so that $\mathbf{n}_\perp = \mathbf{a}/2$.
    The face $(\mathbf{a}, \mathbf{c})$ is centered. Vector $(\mathbf{a} + \mathbf{c})/2$ is a lattice translation. Since face $(\mathbf{a}, \mathbf{b})$ is centered, $(\mathbf{a}+\mathbf{b})/2$ is also a lattice vector and, through addition modulo a lattice vector, so is $(\mathbf{b}+\mathbf{c})/2$. All the faces are centered, and the lattice is face-centered (F) orthorhombic.

3. $p = 0$, $q = 1$, so that $\mathbf{n}_\perp = (\mathbf{a} + \mathbf{b})/4$.

4. $p = 1$, $q = 1$, so that $\mathbf{n}_\perp = (3\mathbf{a} + \mathbf{b})/4$.

Cases (3) and (4) are to be rejected, since they lead to lattices that do not accept a 2-fold axis parallel to **a** or **b**.

In conclusion, the orthorhombic lattice can be primitive (P), base-centered (A, B or C), body-centered (I) or face-centered (F).

The orthorhombic lattice has symmetry mmm, and it is therefore consistent with groups 222, mm2 and mmm.

6.3.6. Group 23

Among the remaining groups, the least symmetric one is group 23. It is a supergroup of 222, which means that it features all the symmetry operations of 222, with additional 3-fold axes. This implies that the lengths of basis vectors **a**, **b** and **c** are equal. We can therefore start out from the results obtained for 222, adding the equality of the lengths of **a**, **b** and **c**. The unit cell is a cube which can be primitive (P), body-centered (I) or face-centered (F). The A, B and C lattices are not consistent with the existence of 3-fold axes, and are therefore ruled out.

The lattice is cubic. The point group of the lattice is m$\bar{3}$m and the cubic lattice is consistent with groups 23, m$\bar{3}$, 432, $\bar{4}$3m and m$\bar{3}$m.

We have now gone through the full list of crystallographic point groups, and the overall result is shown in the last column of Figures 5.17a and 5.17b.

Chapter 7

Space groups

After showing how the space groups are constructed, this chapter aims at enabling the reader to understand what the denomination of a given space group contains, and to efficiently use volume A of the International Tables for Crystallography. This volume lists all the space groups and gives many of their properties. The last section shows some examples of space groups.

7.1. Introduction

We saw that the symmetry operations of a crystal can be expressed in the form $(S, \mathbf{t})$ where S is a rotation or a rotoinversion about an axis going through a given point O, and $\mathbf{t}$ a translation. Point O is not explicitly mentioned in the notation because it is chosen as common to all the operations $(S, \mathbf{t})$. We also saw that the symmetry of the physical properties of a crystal depends only on the operations S, which form a group, and, in Chapter 5, we enumerated the 32 symmetry groups of the operations S, called orientation groups or, more often, point groups.

We must now enumerate the symmetry groups of the operations $(S, \mathbf{t})$. The connection between the point group and the space group was established rigorously in Chapter 4. We recall it briefly here. The group T of the lattice translations $(E, \mathbf{n})$ where E is the identity operation and $\mathbf{n}$ a vector of the crystal lattice, $\mathbf{n} = u\mathbf{a} + v\mathbf{b} + w\mathbf{c}$ ($\mathbf{a}$, $\mathbf{b}$ and $\mathbf{c}$ are the basis vectors of the crystal unit cell and u, v, w any integers), is a subgroup of the group G of the symmetry operations of the crystal. We can decompose this group G into packets (or cosets), each of which is the product of an operation $(S_i, \mathbf{t}_i)$ with all the translation operations, so that:

$$
\begin{aligned}
G &= T + T(S_1, \mathbf{t}_1) + T(S_2, \mathbf{t}_2) + T(S_3, \mathbf{t}_3) + \ldots + T(S_m, \mathbf{t}_m) \\
&= T(E, \mathbf{0}) + T(S_1, \mathbf{t}_1) + T(S_2, \mathbf{t}_2) + T(S_3, \mathbf{t}_3) + \ldots + T(S_m, \mathbf{t}_m).
\end{aligned}
$$

To perform this partition of G by the group T of lattice translations, we start by setting aside the group T. We choose among the remaining elements of G one operation $(S_1, \mathbf{t}_1)$, and we perform all the products of this operation with

© Springer Science+Business Media Dordrecht 2014
C. Malgrange et al., *Symmetry and Physical Properties of Crystals*,
DOI 10.1007/978-94-017-8993-6_7

the lattice translations. This set $T(S_1, \mathbf{t}_1)$ is then put aside, and the process is repeated until all the elements of G have been used.

The vectors $\mathbf{t}_i$ are defined modulo the lattice translation vectors. We can therefore choose them either equal to zero, or equal to $p\mathbf{a} + q\mathbf{b} + r\mathbf{c}$ with p, q, and r smaller than 1.

The set of the cosets $T(S_i, \mathbf{t}_i)$ thus defined forms G_s, the space group or position group of the crystal, which can also be written as:

$$G_s = (\text{E}, \mathbf{0}) + (S_1, \mathbf{t}_1) + (S_2, \mathbf{t}_2) + (S_3, \mathbf{t}_3) + \ldots + (S_m, \mathbf{t}_m) \text{ (modulo T)},^1$$

by replacing each coset $T(S_i, \mathbf{t}_i)$ by $(S_i, \mathbf{t}_i)$ modulo T.

The operations $(S_i, \mathbf{0})$ obtained by replacing, in $(S_i, \mathbf{t}_i)$, the translation $\mathbf{t}_i$ by a zero translation, form the point group or orientation group G_0 of the crystal:

$$G_0 = (\text{E}, \mathbf{0}) + (S_1, \mathbf{0}) + (S_2, \mathbf{0}) + (S_3, \mathbf{0}) + \ldots + (S_m, \mathbf{0}).$$

The 32 point symmetry groups compatible with the existence of a crystal lattice were enumerated in Chapter 5.

The first task in this chapter will be to determine the operations $(S_i, \mathbf{t}_i \neq \mathbf{0})$ which can exist when the existence of the crystal lattice is taken into account. We will thus obtain the symmetry operations associated to screw axes and to glide mirrors. We will then show how the space groups can be determined starting from the point groups, but we will not enumerate them systematically. There are 230 of them. They are described in volume A of the International Tables for Crystallography (I.T. A) [1], presented in Section 7.3.3.

7.2. Enumeration of the operations (S, t)

We considered till now symmetry operations of the form $(S, \mathbf{t})$ where operations S are rotations (or rotoinversions) around an axis that goes through point O, *chosen arbitrarily and common to all the symmetry operations*. We showed (Sect. 4.2.1) that this choice of a common point for all the rotation and rotoinversion axes is possible because we may always replace an operation $(S(O), \mathbf{t})$ by an operation $(S'(O'), \mathbf{t}')$ where O' is any point, if

$$S' = S$$

$$\text{and } \mathbf{t}' = \mathbf{t} + (S - E)\, \mathbf{OO}'.$$

We now reverse the reasoning, in order to determine whether it is possible, for any one of the operations $(S(O), \mathbf{t})$, with common origin O, to displace the

---

1. Adding "modulo T" is mandatory in order for the above operations to form a group because it can happen that the product of two of the operations $(S_i, \mathbf{t}_i)$ is equal to $(S_j, \mathbf{t}_j + \mathbf{n})$ while, as mentioned above, the $\mathbf{t}_i$ are smaller than lattice translations.

origin to a point O' so that we obtain an operation for which the translation part is zero, $(S(O'), \mathbf{0})$. To obtain $\mathbf{t}' = 0$ we require:

$$\mathbf{t} = (E - S)\,\mathbf{OO'}. \tag{7.1}$$

We will see that this equation can have solutions in some cases. When it does not, we will determine the values of $\mathbf{t}$ compatible with the existence of a crystal lattice. We will study separately the cases where S is a rotation (Sect. 7.2.1) and those where S is a rotoinversion, denoted as above with $\bar{S}$ (Sect. 7.2.2).

7.2.1. S is a rotation. Definition of screw axes

It is easily shown that, if S is a rotation, the vector $\mathbf{t}$ deduced from relation (7.1) is perpendicular to the rotation axis. Rotation $S(O)$ transforms vector $\mathbf{OO'}$ into vector $\mathbf{OO''}$ (Fig. 7.1). We obtain:

$$\mathbf{t} = (E - S)\,\mathbf{OO'} = \mathbf{OO'} - \mathbf{OO''} = \mathbf{O''O'} \tag{7.2}$$

which is perpendicular to the rotation axis. Thus the translation part of the operation can be zero only if the initial translation $\mathbf{t}$ is perpendicular to the rotation axis.

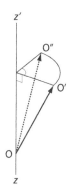

Figure 7.1: Rotation of a vector $\mathbf{OO'}$ around an axis zz' that goes through O

The determination of the new position O' of the axis which cancels the translation will be presented in Section 7.2.3.

We now consider the case $(S, \mathbf{t}_{/\!/})$, where $\mathbf{t}$ is parallel to the rotation axis, so that the translation part can no more be canceled by a change in the position of the rotation axis. Suppose S is a rotation by $2\pi/q$ around a q-fold axis, and determine the operation product of this operation by itself. Using relation (4.8), valid for operations S_1 and S_2 having the same origin, which we reproduce below:

$$(S_2, \mathbf{t}_2)(S_1, \mathbf{t}_1) = (S_2 S_1, S_2 \mathbf{t}_1 + \mathbf{t}_2). \tag{7.3}$$

We obtain: $(S, \mathbf{t}_{/\!/})\,(S, \mathbf{t}_{/\!/}) = (S^2, S\mathbf{t}_{/\!/} + \mathbf{t}_{/\!/})$.

But $St_{/\!/} = t_{/\!/}$ since rotation of a vector parallel to the rotation axis transforms this vector into itself. We obtain $(S, t_{/\!/})^2 = (S^2, 2t_{/\!/})$.

Repeating this operation q times, we obtain:

$$(S, t_{/\!/})^q = (S^q, q\,t_{/\!/}) = (E, q\,t_{/\!/})$$

which is necessarily a symmetry operation of the crystal. Since the rotation part is the identity operation, this symmetry operation is a lattice translation (E, n) and

$$qt_{/\!/} = n.$$

Vector $t_{/\!/}$ is therefore parallel to a crystal row, which is no surprise since it is parallel to a symmetry axis (Sect. 3.3.4). Let c be the smallest lattice vector parallel to the q-fold axis we are discussing.

$$qt_{/\!/} = mc \qquad \text{and} \qquad t_{/\!/} = mc/q$$

where m is any integer.

There are $q - 1$ solutions: $t_{/\!/} = c/q,\ t_{/\!/} = 2c/q,\ \ldots,\ t_{/\!/} = (q-1)c/q$. The next values would be $t_{/\!/} = c$, a crystal lattice vector, and then vectors equal to the above ones modulo c.

The symmetry elements associated to these operations, products of a rotation and a translation parallel to the rotation axis, are called screw axes (Sect. 2.2.4). Table 7.1 summarizes the various screw axes, their short notation, the translation to which they are associated, and the symbol that represents them.

Table 7.1: Screw axes

Order of the axis	2		3			4				6					
Notation	2	2_1	3	3_1	3_2	4	4_1	4_2	4_3	6	6_1	6_2	6_3	6_4	6_5
Translation	0	$c/2$	0	$c/3$	$2c/3$	0	$c/4$	$2c/4$	$3c/4$	0	$c/6$	$2c/6$	$3c/6$	$4c/6$	$5c/6$
Symbol															

Consider a crystal featuring a screw axis, and a given atom of this crystal which is not located on this axis. There are then atoms of the same nature at all the positions which deduce from the position of this atom through the various symmetry operations associated to this axis. The atoms are located on a helix or screw (hence the name screw axis), the axis of which is this screw axis. A helix or screw is a chiral object, $i.e$ an object which is not superimposable to its mirror image.

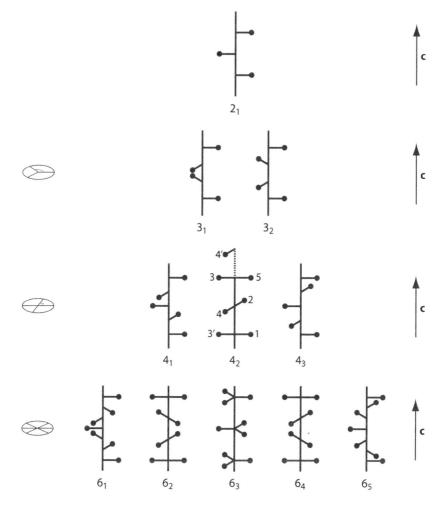

Figure 7.2: Equivalent positions of a general atom through the symmetry operations associated to the various screw axes. We note that only axes 3_1, 3_2, 4_1, 4_3, 6_1, 6_2, 6_4, 6_5 are chiral.

We could therefore assume that the atoms equivalent through the symmetry operations associated to a screw axis form a chiral figure. But we must consider all the atoms equivalent within the unit cell, *i.e.* perform the operation modulo $\mathbf{c}$. As a result, the figure obtained is not always chiral, as Figure 7.2 shows. Consider for example the case of a 4_2 axis. The 4_2 operations performed successively on atom 1 lead to atom 2, then to atom 3, then to atom 4′ rotated by $3\pi/2$ and translated by $3\mathbf{c}/2$ with respect to atom 1. The atom equivalent to 4′ through a lattice translation operation $-\mathbf{c}$ is atom 4. The transform of atom 4 is atom 5. The atom at 3′ is the equivalent of atom 3 through translation $-\mathbf{c}$. We see that the equivalent positions produced, within a unit cell, by these operations 4_2 do not form a helix or screw, and that the figure features mirror symmetry. In the same way, the figures associated to axes 2_1 and 6_3 (*i.e.* axes q_m where $q = 2m$) are not chiral. We note that, among the chiral figures, those associated to axes 3_1, 4_1, 6_1 and 6_2 are respectively symmetric through a mirror operation of the figures associated to axes 3_2, 4_3, 6_5 and 6_4. These figures are said to be enantiomorphic of each other.

7.2.2. S is a rotoinversion, noted $\bar{S}$. Definition of glide mirrors

We proceed in the same way as for rotations, *i.e.* we first look whether a change in position of the rotoinversion axis makes it possible to reduce the translation to zero. Equation (7.1) is

$$(E - \bar{S})\,\mathbf{OO'} = \mathbf{t} \qquad (7.4)$$

which we now solve, not geometrically as was done for rotations, but through matrix algebra. We recall that rotoinversion $\bar{S}$ is a rotation S followed by inversion, $\bar{S} = IS$. We choose a system of axes $Oxyz$ such that axis Oz coincides with the rotation axis, and O is the inversion center. In this axis system:

$$\bar{S} = IS = \begin{pmatrix} -\cos\varphi & \sin\varphi & 0 \\ -\sin\varphi & -\cos\varphi & 0 \\ 0 & 0 & -1 \end{pmatrix},$$

$$E - \bar{S} = \begin{pmatrix} 1+\cos\varphi & -\sin\varphi & 0 \\ \sin\varphi & 1+\cos\varphi & 0 \\ 0 & 0 & 2 \end{pmatrix}.$$

In order for Equation (7.4) to have a solution for all $\mathbf{t}$, the determinant of $E - \bar{S}$, $\det(E - \bar{S})$, must be non-zero.

$$\det(E - \bar{S}) = 2(1 + \cos\varphi)^2 + 2\sin^2\varphi = 2(2 + 2\cos\varphi) = 8\cos^2\varphi/2.$$

This determinant is zero if $\varphi = \pi$, *i.e.* only for a $\bar{2}$ axis. In all other cases, Equation (7.4) has a solution, and zero translation can be obtained through a change in origin.

Let us therefore consider the operations associated to a $\bar{2}$ axis.

We recall that a $\bar{2}$ axis is equivalent to a mirror perpendicular to it, which goes through the inversion center used for the rotoinversion.

The rotation angle φ is equal to π, and Equation (7.4) becomes:

$$\begin{pmatrix} 0 & 0 & 0 \\ 0 & 0 & 0 \\ 0 & 0 & 2 \end{pmatrix} \mathbf{OO'} = \mathbf{t}.$$

This equation has a solution only if $\mathbf{t}$ is parallel to axis Oz, the rotoinversion axis, and therefore perpendicular to the mirror.

Let us note this translation $\mathbf{t}_\perp$, and the associated symmetry operation $(m, \mathbf{t}_\perp)$. In this case $\mathbf{OO'} = \mathbf{t}_\perp/2$. We deduce the following result (Exercise 4.1):

The product of a mirror operation and a translation $\mathbf{t}_\perp$, perpendicular to this mirror, is equivalent to a mirror operation without translation provided the mirror is translated by $\mathbf{t}_\perp/2$.

The mirror operations which cannot transform into $(m, \mathbf{0})$ (pure mirrors) through displacement of the mirror are therefore of the form $(m, \mathbf{t}_{/\!/})$ where $\mathbf{t}_{/\!/}$ is parallel to the mirror. We calculate the product of this mirror operation by itself, using Equation (7.3).

$$(m, \mathbf{t}_{/\!/})^2 = (m, \mathbf{t}_{/\!/})(m, \mathbf{t}_{/\!/}) = (E, m\,\mathbf{t}_{/\!/} + \mathbf{t}_{/\!/}) = (E, 2\mathbf{t}_{/\!/})$$

which must be a symmetry operation of the crystal, and more precisely a lattice translation since the rotation part is the identity operation. Therefore $2\mathbf{t}_{/\!/}$ must be a crystal lattice vector.

We choose in the mirror plane two elementary lattice translations, which we note as $\mathbf{a}$ and $\mathbf{b}$. We must have $2\mathbf{t}_{/\!/} = u\mathbf{a} + v\mathbf{b}$ with u and v integers:

$$\mathbf{t}_{/\!/} = \frac{u}{2}\,\mathbf{a} + \frac{v}{2}\,\mathbf{b}. \tag{7.5}$$

Since the required translations $\mathbf{t}_{/\!/}$ are defined modulo the lattice translations, the solutions are restricted to the following cases:

(a) $\mathbf{t}_{/\!/} = \mathbf{a}/2$ ($u = 1$ and $v = 0$):
the mirror is called an axial mirror and noted as a,

(b) $\mathbf{t}_{/\!/} = \mathbf{b}/2$:
the mirror, called axial, is noted b,

(c) $\mathbf{t}_{/\!/} = \mathbf{a}/2 + \mathbf{b}/2$ ($u = 1$ et $v = 1$) :
the mirror, called diagonal is noted n,

(d) $\mathbf{t}_{/\!/} = \mathbf{0}$:
the mirrors without translation, or pure mirrors, are noted m.

We chose to call $\mathbf{a}$ and $\mathbf{b}$ the basis vectors parallel to the mirror, *i.e.* to consider a mirror parallel to the $(\mathbf{a}, \mathbf{b})$ plane. Clearly, the mirror can be parallel to the $(\mathbf{a}, \mathbf{c})$ plane, with a translation $\mathbf{a}/2$ (a-type mirror) or $\mathbf{c}/2$ (c mirror), or $\mathbf{a}/2 + \mathbf{c}/2$ (n mirror). It can also be parallel to the $(\mathbf{b}, \mathbf{c})$ plane with a translation $\mathbf{b}/2$ (b mirror) or $\mathbf{c}/2$ (c mirror) or $\mathbf{b}/2 + \mathbf{c}/2$ (n mirror).

Tables 7.2 and 7.3 show the symbols used to represent these various mirrors, for two orientations of the projection plane: perpendicular to the mirror for Table 7.2, and parallel to the mirror for Table 7.3. In order to keep the presentation simple, we assumed that the projection plane is parallel to $\mathbf{a}$ and $\mathbf{b}$ and that, in the case of a monoclinic lattice, $\mathbf{c}$ is the basis vector perpendicular to the two others.

When the unit cells are multiple, we must also take into account the vector, or vectors, describing the lattice mode, which we note as $\mathcal{T}$. For example,

Table 7.2: Representation of mirrors perpendicular to the projection plane (**a**, **b**)

Nature of the mirror	Notation	Graphical representation	Translation
Pure mirror	m	————————	none
Axial mirror	a, b	--------------------	1/2 along the projection
	c	··························	1/2 perpendicular to the projection plane
Diagonal translation mirror	n	─·─·─·─·─·─·─·─···	1/2 along the projection + 1/2 perpendicular to the projection plane

Table 7.3: Representation of mirrors parallel to the projection plane (**a**, **b**)

Nature of the mirror	Notation	Graphical representation	Translation
Pure mirror	m	⌐ ⌐ ╱	none
Axial mirror	a, b	⌐↓ ←⌐	1/2 along the arrow
Diagonal translation mirror	n	◺↘	1/2 along the arrow

for an I lattice, the vector $\frac{1}{2}(\mathbf{a} + \mathbf{b} + \mathbf{c})$ and, for an F lattice, the three vectors $\frac{1}{2}(\mathbf{b} + \mathbf{c})$, $\frac{1}{2}(\mathbf{c} + \mathbf{a})$ and $\frac{1}{2}(\mathbf{a} + \mathbf{b})$ which complement the (integral) translations $u\mathbf{a} + v\mathbf{b} + w\mathbf{c}$. Mirrors parallel to these lattice mode vectors, with a glide equal to $\mathcal{T}/2$, are then possible. These mirrors are called diamond mirrors, because they exist in the diamond structure, introduced in Section 7.5.3. Their symbol is d.

7.2.3. Product of a symmetry operation and a translation

We return to the cases where a change of origin can change an operation $(S(O), \mathbf{t})$ into $(S(O'), \mathbf{0})$, whether S is a rotation or a rotoinversion, in order to find the rules for performing this change in origin easily. We thus investigate successively the product of a rotation and a translation perpendicular to the rotation axis, of a mirror symmetry and a translation perpendicular to the mirror, and of a rotoinversion and any translation. In all these cases, the result of the product with the translation is just a displacement of the symmetry element implied (rotation axis, mirror or rotoinversion axis).

Product of a rotation and a translation perpendicular to the rotation axis

Consider a symmetry operation $(S, \mathbf{t}_\perp)$ where S is a rotation by $\varphi = 2\pi/n$ about an axis A_n going through point O, and $\mathbf{t}_\perp$ is a translation perpendicular to this axis. Condition (7.2) can be represented geometrically in the plane perpendicular to the rotation axis going through O (Fig. 7.3). We can search for O' in this plane, since the position of the origin of rotations along an axis is arbitrary. Let $\mathbf{OO''}$ be the transform of $\mathbf{OO'}$ through rotation S. Condition (7.2) shows that $\mathbf{O''O'}$ must be equal to $\mathbf{t}_\perp$. The figure shows a vector $\mathbf{OP}$ equal to $\mathbf{O''O'}$, hence to $\mathbf{t}_\perp$, as well as parallelogram $OPO'O''$. We see that the required point O' is the apex of an isosceles triangle with the vector $\mathbf{t}_\perp$ drawn from O for its base, and with the rotation angle φ for its apical angle. We can thus state the following rule:

Operation $[S(O), \mathbf{t}_\perp]$ where $S(O)$ is a rotation by φ around an axis going through O, and $\mathbf{t}_\perp$ a translation perpendicular to the axis, is equivalent to the pure rotation $[S(O'), \mathbf{0}]$ if point O' is on the perpendicular bisector of the segment $\mathbf{t}_\perp$ drawn from O and if, from this point, vector $\mathbf{t}_\perp$ subtends angle φ.

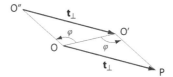

Figure 7.3: Schematic diagram illustrating Equation (7.2). The new axis of rotation is located at O', on the bissector of $\mathbf{OP} = \mathbf{t}_\perp$.

Here are two examples of applications.

1. Consider a crystal with an orthorhombic unit cell, with 2-fold axes parallel to the basis vector $\mathbf{c}$, and choose the origin of the unit cell on one of these 2-fold axes. Figure 7.4a is a projection of the orthorhombic unit cell onto a plane perpendicular to $\mathbf{c}$. Because the crystal is periodic, there is also a 2-fold axis at the other apices of the unit cell. Consider operation $[S(O), \mathbf{a}]$, the product of operation S, a rotation by π around the 2-fold axis going through point O, and the lattice translation $\mathbf{a}$ perpendicular to the axis. This symmetry operation is equivalent to an operation $[S(O'), \mathbf{0}]$.

    According to the above rule, O' lies on the perpendicular bisector of segment $\mathbf{OA} = \mathbf{a}$, and the angle which $\mathbf{OA}$ subtends from O' is π. O' is therefore the middle of segment $\mathbf{OA}$ (Fig. 7.4a). In the same way, performing the product of this rotation by π around the axis going through O and of the translation $\mathbf{a} + \mathbf{b}$ (or $\mathbf{b}$), we obtain a 2-fold axis located at the center S of rectangle OABC (or at O'' such that $\mathbf{OO''} = \mathbf{b}/2$), etc. The set of 2-fold axes present in the unit cell are shown on Figure 7.4b.

2. Consider a crystal with a tetragonal unit cell featuring 4-fold axes parallel to basis vector $\mathbf{c}$. Figure 7.4c is the projection of this unit cell onto a

plane perpendicular to **c**. Let S be the operation rotation by $\pi/2$ around the 4-fold axis parallel to **c** and going through O. Operation [S(O), **a**] is equivalent to [S(O′), **0**] if O′ is at the center of unit cell OABC because segment **OA** subtends an angle $\pi/2$ from O′. Further, a 4-fold axis contains a 2-fold axis which, when combined with the lattice translations, leads to the existence of the 2-fold axes represented on Figure 7.4c.

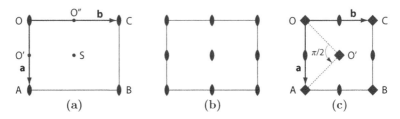

(a)   (b)   (c)

Figure 7.4: (a) Primitive orthorhombic unit cell featuring a 2-fold axis at each lattice node; (b) set of 2-fold axes obtained through the composition of the symmetry operations around these axes with the lattice translations; (c) set of symmetry operations of a tetragonal unit cell of a lattice featuring a 4-fold axis at each lattice node

Product of a mirror operation and a translation perpendicular to the mirror

The symmetry operation (m, $\mathbf{t}_\perp$) *resulting from the combination of a mirror operation and a translation* $\mathbf{t}_\perp$ *perpendicular to it is equivalent to* (m, **0**) *if the mirror is translated by* $\mathbf{t}_\perp/2$.

This rule, derived in Section 7.2.2 and in Exercise 4.1, and the previous one, are widely used to determine all the symmetry operations present in a unit cell starting from the generating elements of the space group and the lattice translations.

Product of a rotoinversion and a translation

We give the results without derivation in order not to make the text too long.

The operation product of an inversion (equivalent to a $\bar{1}$ *axis) and a translation* **t** *is equivalent to an inversion without translation if the center is translated by* **t**/2.

To determine the change in origin which makes it possible to cancel the translation in an operation ($\bar{S}$, **t**) where $\bar{S}$ is a rotoinversion associated to a $\bar{3}$, $\bar{4}$, or $\bar{6}$ axis, the translation must be decomposed into its $\mathbf{t}_{/\!/}$ and $\mathbf{t}_\perp$ components, respectively parallel and perpendicular to the axis. Component $\mathbf{t}_\perp$ is canceled by an operation similar to that of Section 7.2.3: point O is replaced by point O′ located in the plane perpendicular to the axis which goes through O and on the perpendicular bisector of segment **OP** = $\mathbf{t}_\perp$, such that, from O′, **OP** subtends the angle $\varphi - \pi$. Component $\mathbf{t}_{/\!/}$ is canceled by submitting O′ to translation **O′O″** = $\mathbf{t}_{/\!/}/2$. We thus obtain [$\bar{S}$(O), **t**] = [$\bar{S}$(O″), **0**].

7.3. Enumeration of space groups

Starting from a given point group G_0

$$G_0 = (E, \mathbf{0}) + (S_1, \mathbf{0}) + (S_2, \mathbf{0}) + (S_3, \mathbf{0}) + \ldots + (S_m, \mathbf{0}),$$

we want to determine the various space groups G_s which are associated to it, and which can be written as:

$$G_s = (E, \mathbf{0}) + (S_1, \mathbf{t}_1) + (S_2, \mathbf{t}_2) + (S_3, \mathbf{t}_3) + \ldots + (S_m, \mathbf{t}_m) \text{ (modulo T)},$$

taking into account the fact that the operations $(S_i, \mathbf{t}_i)$, where $\mathbf{t}_i$ is non-zero, are associated to screw axes or to glide mirrors. We recall that all the operations $(S_i, \mathbf{t}_i)$ here have a common origin.

This determination can be performed systematically for all the point groups, starting from the simplest ones and using a few theorems deduced from the isomorphism between the quotient group G/T and the point group G_0 to which it is associated. This procedure is tedious, and we will just illustrate the method through an example, that of the space groups deduced from point group mm2.

Point group mm2, associated to an orthorhombic unit cell with basis vectors $\mathbf{a}$, $\mathbf{b}$ and $\mathbf{c}$, contains the following symmetry operations: rotation by π around an axis which we can assume to be parallel to $\mathbf{c}$ for example, and which we note $(A, \mathbf{0})$, symmetry with respect to a mirror perpendicular to $\mathbf{a}$, which we call $(m_1, \mathbf{0})$, symmetry with respect to a mirror perpendicular to $\mathbf{b}$, which we note $(m_2, \mathbf{0})$, and identity $(E, \mathbf{0})$.

$$G_0 = (E, \mathbf{0}) + (A, \mathbf{0}) + (m_1, \mathbf{0}) + (m_2, \mathbf{0}).$$

The mirrors intersect along the 2-fold axis, here parallel to $\mathbf{c}$.

7.3.1. Symmorphic space groups

The first space group that comes to mind is group G_s, associated directly to point group G_0:

$$G_s = (E, \mathbf{0}) + (A, \mathbf{0}) + (m_1, \mathbf{0}) + (m_2, \mathbf{0}) \text{ (modulo T)}.$$

This means that all the symmetry operations of the crystal are the four operations forming G_0 and all their products with lattice translations $\mathbf{n} = u\mathbf{a} + v\mathbf{b} + w\mathbf{c}$.

We now choose an axis system $Oxyz$, parallel to $\mathbf{a}$, $\mathbf{b}$, and $\mathbf{c}$ respectively, and take for a unit, along each of these axes, the length of the respective lattice parameter, a, b or c. If an atom in the unit cell has coordinates x, y, z, the presence of the symmetry elements implies the existence of an atom of the same kind at points with coordinates $-x$, $-y$, z (operation A, $\mathbf{0}$), $-x$, y, z (operation m_1, $\mathbf{0}$) and x, $-y$, z (operation m_2, $\mathbf{0}$). Additionally, there are of course all the positions deduced from the above through the lattice translations, but this goes without saying. Thus the "equivalent positions" of an atom in a general position are defined. The coordinates of these positions are:

$$(x, y, z) \qquad (-x, -y, z) \qquad (-x, y, z) \qquad (x, -y, z).$$

The space group is noted as Pmm2, with P saying that the lattice is primitive.

Consider the cases where the orthorhombic unit cell is a multiple cell. We know it can be base-centered (A, B or C lattice), body-centered (I lattice) or face-centered (F lattice).

(a) Body-centered lattice. The crystal lattice features, apart from the translations $\mathbf{n} = u\mathbf{a} + v\mathbf{b} + w\mathbf{c}$ which form group T, a lattice mode translation $\boldsymbol{\mathcal{T}}$ with coordinates $(\frac{1}{2}, \frac{1}{2}, \frac{1}{2})$. The corresponding space group will be defined by the set of the above operations and of those deduced from them through translation $\boldsymbol{\mathcal{T}}$, so that the new space group is:

$$G_s = (E, \mathbf{0}) + (A, \mathbf{0}) + (m_1, \mathbf{0}) + (m_2, \mathbf{0})$$
$$+ (E, \boldsymbol{\mathcal{T}}) + (A, \boldsymbol{\mathcal{T}}) + (m_1, \boldsymbol{\mathcal{T}}) + (m_2, \boldsymbol{\mathcal{T}}) \quad \text{(modulo T)}.$$

The coordinates of equivalent positions are now:

$$(x, y, z) \qquad (x + \tfrac{1}{2}, y + \tfrac{1}{2}, z + \tfrac{1}{2})$$
$$(-x, -y, z) \qquad (-x + \tfrac{1}{2}, -y + \tfrac{1}{2}, z + \tfrac{1}{2})$$
$$(-x, y, z) \qquad (-x + \tfrac{1}{2}, y + \tfrac{1}{2}, z + \tfrac{1}{2})$$
$$(x, -y, z) \qquad (x + \tfrac{1}{2}, -y + \tfrac{1}{2}, z + \tfrac{1}{2}).$$

This new space group is called Imm2.

(b) Base-centered lattice. For example, in the A lattice, translation $\boldsymbol{\mathcal{T}} \left(0, \frac{1}{2}, \frac{1}{2} \right)$ is added, and the corresponding space group is noted Amm2. The coordinates of equivalent positions will now be:

$$(x, y, z) \qquad (x, y + \tfrac{1}{2}, z + \tfrac{1}{2})$$
$$(-x, -y, z) \qquad (-x, -y + \tfrac{1}{2}, z + \tfrac{1}{2})$$
$$(-x, y, z) \qquad (-x, y + \tfrac{1}{2}, z + \tfrac{1}{2})$$
$$(x, -y, z) \qquad (x, -y + \tfrac{1}{2}, z + \tfrac{1}{2}).$$

(c) Face-centered lattice. The F lattice adds translations $\boldsymbol{\mathcal{T}}_1 \left(0, \frac{1}{2}, \frac{1}{2} \right)$, $\boldsymbol{\mathcal{T}}_2 (\frac{1}{2}, 0, \frac{1}{2})$, $\boldsymbol{\mathcal{T}}_3 \left(\frac{1}{2}, \frac{1}{2}, 0 \right)$. The space group for the F lattice, noted Fmm2, is thus:

$$G_s = (E, \mathbf{0}) + (A, \mathbf{0}) + (m_1, \mathbf{0}) + (m_2, \mathbf{0}) + (E, \boldsymbol{\mathcal{T}}_1) + (A, \boldsymbol{\mathcal{T}}_1)$$

$$+ (m_1, \boldsymbol{\mathcal{T}}_1) + (m_2, \boldsymbol{\mathcal{T}}_1) + (E, \boldsymbol{\mathcal{T}}_2) + (A, \boldsymbol{\mathcal{T}}_2) + (m_1, \boldsymbol{\mathcal{T}}_2)$$

$$+ (m_2, \boldsymbol{\mathcal{T}}_2) + (E, \boldsymbol{\mathcal{T}}_3) + (A, \boldsymbol{\mathcal{T}}_3) + (m_1, \boldsymbol{\mathcal{T}}_3) + (m_2, \boldsymbol{\mathcal{T}}_3) \quad \text{(modulo T)}$$

and the equivalent positions are:

$$(x, y, z) \qquad (x, y + 1/2, z + 1/2)$$
$$(-x, -y, z) \quad (-x, -y + 1/2, z + 1/2)$$
$$(-x, y, z) \qquad (-x, y + 1/2, z + 1/2)$$
$$(x, -y, z) \qquad (x, -y + 1/2, z + 1/2)$$

$$(x + 1/2, y, z + 1/2) \qquad (x + 1/2, y + 1/2, z)$$
$$(-x + 1/2, -y, z + 1/2) \quad (-x + 1/2, -y + 1/2, z)$$
$$(-x + 1/2, y, z + 1/2) \quad (-x + 1/2, y + 1/2, z)$$
$$(x + 1/2, -y, z + 1/2) \quad (x + 1/2, -y + 1/2, z).$$

This provides the enumeration of the so-called symmorphic space groups, which are the space groups where all the symmetry operations are of the form (S, **0**) modulo the lattice translations T for the primitive unit cells, or (S, **0**) and (S, $\mathcal{T}$) modulo T for the multiple unit cells. The number of these space groups for a given crystal system is equal to the number of point groups multiplied by the number of Bravais lattices. We thus obtain a total of 66 symmorphic space groups (Tab. 7.4). Their notation includes a letter which defines the lattice, followed by the symbol for the point group.

Table 7.4: Enumeration of the 66 symmorphic groups, to which 7 supplementary groups must be added

System	Number of point groups	Number of Bravais lattices	Number of space groups
Triclinic	2	1	2
Monoclinic	3	2	6
Orthorhombic	3	4	12
Tetragonal	7	2	14
Rhombohedral	5	1	5
Hexagonal	7+5	1	12
Cubic	5	3	15

Seven other groups must be added to these 66 groups, for two different reasons. On the one hand, there can be various ways of associating the Bravais lattice translation $\mathcal{T}$ to a given point group. This is the case for the base-centered orthorhombic groups, associated to point group mm2, for which it is not equivalent to center the bases parallel or perpendicular to the 2-fold axis: this yields groups Amm2 and Cmm2. On the other hand, there can also be two different ways of positioning the symmetry elements with respect to the elementary lattice translations. For example, crystals with point group $\bar{4}2m$ can have their basis vectors parallel either to the mirrors or to the 2-fold axes, and this results in the existence of two different space groups, P$\bar{4}2m$ and P$\bar{4}m2$ (the notations are explained in Sect. 7.4). For the same reason, there exist I$\bar{4}2m$ and I$\bar{4}m2$, P312 and P321, P3m1 and P31m, P$\bar{3}m1$ and P$\bar{3}1m$, P$\bar{6}m2$ and P$\bar{6}2m$ (Exercise 7.1). We thus obtain the 7 supplementary groups (one for the first reason, and 6 for the second reason).

7.3.2. Non-symmorphic space groups

We again consider point group mm2:

$$G_0 = (E, \mathbf{0}) + (A, \mathbf{0}) + (m_1, \mathbf{0}) + (m_2, \mathbf{0}).$$

Primitive groups

We replace one or more of the $(S, \mathbf{0})$ symmetry operations by compatible operations $(S, \mathbf{t}_i)$, where the $\mathbf{t}_i$ are non-zero, *i.e.* by the operations associated with screw axes or glide mirrors.

We first replace the pure mirror m_2, perpendicular to basis vector $\mathbf{b}$, by a glide mirror. Symmetry operation $(m_2, \mathbf{0})$ is thus successively replaced by the symmetry operations $(m_2, \mathbf{a}/2)$, $(m_2, \mathbf{c}/2)$ and $(m_2, \mathbf{a}/2+\mathbf{c}/2)$ while retaining $(m_1, \mathbf{0})$. The symmetry operations have a common origin, chosen on the 2-fold axis where mirrors m_1 and m_2 intersect.

1. The group contains, instead of $(m_2, \mathbf{0})$, operation $(m_2, \mathbf{a}/2)$ associated to an a-type mirror. It also contains, by assumption, $(m_1, \mathbf{0})$, and therefore the product $(m_2, \mathbf{a}/2)(m_1, \mathbf{0})$, calculated using Equation (7.3). This product is equal to $(A, \mathbf{a}/2)$, where the 2-fold axis goes through the common origin O. Since translation $\mathbf{a}/2$ is perpendicular to the 2-fold axis, a change in the position of this axis leads to zero translation. Applying the rule of Section 7.2.3 shows that the new axis goes through point O' such that $\mathbf{OO'} = \mathbf{a}/4$ (Fig. 7.5a), since segment $\mathbf{OP} = \mathbf{a}/2$ subtends angle π from O'. The space group is

    $$G_s = (E, \mathbf{0}) + (m_1, \mathbf{0}) + (m_2, \mathbf{a}/2) + (A, \mathbf{0}) \qquad \text{(modulo T)} \qquad (7.6)$$

    where the indicated operations now do not all have the same origin. Relation (7.6) must be used in connection with Figure 7.5b, which shows the respective positions of the symmetry elements.

    This group is called Pma2 (group 28 in I.T. A) because we will see, in Section 7.4, that the notation of the space groups in the orthorhombic system is based on the following order for the symmetry elements: first the mirror

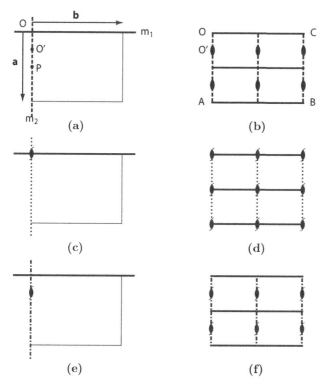

Figure 7.5: An example of the determination of space groups deduced from point group mm2. Operation $(m_2, \mathbf{0})$ is successively replaced by $(m_2, \mathbf{a}/2)$, leading to group Pma2 ((a) and (b)), then by $(m_2, \mathbf{c}/2)$, leading to space group $Pmc2_1$ ((c) and (d)), and finally by $(m_2, \mathbf{a}/2 + \mathbf{c}/2)$, leading to space group $Pmn2_1$ ((e) and (f)).

perpendicular (or the axis parallel) to $\mathbf{a}$, then the mirror perpendicular (or the axis parallel) to $\mathbf{b}$, then the same for $\mathbf{c}$. Figure 7.5b shows the corresponding unit cell OABC, with all the symmetry elements in the unit cell indicated. They are determined by applying the rules of Section 7.2.3 to the products of the operations m_1, m_2 and A with the lattice translations. For example the product of $(m_1, \mathbf{0})$ and the lattice translation $\mathbf{a}$ perpendicular to m_1 is equivalent to a pure mirror translated by $\mathbf{a}/2$ (rule in Sect. 7.2.3). In the same way, the product of $(m_2, \mathbf{a}/2)$ and the lattice translation $\mathbf{b}$ perpendicular to m_2 yields the same glide mirror translated by $\mathbf{b}/2$. Applying the rule of Section 7.2.3 shows that the combination of the 2-fold axis going through O' with a lattice translation $\mathbf{n}$ perpendicular to it (here $\mathbf{a}$ or $\mathbf{b}$ or $(\mathbf{a}+\mathbf{b})$) is equivalent to the existence of a 2-fold axis translated by $\mathbf{n}/2$. Thus there are 2-fold axes going through the points deduced from O' through translations $\mathbf{a}/2$, $\mathbf{b}/2$ and $(\mathbf{a}+\mathbf{b})/2$.

2. The group contains, instead of $(m_2, \mathbf{0})$, operation $(m_2, \mathbf{c}/2)$ associated to a c mirror. The product $(m_2, \mathbf{c}/2)$ $(m_1, \mathbf{0})$ is equal to $(A, \mathbf{c}/2)$, *i.e.* to a screw axis 2_1 at the intersection of the mirrors (Fig. 7.5c) and

$$G_s = (E, \mathbf{0}) + (m_1, \mathbf{0}) + (m_2, \mathbf{c}/2) + (A, \mathbf{c}/2) \text{ (modulo T)}.$$

The space group is noted Pmc2_1 (group 26, I.T. A) and Figure 7.5d shows all the symmetry elements present in the unit cell.

3. The group contains, instead of $(m_2, \mathbf{0})$, operation $(m_2, \mathbf{a}/2 + \mathbf{c}/2)$ associated with an n mirror. The product operation $(m_2, \mathbf{a}/2 + \mathbf{c}/2)(m_1, \mathbf{0})$ is equal to $(A, \mathbf{a}/2 + \mathbf{c}/2)$, which can be replaced by operation $(A, \mathbf{c}/2)$ if the origin of the 2-fold axis is displaced by $\mathbf{a}/4$ (Fig. 7.5e). The space group is

$$G_s = (E, \mathbf{0}) + (m_1, \mathbf{0}) + (m_2, \mathbf{a}/2 + \mathbf{c}/2) + (A, \mathbf{c}/2) \text{ (modulo T)}.$$

It is noted Pmn2_1 (group n° 31, I.T. A) and Figure 7.5f shows all the symmetry elements present in the unit cell.

Other space groups are obtained by substituting glide mirrors for the two mirrors: for example $(m_1, \mathbf{b}/2)$ and $(m_2, \mathbf{a}/2)$. The product of these two operations is equal to $(A, -\mathbf{a}/2+\mathbf{b}/2)$, which becomes $(A, \mathbf{0})$ if the rotation axis is submitted to translation $-\mathbf{a}/4 + \mathbf{b}/4$. The group is noted Pba2 (group 32, I.T. A). In the same way, the reader can, as an exercise, consider the various possible combinations of glide mirrors (Exercise 7.3 is an example). This will show that the combination of operations:

– $(m_1, \mathbf{c}/2)$ and $(m_2, \mathbf{a}/2)$ leads to group Pca2_1 (group 29, I.T. A),

– $(m_1, \mathbf{b}/2 + \mathbf{c}/2)$ and $(m_2, \mathbf{a}/2)$ leads to group Pna2_1(group 33, I.T. A),

– $(m_1, \mathbf{c}/2)$ and $(m_2, \mathbf{c}/2)$ leads to group Pcc2 (group 27, I.T. A),

– $(m_1, \mathbf{b}/2 + \mathbf{c}/2)$ and $(m_2, \mathbf{c}/2)$ leads to group Pnc2 (group 30, I.T. A),

– $(m_1, \mathbf{b}/2 + \mathbf{c}/2)$ and $(m_2, \mathbf{a}/2 + \mathbf{c}/2)$ leads to group Pnn2 (group 34, I.T. A).

We thus enumerated a total of 10 primitive groups associated to point group mm2. Characterizing each of them involves a schematic drawing indicating the position of the various symmetry elements present in the unit cell.

Non-primitive groups

The non-primitive space groups can be enumerated starting from each of these 10 primitive groups, by adding the lattice mode translation or translations $\mathcal{T}$, which transform(s) them into a non-primitive lattice. However, the combination of a glide mirror and a translation $\mathcal{T}$ can yield a pure mirror translated with respect to the first one if translation $\mathcal{T}$ contains a component, parallel to the mirror, equal to the glide vector.

Thus consider group Pma2, the symmetry elements of which are represented on Figures 7.5b and 7.6a. In Figure 7.6a, the origin of the unit cell is chosen on a 2-fold axis to conform to the International Tables for Crystallography A (Sect. 7.3.3). Suppose the group is now C-type, meaning that base $(\mathbf{a}, \mathbf{b})$ is centered, which adds the lattice translation $\mathcal{T} = \frac{1}{2}(\mathbf{a} + \mathbf{b})$. The symmetry operation corresponding to this translation is noted

$$S_{\mathcal{T}} = (\mathrm{E}, \frac{1}{2}\,[\mathbf{a} + \mathbf{b}]).$$

The new space group includes the operations of G_s given in Equation (7.6), which must be associated to Figure 7.6a, and the result of combining them with $S_{\mathcal{T}}$. We investigate them in turn.

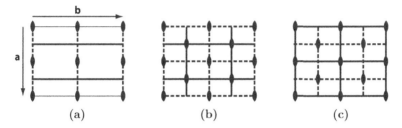

Figure 7.6: Symmetry elements present in the unit cells of (a) group Pma2; (b) group Pma2 with the $(\mathbf{a}, \mathbf{b})$ faces centered, becoming Cmm2; (c) group Cmm2 when the origin is placed on an intersection of two pure mirrors as in I.T. A

The composition of $(m_2, \mathbf{a}/2)$ with $S_{\mathcal{T}}$ is obtained by using relation (7.3), where the rotation or rotoinversion operations are taken to have the same origin.

$$(\mathrm{E},\ \frac{1}{2}\,[\mathbf{a} + \mathbf{b}])\,(m_2, \mathbf{a}/2) = (m_2, [\mathbf{a} + \mathbf{b}/2]) \tag{7.7}$$
$$= (m_2, \mathbf{b}/2) \quad \text{modulo T}.$$

We finally obtain a glide mirror operation, with a translation $\mathbf{b}/2$ perpendicular to the mirror. It transforms into a pure mirror operation by translating the mirror by $\mathbf{b}/4$ (Fig. 7.6b). Relation (7.7) leads to the remark that combination with a pure translation simply adds this translation to the original translation part.

The composition of $(m_1, \mathbf{0})$ with $S_{\mathcal{T}}$ gives $(m_1, \mathbf{a}/2 + \mathbf{b}/2)$. The translation part has two components: $\mathbf{a}/2$, perpendicular to the mirror, which can be canceled by translating the mirror by $\mathbf{a}/4$, and $\mathbf{b}/2$, parallel to the mirror, which turns it into a glide mirror.

Combining $(A, \mathbf{0})$ with $S_{\mathcal{T}}$ displaces the 2-fold axis by $\frac{1}{4}(\mathbf{a} + \mathbf{b})$.

All the symmetry elements present in the unit cell are represented on Figure 7.6b. We see that there are pure mirrors perpendicular to $\mathbf{a}$ and $\mathbf{b}$. The space group is Cmm2. Figure 7.6c shows the unit cell, taking for the origin the intersection of mirrors as in volume A of the International Tables for Crystallography.

In contrast, if a lattice translation $\mathcal{T} = \frac{1}{2}(\mathbf{b} + \mathbf{c})$ is added to group Pma2, the lattice becomes base-centered A, and we obtain group Ama2, the symmetry elements of which are shown on Figure 7.7.

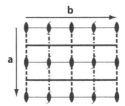

Figure 7.7: Symmetry elements of space group Ama2

The non-primitive space groups associated to point group mm2 are thus fewer than the primitive groups. In International Tables A, the following groups are found: Cmm2, Cmc2$_1$, Ccc2, Amm2, Abm2, Ama2, Aba2, and Fmm2.

The aim of this section was to show one procedure for the enumeration of space groups. Their complete enumeration is beyond the scope of this book. There are 230 crystallographic space groups, perfectly described in volume A of International Tables for Crystallography, which is presented in the following section.

7.3.3. International Tables for Crystallography

The 230 space groups are cataloged in volume A of the International Tables for Crystallography (I.T. A) which, for each group, provide lots of information, as shown on the two pages reproduced on Figures 7.8 a and b, dedicated to space group P4/n, numbered 85.

The first of theses pages is reproduced on Figure 7.8a.

The first line indicates the name of the space group in the Hermann-Mauguin notation (P4/n), then in the Schönflies notation (C_{4h}^3), then the associated point group (4/m), and finally the associated crystal lattice (tetragonal). Since the lattice is primitive, the orders of the space group and the point group are equal, in this case to 8. The 8 symmetry operations of the space group are listed under "Symmetry operations", where each operation is numbered and its nature and position described. For example operation (6), noted n$(\frac{1}{2},\frac{1}{2},0)$ x, y, 0, is the symmetry operation "mirror with diagonal translation n", the glide component of which is the vector $(\frac{1}{2},\frac{1}{2},0)$, with the position of the mirror being given by its equation $z = 0$ or $(x,y,0)$.

The left-hand figure is the projection onto a plane perpendicular to the highest-order axis (here the 4-fold axis) of all the symmetry elements present in the unit cell. The basis vectors $\mathbf{a}$ and $\mathbf{b}$ are oriented as on the earlier figures (7.5 to 7.7). The symbols used for this representation were introduced in Tables 7.1 to 7.3.

$P4/n$ C_{4h}^{3} $4/m$ Tetragonal

No. 85 $P4/n$ Patterson symmetry $P4/m$

ORIGIN CHOICE 1

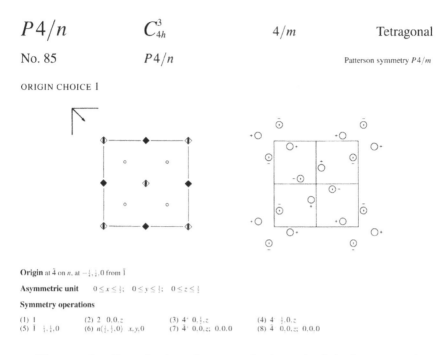

Origin at $\bar{4}$ on n, at $-\frac{1}{4},\frac{1}{4},0$ from $\bar{1}$

Asymmetric unit $0 \le x \le \frac{1}{2}$; $0 \le y \le \frac{1}{2}$; $0 \le z \le \frac{1}{2}$

Symmetry operations

(1) 1	(2) 2 0,0,z	(3) 4^{+} 0,$\frac{1}{2}$,z	(4) 4^{-} $\frac{1}{2}$,0,z
(5) $\bar{1}$ $\frac{1}{4}$,$\frac{1}{4}$,0	(6) $n(\frac{1}{2},\frac{1}{2},0)$ $x,y,0$	(7) $\bar{4}^{+}$ 0,0,z; 0,0,0	(8) $\bar{4}^{-}$ 0,0,z; 0,0,0

Figure 7.8a: Reproduction of a page of volume A of the International Tables for Crystallography: first page describing group P4/n (© IUCr, with permission)

The origin of the chosen unit cell is indicated under the figures. Here it is taken on the center of inversion for the $\bar{4}$ axis, and this is the first choice of the International Tables A. For this group, the Tables also offer another version, not reproduced here, where the origin of the unit cell is chosen on a center of symmetry. The right-hand figure indicates the general positions of the atoms, the signs $+$ and $-$ standing for coordinates $+z$ and $-z$ respectively. The circles with a comma are the images of the circles without comma through one of the improper symmetry operations of the group (mirror, inversion or rotoinversion).

Page 2 (Fig. 7.8b) first presents a choice of generating elements for the space group, then the various possible positions for an atom (indexed through a letter, called the Wyckoff symbol), starting with the general position, for which the coordinates x, y, z are general. The following positions are special positions, because they are located on one or several symmetry elements.

For each position, the table gives the coordinates of the equivalent positions (modulo the lattice translations) through the symmetry operations of the group. For the general position, the atoms in positions equivalent through the symmetry operations are numbered from (1) to (n) where n is the order of the group.

CONTINUED   No. 85 $P4/n$

Generators selected   (1); $t(1,0,0)$; $t(0,1,0)$; $t(0,0,1)$; (2); (3); (5)

Positions

Multiplicity, Wyckoff letter, Site symmetry   Coordinates   Reflection conditions

General:

8	g	1	(1) x,y,z	(2) $\bar{x},\bar{y},z$	(3) $\bar{y}+\frac{1}{2},x+\frac{1}{2},z$	(4) $y+\frac{1}{2},\bar{x}+\frac{1}{2},z$	$hk0: h+k=2n$
			(5) $\bar{x}+\frac{1}{2},\bar{y}+\frac{1}{2},\bar{z}$	(6) $x+\frac{1}{2},y+\frac{1}{2},\bar{z}$	(7) $y,\bar{x},\bar{z}$	(8) $\bar{y},x,\bar{z}$	$h00: h=2n$

Special: as above, plus

4	f	2..	$0,0,z$	$\frac{1}{2},\frac{1}{2},z$	$\frac{1}{2},\frac{1}{2},\bar{z}$	$0,0,\bar{z}$	$hkl: h+k=2n$
4	e	$\bar{1}$	$\frac{1}{4},\frac{1}{4},\frac{1}{4}$	$\frac{1}{4},\frac{3}{4},\frac{1}{4}$	$\frac{3}{4},\frac{3}{4},\frac{1}{4}$	$\frac{3}{4},\frac{1}{4},\frac{1}{4}$	$hkl: h,k=2n$
4	d	$\bar{1}$	$\frac{1}{4},\frac{1}{4},0$	$\frac{1}{4},\frac{3}{4},0$	$\frac{3}{4},\frac{3}{4},0$	$\frac{3}{4},\frac{1}{4},0$	$hkl: h,k=2n$
2	c	4..	$0,\frac{1}{2},z$	$\frac{1}{2},0,\bar{z}$			no extra conditions
2	b	$\bar{4}$..	$0,0,\frac{1}{2}$	$\frac{1}{2},\frac{1}{2},\frac{1}{2}$			$hkl: h+k=2n$
2	a	$\bar{4}$..	$0,0,0$	$\frac{1}{2},\frac{1}{2},0$			$hkl: h+k=2n$

Symmetry of special projections

Along [001] $p4$ Along [100] $p2mg$ Along [110] $p2mm$
$\mathbf{a}'=\frac{1}{2}(\mathbf{a}-\mathbf{b})$ $\mathbf{b}'=\frac{1}{2}(\mathbf{a}+\mathbf{b})$ $\mathbf{a}'=\mathbf{b}$ $\mathbf{b}'=\mathbf{c}$ $\mathbf{a}'=\frac{1}{2}(-\mathbf{a}+\mathbf{b})$ $\mathbf{b}'=\mathbf{c}$
Origin at $0,0,z$ Origin at $x,\frac{1}{4},0$ Origin at $x,x,0$

Maximal non-isomorphic subgroups
I   [2] $P\bar{4}$ (81)   1; 2; 7; 8
      [2] $P4$ (75)   1; 2; 3; 4
      [2] $P2/n$ ($P2/c$, 13)   1; 2; 5; 6
IIa   none
IIb   [2] $P4_2/n$ ($\mathbf{c}'=2\mathbf{c}$) (86)

Maximal isomorphic subgroups of lowest index
IIc   [2] $P4/n$ ($\mathbf{c}'=2\mathbf{c}$) (85); [5] $P4/n$ ($\mathbf{a}'=\mathbf{a}+2\mathbf{b}, \mathbf{b}'=-2\mathbf{a}+\mathbf{b}$ or $\mathbf{a}'=\mathbf{a}-2\mathbf{b}, \mathbf{b}'=2\mathbf{a}+\mathbf{b}$) (85)

Minimal non-isomorphic supergroups
I   [2] $P4/nbm$ (125); [2] $P4/nnc$ (126); [2] $P4/nmm$ (129); [2] $P4/ncc$ (130)
II   [2] $C4/m$ ($P4/m$, 83); [2] $I4/m$ (87)

Figure 7.8b: Reproduction of a page of volume A of the International Tables for Crystallography: the second page describing group P4/n (© IUCr, with permission)

The atom in position (k) deduces from atom in position (1) through the symmetry operation indexed as (k) on the first page. For example, in the space group P4/n described, atom (3) deduces from atom (1) through operation (3), a rotation by $\pi/2$ around the 4-fold axis which goes through the point with coordinates $x = 0$, $y = \frac{1}{2}$.

For each position, the table starts out by indicating its multiplicity (number of equivalent positions), then its Wyckoff symbol (a letter given in alphabetical order, starting with the last position, the one with lowest multiplicity), then the local symmetry of this particular site, and finally the coordinates of the equivalent atomic positions. The local symmetry of a site in a special position deduces from the symmetry elements going through the site. The set of the symmetry operations associated to these symmetry elements which intersect on the site form a point group which is a subgroup of the point group of the crystal.

The asymmetric unit, indicated on page 1, is defined in a general way as a part of space such that, if it is submitted to all the symmetry elements of the space group, all of space is occupied. It is therefore bounded by planes.

However, for a given crystal structure, the asymmetric unit is defined by a group of atoms such that, when complemented by their images through all the symmetry operations of the space group, the motif (or pattern) of the crystal unit cell is retrieved. We note that the set of atoms in the motif of a crystal correspond to a multiple of the chemical formula of the material, while this is not necessarily the case for the atoms in the asymmetric unit. For example, the motif of the unit cell of quartz (chemical formula SiO_2, space group $P3_121$) contains three SiO_2 groups, while the asymmetric unit consists of one Si atom and one O atom. Since the oxygen atom is in a general position with multiplicity 6, there are 6 oxygen atoms in the motif while the Si atoms are in a special position (on a 2-fold axis) with multiplicity $6/2 = 3$, hence 3 silicon atoms in the motif. Starting with one Si atom and one O atom, we retrieve the set Si_3O_6, *i.e.* a motif of three SiO_2 groups.

7.4. Nomenclature

The Hermann-Mauguin space group notation starts with a letter indicating the nature of the Bravais lattice (P for primitive, A, B or C for base-centered, I for body-centered, and F for face-centered). It then refers directly to the notation of the point group which is associated to it (the notation is given in Chap. 5), making explicit the nature of the axes and mirrors involved in the notation for the point group. For example, if an n-fold axis is a screw axis, we add to the n involved in the point group notation a subscript p indicating that the translation is a fraction p/n of the basis vector parallel to the axis (Tab. 7.1). If one of the mirrors is a glide mirror, the letter m is replaced by a, b, c or n depending on the nature of the translation (Tab. 7.2).

When the notation for the point group includes several identical signs, or involves an ambiguity, the following rules are applied, with the understanding that the quoted *directions define, when the subject is an axis, its direction, and, for a mirror, the direction perpendicular to it.*

–   *Orthorhombic system:* the order is **a**, **b**, **c**. Example: group $Cmc2_1$. The unit cell has its (**a**, **b**) bases centered. The point group is mm2. There are pure mirrors perpendicular to **a**, glide mirrors (translation **c**/2) perpendicular to **b**, and screw axes parallel to **c** (translation **c**/2).

–   *Tetragonal system:* the first sign refers to the 4 or $\bar{4}$ axis, which is always parallel to **c**, the second to the directions **a** and **b**, equivalent to each other, and the third to the bisectors of these directions in the (**a**, **b**) plane.

Example: $P\bar{4}2m$ is a space group with a primitive unit cell, with point group $\bar{4}2m$. The $\bar{4}$ axis is parallel to c, there are 2-fold axes parallel to a and b, and pure mirrors at $45°$ to a and b. In group $P\bar{4}m2$, the mirrors are parallel to a and b and the 2-fold axes are at $45°$ to a and b (Exercise 7.1).

– *Hexagonal system*: the first sign refers to the 6-fold axis, parallel to c, the second to the directions of vectors a and b, and the third to the directions perpendicular to a and b.

Example: $P6_3cm$ is a space group with a primitive unit cell, with point group 6mm; the 6-fold axis, parallel to c, is a screw axis (translation $c/2$); there are glide mirrors (glide $c/2$) perpendicular to a and b and pure mirrors parallel to a and b (since they are perpendicular to the normals to a and b).

We now understand, on the basis of these examples, the rationale for the redundancy in the notation for some point groups. For example, in point group 6mm, the notation 6m is sufficient as a descriptor, since the generating elements of the group are the 6-fold axis and a mirror which contains it: the presence of these two generators leads to the existence of 6 mirrors enclosing angles of $30°$ with one another. However, these mirrors form two distinct families with respect to the symmetry operations of the space group. This leads to the existence of two different space groups, $P6_3mc$ and $P6_3cm$.

– *Special case of the hexagonal lattices with rhombohedral point group*: we recall (Sect. 6.3.3) that a crystal with rhombohedral point group (3, $\bar{3}$, 32, 3m or $\bar{3}m$) can have a rhombohedral lattice (noted R) or a hexagonal lattice (noted P). The rule for the hexagonal lattices applies to those with a hexagonal lattice, but with a 1 inserted where there is no symmetry element.

Example of the group $P3_121$: the unit cell is hexagonal, the point group is 32, the 3-fold axis, parallel to c, is a screw axis (translation $c/3$), the 2-fold axes are parallel to a and b, and there is no 2-fold axis perpendicular to a and b. In contrast, for group $P3_112$, the 2-fold axes are perpendicular to a and b.

– *Cubic system*: same order as for the point group, *viz.* directions of the cube sides, then directions of the major diagonals of the cube, then directions of the face diagonals.

First example: $Pn\bar{3}$: primitive unit cell, point group $m\bar{3}$, glide mirrors perpendicular to the cube edges (translation $\frac{1}{2}(b + c)$ for the mirror perpendicular to a), 3-fold axes parallel to the major diagonals of the cube, and center of symmetry.

Second example: $Fd\bar{3}m$: face-centered cubic unit cell, point group $m\bar{3}m$, diamond mirrors (see Sect. 7.2.2) perpendicular to the cube sides (translation $\frac{1}{4}(b + c)$ for the mirror perpendicular to a), 3-fold axes parallel to the major diagonals of the cube and center of symmetry, pure mirrors perpendicular to the face diagonals. This is the space group of diamond and silicon crystals (see Sect. 7.5.3).

7.5. Examples: space groups of some structures

7.5.1. TiO$_2$-type structure (rutile)

This structure[2] is represented in a perspective view on Figure 8.4b. The unit cell is tetragonal primitive. The positions of the atoms in the unit cell are:

Ti atoms : 0, 0, 0 and $\frac{1}{2}, \frac{1}{2}, \frac{1}{2}$;

O atoms: $u, u, 0$; $\frac{1}{2}+u, \frac{1}{2}-u, \frac{1}{2}$; $\frac{1}{2}-u, \frac{1}{2}+u, \frac{1}{2}$; $-u, -u, 0$.

The projection onto plane $(\mathbf{a}, \mathbf{b})$, with the heights z indicated, of the atom positions in two neighboring unit cells is shown on Figure 7.9a. The Ti atoms are shown as full circles and the O atoms as empty circles. This evidences a 4_2 axis which projects at A, glide mirrors (n mirrors) perpendicular to the sides $\mathbf{a}$ and $\mathbf{b}$, with glides respectively $\frac{1}{2}(\mathbf{b}+\mathbf{c})$ and $\frac{1}{2}(\mathbf{a}+\mathbf{c})$, and pure mirrors parallel to the diagonals of the square. There is also a mirror parallel to the plane of the figure at height 0 (and another one at height $\frac{1}{2}$). The point group is thus $4/m\,mm$ and the space group is $P\,4_2/m\,nm$. Since the point group is centrosymmetric, so is the space group. A center of symmetry projecting at A, with height 0, is evidenced.

Figure 7.9b shows the axes and mirrors perpendicular to the projection plane. The 4_2 axis contains a 2-fold axis, and the products of the symmetry operations associated to this 2-fold axis with the lattice translations produce 2-fold axes at the apices and at the center of square $(\mathbf{a}, \mathbf{b})$. The centers of symmetry, represented by empty circles as usual, are also shown. Figure 7.9c shows all the symmetry elements present in the unit cell; this figure is excerpted from International Tables volume A (group 136).

We note that the asymmetric unit here consists of one Ti atom and just one O atom, while the motif, outlined in dotted lines on Figure 7.9a, contains two TiO$_2$ groups. The order of the point group is 16. It is equal to that of the space group since the unit cell is primitive. There are therefore 16 equivalent atoms in general positions. The oxygen atom is in a special position, at the intersection of two pure mirrors, hence on a site with multiplicity $16/4 = 4$ while the Ti atom is at the intersection of 3 pure mirrors, thus on a site with multiplicity $16/8 = 2$.

Many crystals have this structure. Examples are CoF_2, MnF_2, CrO_2, PbO_2. The values of a, c and u vary from compound to compound.

---

2. TiO$_2$ also exists under two other forms: anatase (tetragonal, I4/a md) and brookite (orthorhombic, Pbca).

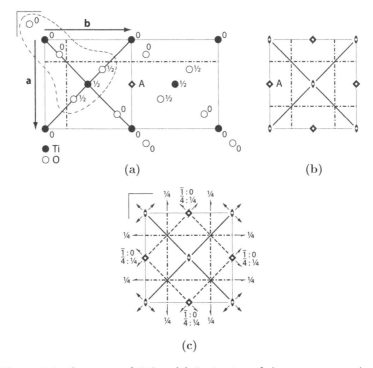

(a)   (b)

(c)

Figure 7.9: Structure of TiO_2. (a) Projection of the atoms onto plane (**a**, **b**), with the heights (z-coordinate) indicated. The Ti atoms are full circles and the O atoms empty circles. The dotted line surrounds the atoms which form the crystal motif. (b) Projection onto the same plane of the axes and mirrors present in the unit cell which are perpendicular to the plane. (c) Projection of all the symmetry elements present in the unit cell. Reproduction of a figure from volume A of the International Tables for Crystallography (© IUCr, with permission).

7.5.2. Metals with close-packed hexagonal structure

This is the case for example of Zn, Na, Be, Mg, Cr. In this structure, which will be described in Chapter 8, the unit cell is hexagonal, with an atom at each of its vertices and another atom at position $(\frac{2}{3}, \frac{1}{3}, \frac{1}{2})$ as shown on Figure 7.10a, a projection of the atomic positions onto the base plane (**a**, **b**) indicating their coordinate parallel to **c**. Several contiguous unit cells were drawn in order to make the determination of the symmetry elements in this structure easier.

By looking at the atom positions, we ascertain the presence of a 6_3 axis projecting at point A, a pure mirror with trace MN and two other mirrors of the same family (enclosing 60° angles between one another) going through A. We also distinguish a glide mirror with translation $\mathbf{c}/2$, with trace uu' (and two other mirrors of the same family going through A). Further, the plane of the figure (as well as the plane parallel to it at height $\frac{1}{2}$) is a symmetry plane, hence

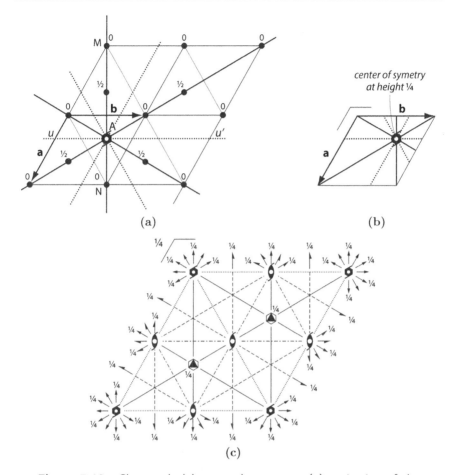

Figure 7.10: Close-packed hexagonal structure: (a) projection of the atomic positions, with the heights indicated, of four adjacent unit cells onto a plane perpendicular to **c**; (b) projection of some of the symmetry elements present in the unit cell; (c) projection of all the symmetry elements present in the unit cell, reproduced from I.T. A (group 194). The origin of the unit cell is at a center of symmetry located on the 6_3 axis.

a pure mirror. We deduce that the space group is $P\,6_3/m\,mc$ (group 194 in I.T. A). The point group is $6/m\,mm$, which has a center of symmetry. There is a center of symmetry at height $\frac{1}{4}$, which projects at A.

Figure 7.10b shows a single unit cell and the symmetry elements we just determined. Figure 7.10c shows all the symmetry elements in the unit cell, as found in International Tables volume A. We note, however, that, in International Tables, the origin of the unit cell is located on a center of symmetry which, on Figure 7.10a, projects at A and has height $\frac{1}{4}$. This explains why the mirror, parallel to the plane of the figure, which was at height 0 (with another one at height $\frac{1}{2}$) on Figure 7.10b, is at height $\frac{1}{4}$ on Figure 7.10c, while the center of symmetry is noted with no further indication on Figure 7.10c since it is in the projection plane.

7.5.3. Structure of diamond

The structure of diamond is shown on Figure 8.6 in a perspective view. This is also the structure of silicon and germanium. The unit cell is face-centered cubic and the projection of the carbon atoms onto the plane perpendicular to one of the sides of the cubic cell, the $(\mathbf{a}, \mathbf{b})$ plane for example, is represented, with height indications, on Figure 7.11a. The motif (or pattern) consists of two atoms (surrounded by a dotted line), with coordinates respectively $(0, 0, 0)$ and $(\frac{1}{4}, \frac{1}{4}, \frac{1}{4})$. This structure features 3-fold axes along the major diagonals of the cubic unit cell.

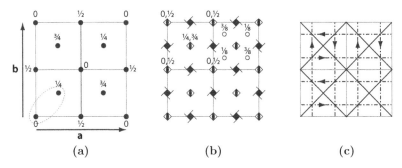

(a)   (b)   (c)

Figure 7.11: Structure of diamond, with face-centered cubic unit cell: (a) projection of the atom positions, with the heights indicated, onto plane $(\mathbf{a}, \mathbf{b})$; (b) position of the symmetry axes parallel to $\mathbf{c}$; (c) projection of the mirrors perpendicular to the $(\mathbf{a}, \mathbf{b})$ plane

Figure 7.11b shows the projection onto the $(\mathbf{a}, \mathbf{b})$ plane of the positions of the symmetry axes perpendicular to this plane. They are 4_1, 4_3 and $\bar{4}$ axes. For the $\bar{4}$ axes, the height of the point used for inversion is indicated. In order not to clutter the figure, they are shown only in the upper left quadrant, because the figure is identical in all four quadrants. In the same way, the position and height of the centers of symmetry are shown only in the upper right quadrant. Figure 7.11c shows the mirrors perpendicular to the plane of the figure. The mirrors parallel to the cube faces are the glide mirrors with translation $\frac{1}{4}(\mathbf{a} + \mathbf{c})$ or $\frac{1}{4}(\mathbf{a} - \mathbf{c})$ for the mirrors perpendicular to $\mathbf{b}$, and with translation $\frac{1}{4}(\mathbf{b} + \mathbf{c})$ or $\frac{1}{4}(\mathbf{b} - \mathbf{c})$ for the mirrors perpendicular to $\mathbf{a}$. The two possibilities are distinguished by an arrow indicating the direction of the component of the translation vector along the projection plane, assuming the component perpendicular to the projection plane is positive. For example, if the translation vector is $\frac{1}{4}(\mathbf{a}+\mathbf{c})$, the arrow will be in the direction of $\mathbf{a}$; if it is $\frac{1}{4}(\mathbf{a}-\mathbf{c})$, the arrow will be along vector $-\mathbf{a}$, since the sign of the translation vector can be changed: if the translation vector is $\mathbf{v}$ to go from atom A to atom B, it is −**v** to go from atom B to atom A. These glide mirrors, for which the translation vector is a quarter of a lattice vector defined by vectors **a**, **b** and **c** but half of a lattice mode vector $\mathcal{T}$ characterizing a multiple unit cell, are called diamond mirrors and noted d (see Sect. 7.2.2).

The diagonal mirrors, perpendicular to the face diagonals, are pure mirrors. According to the rules stated in Section 7.4 (the first symbol shows the nature of the mirrors perpendicular to the cube sides, the second symbol indicates the 3- or $\bar{3}$-fold axis, and the last symbol indicates the nature of the diagonal mirrors), the space group of the diamond structure is noted Fd$\bar{3}$m.

7.6. Exercises

Exercise 7.1

1. (a) Consider point group $\bar{4}$2m. Using a stereographic projection onto a plane normal to the $\bar{4}$ axis, show that the generating elements of this group can be either the $\bar{4}$ axis and a mirror parallel to this axis, or the $\bar{4}$ axis and a 2-fold axis perpendicular to this axis.

    (b) A crystal with this point group has a tetragonal lattice. Comparing the stereographic projections of the point groups of this crystal and its lattice, show that the basis vectors **a** and **b** of this lattice can be parallel either to the 2-fold axes, or to the mirrors. This leads to space groups P$\bar{4}$2m and P$\bar{4}$m2 respectively. Determine, for each of these space groups, the equivalent positions for the general position x, y, z, and all the symmetry elements in the unit cell. Note that the $\bar{4}$ axis is at the intersection of the mirrors, and that the 2-fold axes intersect this $\bar{4}$ axis. Choose as the origin of the unit cell the point common to the various symmetry elements.

2. We consider a crystal with point group 3m, and with a hexagonal lattice.

    (a) By comparing the point group of the crystal and that of the lattice, show that the three mirrors can be either parallel or perpendicular to the basis vectors **a** and **b**.

    (b) Two different space groups are thus defined, P31m and P3m1. Display, through a projection onto a plane perpendicular to the 3-fold axis, all the symmetry elements in the unit cell.

    (c) Why does the same approach fail to apply if the lattice is rhombohedral (space group R3m)?

Exercise 7.2

1. Consider space group P4. Indicate all the symmetry elements present in the unit cell through a projection onto a plane perpendicular to the 4-fold axis.

2. The space group I4 is obtained by adding translation $\frac{1}{2}(\mathbf{a} + \mathbf{b} + \mathbf{c})$ to the symmetry operations of group P4. Indicate all the symmetry elements present in the unit cell using a projection onto a plane perpendicular to the 4-fold axis.

Exercise 7.3

Determine the primitive space group deriving from point group mm2 and containing operations $(m_1, \mathbf{c}/2)$ and $(m_2, \mathbf{a}/2)$ where m_1 is the symmetry operation associated to mirror M_1, perpendicular to $\mathbf{a}$, and m_2 the one associated to mirror M_2, perpendicular to $\mathbf{b}$. Give its notation and the arrangement of the symmetry elements in the unit cell.

Investigation of some structures

The following exercises deal with some crystal structures in the following way. We give the crystal system of the compound, and the position of the atoms in the unit cell. You are asked to determine the corresponding space group. This approach provides a better understanding of the notion of space group, and it is thus well suited to this book. We must however realize that this is not the approach used by a scientist who wants to determine the structure of a given crystal. The researcher starts out from an experimental result: the X-ray diffraction diagram given by this crystal. The accurate investigation of this diagram provides information on the space group of the crystal before the positions of the atoms in the unit cell are known.

Exercise 7.4: structure of neptunium (Np)

Between 278°C and 570°C, the crystal lattice of neptunium is tetragonal and the Np atoms are at the following positions:

$$0, 0, 0 \qquad \tfrac{1}{2}, \tfrac{1}{2}, 0 \qquad \tfrac{1}{2}, 0, u \qquad 0, \tfrac{1}{2}, \bar{u}$$

1. Determine the Bravais lattice.

2. What is the local symmetry of each of the Np atoms? Give the corresponding point group for each of the sites, and deduce the group of which these two point groups are subgroups. This yields the point group of the crystal.

3. Determine the nature of all the symmetry elements which make it possible to determine the space group. Indicate the space group.

Exercise 7.5: structure of NiAs

Nickel arsenide, NiAs, crystallizes in the hexagonal system and the atom positions are:

Ni	$0, 0, 0$	and	$0, 0, \tfrac{1}{2}$,
As	$\tfrac{1}{3}, \tfrac{2}{3}, u$	and	$\tfrac{2}{3}, \tfrac{1}{3}, u + \tfrac{1}{2}$.

1. Draw a diagram, with the heights specified, of the atomic positions projected onto the plane perpendicular to **c**.

2. Determine the symmetry elements present in the unit cell, and deduce the space group of the crystal.

3. What is the symmetry of the Ni and As sites?

Exercise 7.6: structure of FeOCl

Iron oxychloride FeOCl crystallizes in the orthorhombic system, and the atom positions are:

$$
\begin{array}{lll}
\text{Fe} & 0, \tfrac{1}{2}, u & \text{and} \qquad \tfrac{1}{2}, 0, -u \\
\text{Cl} & 0, 0, v & \text{and} \qquad \tfrac{1}{2}, \tfrac{1}{2}, -v \\
\text{O} & 0, 0, v' & \text{and} \qquad \tfrac{1}{2}, \tfrac{1}{2}, -v'
\end{array}
$$

1. Determine the Bravais lattice.

2. Draw a diagram of the projection of the atoms onto a plane perpendicular to **c**, with the heights specified.

3. Determine the symmetry elements present in the unit cell and the space group of this crystal.

4. Give the number of equivalent positions for the general position and the coordinates of these positions.

Complement 7C. Structure determination: outline

Solving a crystal structure means determining on one hand the geometry of the unit cell, and on the other hand the contents of the unit cell. The geometry involves the lengths and relative orientations of the sides of the unit cell that adequately describes the crystal symmetry. The contents of the unit cell involves the nature of the various atoms or ions, their average positions, and their thermal oscillations. The overwhelming majority of crystal structure determination work is based on the diffraction (coherent elastic scattering) of radiation, chiefly X-rays but in some cases neutrons. The replication of the unit cell inherent in crystal periodicity then allows finding the geometric characteristics and contents of the unit cell, whereas this would not be possible for a single unit cell.

As shown by Bragg's law (Sect. 3.2.3), the useful radiations must have wavelength smaller than the interatomic distances, hence the predominant use of X-rays. Neutron and electron diffraction, as discussed in Complement 4C, are also used, but to a much lesser extent. The frequency associated for example with X-rays of wavelength 0.05 nm (0.5 Å) is $\nu = c/\lambda = 3 \times 10^8/5 \times 10^{-11} = 6 \times 10^{18}$ Hz – far above the tracking possibilities of any instrument. As a result, only the intensity of the X-rays, but never their amplitude and phase, can be detected.

Two sets of information must be collected. On the one hand, the geometrical conditions for coherent elastic scattering (diffraction) to occur, and on the other hand the efficiency of diffraction under this condition.

The most convenient description of the geometrical condition for an incident wave, with wave-vector $\mathbf{k}_0$, to be diffracted into a wave with wave-vector $\mathbf{k}$ is that the difference in wave-vector $\mathbf{k} - \mathbf{k}_0$ be equal to a reciprocal lattice vector $\mathbf{h}^*$ of the crystal, as introduced in Section 3.2.1

$$\mathbf{k} - \mathbf{k}_0 = \mathbf{h}^* = h\mathbf{a}^* + k\mathbf{b}^* + l\mathbf{c}^*$$

with $\mathbf{a}^*$, $\mathbf{b}^*$ and $\mathbf{c}^*$ the basis vectors of the reciprocal lattice and h, k and l integers.

The geometrical information thus consists in the orientations and norms of the reciprocal lattice (Sect. 3.2) vectors $\mathbf{h}^*$, which we recall are vectors in reciprocal (or wave-vector) space. Each of them is associated with a Bragg reflection hkl, another name for the coherent elastic diffraction process with scattering vector $\mathbf{h}^*$.

Another approach to diffraction by crystals consists in noting that, as is the case for visible light, the diffraction pattern is the representation of the modulus squared of the diffracted amplitude. The latter is the Fourier transform of the quantity which describes the interaction with matter. In the X-ray case, this is the electronic density $\rho_e(\mathbf{r})$, the number of electrons per unit volume. If the crystal is supposed to be perfectly periodic, the (spatial) Fourier transform is non-zero only for discrete values of the scattering vector $\mathbf{k} - \mathbf{k}_0$, just as the (temporal) Fourier transform of a sound wave (periodic in time) only includes the fundamental frequency and its harmonics. This leads to the reciprocal lattice nodes $\mathbf{h}^* = h\mathbf{a}^* + k\mathbf{b}^* + l\mathbf{c}^*$ shown above and introduced in Section 3.2.1.

Each node of the reciprocal lattice, *i.e.* each vector $\mathbf{h}^*$, is characterized quantitatively in terms of the integrated reflectivity of the Bragg reflection R_{hkl}, the ratio of the area under a curve of diffracted power P_d versus crystal angle (rocking curve) over the intensity incident on the crystal: this quantity is significant because it is independent of the divergence of the incident beam (Fig. 7C.1).

The expression is discussed, without derivation, in the text below, which can be skipped on first reading.

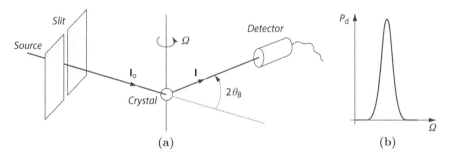

(a)   (b)

Figure 7C.1: (a) A rotating crystal diffraction measurement on a single crystal. The diffracted beam is collected by a detector set at angle $2\theta_B$ with respect to the incident beam, with θ_B the Bragg angle. (b) The variation of the diffracted power P_d with the angle of the crystal, Ω, is the rocking curve. The integral under this curve, scaled with the power P_0 of the incident beam, is the integrated reflectivity.

Integrated reflectivity for X-rays under the kinematical approximation. Structure factor, Lorentz and polarization factors, Debye-Waller factor

The expression for the integrated reflectivity can be shown to be

$$R_{hkl} = \lambda^3 V \, LP \, |\rho_{hkl}|^2$$

where λ is the wavelength of the radiation used, V the volume of the crystal, L is the Lorentz factor, which depends only on the scattering angle and on the diffraction technique used (it is tabulated for example in International Tables for Crystallography, volume C), and P is the polarization factor, which depends only on the scattering angle and on the polarization of the incident beam (it is also tabulated in International Tables for Crystallography, volume C). The central quantity ρ_{hkl} is the hkl Fourier component of the scattering length density for the probe used. In the X-ray case, where the scattering is due to the electrons, $\rho_{hkl} = r_c \rho_{e,hkl}$ where r_c is the length characterizing diffraction by a single free electron and is called the classical electron radius, $r_c = 2.8 \times 10^{-15}$ m and $\rho_{e,hkl}$ the Fourier coefficient hkl of the electron density $\rho_e(\mathbf{r})$ in the crystal. Alternatively, we can write $\rho_{e,hkl} = F_{hkl}/V_c$, where F_{hkl} is called the structure factor of the Bragg reflection used, and V_c the unit cell volume. It is convenient to include in the expression for the structure factor the effect of thermal vibrations of the atoms or ions through the so-called Debye-Waller factor. Then the structure factor can be written as

$$F_{hkl} = \sum_j f_j \exp[i(hx_j + ky_j + lz_j)] \exp\{-W_j\}$$

or

$$F_{hkl} = \sum_j f_j \exp[i(\mathbf{h}^* \cdot \mathbf{r}_j)] \exp\{-W_j\}.$$

In this expression, the sum is over all the atoms, indexed by subscript j, in the unit cell; f_j is the atomic scattering factor (a function of the norm $\|\mathbf{h}^*\|$ of the scattering vector, at least in the approximation where the electron density in the atom has spherical symmetry), $\mathbf{r}_j$ is the average position vector of the atom; $\exp\{-W_j\}$ is the Debye-Waller factor describing the vibration effect; it reduces the scattering more heavily for those atoms which vibrate strongly, as well as for large scattering vectors.

It is important to note that the experimentally measured quantity R_{hkl} is directly related to $|\rho_{hkl}|^2$ and not to ρ_{hkl}. If ρ_{hkl} could be directly measured, determining the structure would be a trivial matter, since the electron density $\rho_e(\mathbf{r})$ could then be reconstructed from its Fourier components. Actually the phase information is missing, and this is known as the phase problem in crystallography. Elaborate strategies, the discussion of which is beyond the scope of this complement, have been developed to obviate this problem.

Structure determination is a largely computer-automated process. It can be performed with single crystal samples or with powders, where the sample consists of many tiny crystals, oriented at random. Powder diffraction, aided by computer files containing the results from a vast number of phases, is also a valuable tool for material identification in industry or geology.

More and more elaborate structures, involving larger and larger unit cells and more and more atoms in the unit cell, are determined routinely. A special case is that of materials of biological interest, where both the challenge and the efforts are big, and where outstanding results in terms of basic science as well as for applications in medicine are obtained.

Further reading

A. Authier, *The Reciprocal Lattice*, International Union of Crystallography, Teaching Pamphlets (1981)

C.A Taylor, *A Non-mathematical Introduction to X-ray Crystallography*, International Union of Crystallography, Teaching Pamphlets (2001);
http://www.iucr.org/_data/assets/pdf_file/0008/3050/1.pdf

J.P. Glusker, *Elementary X-ray Diffraction for Biologists*, International Union of Crystallography, Teaching Pamphlets (2001);
http://www.iucr.org/_data/assets/pdf_file/0008/14399/15.pdf

J.P. Glusker, K.N. Trueblood, *Crystal Structure Analysis: A Primer* (Oxford University Press, 1985)

References

[1] Th. Hahn (ed.), *International Tables for Crystallography*, vol. A, *Space-group Symmetry*, 5th edn. (Kluwer Academic Publishers, Dordrecht, 2005)

Chapter 8

Chemical bonds and crystal structures

This chapter introduces the various types of chemical bonds, and discusses how the nature of the bonds gives an insight into some crystal structures.

8.1. Introduction

The periodic structure of crystals was first evidenced by X-ray diffraction. This technique makes it possible to accurately determine not only the triple periodicity of the atomic arrangement, *i.e.* the crystal lattice, but also the positions of the individual atoms in the crystal motif. One may ask whether we can explain why one atomic arrangement rather than another one is obtained. This cannot be done in a systematic way. However, some rules emerge, and make it possible in many cases, not actually to predict the crystal structure, but rather, as we will see, to interpret the structure once it is determined.

A crystal exists due to the cohesion between atoms, *i.e.* thanks to the existence of interatomic forces. The crystal structure is determined by the balance between the various interatomic forces. In order to understand the rationale for the various crystal structures, we therefore must start out by investigating these forces, *i.e.* the nature of chemical bonding in crystals and the special properties of each of these forces. Whatever the nature of the forces, one general rule applies: the atoms arrange in the most compact (or close-packed) way possible consistent with their size and the special requirements due to the nature of the force.

The distance between two atoms in a solid results from the balance between the attractive and repulsive forces between these two atoms. The physical nature of the attractive force depends on the solid. It can be electrostatic as in ionic solids, or it can be a covalent binding force, or a Van der Waals force, or... Its variation as a function of the distance between the two atoms depends on the nature of these forces. In contrast, the repulsive force, due to the electrostatic repulsion between electrons and to Pauli's "exclusion principle",

© Springer Science+Business Media Dordrecht 2014
C. Malgrange et al., *Symmetry and Physical Properties of Crystals*,
DOI 10.1007/978-94-017-8993-6_8

which forbids two electrons from being in the same quantum state, may in all cases be approximated by a $1/r^{11}$ or $1/r^{12}$ or $\exp(-r/\rho)$ function, where ρ is a characteristic radius. For all types of chemical bonds, the total potential energy looks like the curve shown on Figure 8.1. It has a minimum for a value r_0 of r which is the equilibrium position. Since the repulsive energy varies, in all cases, much faster than the attractive energy, its contribution to the total energy at equilibrium $(r = r_0)$, noted $U(r_0)$, is small with respect to the attractive energy $E_{att}(r_0)$. We may therefore obtain a fair order of magnitude of the total energy at equilibrium by considering only the attractive energy.

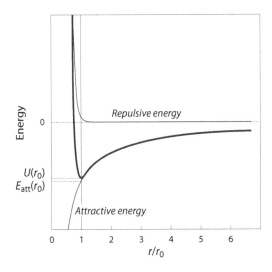

Figure 8.1: Potential energy between two atoms as a function of their distance, expressed in units of the equilibrium distance r_0 of these two atoms (thick line)

In this chapter, we describe the main bonding forces. For each of them, we first show the properties of the corresponding arrangements (such as a definite number of first neighbors, or not; well-defined bond directions, or not...). Then we indicate the order of magnitude of the bonding energy. We recall that the bonding energy of two atoms forming a molecule is the difference in energy between the isolated atoms (*i.e.* when they are located at infinite distance from each other) and the atoms gathered into a molecule. In a general way, the bonding energy, also called cohesive energy, of a group of atoms forming a molecule or forming a crystal is the difference in energy between these atoms when isolated and the atoms in the form of the molecule or crystal. Finally, we give for each bond type a few examples of simple structures involving this type of force. We will mostly restrict this description to crystals containing one or two types of atoms.

8.2. Ionic bonds

8.2.1. Nature and properties

An ion is an atom which lost or gained electrons, so that its outermost electronic shell is complete. Its electronic structure is then usually that of a rare gas.

Example: the Na^+ and Cl^- ions. Their electronic configuration, together with that of the non-ionized atoms, is shown in Table 8.1.

Table 8.1: Electronic configuration of sodium and chlorine atoms and ions

Electronic state → Element or ion	$1s$	$2s$	$2p$	$3s$	$3p$
Na	2	2	6	1	
Na^+	2	2	6		
Cl	2	2	6	2	5
Cl^-	2	2	6	2	6

Ionic crystals are made up of positive and negative ions, and the Coulomb interaction is dominant. Because the crystal is neutral, charge compensation must occur, on the smallest scale possible, which leads to an alternation of positive and negative ions. Each ion is attracted by its nearest neighbors and repelled by its second neighbors.

X-ray diffraction allows to determine accurately not only the positions of the centers of the atoms and thus their distance, but also the distribution $\rho(r)$ of the electronic charge density. The distribution obtained, for example along the axis connecting the Na^+ and the Cl^- ions, in a crystal of NaCl is shown on Figure 8.2b, deduced from Figure 8.2a.

We see that $\rho(r)$ practically goes to zero between the ions (the minimum is less than one thousandth of the maximum density at the center of the ions). This justifies the approach below, where the electrostatic interaction between Na^+ and Cl^- is treated as that of $+q$ and $-q$ point charges (where q is the electron charge), localized at the center of the ions. In fact, the measurement shown on Figure 8.2 can be performed on many ionic crystals, which are different combinations of ions. The results of these accurate measurements lead to the following rule: to a rather good approximation, each species can be described as a "hard sphere" (meaning that it does not deform when it comes to contact), with a well-defined radius called the ionic radius.

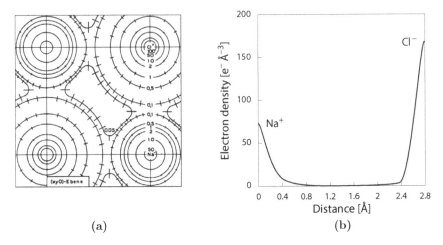

(a)   (b)

Figure 8.2: Electron density in a crystal of NaCl in the (*hk*0) plane:
(a) iso-electron-density curves, after G. Schoknecht [1] (reproduced by
kind permission of Verlag der Zeitschrift für Naturforschung, Tübingen,
Germany); (b) variation in the electron density along the line connecting
a Na^+ ion to its nearest neighbor Cl^-, as deduced from (a). The distance
between the Na^+ and Cl^- ions is 2.815 Å.

The approximation involved in the determination of ionic radii has two origins:

1. The minimum in electron density between the two ions is rather flat, and
    there is therefore some freedom of choice in determining where either of
    the sphere ends.

2. The bonds in the solids considered as ionic are not always purely ionic,
    which can lead to slight variations from one solid to another. The first set
    of ionic radii were published by Goldschmidt *et al.* [2] and Pauling [3], and
    their values are slightly different. Table 8.2 shows the values of some ionic
    radii as determined by Shannon and Prewitt [4]. They showed that the
    values of these ionic radii depend on the coordination number (the number
    of nearest neighbors) of these ions. The values indicated in the table are
    those for coordination number 6.

The consistency of these values can be tested by comparing, as in Table 8.3, the
anion–cation distances derived from these values and the distances measured
in crystals with the NaCl structure, where these ions have coordination 6.

In Table 8.2, the ions on a given row have the same electronic configuration
(the same number Z of electrons) as the rare gas shown on the same row. The
charge of the nucleus increases from left to right, and consequently the attrac-
tion by the nucleus increases, which explains that the ionic radius decreases.
In a given column, the ions have the same charge and, from top to bottom,

the charge of the nucleus increases, which tends to bring the electrons closer to the center, but the increase in the number of electrons tends to increase the volume of the ion.

Table 8.2: Values, in Å, of some ionic radii in crystals, after [4]

Ionic charge	2−	1−	Rare gas	1+	2+	3+
			He (Z = 2)	Li 0.74	Be 0.35	
	O 1.40	F 1.33	Ne (Z = 10)	Na 1.02	Mg 0.72	Al 0.53
	S 1.84	Cl 1.81	A (Z = 18)	K 1.38	Ca 1.00	Sc 0.73
	Se 1.98	Br 1.96	Kr (Z = 36)	Rb 1.49	Sr 1.16	Y 0.90
	Te 2.21	I 2.20	Xe (Z = 54)	Cs 1.70	Ba 1.36	La 1.06

Table 8.3: Comparison of the values of the anion–cation distances calculated from Table 8.2 and measured by X-ray diffraction (these values are from [5])

Compound	Calculated distance	Measured distance
LiF	2.07	2.01
LiCl	2.55	2.565
LiBr	2.70	2.75
LiI	2.94	3.00
NaF	2.35	2.31
NaCl	2.83	2.815
NaBr	2.98	2.985
NaI	3.22	3.215
KF	2.71	2.67
KCl	3.19	3.145
KBr	3.34	3.30
KI	3.58	3.53

The competition between these two effects explains why the ionic radius increases relatively little with atomic number Z. From Li^+ to Cs^+ the number of electrons grows from 2 to 54, a 27-fold increase (*i.e.* 3^3) while the ionic radius is multiplied, not by 3, but only by 2.3.

Note: the forces involved are central, and thus independent of the bond direction, so that the various bonds of a given atom have no preferred relative orientation. The consequence is that the number of bonds between an atom and its first neighbors can have any value, but is restricted by steric effects.

The structure of ionic crystals is thus characterized by

– the alternation of positive and negative ions,

– maximum compacity consistent with the respective ionic radii.

8.2.2. Bonding energy

Consider again the NaCl crystal, known to have a face-centered structure (Fig. 8.3b). The Coulomb interaction energy between two ions Na^+ and Cl^- located a distance r from each other is given by $E_{Coul} = -e^2/r$ where $e^2 = q^2/4\pi\varepsilon_0$, q being the electron charge and ε_0 the electrical permittivity of vacuum. The bonding energy of a Na^+ ion in a NaCl crystal can be calculated using the actual environment of the ion. We see on Figure 8.3 that a Na^+ ion is surrounded by 6 Cl^- ions at a distance r_0 (first neighbors), by 12 Na^+ ions (second neighbors) at distance $r_0\sqrt{2}$, by 8 Cl^- ions at distance $r_0\sqrt{3}$, etc.

The total Coulomb energy of a Na^+ ion is therefore:

$$E_{Coul} = -\frac{e^2}{r_0}\left(6 - \frac{12}{\sqrt{2}} + \frac{8}{\sqrt{3}} - \ldots\right) = -\alpha\frac{e^2}{r_0}$$

where α is called the Madelung constant of NaCl. The series which must be summed to obtain α is an alternating, slowly converging series. Computing it accurately requires special approaches. For NaCl (and for any other face-centered cubic structure of the AX type, where the ions A and X are monovalent) the value of α is 1.748, so that

$$E_{Coul} = -1.748\frac{e^2}{r_0}.$$

For other structures, other values of α are obtained. We note that the same result would be obtained by considering a Cl^- ion and its neighbors. The value of r_0 is obtained from the lattice parameter a determined by X-rays, *viz.* $a = 5.63$ Å for the cubic cell of NaCl and $r_0 = a/2 = 2.815$ Å. The value deduced for E_{Coul} is -8.92 eV, hence an attractive energy of 8.92 eV per ion. If we take into account the repulsive energy, we obtain 7.89 eV. The binding energy measured from the heat of sublimation is 7.86 eV, in very good agreement considering the simplicity of the model.

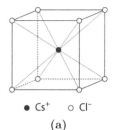
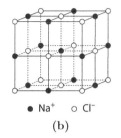
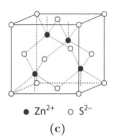

● Cs⁺   ○ Cl⁻   ● Na⁺   ○ Cl⁻   ● Zn²⁺   ○ S²⁻

(a)   (b)   (c)

Figure 8.3: Examples of structures of ionic crystals with formula AX: (a) cesium chloride CsCl, primitive cubic; (b) sodium chloride NaCl (rocksalt), fcc; (c) zinc sulfide ZnS, fcc. In each of these structures, the motif consists of one anion–cation neighbor couple.

8.2.3. Ionic structures with formula AX

The simplest and most frequent of these are the CsCl, NaCl and ZnS-type structures, shown on Figure 8.3.

The CsCl structure is primitive cubic (Fig. 8.3a). The motif consists of a Cs^+ ion and a Cl^- ion. The Cl^- ions are at the cube apices and the Cs^+ ions at the cube centers. It can just as well be described by assuming the Cs^+ ions are at the apices and the Cl^- ions at the centers of the cubes.

The NaCl (rocksalt) structure is face-centered cubic, the motif consisting of one Na^+ and one Cl^- ion (Fig. 8.3b). The Cl^- ions are at the apices of the cube and at the face centers, and the Na^+ ions at the middle of the sides and one at the center of the cube. The structure can also be described by replacing the Cl^- ions by the Na^+ ions and vice versa.

In the zinc blende (or sphalerite) structure,[1] ZnS is face-centered cubic (motif consisting of one Zn^{2+} ion and one S^{2-} ion), with the S^{2-} ions at the vertices of the cubes and at the face centers (Fig. 8.3c). If the cube is thought to be divided into eight small cubes with half the side, we see that the Zn^{2+} ions are at the centers of every other of these cubes. As above, the structure can also be described by substituting the S^{2-} ions for the Zn^{2+} ions and vice versa.

We will attempt to "predict" these simple structures with AX formula using the maximum compacity criterion. We saw that the motif is one AX unit. We will therefore consider that the ions of one type (say the negative A ions, and we will see this is reasonable), with radius r_-, form a compact lattice of ions (simple cubic first, then face-centered cubic), and determine the size of the voids in these structures where the X ions, with radius r_+, could be located. We call a the parameter of the cubic unit cell.

---

1. It is necessary to specify "blende structure" because ZnS can also crystallize in another form, wurtzite (Exercise 8.3).

Primitive cubic lattice

The ions located at the cube apices touch each other and their radius r_- is equal to $a/2$. The ion lattice thus formed exhibits voids centered at the centers of the cubes. The radius r_+ of the sphere which can be placed at the center of a cube while touching the ions with radius r_- is such that $2r_+ = a\sqrt{3} - a$. Hence the ratio of ionic radii should be:

$$\frac{r_+}{r_-} = \sqrt{3} - 1 = 0.73.$$

The positive ion, at the center of the cube, has eight neighbors at the same distance: it has coordination number 8, and the same applies for the negative ions.

Face-centered cubic lattice

The ions touch each other along the face diagonals, and $r_- = a\sqrt{2}/4$. There are two kinds of voids:

1. At the center of the cube: This is called an octahedral site because the figure formed by the six nearest neighbors located at the face centers is an octahedron, the regular polyhedron with eight faces. We would easily see, by considering the whole lattice, that the same type of site exists at the middle of the sides. Such a face-centered cubic lattice, where[2] one type of ions is at the lattice nodes and the other type of ions at the center of the cube and at the middle of the sides, describes the NaCl structure (Fig. 8.3b). The Na^+ and Cl^- ions have coordination number 6.

2. At the centers of the eight cubes with side $a/2$ into which the cube with side a can be divided. These are called tetrahedral sites because the figure formed by the nearest neighbors is a regular tetrahedron. In such a structure, there are twice as many tetrahedral voids as there are atoms at the apices of the cube and at the face centers, and the AX structure uses only every other void. This is the ZnS structure.

A simple geometrical calculation provides the radius ratios for the two kinds of voids:

$$\frac{r_+}{r_-} = 0.41 \qquad \text{for the octahedral site (coordination number 6)},$$

$$\frac{r_+}{r_-} = 0.22 \qquad \text{for the tetrahedral site (coordination number 4)}.$$

Let us now see how the various voids are filled depending on the value of the r_+/r_- ratio. If $r_+/r_- = 0.73$, the negative ions are at the vertices of the cubes and the positive ions at the centers of the cubes: this is the CsCl structure (Fig. 8.3a).

---

2. This description implies that we chose to have a lattice node coincide with one type of ion.

The more r_+/r_- exceeds the value 0.73, the more the positive ion moves the negative ions apart. If r_+/r_- is smaller than 0.73, the positive ion floats about in the void, and the negative ions rearrange into a face-centered cubic lattice where the positive ion sits in the octahedral void, which it enlarges by moving the negative ions apart (NaCl structure on Fig. 8.3b). If $r_+/r_- = 0.41$, the negative ions touch one another. The more r_+/r_- exceeds the value 0.41 (while remaining < 0.73), the more the positive ion pushes the negative ions apart. When r_+/r_- is less than 0.41, the positive ion floats about in the octahedral cavity and the positive ions move to the tetrahedral sites (ZnS-blende structure of Fig. 8.3c), moving apart the negative ions, which come to contact only when r_+/r_- further decreases to the value 0.22. To summarize, we obtain the following rule: if $r_+/r_- > 0.73$, CsCl structure; if $0.41 < r_+/r_- < 0.73$, NaCl structure; if $0.22 < r_+/r_- < 0.41$, ZnS (blende) structure.

This rule is quite well obeyed (Table 8.4). The deviations originate mainly from the fact that the bonds are but rarely purely ionic, in contrast with the starting assumption. There can be for example a covalent contribution, which leads to a tendency for the external electrons not to be distributed with spherical symmetry, so that the starting hypothesis of a spherical ion is no more perfectly satisfied.

Table 8.4: r_+/r_- ratios for some crystals with various cubic structures

CsCl type	r_+/r_-
CsCl	0.94
CsBr	0.87
CsI	0.77

ZnS type	r_+/r_-
AgI	0.52
ZnS	0.40
BeS	0.19

NaCl type	r_+/r_-
KF	1.04
CaO	0.71
KBr	0.70
LiF	0.56
NaCl	0.56
CaS	0.54
MgO	0.51
LiBr	0.38

8.2.4. Other cubic structures

It appears useful to introduce a few other ionic structures, considered as "classic" because many crystals share one or the other of these structures, which are so to speak canonical structures. We chose fluorite CaF_2, rutile TiO_2,[3] and the perovskite structure $CaTiO_3$.

---

3. TiO_2 also exists in two other forms: anatase (tetragonal) and brookite (orthorhombic).

The CaF_2 structure is face-centered cubic (Fig. 8.4a). The Ca^{2+} ions occupy the apices and the face centers of the cube, and the F^- ions the eight tetrahedral sites. This structure can also be described as a set of cubes, with sides half the side of the unit cell, where the apices are occupied by a fluorine ion and the center of every other cube by a Ca^{2+} ion. The ratio of the ionic radii of Ca^{2+} and F^- is equal to 0.74, in good agreement with the coordination number 8 corresponding to a positive ion at the center of a cube whose apices are occupied by a negative ion as for CsCl. However here, since there are twice as many F^- ions as Ca^{2+} ions, the Ca^{2+} ions are at the centers of every other cube only. The result is a face-centered cubic unit cell.

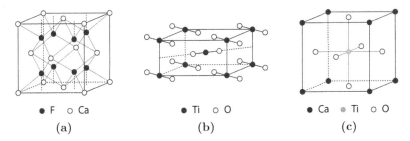

• F	o Ca		• Ti	o O		• Ca	• Ti	o O
(a)			(b)			(c)		

Figure 8.4: Three examples of structures representative of many ionic crystal structures: (a) fluorite CaF_2; (b) rutile TiO_2; (c) $CaTiO_3$, the perovskite structure

Rutile TiO_2 has a tetragonal primitive unit cell (Fig. 8.4b). The Ti atoms are located at the vertices and at the center of the unit cell. Each of the apical Ti atoms is surrounded by two O atoms at the same height, and the Ti atom and its two O neighbors are aligned parallel to one of the diagonals of the square base of the unit cell. The Ti atom at the center of the unit cell also has two O atoms at the same height as itself, but they are aligned parallel to the other diagonal of the square.

The perovskite structure, the model of which is $CaTiO_3$ (Fig. 8.4c), is primitive cubic with Ca atoms at the apices of the cube, O atoms at the face centers, and a Ti atom at the center of the cube. The structure of $CaTiO_3$ actually is slightly distorted with respect to this cubic structure. Many substances have this type of pseudo-cubic unit cell, and they often exhibit a high-temperature phase which is exactly cubic. A good example is barium titanate $BaTiO_3$, which is cubic at high temperature $(T > 120°C)$, and undergoes, when temperature decreases, various transitions toward other structures: first tetragonal, then orthorhombic and finally rhombohedral (Exercise 12.7).

8.3. Covalent bonds

8.3.1. Nature of covalent bonds

In an elementary presentation, we can say that covalent bonding occurs between two atoms when they share a pair of electrons. In contrast to the ionic bonds, the electron density is no more zero between the atoms.

The nature of this bond can only be explained through quantum mechanics. To give a taste of the argument, take the simplest example, that of the H_2^+ ion consisting of two protons and an electron. It can be shown that the wave function of the electron can be written as the combination of the atomic wave functions or "orbitals" Ψ_1 and Ψ_2 of each of the hydrogen atoms: $\Psi^\pm = A_\pm(\Psi_1 \pm \Psi_2)$. The probability density of the electron between the nuclei is non-zero for the Ψ^+ combination (called the bonding solution) while that associated to the Ψ^- combination (called the antibonding solution) is zero. The energies E^+ and E^- associated to these two states vary as a function of the distance ρ between the protons as shown on Figure 8.5. The energy E^+ is always smaller than E^-, and its minimum value $-E_0$ leads to a cohesive energy equal to E_0. Each state of the bonding orbital can be occupied by two electrons with opposite spins (Pauli's principle). In the ground state of the neutral hydrogen molecule (H_2), the electrons of both atoms occupy the bonding state, and they have opposite spins. This is the prototype of the covalent bond, consisting of two electrons with opposite spins, shared between the two bound atoms. If the external electron shell of an atom is several electrons short of being full, this atom can share electrons with several neighboring atoms and thus form several covalent bonds. Each covalent bond then requires the participation of one orbital per bound atom.

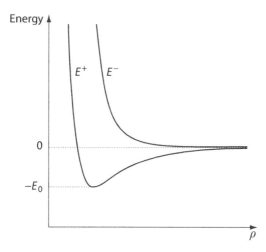

Figure 8.5: Schematic representation of the potential energy of an H_2^+ ion as a function of the distance ρ between the hydrogen nuclei for the bonding orbital, E^+, and the anti-bonding orbital, E^-

8.3.2. Basic property

It can be shown that a covalent bond is the more stable the larger the overlap between the wave functions of the two electrons that take part in the bond. The electronic states of isolated atoms do not all have the spherical symmetry of s states. The p states, with 3-fold orbital degeneracy (the total degeneracy, with spin included, is then 6-fold) have wave functions which point along the three axes of an orthogonal referential. There is thus a strong overlap of the p orbitals with the orbitals of the neighboring atoms if the latter are located along these axes.

Consider the example of the nitrogen–hydrogen bond in the NH_3 molecule. The electronic structure of the nitrogen atom includes two $1s$ electrons, two $2s$ electrons and three $2p$ electrons. The external $2p$ shell of the nitrogen atom lacks three electrons to be complete. The three p orbitals of the nitrogen atom each form a covalent bond with the spherical $1s$ orbital of one of the hydrogen atoms. These bonds are almost at right angles to one another. Actually, the angle between two bonds is $107°$, and the molecule is a pyramid, the base of which is an equilateral triangle and the apical angles are equal to $107°$. The N atom is at the apex of the pyramid and the H atoms at the apices of the equilateral base.

In general, the principle of maximum overlap of wave functions leads to the covalent bonds having preferential directions.

8.3.3. Examples

The following examples are monoatomic crystals. The bonds therefore connect atoms of the same type.

Diamond

It consists of carbon atoms. The electronic structure of the outermost shell of carbon ($n = 2$) consists of two $2s$ electrons and two $2p$ electrons. It lacks four electrons to be full. The four electrons of the $n = 2$ shell form hybrid orbitals consisting of linear combinations of the s, p_x, p_y and p_z orbitals. These orbitals, called sp_3, are oriented along the directions that connect the center of a regular tetrahedron to its four apices. In a diamond crystal, each carbon atom is thus connected to four neighbors by tetrahedral bonds (Fig. 8.6). The diamond crystal can be considered as a giant molecule, with practically infinite extent.

Silicon and germanium crystals have the same structure as diamond. Their outermost shell ($n = 3$ for Si and $n = 4$ for Ge) also includes 4 electrons (two s electrons and two p electrons).

Generally, consider atoms in which the outermost shell would have 8 electrons if it were saturated. If this shell actually contains N electrons, the external shell will be saturated if $8 - N$ covalent bonds are formed. Each bond then uses one of the N electrons of the neighboring atom (an atom of the same type in the monoatomic crystals to which we restrict discussion), and this requires $N \geq 8 - N$, hence $N \geq 4$. The following paragraphs successively discuss a case where $N = 5$, then a case with $N = 6$. An example with $N = 7$ is treated in Section 8.4.2.

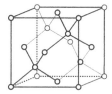

Figure 8.6: Crystal structure of diamond

Arsenic

The outermost electronic shell ($n = 4$) contains two $4s$ electrons and three $4p$ electrons, hence a total of five electrons. Three pseudo-plane bonds are then formed, and the pseudo-planes can be described as corrugated planes. An atom located in the zero height plane is connected to three atoms at height 1.37 Å as shown on Figure 8.7, which represents the projection of these atoms onto the plane at height zero. The arrows indicate the direction of increasing height on the bonds. The angle between bonds is 94.1° (instead of 120° as expected for coplanar bonds). The next corrugated plane is at height 3.516 Å and its projection is displaced with respect to the figure by vector **AB**. The "corrugated" planes are bound to one another by Van der Waals bonds (see next section).

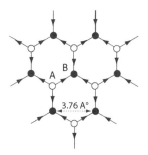

Figure 8.7: Arsenic crystal; structure of a pseudo-plane of atoms. The atoms represented by empty circles are at height zero and those represented by full circles at height 1.37 Å.

Selenium

The outermost electronic shell ($n = 4$) contains six electrons (two $4s$ electrons and four $4p$ electrons), and each Se atom must therefore share two bonds. They are at an angle of 103.1° to each other. The "infinite molecule" is a chain (Fig. 8.8). The positions of the atoms in one chain are deduced from one another by the symmetry elements of a 3_1 screw axis. These helix-shaped chains are stacked parallel to one another and they are bound to one another by Van der Waals forces.

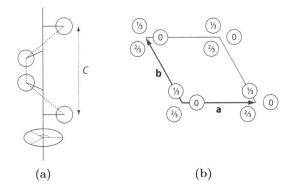

(a)   (b)

Figure 8.8: Selenium crystal: (a) structure of a chain of atoms and its 3_1 screw axis; (b) hexagonal unit cell projected on its basal plane, with the height of the atoms expressed in units of the lattice parameter c

8.3.4. Conclusion

1. Covalent bonds are oriented: this is their main characteristic.

2. The distance between two given atoms engaged in a covalent bond is well defined. For example, the C–C distance in a diamond crystal is equal to that between carbons in the single bond of ethane (C_2H_6), 1.54 Å.

3. The order of magnitude of the energy of these bonds is a few eV. Examples are given in Table 8.5.

Table 8.5: Energy of some covalent bonds. The C–C bond considered is a single bond.

	H–H	C–C	Si–Si	Ge–Ge
Energy (in eV)	4.5	3.6	2.3	1.9

4. There are many solids where the bonds are not totally covalent but partly ionic: they are referred to as iono-covalent bonds. The atoms engaged in these bonds always fill their outermost electronic shell.

8.4. Van der Waals bonds, or molecular bonds

8.4.1. Nature and properties

The simplest example is that of the rare gases, consisting of neutral atoms with their outermost electronic shell full. Although the atoms are neutral, the Van der Waals force is electrostatic in origin. These atoms do not have a permanent electric dipole moment $(\overline{\mathbf{P}(t)} = \mathbf{0})$ but the dipole moment oscillates around its average zero value. At a given time $\mathbf{P}(t)$ is therefore non-zero, and it creates an electric field at the position of the neighboring atom, which it polarizes. The interaction energy between the two atoms can be calculated. It is found to be negative, with absolute value proportional to $1/r^6$, whence an attractive force varying as $1/r^7$.

The order of magnitude of the Van der Waals type bond energies is 0.1 eV.

There is no restriction on the angles between bonds, which may have any orientation.

Van der Waals radii: They can be determined from the contacts between identical atoms bound by Van der Waals forces. Then, knowing a number of these radii, we can determine the others by examining the distances between atoms of different types bound by a Van der Waals force.

8.4.2. Example

Van der Waals bonds appear in all crystals where the elementary building block is a molecule. A very simple example is the chlorine crystal (Fig. 8.9).

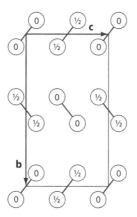

Figure 8.9: Orthorhombic crystal of chlorine: projection of the atoms onto a plane perpendicular to basis vector **a**. The lines connecting two chlorine atoms represent the covalent bonds.

The outermost shell of the chlorine atom has 7 electrons, and the atom can covalently bond only to one neighbor to form a Cl_2 molecule. The crystal consists of Cl_2 molecules connected by Van der Waals forces. This is called a "molecular crystal".

The molecule is extremely simple in this case, but in general the molecules can contain a great number of atoms and have complicated, irregular shapes. Their most compact stacking is not easily determined, all the more since other factors can play a part, for example local charges which create a local electric field in their vicinity.

8.5. Metallic bonds

8.5.1. Nature and properties

In a pure metal, the atoms are all identical and the bond cannot be ionic. Also, the number of electrons in the outermost shell of a metal atom is smaller than 4, and the bond cannot be covalent. The nature of the metallic bond can be understood starting from the crystal structure (as evidenced by X-ray diffraction), and describing quantum mechanically the behavior of an electron in the periodic potential of the metal ions that form the crystal lattice. It can then be shown, for the electrons in a crystal, that there is a set of energy levels forming a continuous band called the conduction band. The electrons in the conduction band behave almost like free electrons, and they also provide the bonding between all the atoms.

The crystal structures observed are well explained by a model of hard spheres with compact (close-packed) stacking. The radius of the spheres is calculated as half the distance between neighboring atoms, as deduced from the crystal structure determined by X-ray diffraction. The radius of the metallic ions thus determined is always larger than the radius of the corresponding ion in an ionic crystal. For example the radius of the Na^+ ion is equal to 1 Å while the radius of Na metal is 1.86 Å. The interpretation is that, in NaCl for example, the Na^+ ions are attracted by the Cl^- ions and therefore get as close as they can to them. In metallic Na, the Na^+ ions are repelled, cohesion being due to the conduction electrons.

The cohesive energy of a metal is of the order of 1 eV per atom (except for the transition metals).

8.5.2. Examples

The maximal compacity of a stacking of identical hard spheres is obtained for the face-centered cubic and close-packed hexagonal structures.

Consider two spheres with radius a, touching each other. A third sphere contacting these two spheres will form with the two first ones an equilateral triangle. A plane of contacting spheres, where the centers are designated as A, form a lattice of plane hexagonal unit cells (Fig. 8.10a), and the lines connecting the centers of the spheres form a set of equilateral triangles. A sphere will position above this plane in the most compact way possible if its center projects at the center of any one of these equilateral triangles. Since there are twice as many centers of equilateral triangles as there are spheres in the compact starting plane, the contacting spheres in the second plane will project on every other of these two sites, which we call B. We will call C sites those which project on the other centers of equilateral triangles. The spheres of the third compact plane can now project either on the C sites or on the A sites. In the first case the crystal will consist of the succession of planes ABCABCABC..., while in the second case the succession will be ABABABAB... It is easy to show (Exercise 8.1) that the atoms corresponding to succession ABCABC... form a face-centered cubic lattice (the motif consisting of a single atom). The planes A, B and C are parallel to the (111) planes of this structure. In the case of the ABABAB... sequence, the atoms form a hexagonal lattice. The **a** and **b** vectors of the hexagonal lattice are the basis vectors of the plane hexagonal starting lattice. Vector **c**, perpendicular to **a** and **b**, has for its length the distance between two successive A (or B) planes. A simple geometrical calculation (Exercise 8.2) shows that the c/a ratio is then $2\sqrt{2/3} = 1.63$.

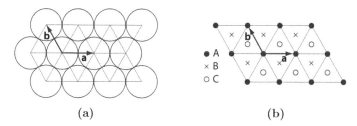

(a)	(b)

Figure 8.10: Compact stacking of spheres: (a) plane of contacting spheres; (b) projection onto this plane of the atom positions in the neighboring planes

Table 8.6 shows that most metal structures are indeed either face-centered cubic or hexagonal, but in the latter case with more or less compacity. However there are also body-centered cubic (I) structures.

Some metals are polymorphic, *i.e.* they feature various structures depending on temperature: this is the case for iron, where γ Fe has an F lattice and α Fe an I lattice. It turns out, when calculated, that the ionic radii are in this case close to each other.

Table 8.6: Structure and lattice parameters of some metals

Face-centered cubic lattice		Body-centered cubic lattice		Hexagonal lattice			
Element	a (Å)	Element	a (Å)	Element	a (Å)	c (Å)	c/a
Ag	4.086	Cr	2.884	Be	2.287	3.583	1.566
Al	4.050	Fe α	2.8665	Co α	2.507	4.069	1.623
Au	4.078	Li	3.509	Cr	2.722	4.427	1.626
Co	3.548	Mo	3.147	Mg	3.209	5.210	1.624
Cu	3.615	Na	4.291	Ni	2.65	4.33	1.634
Feγ	3.591	Nb	3.300	Ti	2.95	4.686	1.588
Ni	3.524	Zr	3.62	Zn	2.665	4.947	1.856

8.6. Remarks and conclusions

We described the various types of bonds between atoms in a crystal. However, bonds are not always of a perfectly well-defined type. In particular, covalent and ionic bonds are only the extreme terms of a set of bonds featuring both an ionic and a covalent character.

Also, several types of bonds can appear together in a crystal: we saw examples of coexistence of covalent binding and Van der Waals binding in crystals of arsenic, selenium and chlorine.

To conclude, we give two examples of simple crystals exhibiting strong anisotropy because they consist of plane layers bound to one another by weak Van der Waals bonds while the binding within the layers have mixed character and higher energy by at least an order of magnitude.

Graphite crystal

It consists of carbon atoms each of which is connected to three other carbon atoms by covalent bonds (sp_2 hybridization between the s, p_x and p_y orbitals), within planes with hexagonal structure (Fig. 8.11). The fourth valence electrons, with orbital p_z, of two neighboring atoms bind to each other through a mainly covalent bond. This bond is partly metallic and is delocalized along the plane. This partly metallic character explains the non-negligible electrical conductivity measured along these planes (see Sect. 9.3.1). The inter-plane binding is Van der Waals type.

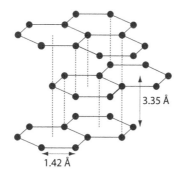

Figure 8.11: Structure of graphite

GeS$_2$ crystal

The layers consist of a set of tetrahedra. The center of each tetrahedron is occupied by a germanium atom, and the apices by sulfur atoms. Within a layer, the Ge atoms form a plane, and the tetrahedra are connected to one another through their apices. Each sulfur atom at the apex of a tetrahedron is thus connected to two Ge atoms. The Ge–S bonds are iono-covalent, and the layers are bound to one another by Van der Waals forces.

8.7. Exercises

Exercise 8.1

Lattice for an ABCABC close-packed stacking.

Comparison of Figures 8.10b and 3.14c shows that the spheres in an ABCABC stacking form a rhombohedral lattice. Show that this rhombohedral lattice is a face-centered cubic lattice.

Exercise 8.2

Lattice for an ABABAB close-packed stacking

1. In this stacking the centers of a sphere in plane B and of its three contacting spheres in plane A form a regular tetrahedron with side a equal to the length of the basis vector **a** of the hexagonal unit cell. Determine, as a function of a, the distance h between planes A and B.

2. Calculate the c/a ratio for the hexagonal unit cell of the ABABAB lattice.

Exercise 8.3

Tetrahedral sites

In the close-packed stacking modes ABCABC and ABABAB, one atom in a given layer and its three nearest neighbors in the adjacent layer form a regular tetrahedron.

1. Calculate the distance d between the center of this tetrahedron and each of its apices.

2. A spherical atom (or ion), with radius r_1, is placed at the center of this tetrahedron so that it contacts the atoms of the close-packed stacking with radius r_2. Calculate the ratio r_1/r_2.

3. Wurtzite is a crystal form of ZnS different from blende. Its lattice is hexagonal, with $a = 3.811$ Å, $c = 6.234$ Å. The atom positions in the unit cell are:

$$
\begin{array}{llll}
\text{Zn} & 0,\, 0,\, 0 & \frac{1}{3},\, \frac{2}{3},\, \frac{1}{2} & \\
\text{S} & 0,\, 0,\, u & \frac{1}{3},\, \frac{2}{3},\, \frac{1}{2} + u & \quad\text{and}\quad u \approx 0.375.
\end{array}
$$

Show that the sulfur atoms are practically in the tetrahedral sites of a structure in which the centers of the Zn atoms form an ABABAB stacking.

Compare this to the blende structure described in Section 8.2.3.

Complement 8C. Magnetic structures

Understanding the physics of magnetic materials requires a knowledge not only of their crystal structure but also of their magnetic structure. The crystal structure indicates the shape and dimensions of the unit cell, the positions of the various individual ions or atoms, and the extent of their thermal vibration. The magnetic structure describes the size and orientation of the magnetic moment carried by some of the ions.

Materials are often designated as magnetic when they react strongly to an applied field. This usually means that they contain permanent magnetic moments which are, below a transition temperature, ordered. The simplest magnetic materials are ferromagnets: below T_C, the Curie temperature, all the magnetic moments are, within a magnetic domain,[4] parallel (Fig. 8C.1a).

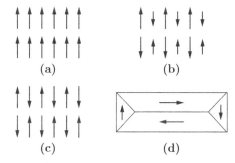

Figure 8C.1: Schematic representation of the magnetic moment arrangements, on a microscopic scale, in ferromagnets (a), ferrimagnets (b) and antiferromagnets (c). In these figures, each arrow represents an atomic magnetic moment. Figure (d) shows, on a macroscopic scale, an arrangement of magnetic domains in a ferromagnet: each arrow then represents the direction of the spontaneous magnetization:[5] the scale of this figure could be anywhere between micrometers and millimeters.

---

4. A magnetic domain is a macroscopic region of the sample where the orientation of the spontaneous magnetization, hence of the magnetic moments, is uniform. There are always at least two opposite (time-reversed) directions for the magnetic moments, as in uniaxial materials. The number of equivalent domains is set by the ratio of the magnetic point symmetry of the material above and below the Curie temperature, and it depends on what crystallographic direction the moments spontaneously take. For example, when the material is cubic above the Curie temperature and the spontaneous magnetization directions are <100>, as in slightly silicon-alloyed iron, there are 6 families of magnetic domains (spontaneous magnetization along ±[100], ±[010], ±[001]) (Fig. 8C.1d). The magnetization process, under the influence of an applied magnetic field, starts with the growth of the favorably oriented domains.

5. The magnetization is the magnetic moment per unit volume. It is analogous to the polarization in dielectrics, discussed in Chapters 9 and 15.

Examples of ferromagnets are iron, cobalt, nickel and many of their alloys or compounds. Another important class of magnetic material comprises the ferrimagnets (Fig. 8C.1b), in which the magnetic moments of one type of ions are "up", and those of another type of ions, with a different value of the moment, are "down", yielding a resulting non-zero (spontaneous) magnetization at the scale of a domain. Examples are the ferrites used in high-frequency transformer cores, or as permanent magnets. The order feature is shared by some materials which do not react strongly to a magnetic field and are therefore not obviously attractive for applications. These have, below the ordering temperature (the Néel temperature T_N), as many "up" as "down" magnetic moments with the same modulus, and therefore zero overall spontaneous magnetization. These materials, the simpler examples of which are NiO, CoO and MnO, or MnF_2, CoF_2 and NiF_2, are called antiferromagnets (Fig. 8C.1c).

The prime tool for investigating the magnetic structure of materials is *neutron diffraction*. Neutrons (see Complement 4C) interact with the nuclei (the nuclear interaction) and, since they carry a magnetic moment, they also sense magnetic fields. In the wavelength range used for diffraction by the microscopic structure of materials, on the order of 10^{-10} m, they are diffracted by the microscopic-scale variation of the magnetic fields in matter, and the probability of magnetic scattering happens to be of the same order of magnitude as for nuclear scattering. From this information, the distribution of magnetic moments that gives rise to these fields can be determined. For example, in some simple antiferromagnets (CoO, NiO or MnO), the magnetic period, in the direction along which the moments alternate, is double that of the chemical structure, because chemically identical ions have alternately up and down orientations. The neutron diffraction pattern will therefore feature new, purely "magnetic" reciprocal lattice nodes and diffraction peaks, which vanish above the Néel temperature.

Just as the symmetry of the crystal structure is described by one of the 230 space groups, the magnetic symmetry is described by one of the 1651 magnetic groups. The 32 crystallographic point groups are replaced by 122 magnetic point groups. The central addition to the toolbox of structure symmetry elements is the operation called "time reversal". The notion under this provocative term is actually simple. Microscopic magnetic moments are represented by axial vectors (see Sect. 10.6). They are due to the equivalent of current loops, *i.e.* of circulating electric charge. Reversing the magnetic moment is tantamount to reversing the velocity of the charges, as if a movie were being played backwards, or if time were flowing backwards. Thus the magnetic symmetry involves operations such as m' (mirror combined with reversal of the magnetic moment, which can be described as time reversal), and the magnetic Bravais lattice description can feature notations such as I_p, meaning anti-centering, *i.e.* the combination of displacement by half the unit cell diagonal and magnetic moment reversal.

Actually, X-rays also interact, albeit weakly, with the magnetic moments. With synchrotron radiation X-rays, valuable complementary information can be extracted on magnetic structures: magnetic X-ray scattering can reveal the two contributions (spin and orbital) to the magnetic moment, whereas magnetic neutron diffraction is sensitive only, via the magnetic field it creates, to the total magnetic moment.

Further reading

D.H. Ryan, *Neutron Powder Diffraction Studies of Magnetic Materials*. Powerpoint document, prepared for the 2009 summer course of the Canadian Institute for Nuclear Scattering; available on the web at www.cins.ca/docs/SS09 (click on DHR_CNBC_June09.ppt)

M. de Graef, *Teaching crystallographic and magnetic point group symmetry using three-dimensional rendered visualizations*, International Union of Crystallography, Teaching Pamphlets;
http://www.iucr.org/_data/assets/pdf_file/0006/13929/final_23.pdf

J. Rodríguez-Carvajal and F. Bourée, Symmetry and magnetic structures, in *Contribution of Symmetries in Condensed Matter*, ed. by B. Grenier, V. Simonet and H. Schober. Proceedings of spring school held at Presqu'île de Giens, France, in May 2009. EDP Web of Conferences, vol. 22 (2012). Available on the web, at
http://dx.doi.org/10.1051/epjconf/20122200010

References

[1] G. Schoknecht, *Z. Naturforsch.* **12A**, 983 (1957)

[2] V.M. Goldschmidt, T. Barth, G. Lunde, W.H. Zachariasen, *Skr. Norske Vidensk. Akad. Skrifter I Mat. Naturv. Kl.* **2**, 1 (1926)

[3] L. Pauling, *J. Am. Chem. Soc.* **49**, 765 (1927)

[4] R.D. Shannon, C.T. Prewitt, *Acta Cryst.* **B25**, 925 (1969)

[5] W.G. Wyckoff, *Crystal Structures*, vol. 1 (Wiley, 1963)

Chapter 9

Crystal anisotropy and tensors

This chapter shows why the anisotropy of crystals requires a representation of their physical properties by tensors, and defines them. It then shows how the symmetry of a given crystal (its point group) influences the form of these material tensors. Curie's and Neumann's principles and their application to the tensor properties of crystals are presented in Section 9.5.

9.1. Introduction

The physical properties of crystals belong to either of two broad categories: intrinsic and extrinsic properties.

The intrinsic physical properties are identical in various samples of a given material because they depend on the very nature of the material: its chemical composition, the nature of the bonds between atoms and their structural arrangement. Examples of intrinsic properties are:

– thermal expansion and elastic properties, which reveal on a macroscopic scale the forces between neighboring atoms,

– the type of electrical conductivity (insulating or conducting or semiconductor), which is related to the nature of the chemical bonds between atoms. A good example is carbon, an insulator when it crystallizes in the form of diamond, while its other (polymorphic) form, graphite, is a conductor.

The extrinsic properties are related to the existence of defects in the crystal, and they can therefore vary extensively between samples. Two examples are:

– plastic deformation: here, in contrast to elastic deformation, the sample does not return to its initial state when the force which created the strain is removed. Plastic deformation is related to the existence and multiplication of linear defects, called dislocation lines (or, for short: dislocations), whose initial number and nature vary from sample to sample depending on the growth process and on the previous deformation. A brief introduction to dislocations is given in Complement 13C.

© Springer Science+Business Media Dordrecht 2014
C. Malgrange et al., *Symmetry and Physical Properties of Crystals*,
DOI 10.1007/978-94-017-8993-6_9

– the value of the electrical conductivity of a semiconductor, which depends on the density of impurities (which means, when it is deliberate, the extent of doping) in the sample.

In this book, we only consider intrinsic physical properties of crystals. In the first part, we saw that crystals are formed from the periodic repetition of the crystal motif, at the microscopic scale, along three non-coplanar directions. A direct consequence of this periodicity is the anisotropy of the physical properties of crystals, which depend on the direction in which a constraint is applied and on the direction in which the resulting effect is measured. The description of this anisotropy involves special mathematical objects, called tensors. This is why this second part of the book starts with two general chapters which analyze the notion of tensor, describe the general properties of tensors, and show how the symmetry of an individual crystal makes it possible to reduce the number of independent coefficients of any tensor representing one of its intrinsic physical properties. The following chapters describe specific physical properties, mainly mechanical and optical.

9.2. Anisotropic continuous medium

The physical properties discussed in this second part of the book are macroscopic properties. In describing them, the crystal is considered as a continuous and anisotropic medium. To satisfy the assumption of continuity, we must assume that any elementary volume V fulfills two conditions. It must contain enough unit cells that it may be considered as homogeneous, but it also must be small enough that the variations of an external applied field within this volume be negligible. When these external fields are sine waves (electric field $\mathbf{E}$ or magnetic field $\mathbf{B}\ldots$), the characteristic length is their wavelength λ. The elementary volume V must therefore have dimensions very much smaller than this wavelength. The two above conditions can be summarized by the double inequality:

$$a \ll \sqrt[3]{V} \ll \lambda$$

where a is the largest one among the parameters of the crystal unit cell.

This means that, if the wavelength of the applied fields is much larger than the dimensions of the unit cell, the variations of the field within the elementary volume can be neglected, and the crystal can be viewed as a continuous medium. Furthermore, the cell parameters can be considered as infinitely small with respect to the relevant volume, and the same applies even more to fractions of these parameters. It is therefore no more justified to distinguish between pure and screw rotation axes, or between pure and glide mirrors, since the translations associated to these symmetry elements are fractions of a lattice parameter (see Sect. 7.2.1 and 7.2.2). In other words, the effect of the crystal's symmetry on its macroscopic properties will be described by its *point group* and not by its space group.

Conversely, if the wavelength of the field applied to the crystal is on the same order of magnitude as its lattice parameter (X-ray range), the crystal can no more be considered as continuous. Its discrete structure must be taken into account, and indeed such probes can be used to investigate the crystal structure and thus the space group of the crystal.

Our discussion of the physical properties will involve either static constraints (mechanical properties), or periodic constraints with wavelengths larger by several orders of magnitude than the lattice parameters (acoustic waves and optical electromagnetic waves). The crystal can thus be considered as an anisotropic continuous medium, and crystal symmetry affects the physical properties through the point group of the crystal.

9.3. Representing a physical quantity by a tensor

Some physical quantities are independent of any notion of orientation: this is the case for temperature or the mass per unit volume of a material. They are expressed by a scalar. Others are expressed by a vector, for example an electric field or the polarization of a crystal under the influence of this field.

Consider two vector-type physical quantities, $\mathbf{A}$ and $\mathbf{B}$, related by a linear law. In a gas or liquid, this relation is written as $\mathbf{A} = \gamma\mathbf{B}$ where γ is a scalar. $\mathbf{A}$ and $\mathbf{B}$ are then parallel. In a crystal, $\mathbf{A}$ is generally not parallel to $\mathbf{B}$, and a tensor must be introduced to describe the physical quantity γ. This will be analyzed in detail on the basis of an example in the next section (Sect. 9.3.1).

Tensors, in the same way as vectors which, as we will see, are just a special form of tensors, are defined by their components in a given system of axes. It is simpler, and this will be done throughout this book, to select for this use not the set of crystal axes (basis $\{\mathbf{a}, \mathbf{b}, \mathbf{c}\}$), but an *orthonormal set of axes* $(Ox_1x_2x_3)$ (basis $\{\mathbf{e}_1, \mathbf{e}_2, \mathbf{e}_3\}$), usually chosen as close as possible to the crystal axes. For example, in the case of the crystal systems with an orthogonal unit cell (orthorhombic, tetragonal, cubic), the *orthonormal set of axes* $(Ox_1x_2x_3)$ in which we express the tensor is chosen parallel to the vectors $\mathbf{a}$, $\mathbf{b}$, $\mathbf{c}$ that define the unit cell. For crystals with a hexagonal unit cell, one of the axes is chosen parallel to the 6-fold axis of the lattice, etc.

9.3.1. Example: electrical conductivity

Electrical conductivity is a good example for understanding the effect of crystal anisotropy on the physical properties of a crystal. An electric field $\mathbf{E}$ applied to a conductor induces a current density $\mathbf{j}$ within the material. The motion of free charges is then characterized by the electrical conductivity σ through

$$\mathbf{j} = \sigma\,\mathbf{E} \tag{9.1}$$

where σ is a scalar if the material is isotropic.

To illustrate this relation, imagine we are performing conductivity measurements on two very different samples of graphite: a compacted graphite powder, and a single crystal of graphite. The crystal structure of graphite, as described in Chapter 8 (Fig. 8.11), consists of a stacking of equidistant dense planes, with hexagonal symmetry, called graphitic planes, connected by Van der Waals bonds.

In the first experiment, performed on compacted powder, we apply an electric field $\mathbf{E}$ in any direction. The current density turns out to be parallel to $\mathbf{E}$ whatever the direction of the applied field $\mathbf{E}$, and proportional to the norm of $\mathbf{E}$. The conductivity σ is then a scalar. In this case, the powder behaves with respect to the applied electric field like an isotropic material. The powder consists of many small crystals, randomly oriented relative to one another, and thus the influence of structural anisotropy on the measured physical property disappears.

Now consider the single crystal of graphite. Due to the special structure of this material, very different values of the current density will be obtained depending on the orientation of the applied electric field. A strong current density will be measured if the field is applied along a direction in the graphitic planes, and a weak current density will be obtained with the field perpendicular to the graphitic planes (Sect. 8.6). The ratio between these values varies between 10^2 and 10^5 depending on crystal quality [1]. If the applied field has a general direction, the current density has no reason to be parallel to the field, and the three components of $\mathbf{j}$ can be expressed as a function of those of $\mathbf{E}$ through the linear relations:

$$
\begin{aligned}
j_1 &= \sigma_{11}E_1 + \sigma_{12}E_2 + \sigma_{13}E_3 \\
j_2 &= \sigma_{21}E_1 + \sigma_{22}E_2 + \sigma_{23}E_3 \\
j_3 &= \sigma_{31}E_1 + \sigma_{32}E_2 + \sigma_{33}E_3,
\end{aligned}
\tag{9.2}
$$

where j_i and E_i with $i = 1, 2, 3$ are the components of $\mathbf{j}$ and $\mathbf{E}$ respectively, in an orthonormal set of axes $(Ox_1x_2x_3)$. These three relations can also be written

$$
j_i = \sum_{k=1}^{3} \sigma_{ik}E_k
$$

or, using Einstein's condensed notation, also referred to as *Einstein's convention*:

$$
j_i = \sigma_{ik}E_k.
$$

This convention consists in suppressing the sum sign, which becomes implicit, before a product of factors when the sum is performed over a subscript that appears twice in the product. Here the summation is performed over subscript k (called the dummy subscript, or dummy index), which appears in σ_{ik} and in E_k. The subscript i, which is not summed over, is called the free subscript, and it must appear on both sides of the equal sign. We will always use this notation in what follows.

Equations (9.2) can also be written in the following matrix form:

$$
\begin{pmatrix} j_1 \\ j_2 \\ j_3 \end{pmatrix} = \begin{pmatrix} \sigma_{11} & \sigma_{12} & \sigma_{13} \\ \sigma_{21} & \sigma_{22} & \sigma_{23} \\ \sigma_{31} & \sigma_{32} & \sigma_{33} \end{pmatrix} \begin{pmatrix} E_1 \\ E_2 \\ E_3 \end{pmatrix}.
$$

Electrical conductivity clearly appears as a table of $3 \times 3 = 9$ components of a quantity described as a tensor quantity. The nine numbers that define it depend on the axis system chosen to express the components of vectors $\mathbf{E}$ and $\mathbf{j}$. It is therefore essential to understand how this table transforms when the vectors $\mathbf{E}$ and $\mathbf{j}$ are expressed in another orthonormal coordinate system. In other words, we must determine the relation that makes it possible to go over from table σ_{ik}, obtained when the vectors $\mathbf{E}$ and $\mathbf{j}$ are expressed in a given orthonormal frame, to the table σ'_{ik} obtained when the same vectors are expressed in another orthonormal frame.

9.3.2. A refresher on orthonormal frame changes

Let R be a frame defined by the orthonormal basis $\{\mathbf{e}_1, \mathbf{e}_2, \mathbf{e}_3\}$ and R' a new frame defined by the orthonormal basis $\{\mathbf{e}'_1, \mathbf{e}'_2, \mathbf{e}'_3\}$. The new basis vectors can be expressed as a function of the old ones through the relations:

$$
\mathbf{e}'_i = a_{ij}\mathbf{e}_j \tag{9.3}
$$

where we use Einstein's convention as defined in the section above: summation over the subscript j which appears twice is implicit.

We immediately see that:

$$
a_{ij} = \mathbf{e}'_i \cdot \mathbf{e}_j. \tag{9.4}
$$

Equation (9.3) can be represented through the table below:

	$\mathbf{e}_1$	$\mathbf{e}_2$	$\mathbf{e}_3$
$\mathbf{e}'_1$	a_{11}	a_{12}	a_{13}
$\mathbf{e}'_2$	a_{21}	a_{22}	a_{23}
$\mathbf{e}'_3$	a_{31}	a_{32}	a_{33}

In this table $a_{kl} = \mathbf{e}'_k \cdot \mathbf{e}_l$. We note that $a_{ji} = \mathbf{e}'_j \cdot \mathbf{e}_i$ is the projection of $\mathbf{e}_i$ onto $\mathbf{e}'_j$, so that:

$$
\mathbf{e}_i = a_{ji}\mathbf{e}'_j. \tag{9.5}
$$

A vector $\mathbf{x}$ can be expressed in basis $\{\mathbf{e}_1, \mathbf{e}_2, \mathbf{e}_3\}$ through relation:

$$\mathbf{x} = x_1\mathbf{e}_1 + x_2\mathbf{e}_2 + x_3\mathbf{e}_3 \qquad \text{or} \qquad \mathbf{x} = x_i\mathbf{e}_i \tag{9.6}$$

and in basis $\{\mathbf{e}'_1, \mathbf{e}'_2, \mathbf{e}'_3\}$ through

$$\mathbf{x} = x'_1\mathbf{e}'_1 + x'_2\mathbf{e}'_2 + x'_3\mathbf{e}'_3 \qquad \text{or} \qquad \mathbf{x} = x'_j\mathbf{e}'_j. \tag{9.7}$$

Taking relation (9.5) into (9.6), we obtain $\mathbf{x} = x_i a_{ji}\mathbf{e}'_j$ and, comparing with (9.7), we obtain

$$x'_j = a_{ji}x_i. \tag{9.8}$$

In the same way, taking (9.3) into (9.7) and comparing with (9.6), we obtain

$$x_j = a_{ij}x'_i. \tag{9.9}$$

Transformation from one orthonormal frame to another orthonormal frame leads to the two following relations, (9.10) and (9.11):

$$a_{ih}a_{ik} = \delta_{hk} \tag{9.10}$$

where δ_{hk} is Kronecker's symbol, equal to 1 if $h = k$ and to 0 if $h \neq k$. The proof is that

$$\mathbf{e}_h \cdot \mathbf{e}_k = \delta_{hk} = a_{ih}\mathbf{e}'_i a_{jk}\mathbf{e}'_j = a_{ih}a_{jk}\delta_{ij} = a_{ih}a_{ik}.$$

In the same way, using the fact that $\mathbf{e}'_h \cdot \mathbf{e}'_k = \delta_{hk}$, we obtain

$$a_{hi}a_{ki} = \delta_{hk}. \tag{9.11}$$

Let us designate as A the matrix $\{a_{ij}\}$, where i is the subscript for the row and j for the column. The i-th row of this matrix consists of the components of the new basis vector $\mathbf{e}'_i$ in terms of the old basis vectors.

This matrix A is called the transformation matrix for going over from the orthonormal coordinate system R, with basis $\{\mathbf{e}_1, \mathbf{e}_2, \mathbf{e}_3\}$, to the orthonormal system R' with basis $\{\mathbf{e}'_1, \mathbf{e}'_2, \mathbf{e}'_3\}$.

Two simple examples of transformation matrices are given below:

1. The new frame R' is the symmetric of R with respect to a plane perpendicular to axis Ox_2 (Fig. 9.1a). In this case, the vectors $\mathbf{e}'_1$, $\mathbf{e}'_2$, $\mathbf{e}'_3$ are given by

$$\mathbf{e}'_1 = \mathbf{e}_1, \qquad \mathbf{e}'_2 = -\mathbf{e}_2, \qquad \mathbf{e}'_3 = \mathbf{e}_3$$

and the transformation matrix for this change of frame is given by

$$A = \begin{pmatrix} 1 & 0 & 0 \\ 0 & -1 & 0 \\ 0 & 0 & 1 \end{pmatrix}.$$

2. The new frame R' is deduced from frame R through a rotation by angle φ around axis Ox_2 (Fig. 9.1b). In this case, the vectors $\mathbf{e}'_1$, $\mathbf{e}'_2$, $\mathbf{e}'_3$ are given by:

$$\mathbf{e}'_1 = \cos\varphi \, \mathbf{e}_1 - \sin\varphi \, \mathbf{e}_3, \qquad \mathbf{e}'_2 = \mathbf{e}_2, \qquad \mathbf{e}'_3 = \sin\varphi \, \mathbf{e}_1 + \cos\varphi \, \mathbf{e}_3$$

and the transformation matrix for this change of frame is given by:

$$A = \begin{pmatrix} \cos\varphi & 0 & -\sin\varphi \\ 0 & 1 & 0 \\ \sin\varphi & 0 & \cos\varphi \end{pmatrix}.$$

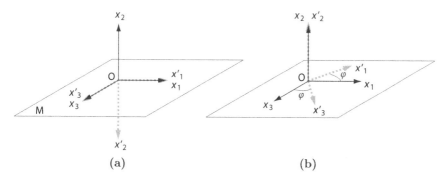

Figure 9.1: Change in coordinate system. The new frame R' $(Ox'_1x'_2x'_3)$, drawn in dotted lines, is obtained from the initial frame R $(Ox_1x_2x_3)$, drawn in solid lines, through: (a) symmetry with respect to a plane M perpendicular to axis Ox_2, (b) rotation by angle φ around axis Ox_2.

9.3.3. Application to electrical conductivity

In frame R $\{\mathbf{e}_1, \mathbf{e}_2, \mathbf{e}_3\}$, the components j_k of $\mathbf{j}$ can be written as a function of the components E_l of $\mathbf{E}$:

$$j_k = \sigma_{kl} E_l.$$

We want to express the coordinates j'_i of $\mathbf{j}$ in frame R' as a function of the coordinates E'_j of $\mathbf{E}$ in this frame.

Using relations (9.8) and (9.9), we obtain:

$$j'_i = a_{ik} j_k = a_{ik} \sigma_{kl} E_l = a_{ik} \sigma_{kl} a_{jl} E'_j.$$

The relation we want is $j'_i = \sigma'_{ij} E'_j$. Therefore

$$\sigma'_{ij} = a_{ik} a_{jl} \sigma_{kl}. \tag{9.12}$$

We note here for the last time that Einstein's convention implies summing over the subscripts k and l which appear twice in the product.

Let $\mathbf{x}$ and $\mathbf{y}$ be two vectors with coordinates respectively x_i and y_i in basis $\{\mathbf{e}_1,\mathbf{e}_2,\mathbf{e}_3\}$. Their coordinates x_i' and y_i' in basis $\{\mathbf{e}_1', \mathbf{e}_2', \mathbf{e}_3'\}$ are given by

$$x_i' = a_{ik}x_k,$$
$$y_j' = a_{jl}y_l.$$

The product of two coordinates x_i' and y_j' takes the form

$$x_i'y_j' = a_{ik}a_{jl}x_ky_l,$$

a relation similar to that of Equation (9.12), connecting components σ_{ij}' and σ_{ij}. Thus, a coefficient σ_{ij} transforms in a change of axes like the product of two vector components.

9.4. Tensors

9.4.1. Definition of a tensor

The last remark above allows us to understand the following definition of tensors:

A tensor of rank n is (when three-dimensional space is considered) a set of 3^n components, characterized by n subscripts, which transform in the change from one orthonormal frame to another orthonormal frame like the product of n vector components.

For example, a tensor $[T]$ of rank 2 (also designated as a second-rank tensor), is a set of 9 coefficients T_{ij} defined in an orthonormal frame R, and called the components of the tensor. In another orthonormal frame R', deduced from R through the transformation matrix $\{a_{ij}\}$, they become the components T_{ij}' such that

$$T_{ij}' = a_{ik}a_{jl}T_{kl} \tag{9.13a}$$

and conversely:

$$T_{ij} = a_{ki}a_{lj}T_{kl}'. \tag{9.13b}$$

The tensor as an entity is denoted by a letter, set between square brackets, *e.g.* $[T]$.

In the same way, a third-rank tensor $[T]$ has 27 components T_{ijk} which transform in the same frame change according to relation

$$T_{ijk}' = a_{il}a_{jm}a_{kn}T_{lmn} \tag{9.14a}$$

and conversely:

$$T_{ijk} = a_{li}a_{mj}a_{nk}T_{lmn}'. \tag{9.14b}$$

Again in the same way, a fourth-rank tensor $[T]$ transforms in a change of frame according to

$$T'_{ijkl} = a_{im}a_{jn}a_{kp}a_{lq}T_{mnpq} \qquad (9.15a)$$

and conversely:

$$T_{ijkl} = a_{mi}a_{nj}a_{pk}a_{ql}T'_{mnpq}. \qquad (9.15b)$$

We deduce immediately that a scalar is a tensor of rank 0 and that a vector is a tensor of rank 1.

In practice, ascertaining the tensor character of a quantity defined by 3^n components implies checking that, in a change of frame, the components transform like the product of n components of vectors.

The above definition restricts the general definition of tensors (see *e.g.* the books by Sands [2] or by Schwartz [3]) to the case of orthonormal frames.[1] It makes the presentation of tensor physical properties of crystals much simpler while retaining the physics of the phenomena. In practice, problems of crystal physics are almost invariably treated with orthonormal frames.

9.4.2. An important property

The derivation used to find how the components of the electrical conductivity transform in a change of frame depends only on the vector nature of the quantities $\mathbf{E}$ and $\mathbf{j}$ and not on their physical nature. We can thus conclude that:

If two vector physical quantities $\mathbf{A}$ and $\mathbf{B}$, with components A_i and B_j in a given orthonormal axis system, are related by $B_i = T_{ij}A_j$, then the physical quantity represented by the 9 coefficients T_{ij} is a tensor. With such a relation, we can state that T_{ij} is a second-rank tensor $[T]$, and we can write $\mathbf{B} = [T]\mathbf{A}$. We use the notation between square brackets to emphasize the tensor nature of the coefficients T_{ij}.

More generally, it can be shown that

A physical quantity relating a tensor physical quantity of rank p (which will be called the cause) to another tensor physical quantity of rank q (called the effect) is represented by a tensor of rank n = p + q.

Let us demonstrate this using an example.

Suppose relation $T_i = U_{ikl}V_{kl}$ connects, in an orthonormal axis system $(Ox_1x_2x_3)$, tensor cause $[V]$, with rank 2, to tensor effect $\mathbf{T}$ (a rank-one tensor, hence actually a vector) through the coefficients U_{ikl}. In another orthonormal axis system $(Ox'_1x'_2x'_3)$, deduced from the former through the transformation

---

1. Using non-orthonormal axes requires the use of covariant and contravariant components, which we do not find necessary here.

matrix $\{a_{ij}\}$, the components of $\mathbf{T}$ and $[V]$ are T_i' and V_{kl}', related to the former ones through:

$$T_m' = a_{mi}T_i, \qquad V_{kl} = a_{nk}a_{pl}V_{np}'$$

whence we deduce:

$$T_m' = a_{mi}T_i = a_{mi}U_{ikl}V_{kl} = a_{mi}U_{ikl}a_{nk}a_{pl}V_{np}'$$

so that

$$
\begin{aligned}
T_m' &= a_{mi}a_{nk}a_{pl}U_{ikl}V_{np}' \\
&= U_{mnp}'V_{np}'
\end{aligned}
$$

with $U_{mnp}' = a_{mi}a_{nk}a_{pl}U_{ikl}$. This relation shows that U_{ikl} is a tensor of rank $3 = 2+1$.

The method we used applies whatever the ranks p and q of the cause and effect tensors are.

In what follows, we will frequently use this property to assert that a set of coefficients indeed represents a tensor.

We note that the relative electrical permittivity $[\varepsilon]$ of a material is a second-rank tensor since it relates the induction $\mathbf{D}$ in the material to the electric field $\mathbf{E}$, i.e.

$$\mathbf{D} = \varepsilon_0[\varepsilon]\mathbf{E} \text{ or } D_i = \varepsilon_0\varepsilon_{ij}E_j$$

where ε_0 is the permittivity of vacuum.

9.4.3. Field tensors and material tensors

Let us return to the example of electrical conductivity

$$j_i = \sigma_{ij}E_j.$$

This relation connects a cause, here the electric field $\mathbf{E}$, to an effect, the current density $\mathbf{j}$, through the electrical conductivity tensor $[\sigma]$.

The electric field $\mathbf{E}$ is an external action applied to the material, and it can be varied at will. It is a field tensor. The same applies to the current density, which varies as $\mathbf{E}$ varies. On the other hand, the conductivity $[\sigma]$ does not vary when the field $\mathbf{E}$ varies. It characterizes a physical property of the crystal, the electrical conductivity: it is called a material tensor because it characterizes the material. Material tensors represent physical properties of the material. Their form largely depends on the isotropic or anisotropic character of the material. We saw, for example, that the electrical conductivity of a material becomes a scalar when the material is isotropic.

Whatever their rank, the number of independent coefficients necessary to define material tensors depends on the crystal symmetry of the material, as we will see in Sections 9.5.2 and 9.6.

In contrast, field tensors are defined in the same way for isotropic and anisotropic materials. We just saw the examples of the electric field and of the electrical current density, which are rank-1 tensor. We will define in Chapter 11 the stress tensor and in Chapter 12 the strain tensor: they both are second-rank field tensors. In Chapter 13, we will see that these two second-rank field tensors are related, in a material, by a material tensor of rank 4 which caracterizes the elastic properties of the material.

9.5. Symmetry properties of tensors

9.5.1. Internal symmetry. Symmetric and antisymmetric tensors

A tensor of rank n is symmetric with respect to two of its subscripts if the following relation applies:

$$T_{ab...kl...n} = T_{ab...lk...n}.$$

A tensor of rank n is antisymmetric with respect to two of its subscripts if the following relation applies:

$$T_{ab...kl...n} = -T_{ab...lk...n}.$$

Most material tensors have intrinsic symmetry (and, more rarely, antisymmetry) properties which result from thermodynamic considerations, independent of the nature of the crystal. They are called internal symmetry properties.

On the other hand, some field tensors also feature an internal symmetry, like the stress tensor whose symmetry is related to the balance condition for the solid to which it is applied (Sect. 11.4).

9.5.2. External symmetry of material tensors. Curie's and Neumann's principles

The influence of crystal symmetry on material tensors can be described as a direct consequence of Curie's principle. This principle was stated in 1894 by Pierre and Jacques Curie when studying piezoelectricity, which they had discovered in 1880. It is expressed as follows:

When certain causes produce certain effects, the symmetry elements of the causes must be present in the effects. In the same way, when some effects reveal some dissymetry, this dissymetry must be present in the causes which produce them.

Consider once more an electric field $\mathbf{E}$ applied to a material, and the current density $\mathbf{j}$ which results because of the electrical conductivity of the material $[\sigma]$: $\mathbf{j} = [\sigma]\mathbf{E}$. The causes here are the electric field (a polar vector with symmetry ∞ m as shown by Fig. 9.2) and the crystal, with point group G_p (we saw in Sect. 9.2 that the group to be considered in discussing macroscopic properties is the point group of the crystal). The effect is the current density (a polar vector with symmetry ∞ m). Thus the symmetry of the causes is the intersection ∞ m $\cap$ G_p of the two point groups, which must be less than or equal to the symmetry of the current density ∞ m.

The symmetry of the causes here depends on the symmetry of the crystal and on the orientation of the electric field with respect to the symmetry elements of the crystal. In the most general case, when the electric field has a general direction with respect to the symmetry elements of the crystal, there is no symmetry element common to the crystal and the electric field. The symmetry ∞ m of the current density is then larger than the intersection ∞ m $\cap$ G_p, and nothing special can be stated. The interesting cases are those for which there is an intersection between ∞ m and G_p. For example, if G_p contains a q-fold axis and if the electric field is parallel to the q-fold axis, then the intersection of the symmetry elements is the q-fold axis. The effect, which must be at least as symmetric, will therefore be a current density parallel to the q-fold axis. Carrying this argument further, and using the properties of rank-2 tensors which we will see in the next chapter, precise information can be gained on the symmetry of the conductivity tensor in relation to the symmetry of the crystal.

Curie's principle is very general. It applies to all areas of physics. For crystallographic applications, one may prefer Neumann's principle, which is less general and was laid down earlier (1885). It is expressed as follows:

The symmetry elements of any macroscopic physical property of a crystal must include the symmetry elements of the point group of the crystal.

Expressed differently, the symmetry group of any physical property of the crystal includes that of the crystal. Thus the point symmetry group of the crystal is either identical to the symmetry group of the physical property, or it is a subgroup of the latter.

Consider for an illustration the example of pyroelectricity. Some crystals, called pyroelectrics (see Chap. 15), exhibit a spontaneous electric polarization **P**. This quantity is difficult to evidence, because the distribution of electric charges which it entails are rapidly neutralized by free charges in air or by migration of charges within the crystal. However, since this polarization varies with temperature, a rapid variation of temperature ΔT entails a rapid variation, Δ**P**, of the polarization, which in turns produces a current which is easy to detect. Δ**P** and ΔT are related by:

$$\Delta \mathbf{P} = \mathbf{p}\Delta T$$

where **p** is a vector (rank-1 tensor) which represents the property of pyroelectricity in the material. The symmetry group of a polar vector is ∞ m (Fig. 9.2). Therefore, according to Neumann's principle, the point group of a pyroelectric crystal must be a subgroup of this group. The only possibilities are the groups:

$$1 \quad 2 \quad 3 \quad 4 \quad 6$$

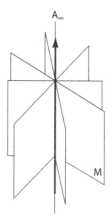

Figure 9.2: Symmetry elements of a polar vector: an A_∞ axis parallel to the vector, and an infinity of mirrors containing this axis

and those obtained by adding to each of these groups a mirror that goes through the n-fold axis, hence the groups

$$m \qquad mm2 \qquad 3m \qquad 4mm \qquad 6mm$$

thus in total 10 point groups. Crystals with one of these 10 point groups can feature pyroelectricity. It is clear that, if the crystal belongs to one of these groups, the spontaneous polarization will be parallel to axis A_n for all the groups that have one. For group m, it will be parallel to the mirror and for group 1, its direction is not predetermined by symmetry.

We just applied Neumann's principle directly to a vector physical property for which the symmetry (∞ m) is known. It can also be applied to a symmetric second-rank tensor, whose symmetry is that of an ellipsoid, mmm, as we will see in Section 10.5. For higher-rank tensors, it is convenient to use an equivalent form of Neumann's principle: if a crystal is invariant under a given symmetry operation, all of its physical properties are invariant under this symmetry operation. Imagine the following experiment. We first measure a physical property in a given reference system for a given position of the crystal. Then we submit the crystal to one of its point symmetry operations, and we again measure the physical property in the same reference system. The result obtained must be identical to the former one.

Since it is equivalent to perform the symmetry operation on the crystal or the inverse operation (necessarily also a member of the group) on the coordinate system, we will retain the second definition. Thus:

To determine the number of independent components of a tensor representing a physical property of a given crystal, it is sufficient to express the invariance of the tensor when all the symmetry operations of the crystal's point group are applied to the coordinate system.

It is clearly sufficient to consider only those symmetry operations which are generating all the elements of the point group.

9.6. Reduction in the number of independent coefficients of a material tensor

We present two methods for reducing the number of components of a tensor, both using the invariance property stated above.

9.6.1. Method using the transformation matrix

We apply the very definition of a tensor based on the transformation of its components in a change of reference frame (Sect. 9.4.1), *viz.*

$$T'_{ijk...} = a_{il}a_{jm}a_{kn}...T_{lmn...}$$

The $T_{ijk...}$ are the coefficients of a tensor expressed in a first, general, orthonormal frame and the $T'_{ijk...}$ are the coefficients of this same tensor expressed in another frame, deduced from the former by a symmetry operation of the crystal's point group. The a_{ij} are the coefficients of the transformation matrix from the first frame to the second.

The invariance of the components of the tensor under this special change in frame implies that

$$T'_{ijk...} = T_{ijk...}.$$

As an example, we consider a physical property represented by a second-rank tensor, and examine in turn the case of crystals with point groups 2, m and 2/m.

If the point group is 2, we can choose the axis system so that axis Ox_2 is parallel to the two-fold axis. A rotation by π around axis Ox_2 changes the frame R $\{e_1, e_2, e_3\}$ into a frame R' $\{e'_1, e'_2, e'_3\}$ such that

$$e'_1 = -e_1, \qquad e'_2 = e_2, \qquad e'_3 = -e_3.$$

The only non-zero coefficients of the frame change matrix A are thus:

$$a_{11} = -1, \qquad a_{22} = 1 \qquad \text{and} \qquad a_{33} = -1$$

and we deduce that in this transformation

$$
\begin{array}{lll}
T'_{11} = T_{11} & T'_{22} = T_{22} & T'_{33} = T_{33} \\
T'_{12} = -T_{12} & T'_{13} = T_{13} & T'_{23} = -T_{23} \\
T'_{21} = -T_{21} & T'_{31} = T_{31} & T'_{32} = -T_{32}.
\end{array}
\tag{9.16}
$$

Furthermore, we must have

$$T'_{ij} = T_{ij}. \tag{9.17}$$

We thus deduce for instance that $T_{12} = 0$ since $T'_{12} = -T_{12}$ from (9.16) and $T'_{12} = T_{12}$ from (9.17). In the same way, $T_{21} = 0$, $T_{23} = 0$ and $T_{32} = 0$, and the tensor finally is:

$$\begin{pmatrix} T_{11} & 0 & T_{13} \\ 0 & T_{22} & 0 \\ T_{31} & 0 & T_{33} \end{pmatrix} \tag{9.18}$$

For this representation of the tensor, we use the classical notation of matrices: the first subscript designates the row and the second one the column.

If the point group is m, we can choose an axis system such that axis Ox_2 is perpendicular to the mirror. This mirror operation changes frame R into a frame R' such that

$$\mathbf{e}'_1 = \mathbf{e}_1 \qquad \mathbf{e}'_2 = -\mathbf{e}_2 \qquad \mathbf{e}'_3 = \mathbf{e}_3.$$

The non-zero coefficients in the transformation matrix are thus $a_{11} = 1$, $a_{22} = -1$ and $a_{33} = 1$. We easily see that we obtain the same tensor form as above in (9.18).

Thus the presence of a two-fold axis or of a mirror perpendicular to this axis reduces the number of components of a second-rank tensor in the same way. The form given in (9.18) is therefore that for a second-rank tensor representing a physical property of a crystal belonging to any one of the point groups 2, m, 2/m (monoclinic groups) when the axis Ox_2 is chosen parallel to the 2-fold axis or perpendicular to the mirror.

9.6.2. "Direct inspection" method

This method, due to Fumi [4], uses another aspect of the definition of tensors, *viz.* the fact that the components of a tensor transform in an axis change like a product of coordinates. We thus consider directly the products of coordinates. Let us apply this method to the example above.

A rotation by π of the coordinate system around axis Ox_2 changes the coordinates x_i of a point M into coordinates x'_j such that

$$x'_1 = -x_1, \qquad x'_2 = x_2, \qquad x'_3 = -x_3.$$

We immediately deduce that, for a second-rank tensor, only the components that have one subscript, and only one, equal to 2 will be zero since, in these cases, $x'_i x'_j = -x_i x_j$, whence $T'_{ij} = -T_{ij}$ while $T'_{ij} = T_{ij}$.

In the same way, the operation of a mirror perpendicular to Ox_2 changes the coordinates x_i of a point M into coordinates x'_j such that

$$x'_1 = x_1, \qquad x'_2 = -x_2, \qquad x'_3 = x_3.$$

We thus retrieve the preceding results.

This method is easy to apply, and it is the one used in practice for all the symmetry operations for which the new components x'_i are expressed as functions of a single component x_i. This is the case for all symmetry operations associated to 2- and 4-fold axes, to centers of symmetry and to mirrors. For 3- and 6-fold axes, the preceding method, using the transformation matrix from one frame to another, is preferred, except for cubic crystals, where the 3-fold axes are parallel to the $< 111 >$ directions. In the latter case, one chooses an orthonormal coordinate system parallel to the basis vectors of the crystal system. A rotation of the coordinate system by $2\pi/3$ around a $[111]$ direction transforms the coordinates x_i of a point P into coordinates x'_i such that

$$x'_1 = x_2, \qquad x'_2 = x_3, \qquad x'_3 = x_1$$

and the direct inspection method is easy to use.

We note that, in all cases, the reduction in the number of components is performed with an orthonormal axis system in which one axis (or several) is parallel or perpendicular to a symmetry element of the crystal.

9.6.3. Special case: central symmetry (inversion symmetry)

In this case both methods are equally convenient. The transformation matrix is such that $a_{ij} = -\delta_{ij}$, and we obtain, for a rank-n tensor

$$T'_{ijk...} = (-1)^n T_{ijk...} = T_{ijk...} \tag{9.19}$$

1. If n is odd, we conclude that all the tensor's components are zero. Thus an odd-rank tensor representing a physical property of a crystal has non-zero components only if the crystal has no center of symmetry. We will see that piezoelectricity, which is described by a third-rank tensor, exists only in crystals that do not have a center of symmetry.

2. If n is even, the tensor is unchanged by the existence of a center of symmetry, and adding a center of symmetry changes nothing to the number of independent coefficients of the tensor. Hence, for the physical properties represented by even-rank tensors, the number of independent coefficients is the same for all point groups of a Laue class (see Sect. 5.7).

We note that relation (9.19) applied to even-rank tensors shows that a central symmetry operation applied to the coordinate system in which the tensor components are expressed leaves its components invariant. This property can be considered without any reference to a crystal. It is general, and the property obtained is therefore also valid for field tensors. We can thus state this result in the following form: a central symmetry operation applied to any even-rank tensor (be it a field tensor or a material tensor) leaves this tensor invariant.

9.7. Exercises

Exercise 9.1

Consider a symmetrical rank-2 material tensor $[T]$ representing a physical property of a crystal with point group 3. Let T_{ij} be the components of this tensor in an orthonormal axis system $(Ox_1x_2x_3)$ where Ox_3 is parallel to the 3-fold axis.

1. Let $(Ox'_1x'_2x_3)$ be an axis system deduced from $(Ox_1x_2x_3)$ through a rotation by $2\pi/3$ around axis Ox_3. Determine the transformation matrix $\{a_{ij}\}$ from axis system $(Ox_1x_2x_3)$ to axis system $(Ox'_1x'_2x_3)$.

2. Express the components T'_{11}, T'_{12} and T'_{22} of $[T]$ in the axis system $(Ox'_1x'_2x_3)$, and deduce that $T_{11} = T_{22}$ and $T_{12} = 0$.

3. Calculate T'_{13}, T'_{23} and T'_{33}, and show that $T_{13} = T_{23} = 0$ and $T_{33} \neq 0$.

Exercise 9.2

Use the direct inspection method to determine the number of non-zero coefficients for a material rank-2 tensor of a crystal with point group mmm if its components are expressed in a frame with axes perpendicular to the mirrors.

References

[1] J.L. Spain, in *Chemistry and Physics of Carbon*, vol. 8, ed. by P.L. Walker (Marcel Dekker Publisher, New York, 1973), pp. 130–131

[2] D.E. Sands, *Vectors and Tensors in Crystallography* (Dover Publications, Mineola, New York, 1995)

[3] L. Schwartz, *Les tenseurs* (Hermann, Paris, 1975)

[4] F.G. Fumi, Physical properties of crystals: the direct inspection method, *Acta Crystallographica* **5**, 44–48 (1952)

Chapter 10

Second-rank tensors

This chapter is devoted to the study of the characteristic proper-
ties of symmetric tensors of rank 2. They represent many physical
properties which, in isotropic materials, are described by a simple
scalar. The end of this chapter introduces axial vectors, which are
antisymmetric tensors of rank 2, and gives examples.

10.1. Introduction

Second-rank tensors are very important in physics. Because most of them are
symmetric, it is useful to investigate their special properties, and in particular
the quadric which is associated to them (Sect. 10.2, 10.3 and 10.4). Section 10.5
investigates the effect of crystal symmetry on material tensors of rank 2. Section 10.6 defines axial vectors, and shows that they are antisymmetric tensors
of rank 2.

10.1.1. Symmetric and antisymmetric tensors

A second-rank tensor is symmetric if $T_{ij} = T_{ji}$ whatever i and j, and the
matrix which represents it in a general axis system has the form

$$\begin{pmatrix} T_{11} & T_{12} & T_{13} \\ T_{12} & T_{22} & T_{23} \\ T_{13} & T_{23} & T_{33} \end{pmatrix}.$$

There are thus 6 independent coefficients.

A second-rank tensor is antisymmetric if $T_{ij} = -T_{ji}$, hence $T_{11} = T_{22} = T_{33}$
$= 0$, so that, in matrix representation:

$$\begin{pmatrix} 0 & T_{12} & T_{13} \\ -T_{12} & 0 & T_{23} \\ -T_{13} & -T_{23} & 0 \end{pmatrix}.$$

There are thus 3 independent coefficients.

© Springer Science+Business Media Dordrecht 2014
C. Malgrange et al., *Symmetry and Physical Properties of Crystals*,
DOI 10.1007/978-94-017-8993-6_10

Any second-rank tensor $[T]$ can be written as the sum of a symmetric tensor $[S]$ and an antisymmetric tensor $[A]$: $[T] = [S] + [A]$, with

$$S_{ij} = \frac{T_{ij} + T_{ji}}{2}$$

and

$$A_{ij} = \frac{T_{ij} - T_{ji}}{2}.$$

This decomposition is independent of the frame in which tensor $[T]$ is expressed. We demonstrate this for the symmetric part. In another orthonormal axis system $(Ox_1' x_2' x_3')$, the components S_{ij}' of tensor $[S]$ become $S_{ij}' = a_{ik}a_{jl}S_{kl}$ (Eq. (9.13a)). If $S_{kl} = S_{lk}$, it is clear that $S_{ij}' = S_{ji}'$.

10.1.2. Matrix form of second-rank tensors

We saw that it is customary to represent the components of a rank-2 tensor, in a given system of axes, by a matrix, extending to tensors the usual rule for writing matrices. Thus the component T_{kl} of the tensor is at the intersection of the k-th row and the l-th column of matrix T. In another system of axes, deduced from the first one through the transformation matrix A, the new component T_{kl}' takes the form

$$T_{kl}' = a_{ki}a_{lj}T_{ij}.$$

If A^T is the transposed matrix of A, $a_{ij}^T = a_{ji}$ and

$$T_{kl}' = a_{ki}T_{ij}a_{jl}^T$$

so that, in matrix notation

$$T' = ATA^T. \tag{10.1}$$

10.1.3. Trace

The trace of a second-rank tensor $[T]$, represented in a given system of axes, is defined by:

$$\mathrm{Tr}\,[T] = T_{11} + T_{22} + T_{33} = T_{ii}.$$

It is independent of the referential chosen to represent the tensor. The trace of tensor $[T]$ in a new axis system, $\mathrm{Tr}\,([T])'$, is expressed, using relation (9.13a), by:

$$(\mathrm{Tr}\,[T])' = T_{ii}' = a_{ik}a_{il}T_{kl}.$$

From (9.10): $a_{ik}a_{il} = \delta_{kl}$ and

$$(\mathrm{Tr}\,[T])' = \delta_{kl}T_{kl} = T_{kk}.$$

Hence $(\mathrm{Tr}\,[T])' = \mathrm{Tr}\,[T]$.

10.2. Representative quadric for a symmetric rank-2 tensor

10.2.1. Characteristic surface

Consider the following equation:

$$S_{ij}x_ix_j = \pm 1 = \varepsilon \tag{10.2}$$

where, for the moment, the coefficients S_{ij} are general, but satisfy the condition $S_{ij} = S_{ji}$. When expanded, this equation takes the form

$$S_{11}x_1^2 + S_{22}x_2^2 + S_{33}x_3^2 + 2S_{12}x_1x_2 + 2S_{13}x_1x_3 + 2S_{23}x_2x_3 = \varepsilon. \tag{10.3}$$

The locus of the points M with coordinates x_1, x_2, x_3 in an orthonormal frame is a second-degree surface, also called quadric (Q) (an ellipsoid or a hyperboloid), having for its center the coordinate origin O. We want the equation of this quadric in another frame $(Ox_1'x_2'x_3')$ obtained from $(Ox_1x_2x_3)$ through the transformation matrix $A = \{a_{ij}\}$.

From Equation (9.9): $x_i = a_{ki}x_k'$

and Equation (10.2) becomes:

$$S_{ij}a_{ki}x_k'a_{lj}x_l' = \varepsilon$$

which can be written as: $S_{kl}'x_k'x_l' = \varepsilon$ if $S_{kl}' = a_{ki}a_{lj}S_{ij}$.

Thus the coefficients S_{ij} which define the quadric (Q) transform in a change of orthonormal frame like the components of a second-rank tensor.

If the coefficients S_{ij} in Equation (10.2) are the coefficients of a symmetric tensor, the quadric (Q) is a geometrical representation of this tensor. We have thus defined the representation quadric of a symmetric tensor $[S]$.

10.2.2. Principal axes and principal coefficients

The quadrics (ellipsoids or hyperboloids on Fig. 10.1) have as symmetry elements three mutually perpendicular mirrors, three 2-fold axes at the intersections of these mirrors, and a center of symmetry. In an orthonormal set of axes $(Ox_1'x_2'x_3')$ parallel to these 2-fold axes, called the principal axes of the quadric, the equation of the quadric becomes:

$$S_{11}'(x_1')^2 + S_{22}'(x_2')^2 + S_{33}'(x_3')^2 = \varepsilon. \tag{10.4}$$

In this new system of axes, called the principal axes of the tensor, the tensor takes a diagonal form:

$$\cdot \begin{pmatrix} S_{11}' & 0 & 0 \\ 0 & S_{22}' & 0 \\ 0 & 0 & S_{33}' \end{pmatrix}$$

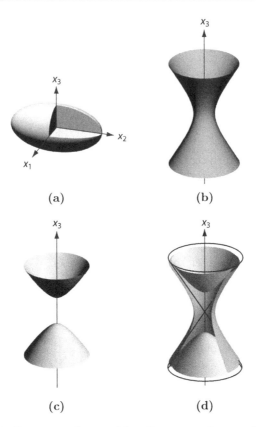

Figure 10.1: Representation quadrics of a symmetric second-rank tensor: (a) an ellipsoid, if the three principal values have the same sign; (b) one-sheeted hyperboloid; (c) two-sheeted hyperboloid; (d) the set of two hyperboloids representing the tensor when one of the principal values has a sign opposite to the two others

and S'_{11}, S'_{22} and S'_{33} are called the principal values or the principal coefficients of tensor $[S]$.

To determine these principal axes (of the quadric and of the tensor), we note that, along these special directions, the normal to the quadric is parallel to the radius vector of the quadric. We know that, if the equation of a surface is $f(x_1, x_2, x_3) = 0$, the components X_i of a vector perpendicular to this surface are $X_i = \partial f/\partial x_i$.

A vector normal to the quadric with Equation (10.3) thus has for its components X_i at a point with coordinates x_j:

$$X_i = 2S_{ij}x_j. \qquad (10.5)$$

It is parallel to the radius vector with components x_i if:

$$S_{ij}x_j = \lambda x_i$$

where λ is any scalar. This system of three equations is just the tool needed to determine the eigenvectors and eigenvalues of matrix $\{S_{ij}\}$. Since this matrix is real and symmetric, it has three real and orthogonal eigenvectors, the eigenvalues of which are real.

We can thus determine the directions of the symmetry axes of the quadric; they coincide with the principal axes of the tensor.

10.2.3. Shape of the quadric

In the frame $(Ox_1' x_2' x_3')$ of its principal axes, the quadric is defined by the three coefficients S_{11}', S_{22}' and S_{33}'. Its geometrical shape depends on the signs of these three coefficients.

If S_{11}', S_{22}' and S_{33}' are positive, we choose $\varepsilon = 1$, and the equation of the quadric can be expressed in the form:

$$\frac{x_1'^2}{a^2} + \frac{x_2'^2}{b^2} + \frac{x_3'^2}{c^2} = 1,$$

the equation of an ellipsoid (Fig. 10.1a) with semi-axes a, b and c such that:

$$a = \frac{1}{\sqrt{S_{11}'}}, \qquad b = \frac{1}{\sqrt{S_{22}'}}, \qquad c = \frac{1}{\sqrt{S_{33}'}}.$$

The principal values of the tensor are thus the reciprocals of the squares of the semi-axes of the ellipsoid.

If S_{11}', S_{22}' and S_{33}' are negative, the associated quadric would be an imaginary ellipsoid if ε was chosen as equal to 1. We choose $\varepsilon = -1$ so that the quadric is an ellipsoid. The principal values of the tensor are negative, and their absolute values are the reciprocals of the squares of the semi-axes of the ellipsoid.

If two coefficients are positive and the third one negative (for example $S_{33}' < 0$), and if we choose $\varepsilon = 1$, the equation of the quadric has the form:

$$\frac{x_1'^2}{a^2} + \frac{x_2'^2}{b^2} - \frac{x_3'^2}{c^2} = 1,$$

with

$$a^2 = \frac{1}{S_{11}'}, \qquad b^2 = \frac{1}{S_{22}'}, \qquad c^2 = -\frac{1}{S_{33}'}.$$

This is the equation of a one-sheet hyperboloid (Fig. 10.1b). The sections of this hyperboloid by the planes $x_3' = \mathrm{const}$ are ellipses, while the sections by the planes $x_1' = \mathrm{const}$ and $x_2' = \mathrm{const}$ are hyperbolas. If we choose $\varepsilon = -1$, the equation becomes:

$$-\frac{x_1'^2}{a^2} - \frac{x_2'^2}{b^2} + \frac{x_3'^2}{c^2} = 1$$

and the quadric is a two-sheeted hyperboloid (Fig. 10.1c). In order to obtain a representation of the tensor in all directions of space, the representation quadric of the tensor consists of the two hyperboloids tangent to the same cone, whose apex is the origin of the axes (Fig. 10.1d).

If two coefficients are negative and the third one positive (for example $S'_{33} > 0$), we retrieve the preceding case if we set:

$$a^2 = -\frac{1}{S'_{11}}, \qquad b^2 = -\frac{1}{S'_{22}}, \qquad c^2 = \frac{1}{S'_{33}}.$$

The one-sheet hyperboloid is obtained with $\varepsilon = -1$ and the two-sheeted one with $\varepsilon = 1$. We will see an application in Section 12.6.

In practice, the principal values of a tensor are usually all positive, and the quadric is then an ellipsoid, the equation of which is given by (10.4) with $\varepsilon = 1$. This is, for simplicity, what we will assume in the next section.

10.3. Properties of the quadric

Consider a second-rank tensor $[T]$ which relates vector $\mathbf{A}$ to vector $\mathbf{B}$ through relation $\mathbf{B} = [T]\mathbf{A}$, *i.e.* $B_i = T_{ij}A_j$. We will, to simplify the presentation, call vector $\mathbf{A}$ the cause and vector $\mathbf{B}$ the effect even if tensor $[T]$ is a field tensor for which this denomination is not justified.

10.3.1. Normal to the quadric

We saw (Eq. (10.5)) that the normal to point M (x_1, x_2, x_3) of the quadric $T_{ij}x_ix_j = 1$ is a vector with components $X_i = 2T_{ij}x_j$, *i.e.* a vector parallel to the vector $[T]\mathbf{OM}$ with components $T_{ij}x_j$. We deduce that, if the cause vector is parallel to radius vector $\mathbf{OM}$, the associated effect vector is parallel to the normal to the quadric at point M. Figure 10.2 illustrates this result. We chose the electric field $\mathbf{E}$ as the cause vector, and its effect is the current density $\mathbf{j}$, which is related to it by the electrical conductivity $[\sigma]$: $\mathbf{j} = [\sigma]\mathbf{E}$.

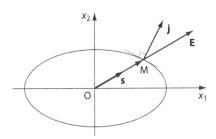

Figure 10.2: If the direction of the cause vector (electric field $\mathbf{E}$) is along $\mathbf{OM}$, the effect vector (electric current density $\mathbf{j}$) is along the normal at point M to the quadric representing the electrical conductivity tensor.

10.3.2. Length and physical meaning of the radius vector

Let M be a point of the quadric representing tensor $[T]$ which connects cause vector $\mathbf{A}$ to effect vector $\mathbf{B}$ through relation $\mathbf{B} = [T]\mathbf{A}$. Set:

$$\mathbf{OM} = r\mathbf{s}$$

where $\mathbf{s}$ is a unit vector. We deduce:

$$x_i = rs_i.$$

The equation of quadric $T_{ij}x_ix_j = 1$ becomes:

$$T_{ij}r^2s_is_j = 1$$

so that
$$\frac{1}{r^2} = s_iT_{ij}s_j \qquad (10.6)$$

which can be written in matrix form:

$$\frac{1}{r^2} = \begin{pmatrix} s_1 & s_2 & s_3 \end{pmatrix} \begin{pmatrix} T_{11} & T_{12} & T_{13} \\ T_{12} & T_{22} & T_{23} \\ T_{13} & T_{23} & T_{33} \end{pmatrix} \begin{pmatrix} s_1 \\ s_2 \\ s_3 \end{pmatrix}. \qquad (10.7)$$

If the cause vector $\mathbf{A}$, with norm A, is parallel to $\mathbf{OM}$, we have $\mathbf{A} = A\mathbf{s}$, $i.e.$ $A_j = As_j$.

The components B_i of the effect vector are given by:

$$B_i = T_{ij}A_j = T_{ij}s_j A$$

and the component of $\mathbf{B}$ parallel to $\mathbf{A}$, $i.e.$ $\mathbf{B} \cdot \mathbf{s}$, becomes:

$$\mathbf{B} \cdot \mathbf{s} = B_is_i = s_iT_{ij}s_jA = A\left(\frac{1}{r^2}\right). \qquad (10.8)$$

We deduce that, for a cause vector with unit norm $(A = 1)$, parallel to a given radius vector of the quadric, the reciprocal of the square of the length of this radius vector is equal to the projection of the corresponding effect vector on the direction of the cause vector.

10.3.3. Intensity of a physical property in a given direction

If tensor T_{ij} is a material tensor representing a given physical property, $s_is_jT_{ij}$ represents the value of this physical quantity in direction $\mathbf{s}$. We demonstrate this by taking the example of electrical conductivity $[\sigma]$, with components σ_{ij} in a given axis system. A measurement of the conductivity of a crystal can be performed by using a parallel-faced plate of this crystal, with thickness e small with respect to the sides of the main face. A voltage V is applied between the metallized major faces. It induces, except for edge effects which we neglect, an electric field $\mathbf{E}$ perpendicular to the plate, so that $\mathbf{E} = E\mathbf{s}$ where $\mathbf{s}$ is a unit vector perpendicular to the plate and $E = V/e$. We then measure the intensity in the circuit, and deduce the current density (the intensity of the current in the circuit, divided by the area of the crystal perpendicular to the electric field, hence in this case the area of the major faces). The latter is the component $\mathbf{j}_s$ of the current density along the normal $\mathbf{s}$ to the crystal (Fig. 10.3), so that:

$$j_s = \mathbf{j} \cdot \mathbf{s} = j_i s_i = \sigma_{ij} s_j s_i E.$$

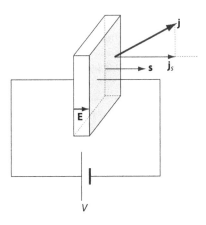

Figure 10.3: Schematic diagram of the principle of the experimental device for measuring the electrical conductivity of a crystal plate, the major faces of which were metallized

The conductivity in direction $\mathbf{s}$ is defined as the quantity σ_s such that:

$$\sigma_s = \frac{j_s}{E} = \sigma_{ij} s_i s_j.$$

In a general way, the value T_s in a direction with unit vector $\mathbf{s}$ of a physical property represented by tensor $[T]$ is given by $s_i s_j T_{ij}$.

From (10.6),

$$T_s = \frac{1}{r^2}$$

and

$$r = \frac{1}{\sqrt{T_s}}. \tag{10.9}$$

Knowing the length of the radius vector $\mathbf{OM}$ of the quadric thus provides the value of the physical property in direction $\mathbf{OM}$.

10.4. Geometrical determination of the principal axes and principal coefficients: the Mohr circle construction

We just saw that a symmetric tensor $[T]$ can be written in the referential of its principal axes $(Ox_1x_2x_3)$:

$$T = \begin{pmatrix} T_{11} & 0 & 0 \\ 0 & T_{22} & 0 \\ 0 & 0 & T_{33} \end{pmatrix}.$$

Consider a new axis system $(Ox'_1x'_2x'_3)$ deduced from $(Ox_1x_2x_3)$ through rotation by angle θ around Ox_3. In this new axis system, the components T'_{ij} of $[T]$ are expressed as:

$$T'_{ij} = a_{ik}a_{jl}T_{kl}$$

where the transformation matrix $A = \{a_{ij}\}$ is:

$$A = \begin{pmatrix} \cos\theta & \sin\theta & 0 \\ -\sin\theta & \cos\theta & 0 \\ 0 & 0 & 1 \end{pmatrix}.$$

We obtain:

$$T'_{11} = a_{1k}a_{1l}T_{kl} = a_{11}a_{11}T_{11} + a_{12}a_{12}T_{22} + a_{13}a_{13}T_{33},$$
$$T'_{11} = \cos^2\theta\, T_{11} + \sin^2\theta\, T_{22},$$
$$T'_{22} = a_{2k}a_{2l}T_{kl} = a_{21}a_{21}T_{11} + a_{22}a_{22}T_{22} + a_{23}a_{23}T_{33},$$
$$T'_{22} = \sin^2\theta\, T_{11} + \cos^2\theta\, T_{22},$$
$$T'_{33} = T_{33}.$$

In the same way:

$$T'_{13} = 0,$$
$$T'_{32} = 0,$$
$$T'_{12} = T'_{21} = -\sin\theta\cos\theta\, T_{11} + \sin\theta\cos\theta\, T_{22}.$$

Thus, tensor $[T]$ is expressed, in the frame $(Ox'_1x'_2x'_3)$, as:

$$\begin{pmatrix} T'_{11} & T'_{12} & 0 \\ T'_{12} & T'_{22} & 0 \\ 0 & 0 & T_{33} \end{pmatrix}.$$

We now express T'_{11}, T'_{22} and T'_{12} as a function of $\cos 2\theta$ and $\sin 2\theta$:

$$T'_{11} = \frac{1}{2}(T_{11} + T_{22}) - \frac{1}{2}(T_{22} - T_{11})\cos 2\theta$$

$$T'_{22} = \frac{1}{2}(T_{11} + T_{22}) + \frac{1}{2}(T_{22} - T_{11})\cos 2\theta \qquad (10.10)$$

$$T'_{12} = \frac{1}{2}(T_{22} - T_{11})\sin 2\theta.$$

The set of relations (10.10) can be expressed graphically by using the Mohr circle representation (Fig. 10.4). Points A and B are placed along axis Ox_1 so that $\overline{OA} = T_{11}$ and $\overline{OB} = T_{22}$. We draw circle $\mathcal{C}$, with its center C at the middle of segment AB, and with diameter d equal to the norm of **AB**.

Relations (10.10) become:

$$T'_{11} = \overline{OC} - \frac{\overline{AB}}{2} \cos 2\theta,$$

$$T'_{22} = \overline{OC} + \frac{\overline{AB}}{2} \cos 2\theta,$$

$$T'_{12} = \frac{\overline{AB}}{2} \sin 2\theta.$$

We draw the axis Cx_2 such that angle (Cx_1, Cx_2) is equal to $\pi/2$, then the radius CP of circle $\mathcal{C}$ enclosing angle 2θ with axis Ox_1. Let P' be the point diametrically opposite to P. Let H and J be the respective projections of P' and P on axis Cx_1, and K the projection of P on axis Cx_2. We see that: $T'_{11} = \overline{OH}$, $T'_{22} = \overline{OJ}$ and $T'_{12} = \overline{CK}$.

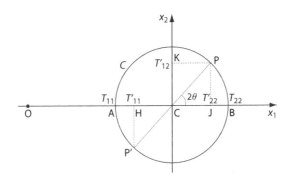

Figure 10.4: Mohr circle construction

This construction makes it possible to easily determine how the tensor coefficients vary when the coordinate axes are rotated.

Furthermore, this construction, when performed the other way, makes it possible to quickly determine the principal axes of a second-rank tensor for which only one principal axis is known.

The procedure that provides the principal axes and principal values of a tensor which, expressed in its initial frame, has the form:

$$\begin{pmatrix} T'_{11} & T'_{12} & 0 \\ T'_{12} & T'_{22} & 0 \\ 0 & 0 & T_{33} \end{pmatrix}$$

is given below, where ψ is the angle by which we must rotate the initial frame around axis Ox_3 to obtain the diagonal form (the angle opposite to the preceding angle θ), and T_{11} and T_{22} are the principal values associated to axes Ox_1 and Ox_2 respectively.

Points H and J are placed on axis Ox_1 so that $\overline{OH} = T'_{11}$ and $\overline{OJ} = T'_{22}$. Let C be the middle of HJ. Point K is placed on axis Cx_2, perpendicular to Ox_1, so that $\overline{CK} = T'_{12}$. Points J and K define respectively the abscissa and the ordinate of a point P in the axis system (Cx_1, Cx_2). Angle 2ψ is then the angle $(\mathbf{CP}, Cx_1)$. Let R be the length of vector CP. We obtain $T_{11} = \overline{OC} - R$ and $T_{22} = \overline{OC} + R$.

Notes:

1. It is evident that, if angle 2ψ is larger than π, it will be simpler to choose an angle equal to $2\psi - \pi$, *i.e.* a rotation angle $\psi - \pi/2$, and to interchange T_{11} and T_{22}.

2. If the subspace to be diagonalized is (Ox_1, Ox_3), then axis Ox_1 will be substituted by Ox_3 and axis Ox_2 by Ox_1 in the above construction.

10.5. Effect of crystal symmetry

To determine the number of independent coefficients of a symmetric second-rank material tensor (a tensor representing a physical property of a crystal), we can use either of the methods presented in Sections 9.6.1 and 9.6.2. A more elegant and quicker method consists in using the representative quadric of the tensor: according to Neumann's principle, it must be at least as symmetric as the crystal. The symmetry elements of a general quadric are:

– one symmetry center,

– three A_2 axes, which are the principal axes,

– three mirrors perpendicular to the A_2 axes.

Hence the point group of the quadric is mmm.

We deduce for example that:

– if the crystal has a 2-fold axis, one of the principal axes of the quadric is necessarily parallel to this 2-fold axis.

– if the crystal has an axis with order n larger than 2 ($n = 3$, 4 or 6), the quadric must also have an axis with order larger than 2. For a quadric, this can only be an axis of revolution (the symmetry of a revolution quadric is ∞/mm). This axis of revolution is then parallel to the n-fold axis.

We now apply these results to the various crystal groups, assuming for simplicity that the representation quadric of the tensor is an ellipsoid. This assumption does not affect the generality of the results to be obtained. We will also see that, finally, the number of independent coefficients of a symmetric second-rank tensor is the same for all the groups of a given crystal system.

10.5.1. Triclinic system

A crystal with point group 1 has no symmetry element. The directions of the principal axes of the quadric then have no special connection with those of the basis vectors **a**, **b** and **c** of the crystal. The tensor representing the physical property of this crystal then has 6 independent coefficients. If the point group of the crystal is $\bar{1}$, the result is the same since the quadric has a center of symmetry. This result is no surprise since we showed (Sect. 9.6.3) that even-rank tensors are invariant under a central symmetry operation.

The tensor has the most general form:

$$\begin{pmatrix} T_{11} & T_{12} & T_{13} \\ T_{12} & T_{22} & T_{23} \\ T_{13} & T_{23} & T_{33} \end{pmatrix}.$$

Its principal axes can now be determined. The six independent coefficients are retrieved: the three principal values, and the three parameters which define the orientation, with respect to the crystal itself, *i.e.* with respect to the basis vectors of its unit cell, of the orthonormal trihedron parallel to the principal axes.

10.5.2. Monoclinic system

Such a crystal has at least a 2-fold axis (group 2) or a mirror (group m). One of these symmetry elements therefore necessarily coincides with one of the symmetry elements of the ellipsoid. If an orthonormal axis system is chosen so that axis Ox_3 be parallel to axis A_2, or perpendicular to mirror m, the tensor takes the form:

$$\begin{pmatrix} T_{11} & T_{12} & 0 \\ T_{12} & T_{22} & 0 \\ 0 & 0 & T_{33} \end{pmatrix}.$$

If the crystal group is $2/m$, the same result is obtained, since the quadric has a mirror perpendicular to each of the A_2 axes.

We retrieve the result of the example discussed in Section 9.6.1, except that we here chose axis Ox_3 parallel to the 2-fold axis and perpendicular to the mirror.

10.5.3. Orthorhombic system

In this system, a crystal has at least three mutually orthogonal 2-fold axes (group 222) or two mutually orthogonal mirrors (group mm2). These symmetry elements must be common to the crystal and to the ellipsoid representing the physical property. The axes of the ellipsoid are therefore parallel to the basis vectors of the orthorhombic unit cell: thus the orientation of the ellipsoid is fixed. In a system of orthonormal axes parallel to the basis vectors $\mathbf{a}$, $\mathbf{b}$ and $\mathbf{c}$, the tensor has the form:

$$\begin{pmatrix} T_{11} & 0 & 0 \\ 0 & T_{22} & 0 \\ 0 & 0 & T_{33} \end{pmatrix}.$$

This result is valid also for group mmm, which is the symmetry group of the ellipsoid.

10.5.4. Uniaxial systems: rhombohedral, trigonal and hexagonal

A crystal belonging to any one of these systems has at least one n-fold axis, with n larger than 2 (except for groups $\bar{4}$ and $\bar{4}2m$). The same must apply to the ellipsoid, which is therefore of revolution around this axis. If Ox_3 is chosen parallel to axis A_n, the tensor has the form:

$$\begin{pmatrix} T_{11} & 0 & 0 \\ 0 & T_{11} & 0 \\ 0 & 0 & T_{33} \end{pmatrix}.$$

The additional symmetry elements in any of these crystal groups do not contribute additional relations among the coefficients, because the ellipsoid already has all the symmetry elements which can be added.

The same result is obtained for groups $\bar{4}$ and $\bar{4}2m$, for example using the direct inspection method (Exercise 10.4) or, alternatively, by noting that the number of independent coefficients for even-rank material tensors is the same for all groups in a given Laue class (Sect. 9.6.3).

10.5.5. Cubic system

All groups in this crystal system have four A_3 axes. The only ellipsoid compatible with the presence of four A_3 axes is a sphere. The quadric is therefore a sphere. The tensor is diagonal, with its three coefficients equal:

$$\begin{pmatrix} T & 0 & 0 \\ 0 & T & 0 \\ 0 & 0 & T \end{pmatrix}$$

whatever the axis system in which it is represented. We conclude that:

The physical properties represented by a rank-2 tensor are isotropic in a cubic crystal.

10.6. Axial vectors, or antisymmetric rank-2 tensors

10.6.1. Polar vectors, axial vectors

Two families of vector-type physical quantities must be distinguished:

- the quantities obviously represented by vectors (displacement, velocity, acceleration, force...): these are called polar vectors.

- the quantities which describe a rotation, such as angular velocity, the moment of a force, magnetic field... These are called axial vectors.

They are denoted as $\overset{\smile}{\mathbf{V}}$ and represented by a line and a sense of rotation around this line. It is however convenient to also represent them through vectors. This requires choosing a convention to connect the direction of the vector and the sense of rotation. The convention chosen is the following: the vector points so that the rotation associated to the axial vector be, for this vector, in the direct direction (Fig. 10.5).

Figure 10.5: Convention for representing axial vectors through vectors

These two vector families transform differently under some symmetry operations. If an axial vector is submitted to a rotation, the sense of rotation of the axial vector is conserved. Therefore a polar vector and an axial vector transform the same way in a rotation operation. In contrast, the behavior is different for symmetry with respect to a center and with respect to a plane.

Consider a given vector **OP**, representing first a polar vector, and then an axial vector, using the above convention. Let x_1, x_2 and x_3 be the components of vector **OP** in a given orthonormal axis system, and let x'_1, x'_2 and x'_3 be those of vector **OP'**, the transform of **OP** in the symmetry operation.

Consider first the operation corresponding to symmetry with respect to plane (Ox_1, Ox_2). The components of the polar vector **OP'** are (Fig. 10.6a):

$$x'_1 = x_1, \qquad x'_2 = x_2, \qquad x'_3 = -x_3.$$

Those of the axial vector **OP'** are (Fig. 10.6b and c):

$$x'_1 = -x_1, \qquad x'_2 = -x_2, \qquad x'_3 = x_3.$$

Now consider the operation corresponding to symmetry with respect to point O. The components of the polar vector **OP'** are (Fig. 10.7a):

$$x'_1 = -x_1, \qquad x'_2 = -x_2, \qquad x'_3 = -x_3.$$

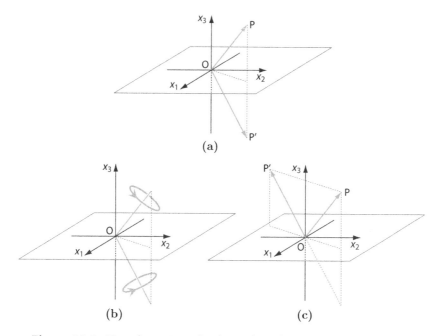

Figure 10.6: Transformation of polar and axial vectors through symmetry with respect to plane (Ox_1, Ox_2): (a) polar vector; (b) axial vector; (c) polar vectors associated to the axial vectors of Figure (b)

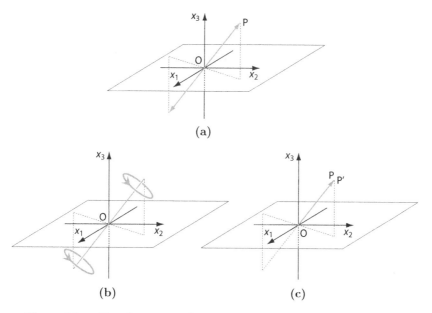

Figure 10.7: Transformation of polar and axial vectors through symmetry with respect to point O: (a) polar vector; (b) axial vector; (c) polar vectors associated to the axial vectors of Figure (b)

and those of the axial vector $\mathbf{OP'}$ are (Fig. 10.7b and c)

$$x'_1 = x_1, \qquad x'_2 = x_2, \qquad x'_3 = x_3.$$

The polar and axial vectors thus transform the same way in direct isometries (rotation), and differently in indirect isometries (symmetry with respect to a plane and central symmetry).

This difference is also encountered when performing a change in coordinate system, since transforming a vector through a symmetry operation is equivalent to performing the inverse symmetry operation on the coordinate system.

In a rotation of the coordinate system (direct isometry), the components of a polar vector and of an axial vector transform the same way: from relation (9.8):

$$x'_i = a_{ij}x_j \qquad (10.11)$$

where the x_i are the old components, the x'_i the new components and $\{a_{ij}\}$ is the transformation matrix for going from the old coordinate system to the new system.

In contrast, if the coordinate system is submitted to a symmetry with respect to a plane or a central symmetry, which are indirect isometries, it can be checked that the new components x'_i of the axial vector are expressed as a function of the old ones x_i through relation:

$$x'_i = -a_{ij}x_j. \qquad (10.12)$$

Further, since the determinant Δ of matrix $\{a_{ij}\}$ is equal to $+1$ for direct symmetry operations and to -1 for indirect symmetry operations, we can, for axial vectors, collect relations (10.11) and (10.12) into a single relation:

$$x'_i = \Delta a_{ij}x_j \qquad (10.13)$$

where Δ is the determinant of the transformation matrix from the old to the new system.

10.6.2. Example of an axial vector: the cross product

Consider a cross product: $\overset{\smile}{\mathbf{W}} = \mathbf{P} \times \mathbf{Q}$ where $\mathbf{P}$ and $\mathbf{Q}$ are two polar vectors. It is an axial vector because it is related to the sense of rotation of vector $\mathbf{P}$ toward vector $\mathbf{Q}$. It can in fact easily be checked that a central symmetry operation on vectors $\mathbf{P}$ and $\mathbf{Q}$ transforms vector $\overset{\smile}{\mathbf{W}}$ into itself. Let p_1, p_2, p_3 and q_1, q_2, q_3 be the components of $\mathbf{P}$ and $\mathbf{Q}$ respectively, in an orthonormal axis system $(Ox_1x_2x_3)$. The components w_i of the cross product $\overset{\smile}{\mathbf{W}} = \mathbf{P} \times \mathbf{Q}$ are:

$$\begin{aligned} w_1 &= p_2q_3 - p_3q_2 \\ w_2 &= p_3q_1 - p_1q_3 \\ w_3 &= p_1q_2 - p_2q_1 \end{aligned} \qquad (10.14)$$

which can be written in a condensed form which will prove useful later on:

$$w_k = \delta_{kij} p_i q_j. \qquad (10.15)$$

δ_{ijk} is the Levi-Civita symbol, defined, for any axis system,[1] as:

$\delta_{ijk} = 0$ if at least two of the subscripts i, j and k are equal,

$\delta_{ijk} = 1$ if the permutation ijk is direct (123, 231, 312),

$\delta_{ijk} = -1$ if it is inverse (132, 213, 321).

Let w'_1, w'_2, w'_3 be the components of vector $\overset{\smile}{\mathbf{W}}'$, the transform of $\overset{\smile}{\mathbf{W}}$ in a symmetry with respect to the origin. Since a central symmetry changes $\mathbf{P}$ into $-\mathbf{P}$ and $\mathbf{Q}$ into $-\mathbf{Q}$, we immediately see on Equations (10.14) that $w'_1 = w_1$, $w'_2 = w_2$ and $w'_3 = w_3$, so that $\overset{\smile}{\mathbf{W}}' = \overset{\smile}{\mathbf{W}}$. We can associate the components of $\overset{\smile}{\mathbf{W}}$ to coefficients V_{ij} defined by:

$$V_{ij} = p_i q_j - q_i p_j \text{ for } i \neq j$$

and
$$V_{11} = V_{22} = V_{33} = 0.$$

We thus have
$$w_1 = V_{23} = -V_{32},$$
$$w_2 = V_{31} = -V_{13}, \qquad (10.16)$$
$$w_3 = V_{12} = -V_{21}.$$

In another axis system $(Ox'_1 x'_2 x'_3)$, deduced from the preceding one through the transformation matrix A with coefficients a_{ij}, the components of vectors $\mathbf{P}$ and $\mathbf{Q}$ become:

$$p'_i = a_{ij} p_j \qquad \text{and} \qquad q'_i = a_{ij} q_j.$$

The coefficients V_{ij} become, in this new axis system, V'_{ij} such that:

$$V'_{ij} = p'_i q'_j - q'_i p'_j = a_{ik} p_k a_{jl} q_l - a_{in} q_n a_{jm} p_m$$
$$= a_{ik} a_{jl} (p_k q_l - q_k p_l)$$
$$= a_{ik} a_{jl} V_{kl}.$$

This shows that V_{ij} is a second-rank tensor. Furthermore, it is antisymmetric since $V_{ij} = -V_{ji}$. Hence:

The components of a cross product (an axial vector or an axial tensor[2] of rank 1) are the components of the polar second-rank antisymmetric tensor $[V]$ defined above.

The axial vector $\overset{\smile}{\mathbf{W}}$ with components w_k and the antisymmetric tensor $[V]$ deduce from each other through relations:

$$w_k = \frac{1}{2} \delta_{kij} V_{ij} \qquad \text{and} \qquad V_{ij} = \delta_{ijk} w_k \qquad (10.17)$$

as can be checked easily.

1. We will show in Appendix 18A that this symbol is a rank-3 axial tensor, also called the permutation tensor.
2. The general definition of an axial tensor is given in Appendix 18A.

10.7. Exercises

Exercise 10.1

Consider a monoclinic crystal with point group 2, where the basis vector **b** of the unit cell is perpendicular to the vectors **a** and **c**, and a rank-2 symmetric tensor $[T]$ representing a physical property of this crystal. The components T_{ij} of this tensor, in an orthonormal axis system $(Ox_1x_2x_3)$ where axis Ox_2 is parallel to **b**, are $T_{11} = 1$, $T_{22} = 2$, $T_{33} = 3$, $T_{13} = T_{31} = 4$ and the other components are zero.

1. Calculate the components T'_{ij} of this tensor in an axis system $(Ox'_1x'_2x_3)$ deduced from $(Ox_1x_2x_3)$ through a rotation by π around Ox_3.

2. Calculate the components T''_{ij} of the same tensor in an axis system $(Ox''_1x_2x''_3)$ deduced from $(Ox_1x_2x_3)$ through a rotation by π around axis Ox_2.

Exercise 10.2

Consider a monoclinic crystal. Its electrical conductivity $[\sigma]$ is represented in an orthonormal axis system $(Ox_1x_2x_3)$, where axis Ox_2 is parallel to the 2-fold axis, by the following matrix:

$$\begin{pmatrix} 7 & 0 & \sqrt{3} \\ 0 & 8 & 0 \\ \sqrt{3} & 0 & 5 \end{pmatrix}$$

where the unit is $10^7 \, \Omega^{-1} \, \mathrm{m}^{-1}$.

1. This crystal is submitted to an electric field **E**. Determine the current density vector for the following cases:

   (a) **E** is parallel to axis Ox_1,

   (b) **E** is parallel to axis Ox_2,

   (c) **E** is parallel to axis Ox_3,

   (d) **E** is parallel to the bisector of the angle between axes Ox_3 and Ox_1,

   (e) **E** is in the plane (Ox_3, Ox_1), and its direction deduces from that of Ox_3 through a rotation by $60°$ around Ox_2. What is your conclusion?

2. A plate with faces parallel to the plane (Ox_1x_2) is cut out of this crystal, and submitted to an electric field **E** parallel to axis Ox_3.

   (a) Calculate the component j_n of **j** along the applied field

   (b) What is the meaning of the ratio j_n/E, where E is the norm of **E**?

3. Express tensor $[\sigma]$ in a new system of axes, rotated by $60°$ around Ox_2 with respect to the preceding one. What do you conclude?

Exercise 10.3

Let σ_{ij} be the electrical conductivity tensor of a crystal exhibiting a 3-fold axis. We choose an orthonormal axis system $(Ox_1x_2x_3)$ such that axis Ox_3 is parallel to this axis.

1. Give the form of this tensor in this axis system.

2. Calculate the conductivity σ_s in a general direction, parallel to the unit vector $\mathbf{s}$ with direction cosines $(s_1,\ s_2,\ s_3)$. Express it as a function of the non-zero components σ_{ij} and of the spherical coordinates $(r = 1,\ \theta,\ \varphi)$ of $\mathbf{s}$. What do you conclude?

Exercise 10.4

1. Using the direct inspection method, determine the form of a symmetric rank-2 tensor representing a physical property of a crystal with point group $\bar{4}$ when it is described in an orthonormal axis system $(Ox_1x_2x_3)$ in which axis Ox_3 is parallel to the $\bar{4}$-fold axis and origin O is at the center of symmetry associated with the rotoinversion by $\pi/2$.

2. Now consider a crystal with point group $\bar{4}2m$, and choose axis Ox_1 perpendicular to the mirror. Show that adding this symmetry element does not change the form of the tensor.

Chapter 11

The stress tensor

The elastic properties of materials relate the applied stress tensor to the resulting strain tensor via the stiffness tensor or its inverse, the compliance tensor. They are the topics of Chapters 11, 12 and 13.

In the present chapter, the stress tensor at a point in the material is defined. The basic relations which must be verified by its spatial derivatives are then determined, and they lead to the condition that this tensor must be symmetric.

11.1. Introduction

In Chapters 11, 12 and 13, we investigate the mechanical properties of crystals, *i.e.* the way they deform under external forces. These external forces induce stresses at each point in the crystal. The relation between the stresses in a material and the external forces applied to its surface is investigated in treatises on the mechanics of continuous media (see for example [1, 2]).

Here, we use a few results of mechanics, in particular to locally define the stress tensor (Chap. 11) and the strain tensor (Chap. 12). At any given point in the crystal, these tensors are related to each other by a material tensor: the compliance tensor, connecting the stress to the strain, or the stiffness tensor, connecting conversely the strain to the stress.

11.2. Stress tensor

11.2.1. Introduction

In an undeformed solid at equilibrium, each atom or molecule is in an equilibrium position which results from the equality of the attractive and repulsive forces exerted on each of them by the neighboring atoms or molecules (Chap. 8).

If the solid is submitted to external forces, it deforms. The external forces are transmitted from atom to atom. Each atom (or molecule) is thus submitted to an additional force, which displaces it to a position where, in the new

© Springer Science+Business Media Dordrecht 2014
C. Malgrange et al., *Symmetry and Physical Properties of Crystals*,
DOI 10.1007/978-94-017-8993-6_11

equilibrium state, this force is balanced by a restoring force which tends to return the atom or molecule to its initial position. The solid is said to be submitted to stress.

Since we are interested in macroscopic properties of the solid, we do not consider an atom but a small volume Δv of the solid, bounded by its surface S (Fig. 11.1). The stresses then lead to forces which the surrounding solid exerts on volume Δv through all elements dS of surface S. More precisely, the solid around Δv exerts on each element dS a force which is proportional to the value of dS:

$$d\mathbf{F} = \mathbf{T}(\mathbf{r}, \mathbf{n})dS.$$

$\mathbf{r}$ is the radius vector to the center P of dS : $\mathbf{r} = \mathbf{O'P}$, and $\mathbf{n}$ a unit vector perpendicular to dS, oriented from within Δv to the outside. $\mathbf{T}(\mathbf{r}, \mathbf{n})$ is a force per unit surface, called the stress at point $\mathbf{r}$; it normally depends not only on $\mathbf{r}$ but also on the orientation, defined by $\mathbf{n}$, of the surface on which it acts.

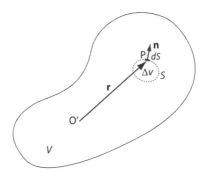

Figure 11.1: Schematic diagram illustrating the surface forces acting on a volume element Δv

For a given elementary surface dS, we can consider the material on one side or the other side of the surface. If the normal is oriented from part 1 toward part 2, *i.e.* if it corresponds to $\mathbf{n}_1$ on Figure 11.2, then $\mathbf{T}(\mathbf{r}, \mathbf{n}_1)dS$ represents the force $d\mathbf{F}_{2/1}$ exerted by part 2 on part 1 of the solid. Conversely, for the opposite normal $\mathbf{n}_2 = -\mathbf{n}_1$, we have $\mathbf{T}(\mathbf{r}, \mathbf{n}_2)dS = d\mathbf{F}_{1/2}$, which represents the force exerted by part 1 on part 2. The action-reaction law (Newton's third law) requires these two forces to be equal and opposite:

$$d\mathbf{F}_{1/2} = -d\mathbf{F}_{2/1} \quad \text{and} \quad \mathbf{T}(\mathbf{r}, \mathbf{n}_2)dS = -\mathbf{T}(\mathbf{r}, \mathbf{n}_1)dS.$$

We set $\mathbf{n}_1 = \mathbf{n}$ and

$$\mathbf{T}(\mathbf{r}, \mathbf{n}) = -\mathbf{T}(\mathbf{r}, -\mathbf{n}).$$

The force per unit area $\mathbf{T}(\mathbf{r}, \mathbf{n})$ is designated as the stress exerted across the small surface element dS, centered at point P such that $\mathbf{O'P} = \mathbf{r}$, with normal unit vector $\mathbf{n}$, by the part of the solid located on the side toward which vector $\mathbf{n}$ is pointing onto the other part. To make the notations simpler, $\mathbf{T}(\mathbf{r}, \mathbf{n})$ is

Figure 11.2: Forces acting on a surface element dS of a solid

written as $\mathbf{T_n}$, and it is enough to designate the point P at the center of the small surface element.

The volume element Δv can also be submitted to forces proportional to its volume, which will be generally noted $\mathbf{f}\Delta v$. The simplest example of such a bulk force is the gravity force, equal to $\rho \mathbf{g}\Delta v$ where ρ is the mass per unit volume of the solid and $\mathbf{g}$ the acceleration of gravity.

11.2.2. Definition of the stress tensor

Consider again the small plane surface element dS centered at P, and its normal vector $\mathbf{n}$. Choose an orthonormal system of axes $(Ox_1x_2x_3)$ whose origin is located at point O very near dS, and on the side opposite to $\mathbf{n}$ (Fig. 11.3). Surface dS intersects axes Ox_1, Ox_2 and Ox_3 at A, B and C respectively. We thus define a small tetrahedron, formed by surface ABC and three faces OBC, OCA and OAB respectively perpendicular to Ox_1, Ox_2 and Ox_3 (Fig. 11.3).

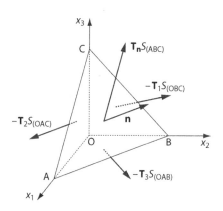

Figure 11.3: Forces acting on the tetrahedron OABC

This tetrahedron, with volume Δv, is submitted, through its four faces, to forces related to the stresses defined above, exerted upon it by the part of the solid which is outside. It can also be submitted to bulk forces $\mathbf{f}\Delta v$. Since this tetrahedron is at equilibrium, the resultant of these forces is zero.

The force acting on surface ABC is $\mathbf{T_n}\,S_{(ABC)}$, in accordance with the definition of $\mathbf{T_n}$ given above.

The forces acting on the other faces of the tetrahedron are:

on OBC: $- \mathbf{T}_1 \, S_{(OBC)}$,

on OCA: $- \mathbf{T}_2 \, S_{(OCA)}$,

on OAB: $- \mathbf{T}_3 \, S_{(OAB)}$.

Here, we call by convention $\mathbf{T}_i$ the stress acting on a surface normal to axis Ox_i, with its normal oriented toward the positive Ox_i (in other words, the stress exerted by the solid lying on the positive x_i side onto the solid lying on the negative x_i side). In the case of the forces applied by the surrounding medium upon tetrahedron OABC, the normals to the surfaces OBC, OAC and OAB are oriented respectively toward the negative Ox_1, Ox_2, Ox_3. This explains the minus signs.

The equilibrium condition for the tetrahedron is given by:

$$-\mathbf{T}_1 S_{(OBC)} - \mathbf{T}_2 S_{(OCA)} - \mathbf{T}_3 S_{(OAB)} + \mathbf{T}_n S_{(ABC)} + \mathbf{f}\Delta v = 0. \qquad (11.1)$$

We now express the areas of triangles OAB, OBC and OCA as a function of the area of triangle ABC. We note that OAB is the projection of area ABC onto plane (Ox_1, Ox_2), which is normal to Ox_3, hence

$$S_{(OAB)} = S_{(ABC)} \cos(\mathbf{n}, Ox_3) = n_3 S_{(ABC)}$$

where we note n_1, n_2, n_3 the direction cosines of vector $\mathbf{n}$. In the same way:

$$S_{(OBC)} = S_{(ABC)} \cos(\mathbf{n}, Ox_1) = n_1 S_{(ABC)},$$
$$S_{(OCA)} = S_{(ABC)} \cos(\mathbf{n}, Ox_2) = n_2 S_{(ABC)}.$$

We can thus write relation (11.1) in the form:

$$\mathbf{T}_n S_{(ABC)} - \mathbf{T}_1 n_1 S_{(ABC)} - \mathbf{T}_2 n_2 S_{(ABC)} - \mathbf{T}_3 n_3 S_{(ABC)} + \mathbf{f}\Delta v = 0.$$

Allowing volume Δv to tend to zero, the ratio $\Delta v / S_{(ABC)}$ tends to zero, and we can write:

$$\mathbf{T}_n = n_1 \mathbf{T}_1 + n_2 \mathbf{T}_2 + n_3 \mathbf{T}_3.$$

Projecting this equation onto the three axes, we obtain

$$(T_n)_1 = T_{11} n_1 + T_{12} n_2 + T_{13} n_3$$
$$(T_n)_2 = T_{21} n_1 + T_{22} n_2 + T_{23} n_3$$
$$(T_n)_3 = T_{31} n_1 + T_{32} n_2 + T_{33} n_3$$

where we denote by T_{ij} the component parallel to Ox_i of the stress $\mathbf{T}_j$.

The three above equations can be written in the following condensed form:

$$(T_n)_i = T_{ij} n_j \qquad (11.2)$$

where $(T_n)_i$ is the component parallel to Ox_i of the stress exerted on the face ABC normal to the unit vector $\mathbf{n}$, with components n_j, by the part of the material located on the side toward which **n** points onto the part located on the other side of ABC.

This relation shows that the coefficients T_{ij} relate vector **n**, normal to face ABC, to the vector $\mathbf{T_n}$ representing the stress acting on this face. We saw (Sect. 9.4.2) that, if two vector quantities **A** and **B**, with components respectively A_i and B_i, are related by $B_i = T_{ij}A_j$, then the T_{ij} are the components of a second-rank tensor $[T]$. Hence we can write:

$$\mathbf{T_n} = [T]\mathbf{n}.$$

The stress tensor $[T]$ describes an external action which depends, not on the crystal itself, but on the forces which are applied on its surface by the external world. It is a field tensor.

The component T_{ij} of this tensor is the component parallel to axis Ox_i of the force per unit area (the stress) exerted on a surface element normal to Ox_j by the part on the positive Ox_j side of the solid onto the part located on the negative x_j side.

We have defined the tensor T_{ij} at point P as $T_{ij}(\mathbf{r})$. In the special case where T_{ij} does not depend on the position of P in the solid, the stress tensor is said to be uniform.

11.2.3. Normal stress and shear stress

The diagonal terms of the stress tensor $[T]$ represent stresses which are called normal because they are the forces perpendicular to the surfaces. For example, T_{11} is the component parallel to axis Ox_1 of the force acting on a surface normal to Ox_1. A positive value of T_{11} represents a tensile stress whereas a negative value represents a compressive stress. The non-diagonal terms T_{ij} (with $i \neq j$) represent the shear components on these faces.

For the same stress tensor, we now express the normal stress $\mathbf{T}_\nu$ and shear stress $\boldsymbol{\tau}$ for a general surface dS, with normal **n**, a unit vector with components n_1, n_2, n_3 (Fig. 11.4).

The components $(T_n)_i$ of the stress force $\mathbf{T_n}$ applied to the surface dS are given by relation (11.2), *viz.*

$$(T_n)_i = T_{ij}n_j.$$

The normal stress $\mathbf{T}_\nu = T_\nu\,\mathbf{n}$ is the projection of $\mathbf{T_n}$ on **n** and

$$T_\nu = \mathbf{n} \cdot \mathbf{T_n} = n_i(T_n)_i,$$
$$T_\nu = n_i n_j T_{ij}$$

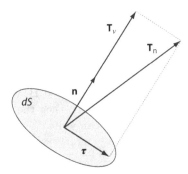

Figure 11.4: Normal stress T_ν and shear stress τ acting on a surface dS, with normal **n**, submitted to stress T_n

which can be written in matrix form:

$$T_\nu = \begin{pmatrix} n_1 & n_2 & n_3 \end{pmatrix} \begin{pmatrix} & & \\ & T_{ij} & \\ & & \end{pmatrix} \begin{pmatrix} n_1 \\ n_2 \\ n_3 \end{pmatrix}.$$

Using result (10.7), we deduce that the normal stress T_ν is equal to the reciprocal of the square of the length of the radius vector of the quadric representing the stress tensor.

The shear stress τ is the projection of $\mathbf{T_n}$ on the surface element dS:

$$\tau = \mathbf{T_n} - T_\nu \mathbf{n}$$

which can also be written as $\tau = \mathbf{n} \times (\mathbf{T_n} \times \mathbf{n})$

since $\mathbf{n} \times (\mathbf{T_n} \times \mathbf{n}) = (\mathbf{n} \cdot \mathbf{n})\mathbf{T_n} - (\mathbf{n} \cdot \mathbf{T_n})\mathbf{n} = \mathbf{T_n} - T_\nu \mathbf{n}$.

11.3. Basic relation

In many cases, the stress tensor is not uniform. It varies from point to point in the solid and is noted $T_{ij}(\mathbf{r})$. We now calculate the resultant of the forces to which a small parallelepiped, with sides parallel to Ox_1, Ox_2 and Ox_3 respectively equal to Δx_1, Δx_2, Δx_3, is submitted, taking into account the possibility of bulk forces **f** per unit volume.

We start by calculating the algebraic sum of the components of the stress forces parallel to Ox_1, $T_{1i}dS$ (Fig. 11.5).

The components acting on the faces perpendicular to Ox_1 give:

$$[T_{11}(x_1 + \Delta x_1) - T_{11}(x_1)]\Delta x_2 \Delta x_3 = \frac{\partial T_{11}}{\partial x_1} \Delta x_1 \Delta x_2 \Delta x_3.$$

Those acting on the faces perpendicular to Ox_2 give:

$$[T_{12}(x_2 + \Delta x_2) - T_{12}(x_2)]\Delta x_1 \Delta x_3 = \frac{\partial T_{12}}{\partial x_2} \Delta x_2 \Delta x_1 \Delta x_3$$

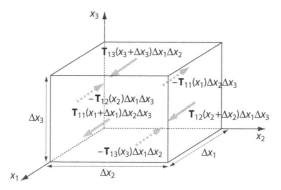

Figure 11.5: Components parallel to axis Ox_1 of the non-uniform stress forces acting on the faces of a rectangle parallelepiped with sides Δx_1, Δx_2 and Δx_3

and those acting on the faces perpendicular to Ox_3 give:

$$[T_{13}(x_3 + \Delta x_3) - T_{13}(x_3)]\Delta x_1 \Delta x_2 = \frac{\partial T_{13}}{\partial x_3}\Delta x_3 \Delta x_1 \Delta x_2$$

hence, in total:

$$\frac{\partial T_{1j}}{\partial x_j}\Delta x_1 \Delta x_2 \Delta x_3.$$

The component parallel to Ox_1 of the resultant of the bulk forces is $f_1 \Delta x_1 \Delta x_2 \Delta x_3$.

Let **F** be the final resultant of the stress forces and of the bulk forces acting on unit volume of the material. Its component parallel to Ox_1, F_1, is:

$$F_1 = \frac{\partial T_{1j}}{\partial x_j} + f_1.$$

The same argument applied to the two other components of **F** makes it possible to finally write:

$$F_i = \frac{\partial T_{ij}}{\partial x_j} + f_i. \tag{11.3}$$

Since the solid is at equilibrium, $F_i = 0$, and we obtain:

$$\frac{\partial T_{ij}}{\partial x_j} + f_i = 0. \tag{11.4}$$

If the bulk forces are negligible with respect to the stresses:

$$\frac{\partial T_{ij}}{\partial x_j} = 0. \tag{11.5}$$

Relation (11.3) will be used in Chapter 14 to investigate the propagation of matter displacements (elastic waves). Since the solid is then no more at equilibrium, it will be applied to Newton's second law.

Relation (11.4) can also be derived in a more formal way by writing the equilibrium condition for a volume V of the solid, submitted to bulk forces $\mathbf{f}$ per unit volume and to surface forces $\mathbf{F}_S$ per unit area, represented by tensor $[T]$:

$$\iiint_V \mathbf{f}dV + \iint_S d\mathbf{F}_S = 0,$$

or

$$\iiint_V f(\mathbf{r})dV + \iint_S \mathbf{T}(\mathbf{r},\mathbf{n})dS = 0.$$

which can be rewritten for each component i, taking relation (11.2) into account:

$$\iiint_V f_i(\mathbf{r})dV + \iint_S T_{ij}(\mathbf{r})n_jdS = 0.$$

Using Ostrogradsky's theorem,[1] we obtain:

$$\iiint_V f_i(\mathbf{r})dV + \iiint_V \frac{\partial T_{ij}}{\partial x_j}dV = 0$$

and

$$\iiint_V \left[f_i(\mathbf{r}) + \frac{\partial T_{ij}}{\partial x_j} \right] dV = 0.$$

Since this relation is valid whatever the volume V selected in the solid, we obtain

$$f_i + \frac{\partial T_{ij}}{\partial x_j} = 0 \qquad (11.6)$$

which is the equilibrium relation we were seeking.

---

1. Ostrogradsky's theorem states that for a vector field $\mathbf{A}$:

$$\iiint \mathrm{div}\mathbf{A}dV = \iint \mathbf{A} \cdot \mathbf{n}dS \text{ or } \iiint \frac{\partial A_j}{\partial x_j}dV = \iint A_k n_k dS.$$

It can be generalized to the components of a tensor as:

$$\iint T_{ik}n_k dS = \iiint \frac{\partial T_{ij}}{\partial x_j}dV.$$

11.4. Symmetry of the stress tensor

We now write that, for a solid at equilibrium, the total moment of the forces applied to it is zero.

The component M_i of the moment $\mathbf{M}$ of a force $\mathbf{F}$ ($\mathbf{M} = \mathbf{r} \times \mathbf{F}$) with components F_k is (Sect. 10.6.2)

$$M_i = \delta_{ijk} x_j F_k$$

where δ_{ijk} is the Levi-Civita symbol, or permutation tensor (cf. Appendix 18A.2). Its value is 0 if two subscripts are equal, 1 if the permutation ijk is direct and -1 if it is inverse.

The condition on the moment is expressed as:

$$M_i = \iiint \delta_{ijk} x_j f_k dV + \iint \delta_{ijk} x_j T_{kl} n_l dS = 0. \tag{11.7}$$

The second term can be written, using Ostrogradsky's theorem:

$$\iint \delta_{ijk} x_j T_{kl} n_l dS = \iiint \frac{\partial}{\partial x_l}(\delta_{ijk} x_j T_{kl}) dV = \iiint \delta_{ijk}\left(x_j \frac{\partial T_{kl}}{\partial x_l} + T_{kj}\right) dV$$

where we used the fact that:

$$\frac{\partial x_j}{\partial x_l} = \delta_{jl}.$$

Further, using (11.6):

$$\frac{\partial T_{kl}}{\partial x_l} = -f_k.$$

We obtain for this second term of (11.7):

$$\iint \delta_{ijk} x_j T_{kl} n_l dS = \iiint \delta_{ijk}(-x_j f_k + T_{kj}) dV$$

and Equation (11.7) becomes:

$$\iiint \delta_{ijk} x_j f_k dV + \iiint \delta_{ijk}(-x_j f_k + T_{kj}) dV = 0$$

or

$$\iiint \delta_{ijk} T_{kj} dV = 0$$

whatever the volume element dV. Hence $\delta_{ijk} T_{kj} = 0$.

For $i = 1$, we obtain $T_{32} - T_{23} = 0$. In the same way, for $i = 2$ we obtain $T_{13} = T_{31}$ and for $i = 3$ we have $T_{21} = T_{12}$. Thus we showed that

$$T_{ij} = T_{ji}.$$

The stress tensor is symmetric.

11.5. Examples of stress tensors

In a material, the stress tensor $T_{ij}(\mathbf{r})$ results from the application of surface forces on the external surface Σ of the material, $(\mathbf{T_n})_\Sigma$. $T_{ij}(\mathbf{r})$ is determined by solving Equations (11.4) with the boundary conditions on surface Σ given by the values of $(\mathbf{T_n})_\Sigma$ at all points of Σ. The solution of this kind of problems is dealt with in treatises on the mechanics of continuous media. In many cases, the solution is a non-uniform stress. It can nevertheless be considered as locally uniform by considering small volume elements, as we will do in the following paragraphs.

In experiments designed to measure the elastic properties of materials, the shape of the objects and the applied force configuration are usually designed so that the stress tensor is almost uniform (Sect. 13.1).

11.5.1. Uniaxial stress

The stress tensor associated to a uniaxial stress is expressed, in a given frame $(Ox_1x_2x_3)$, by:

$$\begin{pmatrix} T & 0 & 0 \\ 0 & 0 & 0 \\ 0 & 0 & 0 \end{pmatrix}.$$

Consider a small cube, with sides parallel to the coordinate axes, and with unit length (Fig. 11.6), and apply relation $\mathbf{T_n} = [T]\mathbf{n}$ to its faces. The faces normal to Ox_2 and Ox_3 are not submitted to any force. Vector $\mathbf{n}$, associated to face ABCD, has the same direction as Ox_1, so that $\mathbf{n} = \mathbf{i}$ and $\mathbf{T_n} = T\mathbf{i}$. If T is positive (as on Fig. 11.6), $\mathbf{T_n}$ is a tensile force[2] and, if T is negative, $\mathbf{T_n}$ is a compressive force. Vector $\mathbf{n}$ associated to the opposite face of the cube, OFGH, is equal to $-\mathbf{i}$, and $\mathbf{T_n} = -T\mathbf{i}$. We retrieve a tensile force if T is positive, and a compressive force if T is negative.

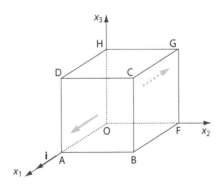

Figure 11.6: Uniaxial tensile stress applied on a small cube with unit sides

---

2. $\mathbf{T_n}$ is a force per unit area but here, for simplicity, we assume the surface areas to be equal to unity.

11.5.2. Pure shear

The stress tensor associated to a pure shear stress is expressed, in a given frame $(Ox_1x_2x_3)$, as:

$$\begin{pmatrix} 0 & T & 0 \\ T & 0 & 0 \\ 0 & 0 & 0 \end{pmatrix}.$$

Consider, as above, a small cube with sides parallel to the coordinate axes, and with unit length. Figure 11.7 shows the forces acting on it if T is positive.

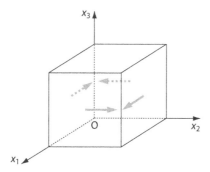

Figure 11.7: Pure shear stress acting on a small cube with unit sides

If we rotate the coordinate system by $45°$ around Ox_3, the tensor, when expressed in this new set of axes, takes the form (Exercise 11.3):

$$\begin{pmatrix} T & 0 & 0 \\ 0 & -T & 0 \\ 0 & 0 & 0 \end{pmatrix}.$$

This is a bi-axial stress acting on the surfaces enclosing an angle of $45°$ wih the surfaces perpendicular to Ox_1 and Ox_2 (Fig. 11.8b).

11.5.3. Hydrostatic pressure

Regardless of the orientation of the surface on which it acts, hydrostatic stress is normal to the surface and has the same value p.

This stress is represented by tensor:

$$\begin{pmatrix} -p & 0 & 0 \\ 0 & -p & 0 \\ 0 & 0 & -p \end{pmatrix}$$

which can be written as: $T_{ij} = -p\delta_{ij}$, where δ_{ij} is the Kronecker symbol. In this case, the three components dF_i of the force $d\mathbf{F}$ acting on a surface element

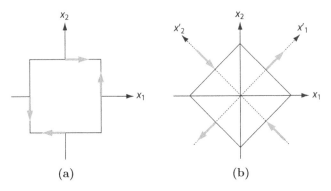

Figure 11.8: (a) Pure shear stress acting on the faces of a small cube with unit sides. (b) This stress is equivalent to a bi-axial stress involving tensile and compressive components, with the same norm, acting on faces at 45° to the initial ones.

dS perpendicular to unit vector $\mathbf{n}$, with direction cosines n_i, are (Eq. 11.2):

$$dF_i = T_{ij}\, n_j dS$$
$$= -p\, \delta_{ij} n_j dS = -p\, n_i dS$$
$$\text{and} \qquad d\mathbf{F} = -p\, \mathbf{n} dS$$

which means a force parallel to $\mathbf{n}$ and with value $p\, dS$ whatever the orientation of $\mathbf{n}$. The minus sign indicates that the pressure force points toward the inside of the volume element (compressive force).

11.6. Effect of gravity

Consider a bar in the shape of a rectangle parallelepiped with square cross-section and height h, resting in equilibrium on a plane surface (Fig. 11.9). This bar is submitted to no external stress on any of its faces, except for the one which rests on the surface (stress from the supporting surface). We choose a coordinate system $(Ox_1 x_2 x_3)$ parallel to the sides of the parallelepiped, with the axis Ox_3 vertical and pointing downward, and the origin O in the plane of the upper face of the bar. The system of Equations (11.4) takes the form:

$$\frac{\partial T_{1j}}{\partial x_j} = 0 \tag{11.8a}$$

$$\frac{\partial T_{2j}}{\partial x_j} = 0 \tag{11.8b}$$

$$\frac{\partial T_{3j}}{\partial x_j} + \rho g = 0 \tag{11.8c}$$

where ρ is the mass per unit volume of the material and g the acceleration of gravity.

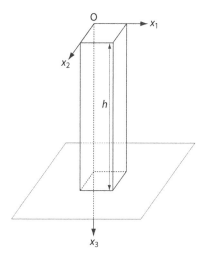

Figure 11.9: Bar in the shape of a rectangle parallelepiped with square cross-section, with height h, resting in equilibrium on a plane surface

The faces perpendicular to Ox_1 and Ox_2 are submitted to no stress, so that $T_{j2} = T_{j1} = 0$ on the faces. In a first approximation, we may therefore consider that, in a cross section of the bar at $x_3 = $ const, the stress is uniform, so that $T_{j2} = T_{j1} = 0$ and $T_{j3} = $ const over the whole surface.

We deduce that:
$$\frac{\partial T_{j2}}{\partial x_2} = \frac{\partial T_{j2}}{\partial x_1} = \frac{\partial T_{j1}}{\partial x_1} = \frac{\partial T_{j1}}{\partial x_2} = 0$$

and in particular
$$\frac{\partial T_{32}}{\partial x_2} = \frac{\partial T_{31}}{\partial x_1} = 0.$$

Equation (11.8c) then reduces to:
$$\frac{\partial T_{33}}{\partial x_3} + \rho g = 0.$$

The solution of this equation is:
$$T_{33} = -\rho g x_3 + C(x_3).$$

C is a constant which depends neither on x_1 nor on x_2 since the stress is supposed to be uniform at constant x_3.

Face $x_3 = 0$ is submitted to no stress, hence $T_{33}(x_3 = 0) = 0$. We deduce that $C(x_3) = 0$ and that
$$T_{33} = -\rho g x_3.$$

On the lower face of the bar ($x_3 = h$), $T_{33} = -\rho g h$.

The force acting on this face of the bar, with area S, is then equal to $-\rho g h S$, *i.e.* to $-P$ if P is the weight of the bar. The surface on which it rests exerts upon the bar a force equal and opposite to its weight, as expected.

In metallurgy, test specimens have height around 100 mm. The stress due to gravity therefore varies from zero to the maximum value $\rho g h$ which, in the case of a steel specimen for example ($\rho = 7.8$ kg dm^{-3}) is equal to $(7.8 \times 10^3) \times 9.81 \times 0.1 = 7.7 \times 10^3$ Pa. This stress is therefore completely negligible with respect to the stresses, of the order of a MPa, applied to the specimens in the elastic range. Gravity can thus be neglected.

11.7. Exercises

Exercise 11.1

Consider a uniaxial stress T acting on a surface perpendicular to the vector $\mathbf{n}$, with direction cosines n_1, n_2, n_3 in an orthonormal axis system $(Ox_1x_2x_3)$.

Determine the components of the stress tensor in the axis system $(Ox_1x_2x_3)$.

First express the tensor in a system of axes where one of the axes, say Ox_1', is parallel to $\mathbf{n}$, and then perform the axis change.

Exercise 11.2

Consider a stress tensor which, in a given orthonormal axis system $(Ox_1x_2x_3)$, takes the form:

$$\begin{pmatrix} 0 & 0 & 0 \\ 0 & 0 & 0 \\ 0 & 0 & T \end{pmatrix}.$$

1. Calculate the moduli T_ν and τ of the normal and shear stresses respectively, to which a surface perpendicular to the unit vector $\mathbf{n}$ with direction cosines n_1, n_2, n_3 is submitted.

2. For what directions of vector $\mathbf{n}$ is the shear stress maximum?

Exercise 11.3

Consider a pure shear tensor represented in an orthonormal axis system $(Ox_1x_2x_3)$ by:

$$\begin{pmatrix} 0 & T & 0 \\ T & 0 & 0 \\ 0 & 0 & 0 \end{pmatrix}.$$

Give the form of this tensor in an axis system deduced from $(Ox_1x_2x_3)$ by a 45° rotation around Ox_3.

Exercise 11.4

The stress tensor $[T]$ at a point M in a material is given, in an orthonormal axis system $(Ox_1x_2x_3)$, by:

$$\begin{pmatrix} 0.7 & 3.6 & 0 \\ 3.6 & 2.8 & 0 \\ 0 & 0 & 7.6 \end{pmatrix} 10^7 \text{ N m}^{-2}.$$

1. Determine the principal stresses and their directions.

2. Retrieve these results using the Mohr circle method.

3. Calculate the stress T acting on a small surface, centered at M, whose normal has direction cosines $(\sqrt{3}/2,\ 1/2,\ 0)$. Find the normal stress and shear stress components.

References

[1] D. Calecki, *Physique des milieux continus*, vol. 1, *Mécanique et thermo-dynamique* (Hermann, Paris, 2007)

[2] L.D. Landau and E.M. Lifshitz, *Course of Theoretical Physics*, vol. 7, *Theory of Elasticity*, 3rd edn. (Elsevier, Oxford, 1986)

Chapter 12

Deformation of a solid. The strain tensor

This chapter introduces the displacement gradient tensor, or distortion tensor, which describes the most general distortion of a crystal. This field tensor of rank 2 is shown to decompose into the strain tensor, which is symmetric, and an antisymmetric tensor associated to a rotation. The last section is devoted to the thermal expansion of crystals.

12.1. Distortion tensor (displacement gradient tensor)

12.1.1. Definition

We consider a material submitted to a deformation. This deformation displaces point P, defined by its position vector $\mathbf{OP} = \mathbf{r}$, to point P' such that $\mathbf{PP'} = \mathbf{u(r)}$ (Fig. 12.1). Vector $\mathbf{u(r)}$ is the displacement vector at point P. It is not uniform, because if it were, the whole material would simply be translated and it would not be deformed. A neighboring point Q such that $\mathbf{PQ} = \Delta\mathbf{r}$ displaces to Q' such that $\mathbf{QQ'} = \mathbf{u(r + \Delta r)}$.

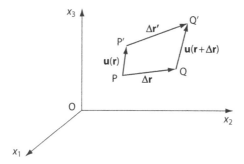

Figure 12.1: Transformation of vector $\mathbf{PQ} = \Delta\mathbf{r}$ into vector $\mathbf{P'Q'} = \Delta\mathbf{r'}$ after deformation

© Springer Science+Business Media Dordrecht 2014
C. Malgrange et al., *Symmetry and Physical Properties of Crystals*,
DOI 10.1007/978-94-017-8993-6_12

Vector $\mathbf{PQ}$ becomes, after transformation, vector $\mathbf{P'Q'} = \Delta\mathbf{r}'$ such that:

$$\mathbf{P'Q'} = \mathbf{P'P} + \mathbf{PQ} + \mathbf{QQ'}$$

so that
$$\Delta\mathbf{r}' = \Delta\mathbf{r} + \mathbf{u}(\mathbf{r} + \Delta\mathbf{r}) - \mathbf{u}(\mathbf{r})$$

or
$$\Delta\mathbf{r}' = \Delta\mathbf{r} + \Delta\mathbf{u} \qquad (12.1)$$

if we note $\Delta\mathbf{u} = \mathbf{u}(\mathbf{r} + \Delta\mathbf{r}) - \mathbf{u}(\mathbf{r})$.

We can replace $\mathbf{u}(\mathbf{r} + \Delta\mathbf{r})$ by its expansion to first order, and thus obtain for the three components of $\Delta\mathbf{u}$:

$$\Delta u_i = \frac{\partial u_i}{\partial x_j} \Delta x_j. \qquad (12.2)$$

Setting
$$\frac{\partial u_i}{\partial x_j} = e_{ij},$$

we obtain:
$$\Delta u_i = e_{ij} \Delta x_j. \qquad (12.3)$$

Coefficients e_{ij} connect the components of a vector $\Delta\mathbf{r}$ to the components of vector $\Delta\mathbf{u}$. They are therefore the components of a tensor (see Sect. 9.4.2) called the displacement gradient tensor, or distortion tensor, and we can write:

$$\Delta\mathbf{u} = [e]\Delta\mathbf{r}. \qquad (12.4)$$

Thus
$$\Delta\mathbf{r}' = \Delta\mathbf{r} + [e]\Delta\mathbf{r} = ([E] + [e])\Delta\mathbf{r} \qquad (12.5)$$

or
$$\Delta x_i' = \Delta x_i + e_{ij}\Delta x_j = (\delta_{ij} + e_{ij})\Delta x_j$$

where $[E]$ is the unit tensor.

If components e_{ij} do not depend on the position of the point, the distortion is uniform, and there are no additional terms in the expansion (12.2). Relations (12.4) and (12.5) are then valid even if points P and Q are not near each other, and the linear relation between $\Delta\mathbf{r}'$ and $\Delta\mathbf{r}$ shows that a line remains a line and a plane remains a plane after distortion.

If the e_{ij} vary from point to point, the distortion is no more uniform, but Equations (12.4) and (12.5) remain valid locally, which implies small $\Delta\mathbf{r}$. We can thus consider, as is done in this chapter, that the distortions are uniform: if they are not, the properties obtained are valid locally.

Further, we will only consider, in what follows, very small displacement gradients (distortions), i.e. $e_{ij} \ll 1$.

12.1.2. Physical meaning of components e_{ij}

We will simplify the interpretation of components e_{ij} by considering a two-dimensional system, in which we choose an orthonormal axis system (Ox_1x_2).

Equations (12.3) now become the following detailed equations:

$$\Delta u_1 = e_{11}\Delta x_1 + e_{12}\Delta x_2, \tag{12.6}$$
$$\Delta u_2 = e_{21}\Delta x_1 + e_{22}\Delta x_2. \tag{12.7}$$

Consider a rectangle whose sides PQ and PS, respectively parallel to Ox_1 and Ox_2, have components $(\Delta x_1, 0)$ and $(0, \Delta x_2)$ respectively. Vectors **PQ** and **PS** become, after deformation, **P′Q′** and **P′S′** respectively. Consider for a start vector **PQ**, which turned into **P′Q′**, and apply relations (12.6) and (12.7):

$$\Delta u_1 = e_{11}\Delta x_1,$$
$$\Delta u_2 = e_{21}\Delta x_1.$$

PQ has rotated by angle θ, and its length has changed. As we immediately see on Figure 12.2, $\tan\theta = \Delta u_2/(\Delta x_1 + \Delta u_1)$. Since we assume that $e_{ij} \ll 1$, Δu_1 and Δu_2 are much smaller than Δx_1 and $\tan\theta \cong \Delta u_2/\Delta x_1 = e_{21}$, and $\tan\theta \cong \theta$. We can write:

$$\theta = e_{21}.$$

Thus e_{21} is the angle by which **PQ**, parallel to Ox_1, has rotated. It is counted as positive in the sense from Ox_1 to Ox_2 (direct sense). The same argument applied to **PS** shows that e_{12} is the angle θ' by which **PS**, parallel to Ox_2, has turned, counting it as positive from Ox_2 to Ox_1 (clockwise).[1]

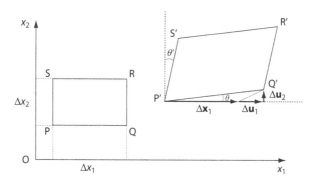

Figure 12.2: Distortion of rectangle PQRS, which becomes P′Q′R′S′. The lengths Δu_1 and Δu_2 are hugely magnified in order to make angle θ visible.

---

1. It is obvious that, to keep Figure 11.2 and the following ones clear, it is necessary to show lengths Δu_1 and Δu_2 much larger than in reality. Therefore, they do not satisfy the condition $e_{ij} \ll 1$, and angles θ and θ' are also too large.

We now calculate the variation in length Δx_1 of **PQ**, which became **P'Q'**: we have $P'Q' - PQ = P'Q' - \Delta x_1$. We see on Figure 12.2 that $\Delta x_1 + \Delta u_1 = P'Q' \cos \theta \cong P'Q'$ since $\theta \ll 1$.

Thus $P'Q' \cong PQ + \Delta u_1$, and we obtain

$$\frac{P'Q' - PQ}{PQ} = \frac{\Delta u_1}{\Delta x_1} = e_{11}.$$

The relative elongation of **PQ** is thus equal to e_{11}.

The same argument applied to **PS** shows that e_{22} is the relative elongation of **PS**, parallel to Ox_2.

These results are easily generalized to three dimensions. We can thus state that:

– *each diagonal term[2] e_{ii} in the distortion tensor represents the relative elongation of a segment parallel to axis Ox_i,*

– *the components e_{ij}, where $i \neq j$, describe the rotation of these segments.*

12.2. Decomposition of the distortion tensor into a rotation and a strain

12.2.1. Introduction using a simple example

Consider the very simple distortion tensor $[T]$ represented by matrix

$$\begin{pmatrix} 0 & -\alpha & 0 \\ \alpha & 0 & 0 \\ 0 & 0 & 0 \end{pmatrix}.$$

This distortion does not affect a vector parallel to Ox_3, and we can therefore focus on the distortion in plane (Ox_1, Ox_2) and investigate what a small square PQRS becomes after distortion (Fig. 12.3). The diagonal terms being zero, the segments parallel to Ox_1 and Ox_2 retain their lengths. Segment **PQ** undergoes a rotation by α in the direct sense and segment **PS** a rotation $-\alpha$ in the clockwise direction, *i.e.* a rotation by α in the direct sense. Thus the square has been transformed into an identical square, rotated by angle α in the direct sense. This shows that there are distortions which do not deform the solid, but just rotate it.

---

2. Here, Einstein's convention does not apply to notation e_{ii}, *i.e.* no sum is implied.

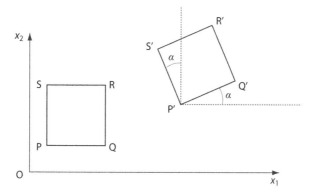

Figure 12.3: Transformation of square PQRS when submitted to a special distortion involving a simple rotation

12.2.2. Expressing the distortion associated to small rotations

Consider a small rotation by angle $d\alpha$ around an axis with unit vector $\mathbf{n}$ going through point O. Let M′ be the transform of point M in this rotation (Fig. 12.4).

$$\mathbf{OM'} = \mathbf{OM} + \mathbf{MM'}.$$

The plane perpendicular to the rotation axis, going through M, intersects this axis at H.

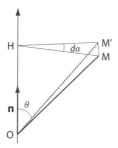

Figure 12.4: Transformation of vector $\mathbf{OM}$ into vector $\mathbf{OM'}$ through a small rotation, by angle $d\alpha$, around an axis with unit vector $\mathbf{n}$ going through point O

Since we assume angle $d\alpha$ to be very small, arc $\mathbf{MM'}$ and its tangent at M are very close to each other. $\mathbf{MM'}$ is perpendicular to $\mathbf{n}$ and also to $\mathbf{HM}$ since the rotation angle $d\alpha$ is small. $\mathbf{MM'}$, being perpendicular to plane $(\mathbf{OH}, \mathbf{HM})$, is thus perpendicular to $\mathbf{OM}$ and to $\mathbf{n}$, hence parallel to vector $\mathbf{n} \times \mathbf{OM}$. Its length MM′ is

$$MM' = HM\, d\alpha = OM \sin\theta\, d\alpha$$

where θ is the angle between **OM** and **n**. We thus obtain

$$MM' = (\mathbf{n} \times \mathbf{OM})\, d\alpha$$

and $\quad \mathbf{OM}' = \mathbf{OM} + (\mathbf{n} \times \mathbf{OM})\, d\alpha. \hspace{2cm} (12.8)$

If we denote as n_i the components of **n** and x_i the components of **OM** in any given orthonormal axis system $(Ox_1x_2x_3)$, the vector $\mathbf{n} \times \mathbf{OM}$ has for components, in this axis system

$$\begin{cases} n_2x_3 - n_3x_2 \\ n_3x_1 - n_1x_3 \\ n_1x_2 - n_2x_1 \end{cases}$$

and Equation (12.8) can be written as

$$\mathbf{OM}' = ([E] + [\Omega])\mathbf{OM}$$

where $[\Omega]$ is a tensor (since it relates vector $\mathbf{OM}' - \mathbf{OM}$ to vector **OM**) which, in the orthonormal basis chosen, is expressed as

$$[\Omega] = \begin{pmatrix} 0 & -n_3 & n_2 \\ n_3 & 0 & -n_1 \\ -n_2 & n_1 & 0 \end{pmatrix}$$

and $[E]$ is the unit tensor.

We deduce that any vector **r** transforms, under rotation by a small angle $d\alpha$ around **n**, into vector $\mathbf{r}'$ such that

$$\mathbf{r}' = ([E] + [\Omega])\mathbf{r} \hspace{2cm} (12.9)$$

where $[\Omega]$ is an antisymmetric tensor. We note that the special case we discussed in Section 12.2.1 satisfies this result.

Since this relation is linear, it applies to any vector $\Delta\mathbf{r}$, irrespective of whether its origin is on the rotation axis.

Relation (12.9) then becomes:

$$\Delta\mathbf{r}' = ([E] + [\Omega])\Delta\mathbf{r}$$

or $\quad \Delta\mathbf{u} = \Delta\mathbf{r}' - \Delta\mathbf{r} = [\Omega]\Delta\mathbf{r}.$

$[\Omega]$, an antisymmetric tensor, is associated to the rotation vector $\mathbf{n}d\alpha$. We thus find a second example of an axial vector, the rotation vector $\mathbf{n}d\alpha$ being equivalent to the antisymmetric second-rank tensor $[\Omega]d\alpha$ (see Sect. 10.6.2).

12.2.3. Strain tensor

A general distortion tensor can be decomposed into the sum of a symmetric tensor $[S]$ and an antisymmetric tensor $[A]$ (Sect. 10.1.1), which we will denote here as $[\Omega]$, such that:

$$S_{ij} = \frac{e_{ij} + e_{ji}}{2} \hspace{2cm} \Omega_{ij} = \frac{e_{ij} - e_{ji}}{2},$$

so that

$$S_{ij} = \frac{1}{2}\left(\frac{\partial u_i}{\partial x_j} + \frac{\partial u_j}{\partial x_i}\right) \tag{12.10}$$

$$\Omega_{ij} = \frac{1}{2}\left(\frac{\partial u_i}{\partial x_j} - \frac{\partial u_j}{\partial x_i}\right). \tag{12.11}$$

Relation (12.5) becomes

$$\Delta \mathbf{r}' = ([E] + [e])\Delta \mathbf{r} = ([E] + [S] + [\Omega])\Delta \mathbf{r}.$$

The distortions we consider here are always small, hence $e_{ij} \ll 1$, and we can write:

$$([E] + [S] + [\Omega]) \cong ([E] + [S])([E] + [\Omega]),$$

and $\quad \Delta \mathbf{r}' = ([E] + [S])([E] + [\Omega])\Delta \mathbf{r}.$

The transformation of vector $\Delta \mathbf{r}$ into $\Delta \mathbf{r}'$ is the product of a rotation $([E]+[\Omega])$ and a deformation $([E] + [S])$.[3] $[S]$ is the symmetric part of the distortion tensor: it is called the strain tensor. $[\Omega]$, its antisymmetric part, is called the rotation tensor.

Generally, interest focuses only on the change in shape of the object, hence on the strain tensor $[S]$, a symmetric tensor. The diagonal terms in $[S]$ are equal to those of $[e]$, and they thus represent relative elongations (see Sect. 12.1.2). Thus S_{11} represents the relative elongation of a segment parallel to axis Ox_1, S_{22} that of a segment parallel to Ox_2 and S_{33} that of a segment parallel to Ox_3.

The strain tensor is a field tensor. It usually results from another field tensor being applied to the crystal, for example a stress, defined in Chapter 11, an electric field if the crystal is piezoelectric (Chap. 15), or a variation in temperature as we will see in Section 12.6.

12.3. Elongation in a given direction

There is often interest in the relative elongation, in a given direction parallel to unit vector $\mathbf{s}$, of a material submitted, in the vicinity of point P, to a given strain tensor $[S]$. If $[S]$ is uniform, $\Delta \ell / \ell|_\mathbf{s}$ represents the relative elongation of a bar parallel to $\mathbf{s}$. The tensor is given through its components S_{ij} in a given orthonormal axis system $(Ox_1x_2x_3)$. Let $(Ox_1'x_2'x_3')$ be a new orthonormal axis system such that Ox_1' is parallel to $\mathbf{s}$. We saw in last section that the relative elongation in direction Ox_1' is given by S_{11}'. Thus

$$\left.\frac{\Delta \ell}{\ell}\right|_\mathbf{s} = S_{11}' = a_{1i}a_{1j}S_{ij},$$

---

3. The order of the operations is irrelevant: since $\Omega_{ij} \ll 1$ and $S_{ij} \ll 1$, $([E] + [S])([E] + [\Omega]) \cong ([E] + [\Omega])([E] + [S])$.

where $\{a_{ij}\}$ is the transformation matrix leading from axis system $(Ox_1x_2x_3)$ to axis system $(Ox_1'x_2'x_3')$. The basis vector $\mathbf{e}_1'$ of the new system is none other than $\mathbf{s}$, hence $a_{1i} = s_i$. We obtain

$$\left.\frac{\Delta\ell}{\ell}\right|_{\mathbf{s}} = s_i s_j S_{ij} \tag{12.12}$$

or, in matrix form,

$$\left.\frac{\Delta\ell}{\ell}\right|_{\mathbf{s}} = (\, s_1 \; s_2 \; s_3 \,) \left(\begin{array}{c} \\ S_{ij} \\ \\ \end{array}\right) \left(\begin{array}{c} s_1 \\ s_2 \\ s_3 \end{array}\right).$$

This relative elongation is equal to the reciprocal of the square of the radius vector of the quadric associated to tensor $[S]$ (Sect. 10.3.2).

12.4. Volume expansion (volumetric strain)

When described in the axis system $(Ox_1x_2x_3)$ of its principal axes, a strain tensor $[S]$, with components S_{ij}, is diagonal. Consider a small rectangle parallelepiped with sides parallel to Ox_1, Ox_2 and Ox_3 and with lengths respectively Δx_1, Δx_2, Δx_3 (Fig. 12.5).

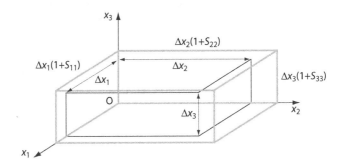

Figure 12.5: A rectangle parallelepiped with sides parallel to principal axes Ox_1, Ox_2 and Ox_3, with lengths respectively Δx_1, Δx_2, Δx_3, shown before and after deformation

This rectangle parallelepiped, with volume ΔV equal to $\Delta x_1 \Delta x_2 \Delta x_3$, becomes, under this deformation, another rectangle parallelepiped since only the diagonal terms of the tensor are non-zero. Its sides have lengths respectively $\Delta x_1(1 + S_{11})$, $\Delta x_2(1 + S_{22})$ and $\Delta x_3(1 + S_{33})$.

The volume $\Delta V'$ of the deformed parallelepiped is:

$$\Delta V' = \Delta V(1 + S_{11})(1 + S_{22})(1 + S_{33}) \approx \Delta V(1 + S_{11} + S_{22} + S_{33}).$$

The relative variation in volume is:

$$\frac{\Delta V' - \Delta V}{\Delta V} = S_{11} + S_{22} + S_{33},$$

equal to the trace of the matrix representing tensor $[S]$ in the coordinate system of its principal axes. We saw (Sect. 10.1.3) that the trace of a tensor is independent of the coordinate system in which it is expressed. Therefore the relative variation in volume (volume expansion, or volumetric strain) is equal to the trace of the strain tensor whatever the coordinate system in which the tensor is expressed.

12.5. Special cases of strain

12.5.1. Simple elongation

The strain tensor associated to simple elongation, expressed in a given axis system $(Ox_1x_2x_3)$, has a single non-zero term, located on the diagonal; for example:

$$\begin{pmatrix} 0 & 0 & 0 \\ 0 & 0 & 0 \\ 0 & 0 & S \end{pmatrix}.$$

A small rectangle parallelepiped with sides parallel to the axes and with lengths respectively Δx_1, Δx_2, Δx_3 remains a rectangle parallelepiped, in which only the side parallel to Ox_3 has suffered a relative elongation equal to S.

12.5.2. Pure shear strain

The associated strain tensor is expressed, in a given coordinate system $(Ox_1x_2x_3)$, as:

$$\begin{pmatrix} 0 & S & 0 \\ S & 0 & 0 \\ 0 & 0 & 0 \end{pmatrix}.$$

It is easy to show (Exercise 12.2) that, in the coordinate system $(Ox_1'x_2'x_3)$ rotated by $45°$ around axis Ox_3, this tensor takes on the form:

$$\begin{pmatrix} S & 0 & 0 \\ 0 & -S & 0 \\ 0 & 0 & 0 \end{pmatrix}$$

which corresponds to an expansion along Ox_1' and a contraction along Ox_2'.

A small square PQRT, with sides parallel to Ox_1 and Ox_2 respectively, becomes after deformation a lozenge (Fig. 12.6). The length of the sides of the lozenge is equal to that of the square since, in the initial coordinate system, $S_{11} = S_{22} = 0$.

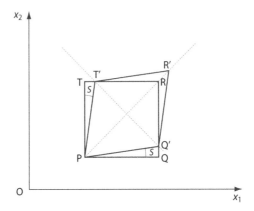

Figure 12.6: Pure shear of the small square PQRT, with sides parallel to Ox_1 and Ox_2, which becomes the lozenge PQ'R'T'

12.5.3. Simple shear strain

The distortion tensor $[e]$ in simple shear is expressed in a given frame $(Ox_1x_2x_3)$ as:

$$\begin{pmatrix} 0 & \gamma & 0 \\ 0 & 0 & 0 \\ 0 & 0 & 0 \end{pmatrix}.$$

A small square PQRS with sides parallel to Ox_1 and Ox_2 becomes, after this distortion, a lozenge PQR''S'' because the tensor has no diagonal terms (Fig. 12.7a). The side parallel to Ox_2 rotates by $-\gamma$ (see Sect. 12.2.1). Because the tensor is not symmetric, we decompose it into a strain and a rotation:

$$[S] = \begin{pmatrix} 0 & \gamma/2 & 0 \\ \gamma/2 & 0 & 0 \\ 0 & 0 & 0 \end{pmatrix} \quad \text{and} \quad [\Omega] = \begin{pmatrix} 0 & \gamma/2 & 0 \\ -\gamma/2 & 0 & 0 \\ 0 & 0 & 0 \end{pmatrix}.$$

Simple shear thus consists in the combination of a rotation by angle $-\gamma/2$ around Ox_3 and pure shear (Fig. 12.7b).

12.6. Thermal expansion

Suppose we submit a bar of an isotropic material, with length ℓ_0 at temperature T_0, to a temperature change ΔT. Its length becomes ℓ such that:

$$\ell = \ell_0(1 + \alpha \Delta T)$$

where α is the linear thermal expansion coefficient of the material. In a general way, an isotropic solid submitted to a uniform rise in temperature expands while retaining its shape, and vector $\Delta \mathbf{r}$ in the material becomes, after a change in temperature ΔT, a vector $\Delta \mathbf{r}'$ parallel to $\Delta \mathbf{r}$ and such that:

$$\Delta \mathbf{r}' = \Delta \mathbf{r}(1 + \alpha \Delta T) = \Delta \mathbf{r} + \Delta \mathbf{u}$$

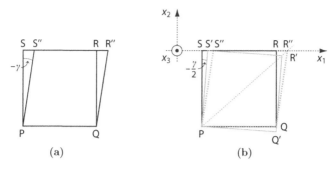

Figure 12.7: (a) Simple shear of the small square PQRS, which becomes lozenge PQR″S″. (b) This transformation can be decomposed into two successive transformations: rotation by $-\gamma/2$ (the square PQRS becomes square PQ′R′S′), and then a deformation of PQ′R′S′ analogous to that of Figure 12.6.

with
$$\Delta\mathbf{u} = \alpha\Delta T\Delta\mathbf{r}.$$

If the solid is not isotropic, the vector $\Delta\mathbf{u}$ is, in general, no more parallel to $\Delta\mathbf{r}$ and its expression at a function of $\Delta\mathbf{r}$ is given by a strain tensor $[S]$ proportional to the change in temperature ΔT:

$$\Delta\mathbf{u} = [S]\Delta\mathbf{r} = [\alpha]\Delta\mathbf{r}\Delta T$$

or $\quad \Delta u_i = \alpha_{ij}\Delta x_j\Delta T,$

and $\quad S_{ij} = \alpha_{ij}\Delta T.$ $\qquad(12.13)$

Tensor $[\alpha]$ thus defined is the thermal expansion tensor.

This second-rank tensor is symmetric and represents a physical property of the material. It relates an external applied constraint (the change in temperature) to the effect it induces (strain). It is a material tensor, which must be invariant with respect to the crystal's symmetry elements. The number of independent coefficients of symmetric rank-2 tensors was determined (Sect. 10.5) for the various point groups. This number depends only on the crystal system because it is found to be identical for all the point groups of a given crystal system. We thus obtain for the various crystal systems the following results:

Triclinic system: 6 independent coefficients.

Monoclinic system: one principal axis of the tensor is parallel to the 2-fold axis (groups 2 and 2/m) or perpendicular to the mirror (group m). There are four independent coefficients. The ellipsoid representing the tensor has one of its axes parallel to the 2-fold axis or perpendicular to the mirror.

Orthorhombic system: the principal axes are parallel to the 2-fold axes or perpendicular to the mirrors. They are the axes of the ellipsoid representing the tensor. There are three independent coefficients.

Tetragonal, rhombohedral and hexagonal systems: the ellipsoid representing the tensor is an ellipsoid of revolution around the axis with order larger than 2. There are two independent coefficients.

Cubic system: the representation ellipsoid is a sphere. The crystal behaves with respect to thermal expansion as an isotropic material. There is only one coefficient.

We note that the strain induced by a change in temperature is uniform if the temperature is uniform.

The coefficients α_{ij} are in general all positive. Nevertheless there are a few cases where one of the coefficients is negative (calcite or beryl for example). In these special cases, there are directions in which the expansion coefficient is zero. Consider for example calcite, with point group $\bar{3}$m, for which the thermal expansion tensor $[\alpha]$ is expressed as:

$$\begin{pmatrix} -5 & 0 & 0 \\ 0 & -5 & 0 \\ 0 & 0 & 27 \end{pmatrix} 10^{-6} \, {}^\circ C^{-1}$$

when expressed in an axis system where Ox_3 is parallel to the 3-fold axis.[4] The relative elongation $\Delta\ell/\ell|_s$, in a given direction with unit vector s, is given by Equation (12.12):

$$\left.\frac{\Delta\ell}{\ell}\right|_s = s_i \, S_{ij} \, s_j = s_i \, s_j \, \alpha_{ij} \, \Delta T.$$

If $\Delta T = 1^\circ C$, we obtain for calcite:

$$\left.\frac{\Delta\ell}{\ell}\right|_s = [-5(s_1^2 + s_2^2) + 27s_3^2] \, 10^{-6}$$

which, expressed in spherical coordinates, becomes:

$$\left.\frac{\Delta\ell}{\ell}\right|_s = [-5\sin^2\theta + 27\cos^2\theta] \, 10^{-6}.$$

The surface representing this relative elongation is obtained by giving any segment parallel to s drawn from a given point O the length $\Delta\ell/\ell|_s$. Here it is a surface of revolution around Ox_3. Figure 12.8a shows a section of this surface by a plane going through this axis of revolution.

We see that

$$\left.\frac{\Delta\ell}{\ell}\right|_s = 0 \qquad \text{if} \qquad \tan^2\theta = \frac{27}{5}$$

thus for all direction at angle $\theta = \theta_0 = 66.7^\circ$ (or its supplement $\theta = 113.3^\circ$) to axis Ox_3.

---

4. We recall that a $\bar{3}$-fold axis is equivalent to a 3-fold axis and a center of symmetry.

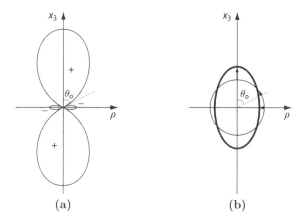

(a)   (b)

Figure 12.8: Thermal expansion of a crystal of calcite. The 3-fold axis, chosen for the axis Ox_3, is an axis of revolution and the figures represent any plane ($O\rho$, Ox_3) going through Ox_3. (a) Relative elongation for all directions of a plane going through axis Ox_3. (b) Transformation of a sphere with unit radius, which becomes an ellipsoid of revolution around Ox_3. The deformations are hugely exaggerated in order to be visible.

We saw (Eq. (10.6)) that, for a given tensor with components T_{ij}, the quantity $T_{ij}s_is_j$ is equal to $1/r^2$ where r is the length of the radius vector of the representation quadric for the tensor, defined by $T_{ij}s_is_j = 1$. This remark makes it possible to retrieve the above result, *viz.* the fact that in the directions of the cone asymptotic to the hyperboloids $-5(x_1^2 + x_2^2) + 27x_3^2 = \pm 1$, the relative elongation is zero since $r \to \infty$. For those directions that enclose with Ox_3 an angle smaller than the half-angle of the cone, the representation of the quadric is a two-sheet hyperboloid with equation $\alpha_{ij}x_ix_j = 1$ and the relative elongation is then equal to $1/r^2$. For directions that are outside this cone, the quadric is a one-sheet hyperboloid with equation $\alpha_{ij}x_ix_j = -1$ and the elongation is equal to $-1/r^2$. Figure 12.8b shows that a sphere with unit radius becomes, through thermal expansion, an ellipsoid with an expansion in the direction Ox_3 and a contraction in all directions of plane (Ox_1, Ox_2). We note that, in direction θ_0, although the radius vector does not expand, it nevertheless undergoes a transformation which is then only a rotation (Exercise 12.5).

Any change in temperature of a solid leads to a relative variation $\Delta V/V$ of its volume V. As we saw in Section 12.4, $\Delta V/V$ is the trace of the strain tensor:

$$\frac{\Delta V}{V} = S_{ii} = \alpha_{ii}\Delta T = (\alpha_{11} + \alpha_{22} + \alpha_{33})\Delta T$$

The bulk thermal expansion is α_{ii}.

If the material is isotropic, with linear expansion coefficient α, we obtain $\Delta V/V = 3\alpha\Delta T$, as expected.

When it is expressed in the coordinate system of its principal axes, tensor $[\alpha]$ is diagonal, and its diagonal terms α_{11}, α_{22}, α_{33} are called the principal expansion coefficients. For simplicity, they are noted α_1, α_2, α_3 respectively. A few examples of the numerical values of these coefficients are given in Table 12.1, in units of 10^{-6} °C^{-1}.

Table 12.1: Principal expansion coefficients for some crystals, in units of 10^{-6} °C^{-1}

Crystal	Crystal system	Temperature range	α_1	α_2	α_3
Zinc	Hexagonal	30 to 118°C	10.8	10.8	64.5
		around − 210°C	−2	−2	55
Quartz	Rhombohedral	around 25°C	13.4	13.4	7.97
Calcite	Rhombohedral	around 40°C	−5	−5	27
Copper	Cubic	around 25°C	16	16	16
Silicon	Cubic	around 25°C	2.35	2.35	2.35

12.7. Exercises

Exercise 12.1

A material is distorted in such a way that the components of the displacement $\mathbf{u}(P)$ at a point P with coordinates (x_1, x_2, x_3) in an orthonormal axis system are:

$$u_1 = 2 \times 10^{-5} x_1 - 10^{-5} x_2$$
$$u_2 = 2 \times 10^{-5} x_1 - 2 \times 10^{-5} x_2$$
$$u_3 = 0$$

1. Determine the components $e_{ij} = \partial u_i / \partial x_j$ of the distortion tensor $[e]$. Decompose tensor $[e]$ into the strain tensor and the rotation tensor.

2. Determine the relative elongation of a segment in a direction parallel to the vector with components $(1, 1, 1)$.

Exercise 12.2

A uniform strain tensor is represented in an orthonormal axis system $(Ox_1x_2x_3)$ by the matrix:

$$\begin{pmatrix} 0 & S & 0 \\ S & 0 & 0 \\ 0 & 0 & 0 \end{pmatrix}.$$

1. How does this strain transform a square with sides being parallel to Ox_1 and Ox_2 respectively and with unit length?

2. Determine the principal values and the principal axes of this tensor.

Exercise 12.3

Consider a homogeneous bar with the shape of an elongated rectangle parallelepiped (see figure below). Choose an axis system such that Ox_1 is parallel to the long side of the bar, Ox_2 and Ox_3 being parallel to the two other sides. This bar undergoes a uniform strain, the principal axes of which are Ox_1, Ox_2 and Ox_3, and the principal values $S_{11} = \eta$, $S_{22} = -0.3\eta$, $S_{33} = -0.3\eta$ with $\eta = 10^{-4}$. A line with length L, drawn on the bar in plane (Ox_1, Ox_2) in the undeformed state, encloses angle α with Ox_1 before deformation.

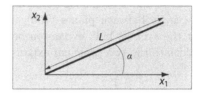

1. Calculate the relative elongation of the line.

2. After deformation, the line encloses with Ox_1 the angle $\alpha' = \alpha + \varepsilon$. Calculate the small angle ε.

Exercise 12.4

A small distortion of a crystal is defined by the following tensor:

$$
e_{ij} = \begin{pmatrix} 4 & 2 & -2 \\ -4 & 1 & 0 \\ 6 & 0 & 3 \end{pmatrix} 10^{-6}.
$$

Determine the strain tensor S_{ij} and the rotation tensor Ω_{ij}. Find the principal axes of tensor $[S]$, the rotation axis and the angle of rotation $\Delta\alpha$ corresponding to tensor $[\Omega]$.

Exercise 12.5

Graphite is hexagonal, with point group 6mm. In order to determine the coefficients of the thermal expansion tensor, the lattice parameters of a graphite crystal were measured at $20°C$ and at $100°C$. At $20°C$, the values are: $a_{20} = 2.4560$ Å, $c_{20} = 6.6960$ Å. At $100°C$, they are: $a_{100} = 2.4557$ Å, $c_{100} = 6.7110$ Å.

1. Calculate, in this temperature range, the average thermal expansion tensor[5] of graphite. Indicate what axis system it is expressed in.

2. Calculate the average thermal expansion coefficient in a general direction at angle θ to the 6-fold axis. Show that there are directions in which the elongation is zero, and that these directions form a cone; determine its apical angle θ_0. Give a graphical interpretation of this result using the representation quadric of this tensor.

3. Show that this special direction undergoes, in a temperature variation, a small rotation; calculate it.

Exercise 12.6

Consider a crystal of rutile TiO_2, with tetragonal symmetry and lattice parameters $a = b = 4.5933$ Å and $c = 2.9592$ Å at $0°C$. When it is heated from $0°C$ to $100°C$, the angle between planes (100) and (101) decreases by 0.095 mrad. Given that the average bulk expansion between $0°C$ and $100°C$ is $23.4 \times 10^{-6}\,°C^{-1}$, calculate the two principal expansion coefficients of this crystal.

Exercise 12.7

Barium titanate $BaTiO_3$ undergoes several phase changes (or phase transitions) at various temperatures: above $120°C$, $BaTiO_3$ is cubic. At $120°C$ it becomes tetragonal, and remains so down to $0°C$. At $0°C$, $BaTiO_3$ becomes orthorhombic and remains so down to $-100°C$, when it becomes rhombohedral. The point symmetry group and the lattice parameters of the various phases are given in the table below.

System	Cubic	Tetragonal	Orthorhombic	Rhombohedral
P. G.	m$\bar{3}$m	4mm	mm2	3m
T		120°C	0°C	$-100°C$
	120°C	**120°C**	**0°C**	**−100°C**
	$a = 4.010$ Å	$a = b = 4.002$ Å	$b_m = c_m = 4.012$ Å	$a = 3.996$ Å
		$c = 4.021$ Å	$a_m = 3.988$ Å	$\alpha = 89.86°$
		0°C	$\alpha = 90.20°$	
		$a = b = 3.991$ Å	**−100°C**	
		$c = 4.034$ Å	$b_m = c_m = 4.012$ Å	
			$a_m = 3.978$ Å	
			$\alpha = 90.40°$	

---

5. The lattice parameters do not vary linearly with temperature. The data given only allow calculating the average thermal expansion coefficients.

In the cubic $\leftrightarrow$ tetragonal transformation, the directions of the basis vectors remain unchanged.

In the tetragonal $\leftrightarrow$ orthorhombic transformation, the direction of $\mathbf{a}$ is unchanged. The rectangular face $(\mathbf{b}, \mathbf{c})$ becomes a lozenge with side $b_m = c_m \neq a$ and with angle $\alpha \neq 90°$, defining a monoclinic pseudo-unit cell. The orthorhombic unit cell is a unit cell with twice the volume of this one. The basis vector $\mathbf{a}_o$ is parallel to the vector $\mathbf{a}_c$ of the cubic unit cell. The vectors $\mathbf{b}_o$ and $\mathbf{c}_o$ are respectively parallel to the diagonals of the lozenge faces of the monoclinic pseudo-unit cell. The 3-fold axis of the rhombohedral phase coincides with one of the 3-fold axes of the cubic phase.

1. Calculate the components of the average[6] thermal expansion tensor of $BaTiO_3$ in the tetragonal phase. Deduce the expansion coefficient for a general direction. Does it vanish for some directions? Which ones?

2. Calculate the norms of the basis vectors of the orthorhombic unit cell at $0°C$ and at $-100°C$. Determine the thermal expansion tensor in this phase (express it in the orthonormal axis system parallel to the basis vectors of the orthorhombic unit cell).

3. Calculate the components of the strain tensor corresponding to the transition from the cubic cell (at $120°C$) to:

   (a) the tetragonal unit cell at $120°C$,

   (b) the orthorhombic unit cell at $0°C$,

   (c) the rhombohedral unit cell at $-100°C$.

   Indicate, in each case, the axis system in which the tensors are expressed.

---

6. See the note for Exercise 12.5.

Chapter 13

Elasticity

The stiffness and compliance tensors, which relate the stress tensor to the resulting strain tensor, are symmetric tensors of rank 4. This chapter shows how the intrinsic symmetry of the strain and stress tensors reduces the number of independent components of the compliance and stiffness tensors, and then how the symmetry of the crystal, described by its point group, can further reduce this number. A short notation, called Voigt's notation, is often used for these tensors. It represents the stress and the strain tensors through 6-component vectors, and the compliance and stiffness tensors by symmetric 6 × 6 matrices which are inverse to each other.

13.1. Introduction

In this chapter, we investigate the deformation of a crystalline material under the influence of a given stress. We first need to know the stress acting at any point in the material. We saw (Sect. 11.5) that the stress tensor $T_{ij}(\mathbf{r})$ in a material results from the application of surface forces on the external surface Σ of the material. For measurements to have an easy interpretation, the shape of the material and the configuration of the external forces to be applied are designed so that the stress tensor is practically uniform. The most usual experiments consist in measuring the strain resulting from a uniaxial stress. The typical specimen shape for this use is shown on Figure 13.1.

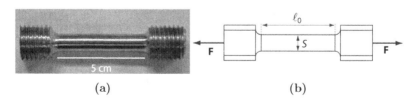

(a)   (b)

Figure 13.1: (a) Photograph of a tensile test specimen; (b) schematic diagram of a traction experiment. The uniaxial stress T in the central part, with cross section S, is $T=F/S$.

© Springer Science+Business Media Dordrecht 2014
C. Malgrange et al., *Symmetry and Physical Properties of Crystals*,
DOI 10.1007/978-94-017-8993-6_13

It consists of a long central cylindrical part with constant cross section, and of two ends with a cross section larger than the central part. The transition between the central part and the ends occurs progressively. The ends of these specimens are secured to the grips. One of them is fixed and the other one submitted to a traction force **F**.

The stress tensor in the central part, with cross section S, can then be shown to be practically uniform. The stress is uniaxial and its value is $T = F/S$. As a result, the length of the constant cross section part of the specimen increases from the zero-stress value ℓ_0 to $\ell = \ell_0 + \Delta\ell$. The elongation $\Delta\ell$ depends on the value of the stress.

The deformation curves are usually represented by plotting the relative elongation $\Delta\ell/\ell$ horizontally and the stress vertically. The general trend is given on Figure 13.2, where we note that the initial part is a straight line.

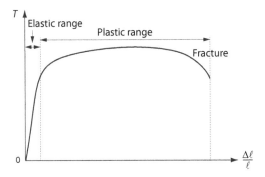

Figure 13.2: Deformation curve of a material submitted to a tensile stress. The linear part of this curve is the elastic deformation range.

This diagram $(T, \Delta\ell/\ell)$ shows several stages in the deformation process:

1. **The elastic deformation range**

   For small values of T, the material has a linear behavior. The elongation is proportional to the applied stress:

   $$\frac{\Delta\ell}{\ell} = \frac{1}{E}T.$$

   This is Hooke's law, and E is Young's modulus of the material, assumed to be isotropic. In this stress range, the material recovers its initial dimensions when the stress is removed: the transformation is reversible. Young's modulus is a measure of the mechanical property called stiffness or rigidity. The larger Young's modulus, the more rigid is a material, and therefore the smaller is its elastic deformation for a given value of the stress.

   Actually, Young's modulus is not sufficient to completely characterize the deformation of the bar, because the cross section of the bar also changes. If the cross section is circular, it remains circular in an isotropic material, and its diameter d undergoes a relative contraction, noted $\Delta d/d$ (with $\Delta d < 0$), also proportional to the stress.

The absolute value of the ratio between this relative contraction and the relative elongation of the bar is called Poisson's ratio, noted ν:

$$\nu = -\frac{\Delta d/d}{\Delta \ell/\ell}$$

so that we can write:

$$\frac{\Delta d}{d} = -\frac{\nu}{E}T.$$

We note that, in this range, the volume of the material changes because the elongation is generally not totally compensated by the decrease in cross section. This volume change is noted as:

$$\frac{\Delta V}{V} = \frac{\Delta \ell}{\ell} + \frac{2\Delta d}{d} = \frac{1-2\nu}{E}T.$$

The order of magnitude of Young's modulus E for metals is some 10^{10} N m^{-2}, and Poisson's coefficient ν is generally close to 0.3 for usual metals.

2. The plastic range

Beyond the elastic range, the strain-stress curve shows a plateau which can be more or less extended depending on the material. This is the plastic deformation range, where deformation occurs at almost constant volume. In this range, which involves the displacement and multiplication of defects called dislocation lines,[1] the material retains some permanent deformation when the stress is released. The deformation in this range is thus not reversible.

The boundary between the two ranges we just defined is called the elastic limit, or yield strength. It is related to the defects in the solid, and is therefore an extrinsic property of the material, variable from one specimen to another. The order of magnitude of this elastic limit for metals is some 10^7 to some 10^8 N m^{-2}. In practice, this limit is often difficult to pinpoint because the transition from the elastic to the plastic range is continuous.

3. The breaking range

In this range, the deformation is localized: the cylinder necks at a point and breaks.

This type of diagram assumes that the stress and strain are simultaneous, *i.e.* that the deformation sets in as soon as the stress is applied. This is the case for metals and many solids. These materials are called elastic materials. Their behavior in a deformation experiment is not influenced by the stress rate, and

---

1. A brief introduction to dislocations is given in the complement to this chapter.

it is well described by Hooke's law, independent of time:

$$\frac{\Delta \ell}{\ell} = \frac{1}{E}T \qquad \text{or} \qquad T = E\frac{\Delta \ell}{\ell}.$$

Some materials, called viscoelastic, feature a time-dependent stress-strain relation. We will not discuss them here. We will therefore restrict discussion to elastic materials and their mechanical behavior in the elastic range.

13.2. Compliance and stiffness tensors

13.2.1. Generalized Hooke's law

As we just saw, Young's modulus is not sufficient to characterize the elastic deformation of a bar, since its cross section decreases at the same time as it elongates. Furthermore, we are interested in crystals, and thus in anisotropic materials. We also want to be able to deal with stresses that are not necessarily uniaxial. Hooke's law must then be generalized. It relates the stress tensor T_{ij} to the strain tensor S_{ij}:

$$S_{ij} = s_{ijkl}T_{kl}. \tag{13.1a}$$

Conversely, the stress tensor is related to the strain tensor by relation:

$$T_{ij} = c_{ijkl}S_{kl}. \tag{13.1b}$$

The coefficients s_{ijkl} are called the compliance constants, and the coefficients c_{ijkl} the stiffness constants.[2]

Relations (13.1a) and (13.1b) each represents a set of 9 equations which relate the components of a rank-2 tensor ($[T]$ or $[S]$) to the components of another rank-2 tensor ($[S]$ or $[T]$) through coefficients s_{ijkl} or c_{ijkl}, which are therefore the components of rank-4 tensors (see Sect. 9.4.2), the compliance tensor $[s]$ for s_{ijkl} and the stiffness tensor $[c]$ for c_{ijkl}. It may be useful to show this in this special case for one of the two tensors, s_{ijkl} for example.

In a change of orthonormal axis system, S_{ij} becomes S'_{mn} and T_{kl} becomes T'_{pq} such that (Eq. (9.13a) and (9.13b)):

$$S'_{mn} = a_{mi}a_{nj}S_{ij},$$
$$T_{kl} = a_{pk}a_{ql}T'_{pq}$$

where $\{a_{ij}\}$ is the transformation matrix from the old to the new axis system. We obtain:

$$S'_{mn} = a_{mi}a_{nj}s_{ijkl}T_{kl} = a_{mi}a_{nj}s_{ijkl}a_{pk}a_{ql}T'_{pq}$$

---

2.   The s_{ijkl} and c_{ijkl} are sometimes called elastic moduli and elastic constants respectively.

so that

$$S'_{mn} = s'_{mnpq}T'_{pq}$$

with

$$s'_{mnpq} = a_{mi}a_{nj}a_{pk}a_{ql}s_{ijkl} \tag{13.2}$$

i.e. a transformation identical to that of the product of four components. This proves that s_{ijkl} is a rank-4 tensor.

We can rewrite relations (13.1a) and (13.1b) in the form:

$$[S] = [s][T],$$
$$[T] = [c][S],$$

where $[s]$ is the compliance tensor and $[c]$ the stiffness tensor.

Since the strain tensor $[S]$ is dimensionless, the components of $[s]$ are dimensionally the inverse of a stress, *i.e.* $m^2\,N^{-1}$. Typical orders of magnitude for the compliance tensor components are $10^{-11}\,m^2\,N^{-1}$. The stiffness tensor components c_{ijkl} are expressed in $N\,m^{-2}$ and have values on the order of $10^{11}\,N\,m^{-2}$.

Example: the central part of the tensile specimen of Figure 13.1 is submitted to a uniform uniaxial stress represented by tensor

$$\begin{pmatrix} T & 0 & 0 \\ 0 & 0 & 0 \\ 0 & 0 & 0 \end{pmatrix}$$

in an axis system $(Ox_1x_2x_3)$ where Ox_1 is parallel to the specimen axis. The generalized Hooke's law (Eq. (13.1a)) provides, for the diagonal components:

$$S_{11} = s_{1111}T, \qquad S_{22} = s_{2211}T \qquad \text{and} \qquad S_{33} = s_{3311}T$$

since only $T_{11} = T$ is non-zero. These components represent the (algebraic) relative elongations along the specimen axis (S_{11}) and perpendicular to it (S_{22} and S_{33}). The two latter components have no reason to be equal, in contrast to the isotropic case. The terms obtained for $i \neq j$, $S_{ij} = s_{ij11}T$, represent shear strains.

13.2.2. Symmetry of the compliance and stiffness tensors

We saw that the stress tensor is symmetric (Sect. 11.4). We can therefore, in relation (13.1a), collect the terms containing components T_{kl} and T_{lk} for which k is different from l. The strain S_{ij} thus takes the form:

$$S_{ij} = s_{ijkk}T_{kk} + \sum_{kl=23,31,12}(s_{ijkl} + s_{ijlk})T_{kl}.$$

Therefore s_{ijkl} and s_{ijlk} cannot be distinguished, and they are chosen equal:

$$s_{ijkl} = s_{ijlk}.$$

In the same way, since $S_{ij} = S_{ji}$ (Sect. 12.2.3), we have $s_{ijkl} = s_{jikl}$. Thus

$$s_{ijkl} = s_{ijlk} = s_{jikl} = s_{jilk}.$$

For the same reasons, we also have:

$$c_{ijkl} = c_{ijlk} = c_{jikl} = c_{jilk}.$$

These symmetry relations reduce the number of independent components of $[s]$ and $[c]$ from 81 to 36. There are 9 different ij-couples, and the same applies for the kl, thus in total 81 different values for $ijkl$. Among the ij (or kl) couples, 3 are such that $i = j$ and 6 are such that $i \neq j$. Symmetry with respect to i and j cancels 3 of the latter 6 couples, which brings to 6 instead of 9 the number of ij couples giving independent values of s_{ijkl}. The same applies for the kl couples. There remain in total $6 \times 6 = 36$ components of s_{ijkl} (and c_{ijkl}) which are independent. We will see that thermodynamic arguments further reduce the number of components, which goes down to 21 because of symmetry with respect to ij and kl: $s_{ijkl} = s_{klij}$ and $c_{ijkl} = c_{klij}$.

13.3. Contracted notation (Voigt's notation)

Because of the symmetry properties of tensor s_{ijkl} and c_{ijkl}, their expression can be simplified by replacing each couple ij (and kl) by a single subscript, denoted α for ij and β for kl, α and β ranging from 1 to 6. The correspondence is given in Table 13.1.

Table 13.1: Equivalence between subscripts ij (or kl) and α (or β)

ij or kl	11	22	33	23 and 32	31 and 13	12 and 21
α or β	1	2	3	4	5	6

The same simplification makes it possible to replace the T_{ij} and the S_{ij} by T_α and S_α components (α ranging from 1 to 6). It is however necessary to be careful about the way these replacements are performed.

13.3.1. Stress tensor

We set

$$T_\alpha = T_{ij} = T_{ji}. \tag{13.3}$$

The tensor then takes the form:

$$\begin{pmatrix} T_1 & T_6 & T_5 \\ T_6 & T_2 & T_4 \\ T_5 & T_4 & T_3 \end{pmatrix}.$$

13.3.2. Strain tensor

We set:

$$S_\alpha = S_{ij} \quad \text{if } i = j \quad \text{and} \quad S_\alpha = S_{ij} + S_{ji} \quad \text{if } i \neq j. \quad (13.4)$$

The strain tensor thus has the form:

$$\begin{pmatrix} S_1 & S_6/2 & S_5/2 \\ S_6/2 & S_2 & S_4/2 \\ S_5/2 & S_4/2 & S_3 \end{pmatrix}.$$

The contracted form makes it possible to represent these two tensors in the form of column vectors with six components, T_α and S_α. We will see in the following section that the benefit of using different relations to go over from the 2-subscript notation to the single subscript notation is that the column vectors S_α and T_α are connected by a matrix relation: $S_\alpha = s_{\alpha\beta}T_\beta$, and conversely $T_\alpha = c_{\alpha\beta}S_\beta$.

13.3.3. Compliance tensor and stiffness tensor

This section shows how to go over from tensors s_{ijkl} and c_{ijkl} to their simplified form $s_{\alpha\beta}$ and $c_{\alpha\beta}$ respectively.

We start out by evaluating $S_1 = S_{11} = s_{11kl}T_{kl}$ as a function of the T_α, defined in Section 13.3.1:

$$S_1 = s_{1111}T_1 + s_{1122}T_2 + s_{1133}T_3 + (s_{1123} + s_{1132})T_4$$
$$+ (s_{1131} + s_{1113})T_5 + (s_{1112} + s_{1121})T_6.$$

Using the symmetry of the s_{ijkl} (Sect. 13.2.2), we set

$$s_{11} = s_{1111}, \qquad s_{12} = s_{1122}, \qquad s_{13} = s_{1133},$$
$$s_{14} = 2s_{1123}, \qquad s_{15} = 2s_{1131}, \qquad s_{16} = 2s_{1112}.$$

We can now write: $S_1 = s_{1\alpha}T_\alpha$. The same argument applies to S_2 and S_3.

Now we evaluate S_{23} and S_{32}, and then add them:

$$S_{23} = s_{2311}T_1 + s_{2322}T_2 + s_{2333}T_3 + (s_{2323} + s_{2332})T_4$$
$$+ (s_{2331} + s_{2313})T_5 + (s_{2312} + s_{2321})T_6,$$
$$S_{32} = s_{3211}T_1 + s_{3222}T_2 + s_{3233}T_3 + (s_{3223} + s_{3232})T_4$$
$$+ (s_{3231} + s_{3213})T_5 + (s_{3212} + s_{3221})T_6.$$

Since we set $S_4 = S_{23} + S_{32}$, we can write:

$$S_4 = s_{4\alpha}T_\alpha$$

if we set:
$$s_{41} = 2s_{2311}, \qquad s_{42} = 2s_{2322}, \qquad s_{43} = 2s_{2333},$$
$$s_{44} = 4s_{2323}, \qquad s_{55} = 4s_{3131}, \qquad s_{66} = 4s_{1212}.$$

Thus we can write:

$$S_\alpha = s_{\alpha\beta} T_\beta$$

providing we set:

$s_{\alpha\beta} = s_{ijkl}$ for α and β equal to 1, 2 or 3 ($i = j$ and $k = l$),

$s_{\alpha\beta} = 2\, s_{ijkl}$ when either of the two subscripts α or β is equal to 1, 2 or 3 and the other one is equal to 4, 5 or 6,

$s_{\alpha\beta} = 4\, s_{ijkl}$ for α and β equal to 4, 5 or 6.

We can summarize the above results by splitting Table 13.2, consisting of 9×9 coefficients s_{ijkl}, into 9 tables with 3 rows and 3 columns each.

Table 13.2: Splitting the 9×9 table of components s_{ijkl} into a set of 3×3 tables

$kl\diagdown^{ij}$	1 1	2 2	3 3	2 3	3 1	1 2	3 2	1 3	2 1
1 1									
2 2		I			II			III	
3 3									
2 3									
3 1		IV			V			VI	
1 2									
3 2									
1 3		VII			VIII			IX	
2 1									

The (6×6) matrix of coefficients $s_{\alpha\beta}$ is obtained by superimposing and adding the sectors of the third large column to the sectors of the second large column, and then performing the same operation (superimposing, then adding) for the second and third large rows thus obtained.

We finally obtain the condensed Table 13.3.

Now we find the relations between components c_{ijkl} and $c_{\alpha\beta}$ that allow writing $T_\alpha = c_{\alpha\beta} S_\beta$.

By definition, $T_\alpha = T_{ij} = T_{ji}$, and relation (13.1b) has the form:

$$T_\alpha = T_{ij} = c_{ij11} S_{11} + c_{ij22} S_{22} + c_{ij33} S_{33} + c_{ij23} S_{23} + c_{ij32} S_{32}$$
$$+ c_{ij31} S_{31} + c_{ij13} S_{13} + c_{ij12} S_{12} + c_{ij21} S_{21}.$$

Table 13.3: Condensed form of the table of components $s_{\alpha\beta}$

$\beta\backslash^{\alpha}$	1 2 3	4 5 6
1		
2	I	II + III
3		
4		V + VI
5	IV + VII	+
6		VIII + IX

Because of the symmetry with respect to k and l of components c_{ijkl}:

$$T_\alpha = T_{ij} = c_{ij11}S_{11} + c_{ij22}S_{22} + c_{ij33}S_{33} + c_{ij23}(S_{23} + S_{32})$$
$$+ c_{ij31}(S_{31} + S_{13}) + c_{ij12}(S_{12} + S_{21})$$

and, taking into account the definition (13.4) of the S_α:

$$T_\alpha = c_{\alpha\beta}S_\beta$$

if we set $c_{\alpha\beta} = c_{ijkl}$.

The table of the $c_{\alpha\beta}$ is now obtained starting from the table of c_{ijkl} by simply canceling the sectors that form the third large column and the third large row (Tab. 13.4). The components $c_{\alpha\beta}$ and $s_{\alpha\beta}$ each form a square 6×6 matrix.

Table 13.4: Condensed form of the table of stiffness components $c_{\alpha\beta}$. The gray areas are canceled.

$\beta\backslash^{\alpha}$	1 2 3	4 5 6	
1			
2	I	II	
3			
4			
5	IV	V	
6			

13.3.4. Relation between the compliance and stiffness tensors

Using the simplified notation we just introduced, the generalized Hooke's law can be written in the following matrix form:

$$S_\alpha = s_{\alpha\beta}T_\beta,$$
$$T_\alpha = c_{\alpha\beta}S_\beta.$$

These relations show that matrices $s_{\alpha\beta}$ and $c_{\alpha\beta}$, which we note here s and c, are mutually inverse:

$$s\,c = E$$

where E is the identity matrix.

To calculate the stiffness components $c_{\alpha\beta}$ starting from the compliance components $s_{\alpha\beta}$, it is sufficient to invert the compliance matrix, and vice versa.

To conclude, the simplified notation we just introduced allows representing the stress $\to$ strain (and conversely strain $\to$ stress) relation by a matrix relation where the rank-2 tensors (stress and strain) become six-component column vectors, and the rank-4 tensors (compliance and stiffness) become 6×6 square matrices.

Matrices s and c represent a simplified form of tensors s_{ijkl} and c_{ijkl}, expressed in a given coordinate system. If we want to know the expression of these matrices in another coordinate system, we have to revert to the four-subscript components of tensors s_{ijkl} and c_{ijkl}, then perform the change in axis system on these components (Eq. (13.2)), and finally revert again to the contracted form of the new components s'_{ijkl} and c'_{ijkl} thus obtained.

13.4. Energy of a strained solid

Consider a small volume of crystal which, before deformation, has the shape of a rectangle parallelepiped. Choose a reference system consisting of orthonormal axes $Ox_1x_2x_3$, parallel to its sides, with lengths respectively Δx_1, Δx_2 and Δx_3 (Fig. 13.3). We submit it to a stress, described by the tensor T_{kl} which induces the strain S_{ij}. Then we increase this deformation by a small quantity dS_{ij}. We now show that the elementary work of the stress forces during this change in deformation has the form $\delta W = T_{ij}dS_{ij}\Delta V$ if ΔV is the volume of the parallelepiped. After deformation, the faces of the parallelepiped are not exactly rectangles with the same area any more, but, since the deformations are assumed to be very small, we can consider that the change in area of the faces is negligible and that the faces remain perpendicular to the coordinate axes. If the parallelepiped had been only translated without undergoing a strain, the work of the stress forces would be zero since the forces acting on two opposite faces are equal and have opposite directions. Thus the work is due only to the relative displacement of one face with respect to the other. We may therefore, for calculation purposes, assume that the three parallelepiped faces which go through the origin remained fixed, and that the work is due to the displacement of the three opposite faces. Consider the face perpendicular to axis Ox_1. It gets displaced because of the components dS_{11}, dS_{21} and dS_{31} of tensor dS_{ij}, which represent the relative deformation of a segment parallel to Ox_1. Term dS_{11} describes a displacement of this face parallel to Ox_1 by $\Delta x_1 dS_{11}$ (Fig. 13.3a). During this displacement, only the component T_{11} of the stress tensor performs work, equal to $T_{11}\Delta x_2 \Delta x_3 \Delta x_1 dS_{11} = T_{11}dS_{11}\Delta V$. The strain component dS_{21} leads to a displacement of this face parallel to Ox_2 by $\Delta x_1 dS_{21}$ (Fig. 13.3b), and only the component T_{21} performs work, equal to $T_{21}\Delta x_2 \Delta x_3 \Delta x_1 dS_{21} = T_{21}dS_{21}\Delta V$. In the same way, the strain dS_{31} leads to a displacement parallel to Ox_3 of the face, and only the component T_{31} does work, equal to $T_{31}dS_{31}\Delta V$.

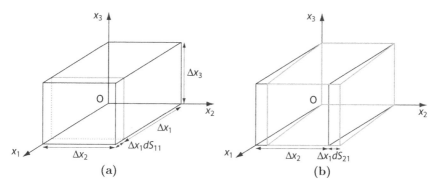

Figure 13.3: (a) Displacement parallel to Ox_1 of the face perpendicular to Ox_1 of the rectangle parallelepiped under the influence of component dS_{11} of the strain tensor; (b) displacement parallel to Ox_2 of the same face under the influence of component dS_{21} of the strain tensor

The total work of the forces acting on the faces perpendicular to Ox_1 is therefore:

$$\delta W_1 = T_{i1}dS_{i1}\Delta V.$$

We would obtain in the same way the works δW_2 and δW_3 of the forces acting respectively on the faces perpendicular to Ox_2 and Ox_3:

$$\delta W_2 = T_{i2}dS_{i2}\Delta V, \qquad \delta W_3 = T_{i3}dS_{i3}\Delta V.$$

The total work δW per unit volume of the material is then equal to:

$$\delta W = (\delta W_1 + \delta W_2 + \delta W_3)/\Delta V \qquad \text{or} \qquad \delta W = T_{ij}dS_{ij} \tag{13.5}$$

or, in detailed form:

$$\delta W = T_{11}dS_{11} + T_{22}dS_{22} + T_{33}dS_{33} + T_{23}dS_{23} + T_{32}dS_{32} + T_{31}dS_{31}$$
$$+ T_{13}dS_{13} + T_{12}dS_{12} + T_{21}dS_{21}.$$

Using relations (13.3) and (13.4) which define the T_α and the S_α components, we see that:

$$\delta W = T_\alpha dS_\alpha$$

and, since $T_\alpha = c_{\alpha\beta}S_\beta$, we can write the above relation in the form:

$$\delta W = c_{\alpha\beta}S_\beta dS_\alpha.$$

The work we just calculated is reversible work, because a small variation in strain with all of its components dS_α equal and opposite to the above variation would take the system back to its initial state.

The variation dF in the free energy $F = U - \Theta S$ per unit volume is such that:

$$dF = dU - \Theta dS - S d\Theta = \delta W + \delta Q - \Theta dS - S d\Theta$$

where S it the entropy and Θ the absolute temperature.

For a reversible transformation:

$$\delta Q_{rev} = \Theta dS \qquad \text{and} \qquad dF = \delta W_{rev} - S d\Theta$$

so that

$$dF = c_{\alpha\beta}S_\beta dS_\alpha - S d\Theta. \qquad (13.6)$$

We deduce that:

$$c_{\alpha\beta} = \left.\frac{\partial^2 F}{\partial S_\beta \partial S_\alpha}\right|_\Theta = \left.\frac{\partial^2 F}{\partial S_\alpha \partial S_\beta}\right|_\Theta = c_{\beta\alpha}.$$

The $c_{\alpha\beta}$ are thus the stiffness coefficients at constant temperature, and this relation shows that the matrix of $c_{\alpha\beta}$ coefficients is symmetric. Therefore, the stiffness tensor c_{ijkl} is symmetric with respect to its two subscript groups ij and kl:

$$c_{ijkl} = c_{klij}.$$

A similar argument using the thermodynamic function Φ, the Gibbs potential or free enthalpy, such that $\Phi = U - \Theta S - T_\alpha S_\alpha$, provides the same results for the compliance tensor.

We have

$$d\Phi = -S d\Theta - S_\alpha dT_\alpha = -S d\Theta - s_{\alpha\beta}T_\beta dT_\alpha$$

and

$$s_{\alpha\beta} = -\left.\frac{\partial^2 \Phi}{\partial T_\beta \partial T_\alpha}\right|_\Theta = -\left.\frac{\partial^2 \Phi}{\partial T_\alpha \partial T_\beta}\right|_\Theta = s_{\beta\alpha}.$$

The $s_{\alpha\beta}$ are the compliance components at constant temperature and furthermore:

$$s_{\alpha\beta} = s_{\beta\alpha} \qquad \text{or} \qquad s_{ijkl} = s_{klij}.$$

These relations bring down to 21 the maximum number of independent components of the elastic and stiffness tensors.

The variation in free energy dF at constant temperature $(d\Theta = 0)$ takes the form, using (13.6):

$$dF = \sum_\alpha c_{\alpha\alpha} S_\alpha dS_\alpha + \sum_{\alpha\beta=23,31,12} c_{\alpha\beta}(S_\alpha dS_\beta + S_\beta dS_\alpha)$$

$$= \sum_\alpha c_{\alpha\alpha} S_\alpha dS_\alpha + \sum_{\alpha\beta=23,31,12} c_{\alpha\beta} d(S_\alpha S_\beta).$$

We obtain after integration:

$$F = \sum_\alpha \frac{1}{2} c_{\alpha\alpha} S_\alpha^2 + \sum_{\alpha\beta=23,31,12} c_{\alpha\beta} S_\alpha S_\beta$$

$$= \frac{1}{2} \sum_\alpha c_{\alpha\alpha} S_\alpha^2 + \frac{1}{2} \sum_{\alpha\neq\beta} c_{\alpha\beta} S_\alpha S_\beta.$$

Thus the elastic free energy per unit volume of a strained solid is given by:

$$F = \frac{1}{2} c_{\alpha\beta} S_\alpha S_\beta.$$

The above derivations were made using the expression of elementary work $\delta W = T_\alpha dS_\alpha$. They could also have been made using the unsimplified forms of the tensors, $\delta W = T_{ij} dS_{ij}$. We would then have obtained $c_{ijkl} = c_{klij}$ and $s_{ijkl} = s_{klij}$, and for the free energy:

$$F = \frac{1}{2} c_{ijkl} S_{ij} S_{kl}.$$

13.5. Effect of crystal symmetry on the form of the elastic tensor

We will use the direct inspection method (Sect. 9.6.2) to determine the number of independent components of the elastic (or stiffness) tensor for some point groups. The results for all point groups are collected in Tables 13.5a and 13.5b. Because the matrices $s_{\alpha\beta}$ and $c_{\alpha\beta}$ are symmetric, the part of the matrix deduced by symmetry is not represented. We recall that, in order to use the direct inspection method or the coordinate system change formula, we must revert to the four-subscript tensor notation.

Table 13.5a: Form of the matrices representing the compliance and stiffness tensors for the various point groups

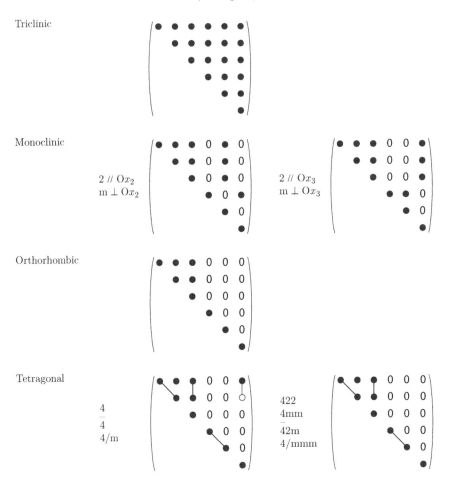

●—● Equal components
●—○ Components with equal absolute values but opposite signs

Table 13.5b: Form of the matrices representing the compliance and stiffness tensors for the various point groups

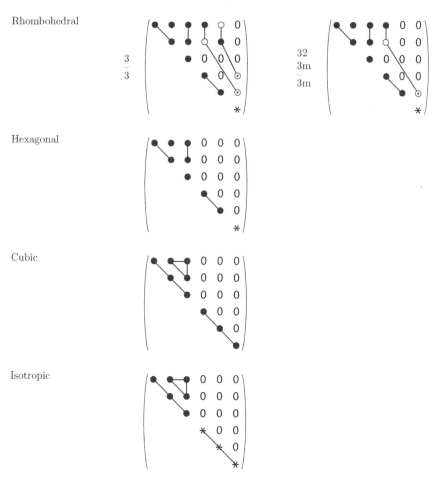

Rhombohedral

$\dfrac{3}{\overline{3}}$

$\begin{matrix}32\\3\text{m}\\\overline{3}\text{m}\end{matrix}$

Hexagonal

Cubic

Isotropic

●—●   Equal components
●—○   Components with equal absolute values but opposite signs

⊙   For s : value twice that of the component shown as a full circle to which it is connected
    For c : value equal to the component shown as a full circle to which it is connected

✱   For s : $2(s_{11} - s_{12})$
    For c : $1/2(c_{11} - c_{12})$

13.5.1. Center of symmetry

Tensors $[s]$ and $[c]$ being even-rank tensors, the existence or non-existence of a center of symmetry among the point symmetry operations of the crystal does not alter the number of independent components of these tensors (see Sect. 9.6.3). Thus, the number of independent components for these tensors is the same for all point groups of a given Laue class.

13.5.2. Groups 2, m and 2/m

We choose axis Ox_3 parallel to the 2-fold axis. Rotating the coordinate system by π around the 2-fold axis transforms components x_1, x_2, x_3 of a vector into components x_1', x_2', x_3' such that:

$$x_1' = -x_1, \qquad x_2' = -x_2, \qquad x_3' = x_3.$$

We deduce that the components s_{ijkl} whose subscripts i, j, k, l contain an odd total number of 1 and 2's and, consequently, an odd number of 3's, are zero. The tensors $[s]$ and $[c]$ shown in their condensed form $\{s_{\alpha\beta}\}$ or $\{c_{\alpha\beta}\}$ reduce to:

$$
\begin{pmatrix}
\bullet & \bullet & \bullet & 0 & 0 & \bullet \\
 & \bullet & \bullet & 0 & 0 & \bullet \\
 & & \bullet & 0 & 0 & \bullet \\
 & & & \bullet & \bullet & 0 \\
 & & & & \bullet & 0 \\
 & & & & & \bullet
\end{pmatrix}
$$

We obtain the same result by considering group m and axis Ox_3 perpendicular to the mirror. The symmetry operation acting on the coordinate system changes the components x_1, x_2, x_3 of a vector into components x_1', x_2', x_3' such that:

$$x_1' = x_1, \qquad x_2' = x_2, \qquad x_3' = -x_3,$$

so that any component of s_{ijkl} (or c_{ijkl}) involving an odd number of 3's is zero.

This result is also valid for group 2/m, as expected since adding a mirror perpendicular to the 2-fold axis does not reduce the number of non-zero coefficients. It is no surprise to obtain the same result for groups 2, m and 2/m since they belong to the same Laue class.

13.5.3. Groups 222, mmm and mm2

These three groups, which belong to the same Laue class, have the same number of independent coefficients, and we will only consider group 222. It has three mutually perpendicular 2-fold axes, which we choose as the axes Ox_1, Ox_2 and Ox_3. We use the results of the preceding section for the 2-fold axis parallel to Ox_3, and we add the symmetry operation associated to the 2-fold axis parallel to Ox_1, a rotation by π of the coordinate system around Ox_1. This operation transforms the components x_1, x_2, x_3 of a vector into components x_1', x_2', x_3' such that

$$x_1' = x_1, \qquad x_2' = -x_2, \qquad x_3' = -x_3.$$

All components involving an odd total number of subscripts equal to 1 are zero. We need not consider the action of the third 2-fold axis, because its existence results from the existence of the two others. We therefore obtain the following matrix:

$$\begin{pmatrix} \bullet & \bullet & \bullet & 0 & 0 & 0 \\ & \bullet & \bullet & 0 & 0 & 0 \\ & & \bullet & 0 & 0 & 0 \\ & & & \bullet & 0 & 0 \\ & & & & \bullet & 0 \\ & & & & & \bullet \end{pmatrix}$$

13.5.4. Groups 422, 4mm and 4/m mm

Each of these groups belongs to the same Laue class and can be deduced from one of the above by replacing a 2-fold axis with a 4-fold axis. For these three groups, axis Ox_3 is chosen parallel to the 4-fold axis. This substitution adds a symmetry operation, the rotation by $\pi/2$ around Ox_3. This operation, performed on the axis system, transforms components x_1, x_2, x_3 of a vector into components x_1', x_2', x_3' such that:

$$x_1' = x_2, \qquad x_2' = -x_1, \qquad x_3' = x_3.$$

We deduce that:

$$s_{1111} = s_{2222}, \qquad s_{1133} = s_{2233} \quad \text{and} \quad s_{1313} = s_{2323}.$$

We finally obtain the following matrix, where lines connecting some dots indicate that the corresponding coefficients are equal:

$$\begin{pmatrix} \bullet & \bullet & \bullet & 0 & 0 & 0 \\ & \bullet & \bullet & 0 & 0 & 0 \\ & & \bullet & 0 & 0 & 0 \\ & & & \bullet & \bullet & 0 \\ & & & & \bullet & 0 \\ & & & & & \bullet \end{pmatrix}$$

This result is also valid for group $\bar{4}$2m, which belongs to the same Laue class.

13.5.5. Cubic system

The least symmetric among the cubic groups is group 23. It has three orthogonal 2-fold axes parallel to the cube sides, and four 3-fold axes parallel to its major diagonals. We can therefore start from the number of independent coefficients determined for group 222 in a system of axes parallel to the 2-fold axes, and add a 3-fold axis, for example the axis parallel to direction [111]. A rotation by $2\pi/3$ around this 3-fold axis, performed on the axis system, transforms the components x_1, x_2, x_3 of a vector into components x'_1, x'_2, x'_3 such that:

$$x'_1 = x_2, \qquad x'_2 = x_3, \qquad x'_3 = x_1.$$

We deduce that:

$$s_{1111} = s_{2222} = s_{3333} \quad \text{and also} \quad s_{1122} = s_{2233} = s_{3311} \text{ etc.}$$

We finally obtain the following matrix:

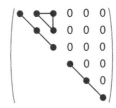

We then check that the additional symmetries in the other cubic groups bring no further simplification.

Thus only 3 independent coefficients are involved in describing the elasticity of cubic crystals.

The order of magnitude of the $s_{\alpha\beta}$ is some 10^{-11} $m^2\,N^{-1}$, that of the $c_{\alpha\beta}$ some 10^{11} $N\,m^{-2}$. Some numerical values of components $s_{\alpha\beta}$ and $c_{\alpha\beta}$ are given in Tables 13.6a and 13.6b respectively.

13.6. Isotropic materials

We saw, in the introduction to this chapter, that only 2 coefficients are required to describe the elastic properties of isotropic materials: Young's modulus E, and Poisson's ratio ν.

Here we will establish the connection between these coefficients E and ν and the components of the compliance and stiffness tensors.

The compliance tensor (as well as the stiffness tensor) of an isotropic material must be invariant under any change in coordinate system. A simple way of expressing this condition is to start from tensor s_{ijkl} for a cubic material, and to perform any change in the coordinate system (for example a rotation by $\pi/4$

Table 13.6a: Numerical values of the compliance coefficients $s_{\alpha\beta}$ for some materials (unit: $10^{-11}\,m^2\,N^{-1}$)

Crystal	P.G.	11	12	44	13	33	14	66
Cu	m$\bar{3}$m	1.5	−0.63	1.32				
Si	m$\bar{3}$m	0.77	−0.22	1.26				
NaCl	m$\bar{3}$m	2.29	−0.47	7.85				
Co	6/mmm	0.47	−0.23	1,32	−0.07	0.32		
Ti	6/mmm	0.97	−0.47	2.15	−0.18	0.69		
SiC	6/mmm	0.21	−0.037	0.59	−0.017	0.18		
ZnS	6mm	1.10	−0.46	3.45	−0.21	0.85		
Al$_2$O$_3$	$\bar{3}$m	0.24	−0.07	0.7	−0.038	0.22	0.049	
LiNbO$_3$	3m	0.57	−0.09	1.67	−0.15	0.49	−0.1	
Quartz (SiO$_2$)	32	1.17	−0.08	1.72	−0.12	0.97	−0.39	
KDP	$\bar{4}$2m	1.49	0.18	7.69	−0.39	1.97		16.7

Table 13.6b: Numerical values of the stiffness coefficients $c_{\alpha\beta}$ for some materials (unit: $10^{11}\,N\,m^{-2}$)

Crystal	P.G.	11	12	44	13	33	14	66
Cu	m$\bar{3}$m	1.69	1.22	0.75				
Si	m$\bar{3}$m	1.65	0.64	0.79				
NaCl	m$\bar{3}$m	0.49	0.13	0.13				
Co	6/mmm	3.07	1.65	0.75	1.03	3.58		
Ti	6/mmm	1.6	0.9	0.46	0.66	1.81		
SiC	6/mmm	5	1	1.7	0.6	5.6		
ZnS	6mm	1.24	0.6	0.29	0.46	1.4		
Al$_2$O$_3$	$\bar{3}$m	4.95	1.6	1.46	1.15	4.97	−0.23	
LiNbO$_3$	3m	2.03	0.53	0.6	0.75	2.45	0.09	
Quartz (SiO$_2$)	32	0.87	0.07	0.58	0.12	1.06	0.18	
KDP	$\bar{4}$2m	0.71	−0.05	0.13	0.13	0.56		0.06

around a 4-fold axis parallel to one of the axes Ox_1, Ox_2 or Ox_3). This must leave all the tensor coefficients unchanged (Exercise 13.2). We thus obtain the result that, for an isotropic material, the three independent components of the compliance and stiffness tensors which characterize the cubic system are connected by the following relations:

$$s_{44} = 2(s_{11} - s_{12}), \tag{13.7a}$$
$$c_{44} = \tfrac{1}{2}(c_{11} - c_{12}). \tag{13.7b}$$

This is of course valid whatever the orthonormal coordinate system used.

There are thus only 2 independent elastic components for isotropic solids. They are expressed as a function of E and ν in the following section.

We note here that cubic materials are different from isotropic materials as far as elastic properties are concerned. This is a general result for all physical properties represented by rank-4 crystals. In contrast, for the properties represented by rank-2 tensors, cubic materials behave as isotropic (Sect. 10.5.5). The deviation from isotropy of the elastic properties of cubic materials can be evaluated by comparing the values of c_{44} and $\tfrac{1}{2}(c_{11} - c_{12})$, or the values of s_{44} and $2(s_{11} - s_{12})$.

13.6.1. Expressing components $s_{\alpha\beta}$ as a function of E and ν

Consider a tensile specimen of an isotropic material, undergoing an elongation under the influence of a uniaxial stress T. We choose Ox_1 parallel to the axis of the specimen and to the stress (Fig. 13.1). The strain is obtained through the generalized Hooke's law:

$$
\begin{pmatrix} S_1 \\ S_2 \\ S_3 \\ S_4 \\ S_5 \\ S_6 \end{pmatrix}
=
\begin{pmatrix}
s_{11} & s_{12} & s_{12} & 0 & 0 & 0 \\
s_{12} & s_{11} & s_{12} & 0 & 0 & 0 \\
s_{12} & s_{12} & s_{11} & 0 & 0 & 0 \\
0 & 0 & 0 & s_{44} & 0 & 0 \\
0 & 0 & 0 & 0 & s_{44} & 0 \\
0 & 0 & 0 & 0 & 0 & s_{44}
\end{pmatrix}
\begin{pmatrix} T \\ 0 \\ 0 \\ 0 \\ 0 \\ 0 \end{pmatrix}
=
\begin{pmatrix} s_{11}T \\ s_{12}T \\ s_{12}T \\ 0 \\ 0 \\ 0 \end{pmatrix}.
$$

Also, $s_{44} = 2(s_{11} - s_{12})$ according to (13.7a).

The bar thus undergoes a relative elongation $S_1 = \Delta\ell/\ell = s_{11}\, T$, and we deduce that $s_{11} = 1/E$ where E is Young's modulus for the material. Along with this elongation, the bar undergoes the same deformation in the two directions perpendicular to the stress: $S_2 = S_3 = s_{12}\, T$. From the definition of Poisson's ratio ν, as given in the introduction, we see that:

$$\frac{s_{12}}{s_{11}} = -\nu$$

so that

$$s_{12} = -\frac{\nu}{E}.$$

Hence:
$$s_{44} = 2\left(\frac{1}{E} + \frac{\nu}{E}\right) = \frac{2(1+\nu)}{E}.$$

The shear (or rigidity) modulus G is defined as the reciprocal of s_{44}, *i.e.* as the ratio of a shear stress and the corresponding shear strain, so that:

$$G = \frac{1}{s_{44}} = c_{44} = \frac{E}{2(1+\nu)}.$$

13.6.2. Stiffness components. Lamé coefficients

For isotropic materials, the stiffness components are not much used in the form of the $c_{\alpha\beta}$. It is customary to use the Lamé coefficients, λ and μ, defined as follows:

$$c_{44} = \mu \text{ and } c_{12} = \lambda \text{ and, using (13.7b): } c_{11} = 2c_{44} + c_{12} = \lambda + 2\mu.$$

The stiffness tensor can thus be written as:

$$\begin{pmatrix} \lambda + 2\mu & \lambda & \lambda & 0 & 0 & 0 \\ \lambda & \lambda + 2\mu & \lambda & 0 & 0 & 0 \\ \lambda & \lambda & \lambda + 2\mu & 0 & 0 & 0 \\ 0 & 0 & 0 & \mu & 0 & 0 \\ 0 & 0 & 0 & 0 & \mu & 0 \\ 0 & 0 & 0 & 0 & 0 & \mu \end{pmatrix}.$$

Hooke's generalized law (Eq. (13.1b)) now takes the form:

for $\alpha = 1, 2, 3$: $T_\alpha = 2\mu S_\alpha + \lambda(S_1 + S_2 + S_3)$,
and for $\alpha = 4, 5, 6$: $T_\alpha = \mu S_\alpha$.

Since $S_\alpha = S_{ij}$ for $\alpha = 1, 2, 3$ and $S_\alpha = S_{ij} + S_{ji} = 2S_{ij}$ for $\alpha = 4, 5, 6$, these two relations can be condensed into a single one:

$$T_{ij} = 2\mu S_{ij} + \lambda(\text{Trace } S)\delta_{ij}.$$

13.7. Representative surface for Young's modulus

It is often valuable to find how the length of a bar of an anisotropic material, with known but general crystallographic direction, changes when it is submitted to a uniaxial stress parallel to the bar, *i.e.* the equivalent of Young's modulus for an isotropic material.

We defined Young's modulus E_w in a given direction with unit vector w as the ratio between the value T of a uniaxial stress in this direction and the relative elongation $\Delta\ell/\ell|_w$ of the material in the same direction, $E_w = T/(\Delta\ell/\ell|_w)$. Choosing an orthonormal axis system $(Ox_1'x_2'x_3')$ where Ox_1' is parallel to vector w, the stress tensor has only one non-zero component in this coordinate system, $T'_{11} = T$. The relative elongation in direction Ox'_1 is given by S'_{11}, so that:

$$\left.\frac{\Delta\ell}{\ell}\right|_w = S'_{11} = s'_{1111}T$$

where s'_{1111} is the compliance component expressed in the axis system $(Ox'_1 x'_2 x'_3)$. We deduce that:

$$E_w = \frac{1}{s'_{1111}}.$$

We thus have to calculate s'_{1111} as a function of the independent components s_{ijkl} of the compliance tensor in the orthonormal axis system Ox_1, Ox_2 and Ox_3 associated to the crystallographic system. The rule for axis system change provides:

$$s'_{1111} = a_{1i}a_{1j}a_{1k}a_{1l}s_{ijkl}$$

where $\{a_{ij}\}$ is the transformation matrix for going over from coordinate system $(Ox_1 x_2 x_3)$ to system $(Ox'_1 x'_2 x'_3)$.

The only coefficients of the matrix actually involved are the a_{1j}, *i.e.* the components of the new basis vector $\mathbf{e}'_1$, which is none other than vector $\mathbf{w}$, in the frame $Ox_1 x_2 x_3$. These are the direction cosines of $\mathbf{w}$ in this frame. They are noted w_1, w_2, w_3, yielding:

$$s'_{1111} = w_i w_j w_k w_l s_{ijkl}.$$

The representative surface for Young's modulus is defined as the locus of the ends of vectors starting from the same origin, with lengths in a given direction $\mathbf{w}$ equal to the corresponding value s'_{1111}. This surface thus represents the reciprocal of Young's modulus for any direction of the applied uniaxial stress.

Example for the cubic system:

$$s'_{11} = s_{11}\left(w_1^4 + w_2^4 + w_3^4\right) + s_{44}\left(w_2^2 w_3^2 + w_3^2 w_1^2 + w_1^2 w_2^2\right)$$
$$+ 2s_{12}\left(w_2^2 w_3^2 + w_3^2 w_1^2 + w_1^2 w_2^2\right).$$

Since $\mathbf{w}$ is a unit vector, we have:

$$\left(w_1^4 + w_2^4 + w_3^4\right) = \left(w_1^2 + w_2^2 + w_3^2\right)^2 - 2\left(w_2^2 w_3^2 + w_3^2 w_1^2 + w_1^2 w_2^2\right)$$
$$= 1 - 2\left(w_2^2 w_3^2 + w_3^2 w_1^2 + w_1^2 w_2^2\right)$$

and

$$s'_{11} = s_{11} - 2\left(s_{11} - s_{12} - \frac{s_{44}}{2}\right)\left(w_2^2 w_3^2 + w_3^2 w_1^2 + w_1^2 w_2^2\right).$$

Thus Young's modulus is not isotropic for cubic crystals.

For direction [100], $s'_{11} = s_{11}$, while for direction [111], we obtain: $s'_{11} = s_{11} - 2/3\left(s_{11} - s_{12} - s_{44}/2\right)$. It can be shown that this value is extremal (Exercise 13.5).

Note: in an isotropic material, $s_{11} - s_{12} = s_{44}/2$, and, as expected, $1/E = s_{11}$ for any direction.

13.8. Compressibility

Two types of compressibility at constant temperature can be defined depending on whether interest focuses, for a solid submitted to a pressure variation Δp, on the change in volume or on the change in length in a given direction. In either case, the pressure variation Δp induces a variation of the stress tensor

$$\Delta T_{ij} = -\delta_{ij}\Delta p$$

which leads to a strain variation ΔS_{ij} such that:

$$\Delta S_{ij} = s_{ijkl}\Delta T_{kl} = -s_{ijkl}\delta_{kl}\Delta p = -s_{ijkk}\Delta p \tag{13.8}$$

13.8.1. Bulk compressibility

The bulk compressibility coefficient χ at constant temperature Θ is defined by:

$$\chi = -\frac{1}{V}\left(\frac{\partial V}{\partial p}\right)_{\Theta}.$$

If ΔV is a small variation in the volume V of the material caused by a small pressure variation Δp at constant temperature:

$$\chi = -\frac{1}{\Delta p}\left(\frac{\Delta V}{V}\right).$$

We saw (Sect. 12.4) that the relative volume variation is such that:

$$\frac{\Delta V}{V} = S_{ii}$$

and, because of Equation (13.8):

$$\frac{\Delta V}{V} = -s_{iikk}\Delta p.$$

Thus

$$\chi = -\frac{1}{\Delta p}\left(\frac{\Delta V}{V}\right) = s_{iikk}$$

i.e. the sum of the nine coefficients in the first quarter of matrix $s_{\alpha\beta}$:

$$\chi = [s_{11} + s_{22} + s_{33} + 2(s_{12} + s_{23} + s_{31})].$$

In the case of an isotropic material, this relation becomes:

$$\chi = 3[s_{11} + 2s_{12}]$$

and, since $s_{11} = 1/E$ and $s_{12} = -\nu/E$, we obtain:

$$\chi = 3\left(\frac{1 - 2\nu}{E}\right).$$

Its reciprocal K is called the bulk modulus:

$$K = \frac{1}{\chi} = \frac{E}{3(1 - 2\nu)}.$$

K and χ are positive quantities, since when the pressure applied on a solid is increased, its volume decreases: $\Delta V/V < 0$. Thus for all materials $\nu < 1/2$. It can also be shown [1] that the Lamé coefficient μ is always positive. If E is expressed as a function of μ and ν in the expression for K, we obtain:

$$K = \frac{2}{3} \frac{\mu(1 + \nu)}{(1 - 2\nu)}$$

whence:

$$-1 < \nu < \frac{1}{2}.$$

In practice, a negative value of ν is exceptional, but it was encountered [2]. For most polycrystalline metals, described as isotropic materials, ν is on the order of 0.3.

13.8.2. Linear compressibility of a bar

The linear compressibility coefficient at constant temperature in a given direction $\boldsymbol{w}$ of a material, $\beta_{\boldsymbol{w}}$, is defined by:

$$\beta_{\boldsymbol{w}} = -\frac{1}{\ell}\left(\frac{\partial \ell}{\partial p}\right)_{\Theta}$$

which can be written:

$$\beta_{\boldsymbol{w}} = -\frac{1}{\ell}\left(\frac{\Delta \ell}{\Delta p}\right) = -\frac{1}{\Delta p}\left(\frac{\Delta \ell}{\ell}\right)$$

if Δp is a small pressure variation, at constant temperature, which induces a small variation in length $\Delta \ell$ in a segment with length ℓ parallel to vector $\boldsymbol{w}$.

We saw (Eq. (12.12)) that the relative elongation of such a segment, submitted to a variation in the strain tensor ΔS_{ij}, is given by:

$$\left.\frac{\Delta \ell}{\ell}\right|_{\boldsymbol{w}} = w_i w_j \Delta S_{ij}$$

and, using Equation (13.8):

$$\left.\frac{\Delta \ell}{\ell}\right|_{\boldsymbol{w}} = -w_i w_j s_{ijkk} \Delta p.$$

Hence the linear compressibility coefficient at constant temperature is given by:

$$\beta_{\boldsymbol{w}} = w_i w_j s_{ijkk}.$$

13.9. Notes on non-uniform stresses and strains

We saw in Section 11.5 that it is necessary, in order to determine tensor $T_{ij}(\mathbf{r})$ in a material submitted to given stresses $(\mathbf{T_n})_\Sigma$ on its external surface Σ, to solve Equations (11.4)

$$\frac{\partial T_{ij}}{\partial x_j} + f_i = 0, \tag{11.4}$$

at all points in the solid, with the boundary conditions:

$$(\mathbf{T_n})_\Sigma = (T_{ij})_\Sigma (n_j)_\Sigma. \tag{13.9}$$

We must therefore determine a tensor $[T]$ satisfying (11.4) and (13.9).

Since tensor $[T]$ is related to tensor $[S]$ through Hooke's law $[T] = [c][S]$, condition (11.4) leads to a condition on the derivatives of $[S]$. But the six components of the symmetric tensor $[S]$ cannot take any value in the case of non-uniform strains, because they are expressed as a function of the derivatives of the three components of the displacement vector $\mathbf{u}$. Consider for example the components S_{11}, S_{22} and S_{12} which imply functions u_1 and u_2. By definition:

$$S_{11} = \frac{\partial u_1}{\partial x_1} \qquad S_{22} = \frac{\partial u_2}{\partial x_2} \qquad S_{12} = \frac{1}{2}\left(\frac{\partial u_1}{\partial x_2} + \frac{\partial u_2}{\partial x_1}\right).$$

We immediately see that these three quantities are not independent. The second derivative of S_{12} with respect to x_1 and x_2 has the form:

$$\frac{1}{2}\left(\frac{\partial^3 u_1}{\partial x_1 \partial x_2^2} + \frac{\partial^3 u_2}{\partial x_2 \partial x_1^2}\right)$$

thus the half-sum of

$$\frac{\partial^2 S_{11}}{\partial x_2^2} \qquad \text{and} \qquad \frac{\partial^2 S_{22}}{\partial x_1^2}.$$

We must therefore have:

$$2\frac{\partial^2 S_{12}}{\partial x_1 \partial x_2} = \frac{\partial^2 S_{11}}{\partial x_2^2} + \frac{\partial^2 S_{22}}{\partial x_1^2}. \tag{13.10}$$

It can be shown that there are six such compatibility relations.

Further, since the basic relation (11.4) must be verified, one way of incorporating the compatibility relations is to substitute, in these equations, the S_{ij} by their expression as a function of the T_{ij}, and to add condition (11.4). We then obtain six coupled equations, called the Beltrami-Michell equations. They can usually not be solved analytically, and their solution is obtained numerically.

Another way of translating Equation (11.4) is to replace in this equation the stresses by their expression as a function of the S_{ij}, and then to express the S_{ij} as a function of the derivatives of displacement $\mathbf{u}$. The calculation is easy in the isotropic case.

Relation (11.4) for $i = 1$ can be written, in the isotropic case:

$$\frac{\partial T_{1j}}{\partial x_j} + f_1 = 0 \qquad \text{or} \qquad \frac{\partial T_1}{\partial x_1} + \frac{\partial T_6}{\partial x_2} + \frac{\partial T_5}{\partial x_3} + f_1 = 0. \qquad (13.11)$$

Hooke's law $T_\alpha = c_{\alpha\beta}S_\beta$ can we written, using the Lamé coefficients (Sect. 13.6.2), as:

$$T_1 = 2\mu S_1 + \lambda(S_1 + S_2 + S_3),$$
$$T_5 = \mu S_5,$$
$$T_6 = \mu S_6$$

and Equation (13.11) becomes:

$$\mu\left(2\frac{\partial S_1}{\partial x_1} + \frac{\partial S_6}{\partial x_2} + \frac{\partial S_5}{\partial x_3}\right) + \lambda\left(\frac{\partial S_1}{\partial x_1} + \frac{\partial S_2}{\partial x_1} + \frac{\partial S_3}{\partial x_1}\right) + f_1 = 0. \qquad (13.12)$$

Also (Eq. (12.10)):

$$S_{ij} = \frac{1}{2}\left(\frac{\partial u_i}{\partial x_j} + \frac{\partial u_j}{\partial x_i}\right).$$

Since, from Equation (13.4), $S_5 = S_{13} + S_{31}$ and $S_6 = S_{12} + S_{21}$,

$$S_5 = \frac{\partial u_1}{\partial x_3} + \frac{\partial u_3}{\partial x_1} \qquad \text{and} \qquad S_6 = \frac{\partial u_1}{\partial x_2} + \frac{\partial u_2}{\partial x_1}.$$

Equation (13.12) thus becomes:

$$\mu\left(2\frac{\partial^2 u_1}{\partial x_1^2} + \frac{\partial^2 u_1}{\partial x_2^2} + \frac{\partial^2 u_2}{\partial x_2\partial x_1} + \frac{\partial^2 u_1}{\partial x_3^2} + \frac{\partial^2 u_3}{\partial x_3\partial x_1}\right)$$
$$+\lambda\left(\frac{\partial^2 u_1}{\partial x_1^2} + \frac{\partial^2 u_2}{\partial x_1\partial x_2} + \frac{\partial^2 u_3}{\partial x_1\partial x_3}\right) + f_1 = 0.$$

Reorganizing the terms, we obtain:

$$\mu\left(\frac{\partial^2 u_1}{\partial x_1^2} + \frac{\partial^2 u_1}{\partial x_2^2} + \frac{\partial^2 u_1}{\partial x_3^2}\right) + (\lambda + \mu)\left[\frac{\partial^2 u_1}{\partial x_1^2} + \frac{\partial^2 u_2}{\partial x_1\partial x_2} + \frac{\partial^2 u_3}{\partial x_1\partial x_3}\right] + f_1 = 0$$

and

$$\mu\left(\frac{\partial^2 u_1}{\partial x_1^2} + \frac{\partial^2 u_1}{\partial x_2^2} + \frac{\partial^2 u_1}{\partial x_3^2}\right) + (\lambda + \mu)\frac{\partial}{\partial x_1}\left[\frac{\partial u_1}{\partial x_1} + \frac{\partial u_2}{\partial x_2} + \frac{\partial u_3}{\partial x_3}\right] + f_1 = 0$$

so that

$$\mu\Delta u_1 + (\lambda + \mu)\frac{\partial}{\partial x_1}(\text{div } \mathbf{u}) + f_1 = 0.$$

We would obtain similar relations for $i = 2$ and $i = 3$ in (11.4), so that finally:

$$\mu\Delta u_i + (\lambda + \mu)\frac{\partial}{\partial x_i}(\text{div } \mathbf{u}) + f_i = 0.$$

The three above relations can be collected into a single vector equation:

$$(\lambda + \mu)\mathbf{grad}(\text{div } \mathbf{u}) + \mu\Delta\mathbf{u} + \mathbf{f} = 0$$

This is Navier's equation.

To determine the strains in the solid, we must then take into account the boundary conditions on **u**. In a few simple cases, the solution is analytical, but in most cases it has to be determined numerically.

13.10. Exercises

Exercise 13.1

Determine the form of the elasticity tensor of a crystal with point group $\bar{4}3m$ by using that for a crystal with point group $\bar{4}2m$ given in Table 13.5a. Express it in an orthonormal axis system $(Ox_1x_2x_3)$ parallel to the basis vectors **a**, **b** and **c** of the cubic unit cell.

Exercise 13.2

Consider a cubic crystal, to which an orthonormal axis system parallel to the basis vectors of the cubic unit cell is associated.

1. It is submitted to a uniform uniaxial stress T in the [100] direction. Calculate the relative elongation in the direction of the applied stress.

2. The uniaxial stress is now applied in the direction [110]. Calculate the relative elongation in the direction of the applied stress.

3. If the material was isotropic, the results obtained in (1) and (2) would have to be identical. Deduce the relation which exists between the elastic coefficients s_{11}, s_{12} and s_{44} in an isotropic material.

Exercise 13.3

A small crystal sample with volume v of tin (point group $4/m\,mm$) is submitted to a hydrostatic stress p with value $10^7\,\mathrm{N\,m^{-2}}$. Calculate the resulting relative variation in volume. The form of the matrix $\{s_{\alpha\beta}\}$ for this group is given in Table 13.5a, and its elements, expressed in units of $10^{-11}\,\mathrm{m^2\,N^{-1}}$, are

$$s_{11} = 1.85 \qquad s_{12} = -0.99 \qquad s_{13} = -0.25$$
$$s_{33} = 1.18 \qquad s_{44} = 5.7 \qquad s_{66} = 13.5.$$

Exercise 13.4

Show that the relative variation in volume of a crystal with cubic symmetry, submitted to a uniaxial stress T, is independent of the direction of the stress and is expressed by $(s_{11} + 2s_{12})T$.

Exercise 13.5

Show that Young's modulus for a cubic material is maximum in the direction [111] in all cases where $s_{11} - s_{12} - s_{44}/2$ is positive.

Exercise 13.6

A crystal of lead telluride PbTe, with point group m$\bar{3}$m, is submitted to a uniform uniaxial tensile stress, with intensity 10^8 N m^{-2}, perpendicular to a (110) surface. The resulting strains are the following:

– a relative increase in length of 24×10^{-4} in the direction of the uniaxial stress [110],

– relative decreases in length along directions [001] and [1$\bar{1}$0], with values 6×10^{-5} and 15×10^{-4} respectively.

Determine the values of the elastic coefficients of PbTe.

Exercise 13.7

A cubic crystal with lattice parameter a is submitted to a uniform uniaxial stress T, parallel to direction [111].

1. Determine the stress tensor in an orthonormal axis system $(Ox_1x_2x_3)$ parallel to the basis vectors of the cubic unit cell.

2. Deduce the resulting strain tensor, in the orthonormal axis system linked to the basis vectors of the cubic unit cell.

3. Show that the cubic unit cell becomes a rhombohedral unit cell. Calculate the lattice parameters ($a' = b' = c'$ and $\alpha' = \beta' = \gamma' = \pi/2 - \varepsilon$) as a function of the uniaxial stress T and of the elastic coefficients $s_{\alpha\beta}$.

Exercise 13.8

Consider test specimens made of copper, with point group m$\bar{3}$m, in the shape of rectangle parallelepipeds with length L and square cross section, with side ℓ. The values are $L = 5$ mm, $\ell = 1$ mm, and the compliance tensor components of copper are: $s_{11} = 1.49 \times 10^{-11}$ m^2 N^{-1}, $s_{12} = -0.63 \times 10^{-11}$ m^2 N^{-1}, $s_{44} = 1.33 \times 10^{-11}$ m^2 N^{-1}.

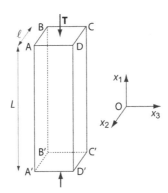

Assume the experimental conditions are such that the stresses to which the specimens are submitted are uniform.

1. The first specimen is cut so that its faces are all $\{100\}$ type. The uniaxial compressive stress T is applied onto the square faces ABCD and A′B′C′D′ (figure above). Determine the variation in length ΔL of the specimen, and the variation $\Delta \ell$ of the sides of the section, for $T = 10^9 \, \mathrm{N\,m^{-2}}$.

2. Consider a second specimen, with the same dimensions as the first one, with faces now having the following crystallographic orientations: faces ABCD and A′B′C′D′ are (110) faces, faces ABB′A′ and DCC′D′ are (001) faces and faces ADD′A′ and BCC′B′ are ($\bar{1}$10) faces. The uniaxial compressive stress is applied on faces ABCD and A′B′C′D′. Determine the shape taken by the specimen, and its new dimensions.

Complement 13C. Plastic deformation and crystal defects

Throughout this book, the assumption is made, for simplicity's sake, that the crystal is perfect and has infinite extension. The latter condition is obviously not to be taken literally, but the assumption of crystal perfection is also most questionable. Actually, it is only through tremendous efforts that crystals of silicon are manufactured (see Complement 1C) with no defect of one kind – the linear defects called dislocations which will be discussed below. Other defects, called point defects (vacancies and interstitials) are present even in these extraordinarily good crystals, because thermodynamics requires them at equilibrium. When it comes to the discussion of X-ray or neutron diffraction by crystals, an introductory, geometrical treatment invokes a perfect and infinite crystal in order to make the reciprocal lattice (Sect. 3.2) consist of point-like nodes. Nevertheless a discussion of the scattering mechanism and scattered intensity leads to the condition that, for the experimental results to be reasonably simple to analyze ("kinematical approximation"), the crystal must be small and imperfect. And an essential physical property of metals, mechanical ductility, is directly linked to dislocations.

The macroscopic strain produced by a stress, for example in a tensile test, disappears once the applied stress is removed if the strain is small enough: this is the *elastic* deformation range. Under a larger stress, part of the distortion remains after the stress is removed. This is *plastic* deformation, and obviously it is vital for metal shaping to produce permanent effects, hence in all technical activities. In terms of the macroscopic stress-strain diagram, the plastic regime starts where the curve deviates from the linear initial part (Sect. 13.1).

The mechanism of plastic deformation was suggested in the 1930s to be the motion and multiplication of crystal defects called dislocation lines, or for short dislocations. This postulate was required to remove the paradox associated with the fact that plastic deformation occurs for stresses much smaller than predicted from the cohesive energy for a homogeneous crystal with no singularity. The existence of dislocations was indeed confirmed through direct observation when the electron microscope was developed, and they were explored in great detail also with several other techniques, notably the X-ray imaging techniques called "diffraction topographic methods" or Bragg-diffraction imaging.

A dislocation can be envisaged as the result of i) performing in an initially perfect crystal a cut ending along a line (the dislocation line); ii) moving one lip of the cut with respect to the other: the displacement defines the Burgers vector **b**; iii) adding or removing some perfect crystal material to fill the gap or remove the overlap; iv) gluing the material together. The Burgers vector must be a lattice translation in order to allow the displaced parts to match, and it is usually, for thermodynamic reasons, the shortest lattice translation possible.

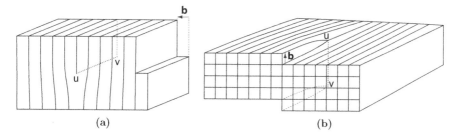

Figure 13C.1: (a) An edge dislocation: schematic representation of the atomic planes. The Burgers vector is perpendicular to the dislocation line uv. (b) A screw dislocation. The line uv and the Burgers vector are parallel.

Figure 13C.1 shows the two pure situations: in (a), the dislocation line uv is perpendicular to the Burgers vector **b**; this is an edge dislocation, and the plane containing the line and the Burgers vector is called the glide plane.

Figure 13C.1b shows the situation where the Burgers vector **b** is parallel to the line uv, defining a *screw* dislocation. In reality, dislocations usually have mixed character, *i.e.* their Burgers vector lies at an angle to the dislocation line.

As Figure 13C.1 suggests, the lattice planes are distorted in the vicinity of the dislocation. The distortion decreases with increasing distance from the core (the center) of the dislocation. The visibility of dislocations in the electron microscope and in the Bragg-diffraction imaging techniques is due to the strain field: the region with inhomogeneous distortion diffracts differently from the perfect-crystal matrix surrounding it.

The simplest case of plastic deformation through the motion of a dislocation is illustrated in Figure 13C.2, which suggests how one part of the crystal will have moved by one Burgers vector when the dislocation line has moved across the crystal. The hatched area generated by the dislocation line is called the glide plane.

Dislocations interact with one another through their stress field. This leads to a preferential arrangement of dislocations into two-dimensional boundaries (subgrain boundaries), separating regions with slightly different orientations (typically minutes of arc).

There are other defects in crystals than these line (one-dimensional) defects. They can be classified according to their geometry as point, or zero-dimensional, defects (vacancies, interstitials); plane, or two-dimensional defects (twin boundaries, stacking faults), or three-dimensional (voids, precipitates). These defects also interact with dislocations, and consequently affect the plastic deformation process, leading to hardening (e.g. precipitation hardening or work hardening) when dislocation displacements are hampered. Dislocations also play an important role in the mechanism of crystal growth.

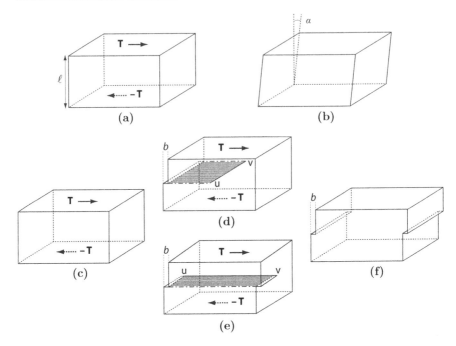

Figure 13C.2: Schematic illustration of elastic and plastic deformation processes. (a) and (c) are identical and show an initial, strain-free state. (b) is the result of elastic deformation. (d) shows the plastic deformation stage in which an edge dislocation uv, which Burgers vector **b**, moves to the right in its glide plane. (e) shows the plastic deformation stage in which a screw dislocation, with the same Burgers vector **b**, moves toward the back of the figure. (f) is the plastically deformed state after the dislocation in (d) or (e) has moved across the sample.

The density of dislocations in crystals is expressed as their total length per unit volume, hence as $cm \cdot cm^{-3}$ or cm^{-2}. It varies from zero in the perfect single crystals of silicon made for the production of microelectronic chips, to typically 10^{11} cm^{-2} in deformed metals.

Further reading

D. Hull, D.J. Bacon, *Introduction to Dislocations*, 5th edn. (Elsevier, 2011)

H. Föll, website on crystal defects (2001)
http://www.tf.uni-kiel.de/matwis/amat/mw1_ge/index.html

References

[1] D. Calecki, *Physique des milieux continus*, vol. 1, *Mécanique et thermo-dynamique* (Hermann, Paris, 2007)

[2] A. Yeganeh-Haerl, D.J. Weidner, J.B. Parise, *Science* **257**, 650–652 (1992)

Chapter 14

Elastic waves in crystals

This chapter introduces an important application of the elasticity tensors: the propagation in crystals of acoustic waves, the velocity of which depends directly on the stiffness tensor. The end of this chapter discusses briefly the relation between this macroscopic property and atom vibrations.

14.1. Introduction

In the preceding chapter, we investigated the elastic behavior of crystalline solids submitted to static forces. These forces induce strains, which are connected to them through the elastic coefficients. These coefficients were first measured using static approaches, which yield limited accuracy because the induced strains are very small. We will see in this chapter that, if the material is submitted to a strain wave, the velocity at which this wave propagates depends on the stiffness coefficients of the material. These propagation velocities can be measured with an accuracy much higher than through static measurements. Therefore the stiffness coefficients, from which the elastic coefficients are deduced, are nowadays measured using techniques based on elastic wave propagation velocity. The elastic waves are excited in the crystal by transducers made from piezoelectric crystals (Sect. 15.3.6): when excited by a sinusoidal electric field, they create a sinusoidal strain wave which is transmitted by contact to the crystal under investigation.

In Sections 14.2, 14.3 and 14.4, the crystal is, as in all preceding chapters starting from Chapter 9, considered as a continuous medium. This assumption only holds as long as the wavelength of the elastic waves is much larger than the dimensions of the crystal's unit cell. The frequency range used in applications is that of sound and ultrasound waves, gathered under the denomination of acoustic waves. Their frequency varies from 1 kHz to about 10 GHz, so that, considering the order of magnitude of the propagation velocity of sound in crystal materials (some 10^3 m s^{-1}), the condition is satisfied. Nevertheless, it is worthwhile understanding the connection between this macroscopic approach and the microscopic treatment which directly takes into consideration

© Springer Science+Business Media Dordrecht 2014
C. Malgrange et al., *Symmetry and Physical Properties of Crystals*,
DOI 10.1007/978-94-017-8993-6_14

the atomic arrangement and the binding forces between atoms. This is done in the last section, using a highly simplified model (a linear chain of equidistant identical atoms) which however provides insight into the behavior of a real crystal (three-dimensional unit cell, and several atoms per unit cell). These results are important, and they are given without derivation in Section 14.5.3.

14.2. Plane elastic waves

Consider a solid submitted to an elastic perturbation, and a small volume element dv of this solid centered at a point with position vector $\mathbf{r}_0$ at rest. Its position at time t becomes:

$$\mathbf{r}(t) = \mathbf{r}_0 + \mathbf{u}(\mathbf{r}_0, t) \tag{14.1}$$

where $\mathbf{u}(\mathbf{r}_0, t)$ is the displacement vector.

The resultant $\mathbf{F}$ of the forces applied to this element dv is, using (11.3), such that:

$$F_i dv = \left(\frac{\partial T_{ij}}{\partial x_j} + f_i \right) dv$$

and, applying Newton's second law:

$$\left(\frac{\partial T_{ij}}{\partial x_j} + f_i \right) dv = (\rho dv) \frac{\partial^2 x_i}{\partial t^2} \tag{14.2}$$

where ρ is the mass per unit volume of the solid.

If the bulk forces $\mathbf{f}$ are zero or considered to be negligible, as we assume in the rest of this chapter, then:

$$\frac{\partial T_{ij}}{\partial x_j} = \rho \frac{\partial^2 x_i}{\partial t^2}. \tag{14.3}$$

Using (14.1),

$$\frac{\partial^2 x_i}{\partial t^2} = \frac{\partial^2 u_i}{\partial t^2}$$

and (14.3) becomes:

$$\frac{\partial T_{ij}}{\partial x_j} = \rho \frac{\partial^2 u_i}{\partial t^2}. \tag{14.4}$$

We can relate the components T_{ij} to the partial derivatives of the components u_i with respect to the position variables. The stress tensor T_{ij} is related to the strain tensor S_{kl} through the stiffness tensor c_{ijkl} (Eq. (13.1b)). Furthermore (Eq. (12.10)),

$$S_{kl} = \frac{1}{2} \left(\frac{\partial u_k}{\partial x_l} + \frac{\partial u_l}{\partial x_k} \right).$$

We thus obtain:

$$T_{ij} = c_{ijkl} S_{kl} = \frac{1}{2} c_{ijkl} \left(\frac{\partial u_k}{\partial x_l} + \frac{\partial u_l}{\partial x_k} \right)$$

and

$$\frac{\partial T_{ij}}{\partial x_j} = \frac{1}{2} c_{ijkl} \left(\frac{\partial^2 u_k}{\partial x_j \partial x_l} + \frac{\partial^2 u_l}{\partial x_j \partial x_k} \right).$$

The two terms on the right-hand side of this equation are equal since we sum over j, k and l. We can thus write:

$$\frac{\partial T_{ij}}{\partial x_j} = c_{ijkl} \left(\frac{\partial^2 u_l}{\partial x_j \partial x_k} \right)$$

and Equation (14.4) becomes:

$$c_{ijkl} \left(\frac{\partial^2 u_l}{\partial x_j \partial x_k} \right) = \rho \frac{\partial^2 u_i}{\partial t^2}. \qquad (14.5)$$

We now show that the general solution to this equation has the form:

$$\mathbf{u}(\mathbf{r},t) = A e F \left(t - \frac{\mathbf{s} \cdot \mathbf{r}}{v} \right) + B e' G \left(t + \frac{\mathbf{s} \cdot \mathbf{r}}{v} \right).$$

For simplicity, we restrict consideration to the first term in this solution:

$$\mathbf{u}(\mathbf{r},t) = A e F \left(t - \frac{\mathbf{s} \cdot \mathbf{r}}{v} \right) = A e F \left(t - \frac{s_j x_j}{v} \right).$$

This is the equation of a plane wave propagating along a direction parallel to unit vector $\mathbf{s}$, with components s_i, because $s_j x_j$, hence the phase of the wave, is constant over any plane perpendicular to vector $\mathbf{s}$ (Fig. 14.1). If we consider any point M with coordinates x_j, the plane P perpendicular to $\mathbf{s}$ and going through M intersects axis $O\ell$, parallel to $\mathbf{s}$, at H the projection of M and of all points of the plane onto $O\ell$. For all points M in the plane, $\mathbf{OM} \cdot \mathbf{s} = x_j s_j = \overline{OH} = \ell$ and we can write:

$$\mathbf{u}(\mathbf{r},t) = A e F \left(t - \frac{\mathbf{s} \cdot \mathbf{r}}{v} \right) = A e F \left(t - \frac{\ell}{v} \right), \qquad (14.6)$$

a plane wave whose wave planes move parallel to $\mathbf{s}$, at velocity v. Unit vector $\mathbf{e}$, whose components are noted e_i, is called the polarization vector of the wave.

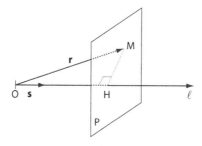

Figure 14.1: Plane P perpendicular to unit vector $\mathbf{s}$. This plane is the locus of points M for which $\mathbf{s} \cdot \mathbf{r} = \mathbf{s} \cdot \mathbf{OM} = \overline{OH} = \mathrm{const} = \ell$.

Consider the form (14.6) of the displacement, and calculate its derivatives with respect to time and space, in order to substitute them into (14.5):

$$\frac{\partial^2 u_i}{\partial t^2} = A e_i F'',$$

$$\frac{\partial u_l}{\partial x_k} = -A e_l \frac{s_k}{v} F' \quad \text{and} \quad \frac{\partial^2 u_l}{\partial x_j \partial x_k} = A e_l \frac{s_j s_k}{v^2} F''$$

where F' and F'' are respectively the first and second derivatives of $F(w)$ with respect to w.

Equation (14.5) becomes:

$$c_{ijkl} e_l \frac{s_j s_k}{v^2} F'' = \rho e_i F''$$

so that

$$c_{ijkl} s_j s_k e_l = \rho v^2 e_i. \tag{14.7}$$

We set:

$$\Gamma_{il} = c_{ijkl} s_j s_k \tag{14.8}$$

and Equation (14.7) can be written as:

$$\Gamma_{il} e_l = \rho v^2 e_i. \tag{14.9}$$

This equation shows that vector $\mathbf{e}$ is an eigenvector of matrix $\{\Gamma_{il}\}$, and that the eigenvalue associated to it is equal to ρv^2.

The terms in matrix $\{\Gamma_{il}\}$ relate a vector to another vector. They are therefore the coefficients of a rank-2 tensor $[\Gamma]$ (Sect. 9.4.2), called the Christoffel tensor. This tensor is symmetric, since:

$$\Gamma_{li} = c_{ljki} s_j s_k = c_{kilj} s_j s_k$$

because $c_{\alpha\beta} = c_{\beta\alpha}$ (Sect. 13.4).

Also, since tensor c_{ijkl} is symmetric with respect to i and j and with respect to k and l (Sect. 13.2.2):

$$\Gamma_{li} = c_{ikjl} s_k s_j \quad \text{and} \quad \Gamma_{li} = \Gamma_{il}.$$

Because matrix Γ is real and symmetric, it has three orthogonal eigenvectors. The eigenvalues are the solutions of the following equation, called the Christoffel equation:

$$\begin{vmatrix} \Gamma_{11} - \rho v^2 & \Gamma_{12} & \Gamma_{13} \\ \Gamma_{12} & \Gamma_{22} - \rho v^2 & \Gamma_{23} \\ \Gamma_{13} & \Gamma_{23} & \Gamma_{33} - \rho v^2 \end{vmatrix} = 0.$$

Once the eigenvalues are known, it is easy to deduce the eigenvectors, *i.e.* the polarization vectors of the corresponding elastic wave.

Expanding the expression of one of the coefficients Γ_{il}, for example Γ_{11}:

$$\Gamma_{11} = c_{1111}s_1^2 + c_{1121}s_1s_2 + c_{1131}s_1s_3 + c_{1211}s_2s_1 + c_{1221}s_2^2 + c_{1231}s_2s_3$$
$$+ c_{1311}s_3s_1 + c_{1321}s_3s_2 + c_{1331}s_3^2.$$

Hence, using for the c_{ijkl} the two-subscript notation with α and β:

$$\Gamma_{11} = c_{11}s_1^2 + c_{66}s_2^2 + c_{55}s_3^2 + 2\left(c_{16}s_1s_2 + c_{15}s_3s_1 + c_{56}s_2s_3\right).$$

The other Γ_{ij} coefficients can be calculated in the same way, and the result is given in the form of Table 14.1. Each row of the table gives, for the coefficient Γ_{il} indicated at the beginning of the row, the elastic coefficients to be associated to each of the six terms s_1^2, s_2^2, s_3^2, s_2s_3, s_3s_1, s_1s_2 in the sum.

Table 14.1: Table showing the form of the components Γ_{ij} of Christoffel's tensor

	s_1^2	s_2^2	s_3^2	s_2s_3	s_3s_1	s_1s_2
Γ_{11}	c_{11}	c_{66}	c_{55}	$2\,c_{56}$	$2\,c_{15}$	$2\,c_{16}$
Γ_{22}	c_{66}	c_{22}	c_{44}	$2\,c_{24}$	$2\,c_{46}$	$2\,c_{26}$
Γ_{33}	c_{55}	c_{44}	c_{33}	$2\,c_{34}$	$2\,c_{35}$	$2\,c_{45}$
Γ_{23}	c_{56}	c_{24}	c_{34}	$c_{44}+c_{23}$	$c_{45}+c_{36}$	$c_{46}+c_{25}$
Γ_{31}	c_{15}	c_{46}	c_{35}	$c_{45}+c_{36}$	$c_{13}+c_{55}$	$c_{14}+c_{56}$
Γ_{12}	c_{16}	c_{26}	c_{45}	$c_{25}+c_{46}$	$c_{56}+c_{14}$	$c_{12}+c_{66}$

In conclusion, three elastic waves can propagate in a crystal. The propagation velocity v_j of each of these waves j, the phase velocity, is deduced from the corresponding eigenvalue $\xi_j = \rho v_j^2$ of the matrix $\{\Gamma_{il}\}$ associated with the Christoffel tensor. The corresponding polarization vector $\mathbf{e}_j$ is the eigenvector associated to ξ_j. The three polarization vectors are mutually orthogonal.

14.3. Application to a cubic crystal

The stiffness matrix of a cubic crystal is expressed, in a system of axes parallel to the basis vectors of the crystal lattice:

$$c = \begin{pmatrix} c_{11} & c_{12} & c_{12} & 0 & 0 & 0 \\ c_{12} & c_{11} & c_{12} & 0 & 0 & 0 \\ c_{12} & c_{12} & c_{11} & 0 & 0 & 0 \\ 0 & 0 & 0 & c_{44} & 0 & 0 \\ 0 & 0 & 0 & 0 & c_{44} & 0 \\ 0 & 0 & 0 & 0 & 0 & c_{44} \end{pmatrix}.$$

The Christoffel tensor coefficients can then be expressed as:

$$\begin{aligned}
\Gamma_{11} &= c_{11}s_1^2 + c_{44}\left(s_2^2 + s_3^2\right) & \Gamma_{12} &= (c_{44} + c_{12})\, s_1 s_2 \\
\Gamma_{22} &= c_{11}s_2^2 + c_{44}\left(s_3^2 + s_1^2\right) & \Gamma_{23} &= (c_{44} + c_{12})\, s_2 s_3 \\
\Gamma_{33} &= c_{11}s_3^2 + c_{44}\left(s_1^2 + s_2^2\right) & \Gamma_{31} &= (c_{44} + c_{12})\, s_3 s_1.
\end{aligned}$$

Their value depends on the propagation direction **s** of the plane wave. The next two sections deal with two special cases.

14.3.1. Propagation of a plane wave along direction [100]

In this case: $\qquad\qquad s_1 = 1, \qquad\qquad\qquad s_2 = s_3 = 0.$

The only non-zero coefficients of the Christoffel tensor are thus:

$$\Gamma_{11} = c_{11}, \qquad \Gamma_{22} = c_{44}, \qquad \Gamma_{33} = c_{44}.$$

The propagation velocities must satisfy equation:

$$\begin{vmatrix}
c_{11} - \rho v^2 & 0 & 0 \\
0 & c_{44} - \rho v^2 & 0 \\
0 & 0 & c_{44} - \rho v^2
\end{vmatrix} = 0,$$

the obvious solutions of which are:

$$v_1 = \sqrt{\frac{c_{11}}{\rho}}$$

with eigenvector (1, 0, 0) so that the wave is longitudinal (Fig. 14.2a), and

$$v_2 = v_3 = \sqrt{\frac{c_{44}}{\rho}}$$

for which the eigenvectors are vectors of the subspace formed by vectors $(0, 1, 0)$ and $(0, 0, 1)$, in other words all vectors perpendicular to vector **s**: the wave is transverse (Fig. 14.2b).

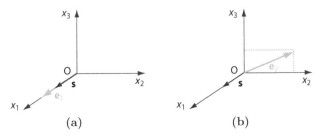

(a)   (b)

Figure 14.2: Polarization vectors for waves propagating in a cubic crystal along direction [100]: (a) longitudinal wave; (b) transverse wave

14.3.2. Propagation along direction [110]

In this case:
$$s_1 = s_2 = \frac{1}{\sqrt{2}}, \qquad s_3 = 0.$$

The only non-zero values of Christoffel's tensor are:

$$\Gamma_{11} = \frac{c_{11} + c_{44}}{2}, \qquad \Gamma_{22} = \frac{c_{11} + c_{44}}{2}, \qquad \Gamma_{33} = c_{44}, \qquad \Gamma_{12} = \frac{c_{44} + c_{12}}{2}.$$

The propagation velocities must satisfy equation:

$$\begin{vmatrix} \dfrac{c_{11} + c_{44}}{2} - \rho v^2 & \dfrac{c_{44} + c_{12}}{2} & 0 \\[2ex] \dfrac{c_{44} + c_{12}}{2} & \dfrac{c_{11} + c_{44}}{2} - \rho v^2 & 0 \\[2ex] 0 & 0 & c_{44} - \rho v^2 \end{vmatrix} = 0.$$

Vector $(0, 0, 1)$ is an obvious eigenvector, and the associated propagation velocity is $v_3 = \sqrt{c_{44}/\rho}$ (Fig. 14.3). This wave is transverse.

The eigenvalues v_1 and v_2 are solutions of:

$$\left[\left(\frac{c_{11} + c_{44}}{2} - \rho v^2 \right)^2 - \left(\frac{c_{44} + c_{12}}{2} \right)^2 \right] = 0$$

hence

$$v_1 = \sqrt{\frac{2c_{44} + c_{12} + c_{11}}{2\rho}} \quad \text{and} \quad v_2 = \sqrt{\frac{c_{11} - c_{12}}{2\rho}}.$$

It is easy to show that the polarization vector of the wave with propagation velocity v_1 is $\mathbf{e}_1 = (1/\sqrt{2})(1, 1, 0)$, so that the wave is longitudinal because $\mathbf{e}_1$ is parallel to $\mathbf{s}$. The polarization vector of the wave with propagation velocity v_2 is $\mathbf{e}_2 = (1/\sqrt{2})(1, -1, 0)$, so that this is a transverse wave.

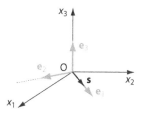

Figure 14.3: Polarization vectors for a wave propagating in a cubic crystal along direction [110]: ($\mathbf{e}_1$) and the wave is longitudinal, ($\mathbf{e}_2$, $\mathbf{e}_3$) and the waves are transverse

Notes:

1. We found longitudinal and transverse waves to be the solutions in the two above cases, but these are very special cases. For a general propagation direction, the vibration modes are neither purely longitudinal

nor purely transverse. The wave whose polarization vector is closest to the propagation direction $\mathbf{s}$ (*i.e.* for which the angle between $\mathbf{e}$ and $\mathbf{s}$ is smallest) is called quasi-longitudinal and the two other waves are called quasi-transverse.

2. The velocity v is a phase velocity. In the case of a sine wave

$$\mathbf{u}(\mathbf{r}, t) = A\mathbf{e}\cos(\omega t - \mathbf{k} \cdot \mathbf{r}) = A\mathbf{e}\cos(\omega t - k\mathbf{s} \cdot \mathbf{r}) = A\mathbf{e}\cos\omega\left(t - \frac{k}{\omega}\mathbf{s} \cdot \mathbf{r}\right),$$

the phase velocity v is equal to ω/k. We found above that the value of v does not depend on the angular frequency ω. Thus the crystal material is non-dispersive for acoustic waves. However it is anisotropic, and this velocity depends on the direction of vector $\mathbf{k}$ since the Christoffel tensor $[\Gamma]$ depends on it. The group velocity $\mathbf{v}_g$, with components $(v_g)_i = \partial\omega/\partial k_i$, which is the propagation direction of energy, is generally no more parallel to the wave vector.

3. We showed (Sect. 12.4) that the relative volume variation of a deformed solid is equal to the trace of the strain tensor:

$$\frac{\partial u_i}{\partial x_i} = \text{div } \mathbf{u}.$$

In the special case of a transverse wave, it is easy to show that div $\mathbf{u} = 0$. If one of the axes, for example Ox_1, is chosen parallel to the propagation vector $\mathbf{s}$, the only non-zero components of $\mathbf{u}$ are u_2 and u_3 and, since they do not depend on x_2 and x_3, div $\mathbf{u} = 0$ and propagation occurs with no change in volume. In contrast, propagation of any other waves, for which div $\mathbf{u} \neq 0$, entails expansion and compression regions in the solid.

14.4. Isotropic solid case

This case is easily deduced from the propagation in a cubic crystal, by considering a general direction and using relation (13.7b) which connects the stiffness coefficients of an isotropic material:

$$c_{44} = \frac{c_{11} - c_{12}}{2}.$$

Consider for example propagation direction [100] in a cubic material. From the results of Section 14.3.1, one longitudinal wave, with propagation velocity v_ℓ, and transverse waves, with propagation velocity v_t, can propagate, with

$$v_\ell = \sqrt{\frac{c_{11}}{\rho}} \qquad v_t = \sqrt{\frac{c_{44}}{\rho}}.$$

Using the Lamé coefficients λ and μ (see Sect. 13.6.2) such that:

$$\lambda + 2\mu = c_{11} \qquad \mu = c_{44},$$

the velocities of the longitudinal and transverse waves, v_ℓ and v_t respectively, become:

$$v_\ell = \sqrt{\frac{\lambda + 2\mu}{\rho}} \qquad \text{and} \qquad v_t = \sqrt{\frac{\mu}{\rho}}.$$

The ratio of these longitudinal and transverse wave velocities is:

$$\frac{v_\ell}{v_t} = \sqrt{\frac{2\mu + \lambda}{\mu}} = \sqrt{2 + \frac{\lambda}{\mu}}.$$

This ratio is always larger than 1. Hence, in an isotropic material, the longitudinal elastic waves propagate faster than the transverse elastic waves.

Of course the same results would be obtained using the results of Section 14.3.2, derived for a propagation direction parallel to [110].

14.5. Microscopic approach – Crystal lattice dynamics

Till now, we considered the crystal to be a continuous and homogeneous medium. It is worthwhile understanding how this macroscopic approach can be connected to a microscopic description involving the vibrations of the atoms. At equilibrium, the average interatomic distance in a crystal is the one which minimizes the crystal's energy. If an atom is moved away from its equilibrium position, it is submitted by its neighbors to restoring forces which cause it to oscillate around its equilibrium position. These oscillations are transmitted to the neighboring atoms, thus creating within the solid an elastic wave which propagates in the crystal.

The full treatment of atomic vibrations in a crystal, crystal lattice dynamics, involves no major difficulty, but it implies a rather cumbersome algebraic approach, with a wealth of subscripts, which is beyond the scope of this book. Nevertheless, the main characteristics of the results can be deduced from the simple model of a one-dimensional crystal, *i.e.* a linear chain of atoms. We will start out by working out the case of a chain of identical atoms, then that of a chain containing two types of atoms. These two examples provide the essential results of the full treatment.

14.5.1. Linear chain of identical atoms

Consider a linear chain of N identical atoms, with mass M, the same distance a apart (Fig. 14.4). The atom displacements in the chain are assumed to be small enough that a series expansion to first order of the energy with respect to the displacement be valid, which is equivalent to considering a harmonic potential. Furthermore, we assume that the forces between atoms are restricted to forces between nearest neighbors. We can thus imagine a chain of atoms connected by springs with the same coefficient K.

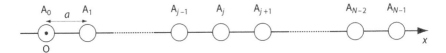

Figure 14.4: One-dimensional chain consisting of a single kind of atoms, with mass M, at the same distance a from one another

The force between two atoms is a restoring force proportional to the difference in distance between the displaced and the undisplaced atoms. There are N atoms, denoted as $A_0, A_1, \ldots, A_j, \ldots, A_{N-1}$. Atom j is, at equilibrium, at position $x_j^0 = ja$, and, if it is displaced by u_j, it is submitted to two forces, with respective algebraic values (Fig. 14.5):

(a) $K(u_{j-1} - u_j)$, the force exerted by atom $j - 1$,

(b) $K(u_{j+1} - u_j)$, the force exerted by atom $j + 1$.

Their resultant F_j is:

$$F_j = K(u_{j+1} + u_{j-1} - 2u_j).$$

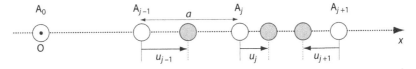

Figure 14.5: Displacement of the atoms in the one-dimensional chain. The positions of the atoms at equilibrium are the empty circles and the positions after displacement are the full circles.

Applying Newton's second law to the j-th atom, we obtain the equation for its motion:

$$M\frac{\partial^2 u_j}{\partial t^2} = K(u_{j+1} + u_{j-1} - 2u_j) \tag{14.10}$$

thus altogether N equations for the N atoms. This system of N equations simplifies considerably if we seek a solution where all the atoms have a sinusoidal movement with the same amplitude, the displacement being in the form of a plane monochromatic wave. For atom j:

$$u_j = A \cos(kx_j^0 - \omega t) = A \cos(kja - \omega t) = A\,\Re\left[\exp i(kja - \omega t)\right] \tag{14.11}$$

where $\Re$ means "real part of".

We set $v_j = \exp i(kja - \omega t)$, and we seek v_j as a solution of Equation (14.10), which becomes

$$M\frac{\partial^2 v_j}{\partial t^2} = K(v_{j+1} + v_{j-1} - 2v_j). \tag{14.12}$$

Since this equation is linear, u_j will be a solution if v_j is. We obtain:

$$\frac{\partial^2 v_j}{\partial t^2} = -\omega^2 v_j,$$

$$v_{j-1} = e^{-ika} v_j \qquad \text{and} \qquad v_{j+1} = e^{ika} v_j.$$

Equation (14.12) becomes:

$$\omega^2 = \frac{K}{M}[2 - (e^{-ika} + e^{ika})] = 2\frac{K}{M}(1 - \cos\ ka)$$

$$\omega^2 = 4\frac{K}{M}\sin^2\left(\frac{ka}{2}\right)$$

and

$$\omega\,(k) = 2\sqrt{\frac{K}{M}}\left|\sin\frac{ka}{2}\right|. \tag{14.13}$$

The $\omega(k)$ curve is called the dispersion curve.

It is periodic, with period equal to $2\pi/a$ (Fig. 14.6). Furthermore it is symmetric. It is therefore sufficient to consider half a period, for example $0 < k < \pi/a$.

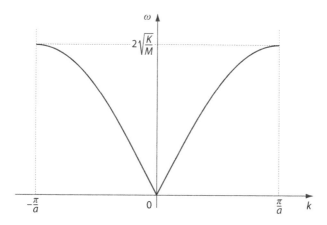

Figure 14.6: Dispersion curve $\omega(k)$ of a linear chain consisting of identical equidistant atoms with mass M

We notice that the dispersion relation (14.13) does not depend on j. Thus the N equations (14.12) lead to a single dispersion relation, which justifies the choice we made of a sinusoidal displacement wave (14.11). However, we started out from N coupled oscillator equations, which means that if a single atom starts vibrating, it will not continue vibrating with the same amplitude. It transfers energy to its neighbors in a complicated way. The sine-wave solutions we sought are collective oscillations, uncoupled (independent) from one another, called the normal modes. Each mode corresponds to one well-defined value of k and one well-defined value of ω; hence, in the language of quantum physics, it is a stationary state. The number of normal modes is equal to the number of equations we started from, N. If we want each atom to have the same environment, we must assume that N is very large, and that the chain is closed on itself (Born and Von Karman's periodic boundary conditions). This implies that atom j coincides with atom $j + N$, so that $u_j = u_{j+N}$, and $e^{ikNa} = 1$, so that:

$$kNa = 2p\pi \text{ with } p \text{ an integer}$$

and
$$k = \frac{2p\pi}{Na}.$$

There are thus N values of k in one period $2\pi/a$, for example between $-\pi/a$ and π/a.

We now look at two limiting cases: k very small, *i.e.* $k \ll \pi/a$ and $k = \pi/a$.

If $k \ll \pi/a$, then $\lambda \gg a$. The wavelength is very large compared to the interatomic distance, and we expect to retrieve the results obtained when we assume the medium to be continuous. We notice that, in this case, many atoms have almost identical motions, and thus the discrete character of the chain vanishes.

Equation (14.13) becomes $\omega \approx ka\sqrt{K/M}$ and the phase velocity $v_\varphi = \omega/k$ is:

$$\frac{\omega}{k} = a\sqrt{\frac{K}{M}}.$$

It is independent of k. The crystal is non-dispersive for such waves, in agreement with the result of the macroscopic treatment, and the velocity of the acoustic waves v_a is:

$$v_a = a\sqrt{\frac{K}{M}}. \tag{14.14}$$

If $k = \pi/a$, Equation (14.11) becomes $u_j = A\cos(\omega t - j\pi)$. There is no propagation any more, and two neighboring atoms, j and $j + 1$, vibrate in phase opposition.

The maximum angular frequency ω_m of the associated waves is given by:

$$\omega_m = 2\sqrt{\frac{K}{M}}$$

whence a cutoff frequency $f_c = \omega_m/2\pi$ which can be related to the above acoustic wave velocity v_a determined in (14.14):

$$f_c = \frac{1}{\pi}\sqrt{\frac{K}{M}} = \frac{v_a}{\pi a}. \tag{14.15}$$

The results obtained with this extremely simple model can be generalized to the case of a real monoatomic crystal. It can be shown that there are three dispersion curves for one propagation direction of the wave. Each curve is associated to one of the polarization vectors determined at the beginning of this chapter, and its slope at origin is the associated propagation velocity.

This one-dimensional model can be used to check that the frequency range of acoustic waves does correspond to the linear part of these dispersion curves. The order of magnitude of the cutoff frequency defined above (Eq. 14.15) can be calculated from the order of magnitude of the acoustic wave velocity v_a, some $10^3 \, \text{m s}^{-1}$: we choose arbitrarily $3 \times 10^3 \, \text{m s}^{-1}$. If we assume an interatomic distance on the order of $2 \, \text{Å}$, we obtain a cutoff frequency on the order of $5 \, \text{THz}$. The acoustic wave range does not extend beyond a few GHz, it is thus indeed in the linear part of the dispersion curve.

14.5.2. Linear chain with two different atoms

We now consider a linear chain of atoms, with period a, containing two types of equidistant atoms, with mass respectively M and m (Fig. 14.7). The coordinate x_j^0 of the j-th atom at equilibrium is $x_j^0 = ja/2$.

Figure 14.7: One-dimensional chain, with period a, consisting of two types of atoms, with mass M and m, a distance $a/2$ apart

During motion, the coordinate x_j of this atom becomes: $x_j = j(a/2) + u_j$. As in the case of the monoatomic chain, we model the interaction force between nearest neighbor atoms with a harmonic restoring force, the equivalent of a spring with stiffness K.

The force on the j-th atom is then given by:

$$F_j = K(u_{j+1} - u_j) + K(u_{j-1} - u_j).$$

The equations of motion are respectively

$$M\frac{\partial^2 u_j}{\partial t^2} = K(u_{j+1} + u_{j-1} - 2u_j), \qquad (14.16a)$$

for atom j with mass M, and

$$m\frac{\partial^2 u_{j+1}}{\partial t^2} = K(u_{j+2} + u_j - 2u_{j+1}) \qquad (14.16b)$$

for atom $j+1$ with mass m.

As in the preceding case, we look for a solution having the form of a sinusoidal wave:

$$u_j = A\Re\left[e^{i(kja/2-\omega t)}\right] \qquad \text{and} \qquad u_{j+1} = \alpha A\Re\left[e^{i(k(j+1)a/2-\omega t)}\right]$$

where α is the ratio of the respective displacement amplitudes of masses m and M. We obtain:

$$-M\omega^2 = K[\alpha(e^{ika/2} + e^{-ika/2}) - 2]$$
$$-m\alpha\omega^2 e^{ika/2} = K[(e^{ika} + 1) - 2\alpha e^{ika/2}]$$

so that

$$M\omega^2 = 2K[1 - \alpha\cos ka/2] \qquad (14.17a)$$
$$m\alpha\omega^2 = 2K[\alpha - \cos ka/2]. \qquad (14.17b)$$

From Equations (14.17a) and (14.17b):

$$\alpha = \frac{2K - M\omega^2}{2K\cos(ka/2)} = \frac{2K\cos(ka/2)}{2K - m\omega^2}.$$

This implies

$$(2K - m\omega^2)(2K - M\omega^2) = 4K^2\cos^2\left(\frac{ka}{2}\right)$$

so that

$$mM\omega^4 - 2K(m + M)\omega^2 + 4K^2\sin^2\left(\frac{ka}{2}\right) = 0.$$

This is a second-degree equation in ω^2, with solutions:

$$\omega^2 = \frac{K(m + M) \pm \left[(m + M)^2 K^2 - 4K^2 mM\sin^2(ka/2)\right]^{1/2}}{mM}$$

so that

$$\omega^2 = \frac{K}{mM}\left[(m + M) \pm \left[(m + M)^2 - 4mM\sin^2(ka/2)\right]^{1/2}\right].$$

The $\omega(k)$ curve has two branches (Fig. 14.8). Their limiting values for $k \ll \pi/a$ and for $k = \pi/a$ can be easily calculated.

If $k \ll \pi/a$, $\sin ka/2 \approx ka/2 \ll 1$ and we obtain:

$$\omega \approx \sqrt{2K\frac{(m + M)}{mM}} \qquad \text{and} \qquad \omega \approx ka\sqrt{\frac{K}{2(m + M)}}.$$

The corresponding values of α are $-M/m$ and 1.

The first solution corresponds to point A in Figure 14.8. The atoms with masses m and M respectively vibrate in phase opposition and their center of mass is at rest (Fig. 14.9a). If the chain consists of an alternation of positive and negative ions, their motion leads to an electric polarization of the material. This gives a possibility of interacting with electromagnetic waves which leads, in particular, to wave absorption in the relevant frequency range. These frequencies happen to be those of optical electromagnetic waves. This is why this branch of the dispersion curve is called the optical branch or the optical mode.

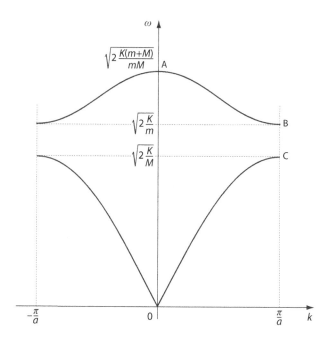

Figure 14.8: Dispersion curve $\omega(k)$ for a periodic linear chain consisting of two types of equidistant atoms, with masses respectively M and m

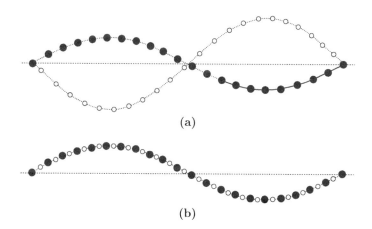

(a)

(b)

Figure 14.9: Vibration modes for a periodic linear chain consisting of two types of equidistant atoms, in the approximation $k \ll \pi/a$, so that $\lambda \gg 2a$. For clarity, (i) we assume the vibrations to be transverse, and not longitudinal, and (ii) the wavelength is not extremely large with respect to period a. (a) Optical branch: the atoms with masses m and M respectively vibrate in phase opposition and their center of mass is at rest. (b) Acoustic branch: the two types of atoms vibrate in phase with the same amplitude.

The second solution corresponds to long wavelength acoustic waves. Both types of atoms vibrate in phase with the same amplitude (Fig. 14.9b). The material is no more dispersive (ω/k is constant) and the propagation velocity v is equal to ω/k:

$$v = a\sqrt{\frac{K}{2(m+M)}}.$$

This is the equivalent of the propagation velocity of sound waves.

If $k = \pi/a$, $\sin ka/2 = 1$ and the two solutions for ω^2 are:

$$\omega = \sqrt{2\frac{K}{m}} \qquad \text{and} \qquad \omega = \sqrt{2\frac{K}{M}}.$$

The corresponding values of α are $\alpha = \infty$ and $\alpha = 0$. This limit corresponds to a half-wavelength equal to a, the distance between atoms of the same type. In the first solution, mass m oscillates and mass M is at rest (point B in Fig. 14.8). It is no surprise that the frequency then depends only on mass m. In the second case, mass M oscillates and mass m is at rest (point C on Fig. 14.8).

14.5.3. Extension to the real crystal case

The simplified theory we just presented makes it possible to understand the results for a real three-dimensional crystal. The dispersion curves $\omega(\mathbf{k})$ are then represented for a given direction of vectors $\mathbf{k}$. The results are as follows:

- The dispersion curves feature 3 acoustic modes, analogous to the single acoustic mode of the linear chain. The slopes at origin of these modes are equal to the propagation velocities of the acoustic waves determined under the macroscopic theory. If two velocities are the same in some propagation direction, there are only two acoustic modes, one of them being degenerate.

- If the crystal contains one single atom per primitive unit cell, there are only acoustic modes. If the crystal contains two atoms per unit cell, there are, additionally, three optical modes, the properties of which are similar to those of the optical mode of the linear chain. The optical modes can also be degenerate.

- If the crystal contains p atoms per unit cell, there are $3p$ modes, 3 of which are acoustic modes and $3(p-1)$ are optical modes.

- In all cases, the linear parts of the acoustic modes, near the origin of the dispersion curves, are compatible with the assumptions of continuity and homogeneity of the material, and consistent with the macroscopic results on the propagation of acoustic waves.

14.6. Exercises

Exercise 14.1

Consider the propagation of elastic waves in a crystal with mass per unit volume ρ.

1. The crystal is monoclinic, and **c** is the basis vector parallel to the 2-fold axis or perpendicular to the mirror. We consider an elastic wave with wave-vector parallel to direction [001].

   (a) Write down Christoffel's matrix.

   (b) Determine the velocity and the polarization of the waves which propagate in the crystal.

2. Same questions for an orthorhombic crystal, and a propagation direction parallel to direction [001].

3. Same questions for a crystal having an n-fold rotation axis, with $n > 2$, and propagation direction parallel to direction [001].

4. The crystal is hexagonal. The propagation direction is a general direction in the plane perpendicular to **c**. Show that a wave with polarization parallel to **c**, therefore transverse, propagates.

Exercise 14.2

The components of the stiffness tensor of diamond are to be determined from the measurement of the propagation velocities of elastic waves. Diamond is cubic and its mass per unit volume is 3.51 g cm^{-3}.

1. Calculate the propagation velocity of longitudinal waves along direction [100].

2. Calculate the propagation velocities along direction [111].

3. The propagation velocities were measured in a diamond crystal for both of the above wave-vector directions. The results are as follows:

   For **k** parallel to [100]:
   – velocity of transverse waves: $v_t = 11{,}196$ m s^{-1}.

   For **k** parallel to direction [111]:
   – velocity of the longitudinal wave: $v_l = 14{,}490$ m s^{-1},
   – velocity of the transverse waves: $v_t = 13{,}239$ m s^{-1}.

   Determine the components of the stiffness tensor of diamond.

Exercise 14.3

Consider a tetragonal crystal with point group 422.

Determine the propagation velocity of elastic waves with wave-vector parallel to direction [110] in this crystal.

Chapter 15

Crystal thermodynamics. Piezoelectricity

The aim of this chapter is to introduce crossed effects, in which quantities related to different areas of physics (thermal, mechanical, electrical) are connected through thermodynamics. It then deals with an example which is important through its many diverse applications: piezoelectricity.

The mechanical properties of crystals were studied in Chapters 11, 12 and 13. We briefly recall below their electrical and thermal properties. Usually, these properties are studied independently of one another. For example, the application of stress leads to a deformation of the material, the magnitude of which depends on the value of its elastic coefficients. In a similar way, an electric field applied to a crystal induces an electrical induction which depends on the dielectric permittivity of the material. In both of these examples, the physical quantities involved are of the same nature, or, to put it another way, belong to the same domain or area (mechanical and electrical respectively), and the effects involved are called principal effects. This is no more the case when, applying a variation in temperature (thermal constraint), we measure a variation in polarization (an electrical quantity, the phenomenon of pyroelectricity) or a strain (a mechanical quantity, the phenomenon of thermal expansion). The response does not belong to the same area, and the effects involved are called crossed effects. Thus, the response of a crystal to a constraint may have several aspects and imply various physical properties.

The aim of this chapter is to examine the various responses of crystals to a given constraint. To introduce these effects and the connections between them, it is necessary to assume the crystal to be in an overall thermodynamic state defined by one parameter in each of the thermal, mechanical and electrical domains. This will be done in the first section, and will make it possible to introduce various crossed effects. We will then study more specifically two special crossed effects, *viz.* the properties of pyroelectricity (Sect. 15.2) and piezoelectricity (Sect. 15.3) featured by some crystals.

© Springer Science+Business Media Dordrecht 2014
C. Malgrange et al., *Symmetry and Physical Properties of Crystals*,
DOI 10.1007/978-94-017-8993-6_15

15.1. Crystal thermodynamics

15.1.1. Conjugate variables

Mechanical quantities

We have seen that, under the effect of a stress represented by a tensor $[T]$, a material undergoes a strain $[S]$ such that

$$[S] = [s][T]$$
$$\text{or} \quad S_{ij} = s_{ijkl}T_{kl}$$

and, in contracted notation (Sect. 13.3):

$$S_\alpha = s_{\alpha\beta}T_\beta. \tag{15.1}$$

$[s]$ is the tensor of elastic compliances, which connects the response $[S]$ of the crystal to the constraint $[T]$.

In a similar way, the stress is related to the strain by the tensor of elastic stiffnesses $[c]$ according to the relation

$$[T] = [c][S]$$
$$\text{or} \quad T_{ij} = c_{ijkl}S_{kl}$$

and in contracted notation

$$T_\alpha = c_{\alpha\beta}S_\beta.$$

During a change in the strain dS_{ij}, the work δW received by unit volume of the system[1] is given by Equation (13.5), *i.e.*

$$\delta W = T_{ij}dS_{ij} \quad \text{or} \quad \delta W = T_\alpha dS_\alpha. \tag{15.2}$$

Tensors $[S]$ and $[T]$ are the conjugate quantities, or conjugate variables, in the mechanical domain.

Electrical quantities

When submitted to an electric field $\mathbf{E}$, a material polarizes, and its polarization $\mathbf{P}$, the dipole moment per unit volume, is $\mathbf{P} = \varepsilon_0[\chi]\mathbf{E}$, where $[\chi]$ is the electrical susceptibility tensor. The electrical induction $\mathbf{D}$ is such that

$$\mathbf{D} = \varepsilon_0\mathbf{E} + \mathbf{P} = \varepsilon_0([E] + [\chi])\mathbf{E} = \varepsilon_0[\varepsilon]\mathbf{E} \tag{15.3}$$

---

1. Here we assume, for simplicity's sake, that the stress and the strain are homogeneous. If this were not the case, we would have to write

$$\delta W = \iiint T_{ij}dS_{ij}\,dv,$$

where the integral is performed over unit volume.

which can also be written as

$$D_i = \varepsilon_0 \varepsilon_{ij} E_j. \tag{15.4}$$

ε_0 is the permittivity of vacuum, the value of which, in the international system (SI) of units, is $10^{-9}/36\pi = 8.8 \times 10^{-12}$, and $[\varepsilon]$ is the electrical permittivity tensor which connects the response $\mathbf{D}$ of the crystal to the electric field $\mathbf{E}$.

Relation (15.4) can be inverted by introducing the impermeability tensor $[\eta]$, the inverse of the relative electrical permittivity tensor $[\varepsilon]$:

$$E_i = \eta_{ij} \frac{D_j}{\varepsilon_o}.$$

During a variation in induction $d\mathbf{D}$, the work received by unit volume of the system is (see e.g. [1])[2]

$$\delta W = E_i dD_i = \mathbf{E} \cdot d\mathbf{D}. \tag{15.5}$$

$\mathbf{D}$ and $\mathbf{E}$ are the conjugate quantities in the electrical domain.

Relation (15.3) must be altered for some materials which additionally feature a polarization $\mathbf{P}_0$, the spontaneous polarization, in the absence of an electrical field. For these materials:

$$\mathbf{D} = \mathbf{P}_0 + \varepsilon_0[\varepsilon]\mathbf{E} \tag{15.6}$$

so that, in zero field,

$$\mathbf{D} = \mathbf{P}_0.$$

These materials are called pyroelectric because, as we will see, this intrinsic polarization varies with temperature. They can only belong to some crystal classes[3] (Sect. 9.5.2).

Thermal quantities

If we apply a reversible change in temperature $d\Theta$ to unit volume of a material, it receives per unit volume a quantity of heat δQ_{rev}:

$$\delta Q_{\mathrm{rev}} = C_{\mathrm{exp}} d\Theta$$

---

2. The exact form given by Landau and Lifshitz in [1] is

$$\delta W = \iiint E_i dD_i dv,$$

where the integral must be taken over the whole volume, including vacuum if the dielectric does not occupy all of the space between conductors. We therefore assume here that the material fills all the space between conductors, and that $\mathbf{E}$ and $\mathbf{D}$ are homogeneous in the material. This makes it possible to refer to the work per unit volume of the material.

3. We recall that a crystallographic class includes all crystals with a given point group.

where C_{exp} is the heat capacity per unit volume of the material under the conditions of the experiment (e.g. at constant volume or at constant pressure).

The quantity of heat is not one of the variables defining the state of a system, and δQ is not a total differential. We must consider the entropy of the system $\mathcal{S}$, defined by its variation $d\mathcal{S}$ during a reversible transformation where the temperature changes from Θ to $\Theta + d\Theta$:

$$dS = \frac{\delta Q_{\text{rev}}}{\Theta} = \frac{C_{\text{exp}}}{\Theta} d\Theta. \tag{15.7}$$

Entropy $\mathcal{S}$ and temperature Θ are the conjugate variables of the thermal domain.

15.1.2. Independent variables

Consider a system with unit volume, defined by three quantities, each of them chosen in one of the above domains, *viz.* the thermal, electrical and mechanical domains. In the thermal domain, we can choose temperature Θ or entropy $\mathcal{S}$, which are scalars. In the electrical domain, the electric field $\mathbf{E}$ or induction $\mathbf{D}$, which are vectors defined by three components. And in the mechanical domain the strain tensor or the stress tensor, both symmetrical tensors with six independent components each. The state of the system is thus defined in total by 10 independent variables. For clarity, we actually consider the tensor form of tensors $[T]$ and $[S]$, with the two-subscript notation T_{ij} and S_{ij} for their coefficients,[4] and will thus consider 13 variables which are only quasi-independent since three of them are equal to three others ($T_{ij} = T_{ji}$ and $S_{ij} = S_{ji}$ for $i \neq j$). We will then revert to the contracted notation and to the 10 independent variables.

When the system changes state, its internal energy per unit volume U changes by dU such that

$$dU = \Theta dS + \delta W$$

where dS is the change in entropy and δW the reversible work, so that here

$$\delta W = E_i dD_i + T_{kl} dS_{kl}$$

and

$$dU = \Theta dS + E_i dD_i + T_{kl} dS_{kl}.$$

If we are interested in the internal energy of the system, the usual choice of independent variables is the entropy $\mathcal{S}$ per unit volume, the three components of induction D_i, and the 9 components of strain S_{kl}. We call these variables, which appear in the differential of the various terms in dU, the thermodynamic coordinates of the system, and we denote them as x_i, with index i ranging

---

4.   This makes it possible to clearly define the transition from the two subscripts ij to the single subscript α for the new tensors which will be introduced (thermal expansion and piezoelectricity).

from 1 to 13. The variables Θ, E_i and T_{kl} are the variables conjugate to the variables x_i chosen to define the system. We note X_i the quantity conjugate to the thermodynamic coordinate x_i, and thus

$$dU = X_i dx_i \qquad (15.8)$$

using Einstein's convention. The conjugate quantities X_i are intensive quantities, whereas the thermodynamic coordinates are often inaccurately described as extensive quantities. The electrical induction $\mathbf{D}$ is in fact homogeneous to a polarization, *i.e.* to a dipole moment per unit volume, and the components of the strain tensor define the relative variations in length. Thus the components D_i and S_{kl} cannot be considered as extensive quantities. Nevertheless they are quasi-extensive quantities because they are the ones to be multiplied by the value of the volume when the latter is no more unity.

In practice, it is often more convenient to choose as independent variables defining the system the conjugate quantities X_i. We are then led to considering another thermodynamic function[5] of the system, the free enthalpy or Gibbs potential G, such that

$$G = U - \Theta \mathcal{S} - E_i D_i - T_{kl} S_{kl}$$

and

$$dG = -\mathcal{S}d\Theta - D_i dE_i - S_{kl} dT_{kl}. \qquad (15.9)$$

We thus conclude that

$$\mathcal{S} = -\left.\frac{\partial G}{\partial \Theta}\right|_{E_j, T_{mn}} \qquad D_i = -\left.\frac{\partial G}{\partial E_i}\right|_{\Theta, (E_j), T_{mn}} \qquad S_{kl} = -\left.\frac{\partial G}{\partial T_{kl}}\right|_{\Theta, E_i, (T_{mn})}. \qquad (15.10)$$

In these formulas, the indexed variables that are taken to be constant during differentiation are indicated in parentheses when all of them are constant except for the one with respect to which differentiation is performed. For example, D_1 is the derivative of G with respect to E_1 at constant (E_i), Θ, T_{kl}, meaning that E_2, E_3, Θ and all T_{kl}'s are held constant.

Using the above notation, with the x_i representing the thermodynamic coordinates (the variables involved in the differential element of the elementary variation in internal energy (15.8)), and X_i their respective conjugate quantities, chosen here as defining the state of the system, the relations (15.9) and (15.10) respectively take on the condensed form

$$dG = -x_i dX_i$$

and

$$x_i = -\left.\frac{\partial G}{\partial X_i}\right|_{(X_j)}. \qquad (15.11)$$

---

5. The use of various thermodynamic functions depending on the independent variables of a problem is particularly well discussed in [2].

We can make a Taylor expansion of the value of each coordinate x_i as a function of its value $x_i(0)$ in the initial state. We denote as $X_j(0)$ the values of the conjugate quantities in the initial state. We obtain

$$x_i - x_i(0) = \sum_j \left. \frac{\partial x_i}{\partial X_j} \right|_0 (X_j - X_j(0)) \tag{15.12}$$

if we restrict the Taylor expansion to its first term. This is equivalent to considering that the first derivatives

$$\left. \frac{\partial x_i}{\partial X_j} \right|_0$$

are constant, or that the variations $(X_j - X_j(0))$ are small. This is an approximation, usually valid, which will be used in this and the following chapters. Only in the last chapter shall we show that some special effects (such as electro-optical and piezo-optical effects) require the inclusion of higher order terms. The same would apply for other effects such as non-linear elasticity, but we will not deal with them in this book.

Using Equations (15.11), relation (15.12) can be written as

$$x_i - x_i(0) = - \sum_j \left. \frac{\partial^2 G}{\partial X_j \partial X_i} \right|_0 (X_j - X_j(0))$$

or

$$x_i - x_i(0) = M_{ij}(X_j - X_j(0)) \tag{15.13}$$

using Einstein's convention for subscript j (ranging from 1 to 13) and setting

$$M_{ij} = - \left. \frac{\partial^2 G}{\partial X_j \partial X_i} \right|_0 . \tag{15.14}$$

Since the order of the differentiations does not matter:

$$M_{ji} = M_{ij}.$$

We thus defined a symmetrical matrix with 13 rows and 13 columns, called the physical property matrix M.

Consider an initial reference state at $\Theta = \Theta_0$, without stress so that $T_{kl} = 0$, and in zero electric field, so that $E_i = 0$, and a final state where $\Theta = \Theta_0 + \Delta\Theta$, $E_i \neq 0$, $T_{kl} \neq 0$. In the reference state, E_i is zero, hence, according to Equation (15.4), induction D_i is also zero, except if the crystal is naturally polarized, which is the case for pyroelectric crystals (Eq. (15.6)). For the latter the reference state has a spontaneous polarization $\mathbf{P}_0$ at Θ_0.

Since the stress tensor T_{kl} is zero in the reference state, so is the strain S_{kl}. Thus, in the reference state, $S = S_0$, $D_i = 0$ (except for pyroelectrics where $\mathbf{D}_0 = \mathbf{P}_0$) and $S_{kl} = 0$. Thus the only thermodynamic quantity with non zero value $x_i(0)$ is $\mathcal{S}_0$. The 13 components of the vector $x_i - x_i(0)$ are therefore $\mathcal{S} - \mathcal{S}_0 = \Delta\mathcal{S}$, D_j and S_{mn}.

A slightly more detailed presentation of Equations (15.13), using Einstein's convention (Sect. 9.3.1), is the following:

$$\Delta\mathcal{S} = -\frac{\partial^2 G}{\partial\Theta^2}\Delta\Theta - \frac{\partial^2 G}{\partial E_i \partial\Theta}E_i - \frac{\partial^2 G}{\partial T_{kl}\partial\Theta}T_{kl}, \tag{15.15a}$$

$$D_j = -\frac{\partial^2 G}{\partial\Theta\partial E_j}\Delta\Theta - \frac{\partial^2 G}{\partial E_i\partial E_j}E_i - \frac{\partial^2 G}{\partial T_{kl}\partial E_j}T_{kl}, \tag{15.15b}$$

$$S_{mn} = -\frac{\partial^2 G}{\partial\Theta\partial T_{mn}}\Delta\Theta - \frac{\partial^2 G}{\partial E_i\partial T_{mn}}E_i - \frac{\partial^2 G}{\partial T_{kl}\partial T_{mn}}T_{kl}. \tag{15.15c}$$

The partial derivatives are calculated for the reference state, assuming the variables other than those with respect to which derivation is performed to be constant.

Equations (15.15) connect the components of a 13-component vector $\mathbf{y}$ ($\Delta\mathcal{S}$, D_j, S_{mn}) to the 13 components of another vector $\mathbf{Y}$ ($\Delta\Theta$, E_i, T_{kl}) via the physical property matrix defined above in (15.14), *i.e.*

$$y_i = M_{ij}Y_j. \tag{15.16}$$

15.1.3. Principal effects *vs* crossed effects

Some coefficients of the matrix M_{ij} are known and represent the principal effects. They are the "diagonal" terms in the broad sense, which we recall below.

The variation in entropy $\Delta\mathcal{S}$ at zero electric field E_i and stress T_{kl}, as a function of the variation in temperature, is

$$\Delta\mathcal{S} = \frac{\Delta Q}{\Theta} = \frac{C}{\Theta}\Delta\Theta$$

where C is the heat capacity per unit volume of the material (at zero electric field and stress).

Similarly, if an electric field E_i is applied (at zero $\Delta\Theta$ and T_{kl}), the induction D_j is given by $D_j = \varepsilon_{ji}E_i$, where ε_{ji} is the relative electrical permittivity tensor. We note that $\varepsilon_{ij} = \varepsilon_{ji}$ since

$$\varepsilon_{ij} = -\frac{\partial^2 G}{\partial E_i \partial E_j}$$

and the order of the derivations is irrelevant.

The tensor $[\varepsilon]$ is symmetrical.

Applying a stress T_{kl} (at zero $\Delta\Theta$ and E_i) gives rise to a strain S_{mn} such that $S_{mn} = s_{mnkl}T_{kl}$. We note that $s_{mnkl} = s_{klmn}$ since

$$s_{mnkl} = -\frac{\partial^2 G}{\partial T_{kl}\partial T_{mn}}.$$

We retrieve the symmetry of tensor $[s]$ which we already derived in Section 13.4.

The non-diagonal terms correspond to the crossed effects, and are pairwise equal since matrix M_{ij} is symmetrical. Thus the system (15.15) can be rewritten as:

$$\Delta S = (C/\Theta)\Delta\Theta + p_i E_i + \alpha_{kl}T_{kl}, \tag{15.17a}$$

$$D_j = p_j\Delta\Theta + \varepsilon_{ji}E_i + d_{jkl}T_{kl}, \tag{15.17b}$$

$$S_{mn} = \alpha_{mn}\Delta\Theta + d_{imn}E_i + s_{mnkl}T_{kl}. \tag{15.17c}$$

In these expressions, we set:

$$p_i = -\frac{\partial^2 G}{\partial E_i \partial\Theta} = -\frac{\partial^2 G}{\partial\Theta\partial E_i}, \tag{15.18a}$$

$$\alpha_{kl} = -\frac{\partial^2 G}{\partial T_{kl}\partial\Theta} = -\frac{\partial^2 G}{\partial\Theta\partial T_{kl}}, \tag{15.18b}$$

$$d_{jkl} = -\frac{\partial^2 G}{\partial T_{kl}\partial E_j} = -\frac{\partial^2 G}{\partial E_j\partial T_{kl}}. \tag{15.18c}$$

The coefficients p_j (called the pyroelectric coefficients) in Equation (15.17b) describe the pyroelectric effect, *i.e.* a variation in spontaneous polarization under the influence of a change in temperature at zero stress and electric field. We recall, from the beginning of this section, that the reference state for $\mathbf{D}$ in pyroelectric crystals is the spontaneous polarization $\mathbf{P}_0$ at the reference temperature Θ_0. The quantity D_j in Equation (15.17b) is then the component ΔP_j of the variation in spontaneous polarization $\Delta\mathbf{P}$ under the influence of the variation in temperature, *i.e.* $\Delta P_j = p_j\Delta\Theta$. The same coefficients describe the electrocaloric effect (Eq. (15.17a)), *i.e.* the production of heat under the influence of an electric field in pyroelectric crystals.

The coefficients α_{mn} (Eq. (15.17c)) describe thermal expansion (strain S_{mn} under the influence of a variation in temperature $\Delta\Theta$, with $S_{mn} = \alpha_{mn}\Delta\Theta$) and also the piezocaloric effect (Eq. (15.17a)), *i.e.* the production of heat $\Delta Q = \Theta\Delta S = \Theta\alpha_{kl}T_{kl}$ under the influence of a stress T_{kl}.

The coefficients d_{jkl} (Eq. (15.17b)) describe the direct piezoelectric effect, *i.e.* the appearance of a polarization under the influence of a stress $D_j = P_j = d_{jkl}T_{kl}$ or, for pyroelectric crystals, the appearance of an additional polarization $\Delta P_j = d_{jkl}T_{kl}$. They also describe (Eq. (15.17c)) the appearance of a strain under the influence of an electric field, $S_{mn} = d_{imn}E_i$. This is called the inverse piezoelectric effect.

15.1.4. Summary of the various effects

The above results are summarized in Table 15.1.

The various effects are represented on Figure 15.1 by a drawing borrowed from [3].

Table 15.1: Table describing the physical effects related to the coefficients of the physical property matrix. The principal effects are shaded.

	Thermodynamic coefficients		
	$\Delta\Theta$	E_i	T_{kl}
ΔS	Heat capacity C/Θ	Electrocaloric effect p_i	Piezocaloric effect α_{kl}
D_j	Pyroelectricity p_j	Electrical permittivity ε_{ji}	Direct piezo-electric effect d_{jkl}
S_{mn}	Thermal expansion α_{mn}	Inverse piezo-electric effect d_{imn}	Elasticity s_{mnkl}

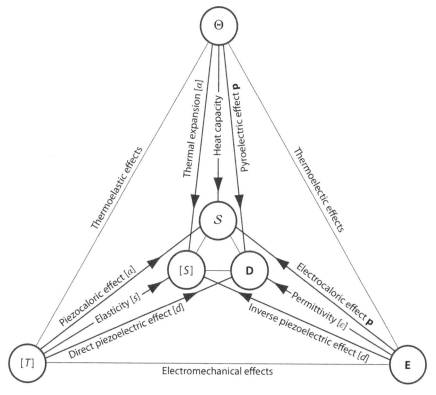

Figure 15.1: Schematic representation of the principal and crossed effects (adapted from [3], by permission of Oxford University Press)

Two equilateral triangles are drawn with the same center and the same orientation. The thermodynamic coordinates (the x_i, which are the entropy, electrical induction and strain) are placed at the vertices of the internal triangle (i). Their conjugate quantities (the X_i, which are the temperature, the electric field and the stress) are placed at the vertices of the external triangle (e). The arrangement is such that two conjugate quantities are located on the same median of the triangles. The arrows indicate the paths from one quantity to another. Those connecting a quantity and its conjugate (arrow from e to i) represent the principal effects. For example, the arrow connecting Θ to $\mathcal{S}$ (thermal domain) corresponds to a change in entropy due to a variation in temperature, *i.e.* the heat required to obtain a variation in temperature, and the physical quantity associated to this effect is the heat capacity. The temperature vertex Θ of the external triangle can be connected to quantities from another domain in the internal triangle, *viz.* the electrical induction or the strain, and the associated arrows (from e to i) correspond to crossed effects (pyroelectric effect and thermal expansion). The straight line segments parallel to the sides of the triangle are paths from i to i or e to e. The area between two such parallel segments, such as the segment connecting Θ to $\mathbf{E}$ and the one connecting $\mathcal{S}$ to $\mathbf{D}$, represent various effects that couple two domains, here the thermal and the electrical domain; the various effects are designated as thermoelectric.

15.1.5. Condensed representation of the physical property matrix

The way we defined it, the physical property matrix connects two 13-component vectors. We know that, because tensors S_{ij} and T_{ij} are symmetric, we can replace the nine components S_{ij} and T_{ij} by the 6 independent components S_α and T_α defined in (13.3) and (13.4) respectively. Thus vectors y_i and Y_i become 10-component vectors and matrix M a 10 by 10 matrix (Fig. 15.2). The "diagonal" terms which connect the S_α and T_α are then the contracted elastic coefficients $s_{\alpha\beta}$. The crossed terms implying the S_α and T_α are the components $d_{i\alpha}$ and α_β which are deduced from components d_{ijk} and α_{ij}. We will show in Section 15.3.2 how to go over from the d_{ijk} to the $d_{i\alpha}$. Let us see how we go over from the components α_{ij} of the thermal expansion tensor (equal to the piezocaloric tensor) to the contracted components α_β. If we apply a temperature variation $\Delta\Theta$ at zero or constant electric field and stress, the components of the strain tensor S_{mn} can be written as: $S_{mn} = \alpha_{mn}\Delta\Theta$, which we want to write as $S_\beta = \alpha_\beta\Delta\Theta$.

It is clear that, for S_{11}, S_{22} and S_{33}, which are respectively equal to S_1, S_2 and S_3, we simply set $\alpha_\beta = \alpha_{mm}$.

The same does not apply for $\beta = 4$, 5 or 6. For example, for $\beta = 4$, $S_4 = S_{23} + S_{32} = (\alpha_{23} + \alpha_{32})\Delta\Theta = \alpha_4\Delta\Theta$ if we set $\alpha_4 = \alpha_{23} + \alpha_{32}$.

$$
\begin{pmatrix} \Delta S \\ D_1 \\ D_2 \\ D_3 \\ S_1 \\ S_2 \\ S_3 \\ S_4 \\ S_5 \\ S_6 \end{pmatrix}
=
\left(
\begin{array}{c|ccc|cccccc}
\frac{C}{T} & p_1 & p_2 & p_3 & \alpha_1 & \alpha_2 & \alpha_3 & \alpha_4 & \alpha_5 & \alpha_6 \\
\hline
p_1 & \varepsilon_{11} & \varepsilon_{12} & \varepsilon_{13} & d_{11} & d_{12} & d_{13} & d_{14} & d_{15} & d_{16} \\
p_2 & \varepsilon_{12} & \varepsilon_{22} & \varepsilon_{23} & d_{21} & d_{22} & d_{23} & d_{24} & d_{25} & d_{26} \\
p_3 & \varepsilon_{13} & \varepsilon_{23} & \varepsilon_{33} & d_{31} & d_{32} & d_{33} & d_{34} & d_{35} & d_{36} \\
\hline
\alpha_1 & d_{11} & d_{21} & d_{31} & s_{11} & s_{12} & s_{13} & s_{14} & s_{15} & s_{16} \\
\alpha_2 & d_{12} & & & s_{12} & s_{22} & s_{23} & s_{24} & s_{25} & s_{26} \\
\alpha_3 & d_{13} & & & s_{13} & & s_{33} & s_{34} & s_{35} & s_{36} \\
\alpha_4 & d_{14} & & & s_{14} & & & s_{44} & s_{45} & s_{46} \\
\alpha_5 & d_{15} & & & s_{15} & & & & s_{55} & s_{56} \\
\alpha_6 & d_{16} & & & s_{16} & & & & & s_{66}
\end{array}
\right)
\begin{pmatrix} \Delta T \\ E_1 \\ E_2 \\ E_3 \\ T_1 \\ T_2 \\ T_3 \\ T_4 \\ T_5 \\ T_6 \end{pmatrix}
$$

Figure 15.2: Physical property matrix

Thus we conclude that

$$\text{for } \beta = 1, 2, 3: \qquad \alpha_\beta = \alpha_{mm},$$

$$\text{for } \beta = 4, 5, 6: \qquad \alpha_\beta = \alpha_{mn} + \alpha_{nm}.$$

It is easy to check that this result remains valid for the piezocaloric effect, which can be written as $\Delta S = \alpha_\beta T_\beta$.

In the following sections, we study more specifically two of the crossed effects we just introduced, *viz.* pyroelectricity and piezoelectricity.

15.2. Pyroelectricity. Pyroelectric crystals

Pyroelectric crystals feature a spontaneous polarization in the absence of an electric field. They are crystals for which the centers of gravity of the positive and of the negative charges in the unit cell do not coincide. For these materials:

$$\mathbf{D} = \mathbf{P}_0 + \varepsilon_0[\varepsilon]\mathbf{E}.$$

In the absence of an applied electric field, and at a given reference temperature Θ_0: $\mathbf{D} = \mathbf{P}_0$.

This polarization is difficult to evidence because the surface charges due to polarization are rapidly neutralized by free charges from the air or through conduction in the crystal. However, its variation as a function of temperature can be measured more easily, and this change (at zero electric field and stress) $\Delta \mathbf{P} = \mathbf{P}(\Theta) - \mathbf{P}(\Theta_0)$ is related, according to Equation (15.17b), to the variation in temperature $\Delta \Theta = \Theta - \Theta_0$ by

$$\Delta P_i = p_i \Delta \Theta \qquad \text{or} \qquad \Delta \mathbf{P} = \mathbf{p} \Delta \Theta.$$

$\mathbf{p}$ is a vector whose components are the pyroelectric coefficients.

We showed (Sect. 9.5.2) that pyroelectricity can occur only in crystals belonging to one of the following point groups: 1, 2, 3, 4, 6, m, mm2, 3m, 4mm and 6mm.

The best known pyroelectric crystal is tourmaline, whose variation in polarization with temperature was evidenced in antiquity. The materials most used in applications (especially for infrared radiation detectors) are ferroelectric materials because their pyroelectric coefficients are usually higher. Ferroelectric crystals are pyroelectric crystals whose spontaneous polarization can be reversed under the influence of high enough an electric field. These crystals generally undergo a phase change at a given temperature, called the Curie temperature Θ_c, above which they are no more pyroelectric. Therefore the pyroelectric coefficients are usually negative in a representation where the components of the spontaneous polarization are positive, since the spontaneous polarization decreases with increasing temperature and finally disappears beyond Θ_c. Among the best known materials are potassium dihydrogen phosphate KH_2PO_4, usually denoted as KDP, and barium titanate $BaTiO_3$. The point group of KDP is mm2 (orthorhombic system) in the ferroelectric phase and $\bar{4}2m$ (tetragonal system) above its Curie temperature Θ_c equal to $-150°C$. Barium titanate has the perovskite structure (Sect. 8.2.4), with point group $m\bar{3}m$ (cubic system) above the Curie temperature $\Theta_c = 120°C$. As temperature is reduced below Θ_c, its point group is 4mm (tetragonal system). It then undergoes more transitions, its point group becoming mm2 (orthorhombic system) at $\Theta = 5°C$, then 3m (rhombohedral system) at $\Theta = -90°C$. It remains ferroelectric in these three phases. Ferroelectric materials are often used in the form of ceramics, *i.e.* as powders submitted to compacting and sintering at high temperatures, because they are easier to manufacture than single crystals and they can be given any shape. Without special treatments, the microcrystals forming the ceramics have random relative misorientation and their polarization can take on all possible orientations. Their average polarization is then zero. To obtain a sizable polarization, the ceramic is submitted to a strong electric field which orients the polarization of each of the domains in the grains in the direction closest to the electric field consistent with the crystal symmetry and the grain orientation.

15.3. Piezoelectric crystals

15.3.1. Direct effect and inverse effect

A piezoelectric crystal polarizes under the action of an applied stress (direct effect) and, conversely, deforms when an electric field is applied (inverse effect). The direct effect was discovered by Pierre and Jacques Curie in 1880. The inverse effect was predicted the following year by Gabriel Lippman on the basis of thermodynamic considerations, and its existence immediately verified by the Curie brothers.

The direct piezoelectric effect connects the polarization induced in the crystal **P** and the applied stress $[T]$. This effect shows up at zero electric field, and we can write, in agreement with Equation (15.17b):

$$D_i = P_i = d_{ijk} T_{jk}. \tag{15.19}$$

The inverse piezoelectric effect connects the quantities conjugate to $[T]$ and **D**, *viz.* the strain $[S]$ and the electric field **E** respectively. Equation (15.17c) shows that

$$S_{ij} = d_{kij} E_k. \tag{15.20}$$

It is worth noticing the difference in the order of the subscripts in the piezo-electric coefficients d_{ijk} for the direct and inverse effects in Equations (15.19) and (15.20). This is due to the definition, in (15.18), of the coefficients d_{ijk}: the first subscript points to a differentiation of free enthalpy with respect to a component of **E**, while the two last ones point to differentiation with respect to a component of $[T]$.

The above conditions, *viz.* zero electric field to describe the direct effect, and zero stress to describe the inverse effect, are fulfilled only in special geometries or for high sample symmetry. Usually, the electric and mechanical quantities are coupled, and we must consider, at constant temperature, the two coupled equations resulting from (15.17b) and (15.17c):

$$P_j = \varepsilon_{ji} E_i + d_{jkl} T_{kl}, \tag{15.21a}$$
$$S_{mn} = d_{imn} E_i + s_{mnkl} T_{kl}. \tag{15.21b}$$

15.3.2. Piezoelectric tensor $[d]$ and its two-subscript notation

Equation (15.19) shows that the coefficients d_{ijk} connect the second-rank tensor $[T]$ with the vector **P**, a first-rank tensor. They are therefore the coefficients of a third-rank tensor (Sect. 9.4.2).

The same conclusions are reached by considering Equation (15.20), where the coefficients d_{kij} connect the first-rank tensor **E** to the second-rank tensor $[S]$.

A third-rank tensor comprises $3^3 = 27$ coefficients. This number can here be reduced by taking into account the symmetry of the stress tensor, allowing the 9 coefficients T_{ij} to be replaced by the six coefficients T_α (Eq. (13.3)).

Suppose we apply a stress for which only the components $T_{23} = T_{32}$ are non-zero. The resulting polarization is $P_i = d_{i23} T_{23} + d_{i32} T_{32} = (d_{i23} + d_{i32}) T_{23}$. Since we cannot distinguish d_{i23} and d_{i32}, we set:

$$d_{ijk} = d_{ikj}$$

which reduces the number of independent components of tensor $[d]$ to 18.

Using the contracted notation, we immediately see that Equation (15.19) can be written as:

$$P_i = d_{i\alpha} T_\alpha \tag{15.22}$$

provided we set

$$d_{i\alpha} = d_{ijk} \quad \text{for } \alpha = 1, 2 \text{ and } 3 \, (j = k), \tag{15.23a}$$

$$d_{i\alpha} = 2d_{ijk} \quad \text{for } \alpha = 4, 5, 6 \, (j \neq k). \tag{15.23b}$$

Now relation (15.22) can be written in matrix form. The components P_i form a column vector with three rows, the components T_α a column vector with six rows, and the table of coefficients $d_{i\alpha}$ a matrix with 3 rows and 6 columns. We thus obtain

$$\begin{pmatrix} P_1 \\ P_2 \\ P_3 \end{pmatrix} = \begin{pmatrix} d_{11} & d_{12} & d_{13} & d_{14} & d_{15} & d_{16} \\ d_{21} & d_{22} & d_{23} & d_{24} & d_{25} & d_{26} \\ d_{31} & d_{32} & d_{33} & d_{34} & d_{35} & d_{36} \end{pmatrix} \begin{pmatrix} T_1 \\ T_2 \\ T_3 \\ T_4 \\ T_5 \\ T_6 \end{pmatrix}.$$

The inverse piezoelectric effect can be written as:

$$S_{ij} = d_{kij} E_k.$$

We see that, using the definitions (15.23) of the coefficients $d_{i\alpha}$, and the contracted notation for tensor $[S]$ (Eq. (13.4)), we can write:

$$S_\alpha = d_{i\alpha} E_i.$$

This relation becomes a matrix relation provided we use the matrix d^T, transposed from the preceding one, $i.e.$

$$d^T_{\alpha i} = d_{i\alpha} \quad \text{and} \quad S_\alpha = d^T_{\alpha i} E_i$$

or

$$\begin{pmatrix} S_1 \\ S_2 \\ S_3 \\ S_4 \\ S_5 \\ S_6 \end{pmatrix} = \begin{pmatrix} d_{11} & d_{21} & d_{31} \\ d_{12} & d_{22} & d_{32} \\ d_{13} & d_{23} & d_{33} \\ d_{14} & d_{24} & d_{34} \\ d_{15} & d_{25} & d_{35} \\ d_{16} & d_{26} & d_{36} \end{pmatrix} \begin{pmatrix} E_1 \\ E_2 \\ E_3 \end{pmatrix}.$$

15.3.3. Effect of crystal symmetry on the form of the tensor

The piezoelectric tensor, a material tensor, must be invariant under all the point symmetry operations of the crystal (Sect. 9.5.2). We remember that, in applying this principle, we must revert to the three-subscript notation of the piezoelectric tensor.

Centrosymmetric crystal

A centrosymmetric crystal cannot be piezoelectric, since odd-rank tensors are identically zero for all crystals featuring a center of symmetry (Sect. 9.6.3). We can retrieve this result directly from Curie's principle. The symmetry of the effect (electrical polarization, with symmetry ∞m) must include the symmetry of the cause (crystal + stress). Stress is centrosymmetric. If the crystal also is centrosymmetric, the intersection of the symmetry groups of the stress and of the crystal features a center of symmetry, and this is incompatible with symmetry group ∞m.

Examples of reduction in the number of non-zero coefficients

We can apply the direct inspection method (Sect. 9.6.2) to all crystallographic groups except for those which contain a 3-fold or 6-fold axis. For the latter, we must revert to the matrix method (Sect. 9.6.1).

We give below a few examples of the use of the direct inspection method, while the result for all crystal groups is given in Table 15.2.

1. **Crystals with group 2**

    We choose axis Ox_3 parallel to the 2-fold axis. Rotating the coordinate system by angle π around Ox_3 changes the components x_i of a vector into components x_i' such that

    $$x_1' = -x_1, \quad x_2' = -x_2, \quad x_3' = x_3.$$

    Coefficients d_{ijk} transform as the product of 3 components, so that for example $d_{111}' = -d_{111}$, and on the other hand they must be invariant in the coordinate system rotation, so that $d_{111}' = d_{111}$, which means that $d_{111} = 0$.

    We understand from this example that, if the subscripts i, j and k include an even number (0 or 2) of the value 3, which implies an odd number of the set of 1's and 2's, $d_{ijk} = 0$. Reverting to the 2-subscript notation, we obtain the table

    $$\begin{pmatrix} 0 & 0 & 0 & d_{14} & d_{15} & 0 \\ 0 & 0 & 0 & d_{24} & d_{25} & 0 \\ d_{31} & d_{32} & d_{33} & 0 & 0 & d_{36} \end{pmatrix}. \qquad (15.24)$$

2. **Crystals with group m**

    Let us choose Ox_3 perpendicular to the mirror. The action of the mirror perpendicular to Ox_3 on the coordinate system changes the components x_i of a vector into components x_i' such that

    $$x_1' = x_1, \quad x_2' = x_2, \quad x_3' = -x_3.$$

This time the coefficients d_{ijk} that contain an odd number of indices equal to 3 are zero. The tensor now has the form:

$$\begin{pmatrix} d_{11} & d_{12} & d_{13} & 0 & 0 & d_{16} \\ d_{21} & d_{22} & d_{23} & 0 & 0 & d_{26} \\ 0 & 0 & 0 & d_{34} & d_{35} & 0 \end{pmatrix}.$$

Adding a 2-fold axis perpendicular to this mirror, we check that, for the centrosymmetric group 2/m, all coefficients are zero.

3. **Crystals with group 222**

   We again consider matrix (15.24), and now add the effect of the existence of another 2-fold axis. Choosing axis Ox_2 along this second 2-fold axis, the zero coefficients are those containing an even number of 2's. The tensor thus becomes:

   $$\begin{pmatrix} 0 & 0 & 0 & d_{14} & 0 & 0 \\ 0 & 0 & 0 & 0 & d_{25} & 0 \\ 0 & 0 & 0 & 0 & 0 & d_{36} \end{pmatrix}.$$

   The effect of a third 2-fold axis would bring nothing more, since the existence of this third axis is the direct consequence of the existence of the two others. Equivalently, we can say that the generating elements of the group 222 are restricted to rotations by π around two perpendicular axes.

4. We note that group 423, although it is not centrosymmetric, has a piezoelectric tensor that is identically zero (Exercise 15.1).

Note: a crystal belonging to a piezoelectric group does not necessarily feature a piezoelectric effect. The values of the components of the piezoelectric tensor depend very much on the crystal itself, and in some cases the coefficients are all zero even when measured with high sensitivity. In this respect, the piezoelectric tensor is different from the elastic or thermal expansion tensors.

Conversely, experimentally detecting a piezoelectric signal indicates beyond doubt the absence of centrosymmetry in a crystal. When possible, the piezoelectric test is therefore a valuable complement to structural diffraction data, for which the conclusion on the presence or absence of centrosymmetry is difficult.

Form of the piezoelectric tensors for all crystal groups

Table 15.2 shows the non-zero coefficients of the piezoelectric tensor for all crystal groups, and Table 15.3 gives some numerical values for the coefficients $d_{i\alpha}$, in units of 10^{-12} C N^{-1}.

Table 15.2: Form of the matrix representing the piezoelectric tensor for the various crystal groups

Triclinic
1

$$\begin{pmatrix} \bullet & \bullet & \bullet & \bullet & \bullet & \bullet \\ \bullet & \bullet & \bullet & \bullet & \bullet & \bullet \\ \bullet & \bullet & \bullet & \bullet & \bullet & \bullet \end{pmatrix}$$

Monoclinic

$2 \mathbin{/\!/} Ox_2$
$$\begin{pmatrix} 0 & 0 & 0 & \bullet & 0 & \bullet \\ \bullet & \bullet & \bullet & 0 & \bullet & 0 \\ 0 & 0 & 0 & \bullet & 0 & \bullet \end{pmatrix}$$

$2 \mathbin{/\!/} Ox_3$
$$\begin{pmatrix} 0 & 0 & 0 & \bullet & \bullet & 0 \\ 0 & 0 & 0 & \bullet & \bullet & 0 \\ \bullet & \bullet & \bullet & 0 & 0 & \bullet \end{pmatrix}$$

$m \perp Ox_2$
$$\begin{pmatrix} \bullet & \bullet & \bullet & 0 & \bullet & 0 \\ 0 & 0 & 0 & \bullet & 0 & \bullet \\ \bullet & \bullet & \bullet & 0 & \bullet & 0 \end{pmatrix}$$

$m \perp Ox_3$
$$\begin{pmatrix} \bullet & \bullet & \bullet & 0 & 0 & \bullet \\ \bullet & \bullet & \bullet & 0 & 0 & \bullet \\ 0 & 0 & 0 & \bullet & \bullet & 0 \end{pmatrix}$$

Orthorhombic

222
$$\begin{pmatrix} 0 & 0 & 0 & \bullet & 0 & 0 \\ 0 & 0 & 0 & 0 & \bullet & 0 \\ 0 & 0 & 0 & 0 & 0 & \bullet \end{pmatrix}$$

mm2
$$\begin{pmatrix} 0 & 0 & 0 & 0 & \bullet & 0 \\ 0 & 0 & 0 & \bullet & 0 & 0 \\ \bullet & \bullet & \bullet & 0 & 0 & 0 \end{pmatrix}$$

Tetragonal

4
$$\begin{pmatrix} 0 & 0 & 0 & \bullet & \bullet & 0 \\ 0 & 0 & 0 & \bullet & \bullet & 0 \\ \bullet & \bullet & \bullet & 0 & 0 & 0 \end{pmatrix}$$

$\bar{4}$
$$\begin{pmatrix} 0 & 0 & 0 & \bullet & \bullet & 0 \\ 0 & 0 & 0 & \bullet & \bullet & 0 \\ \bullet & \bullet & 0 & 0 & 0 & \bullet \end{pmatrix}$$

422
$$\begin{pmatrix} 0 & 0 & 0 & \bullet & 0 & 0 \\ 0 & 0 & 0 & 0 & \bullet & 0 \\ 0 & 0 & 0 & 0 & 0 & 0 \end{pmatrix}$$

4mm
$$\begin{pmatrix} 0 & 0 & 0 & 0 & \bullet & 0 \\ 0 & 0 & 0 & \bullet & 0 & 0 \\ \bullet & \bullet & \bullet & 0 & 0 & 0 \end{pmatrix}$$

$\bar{4}2m$
$2 \mathbin{/\!/} Ox_1$
$$\begin{pmatrix} 0 & 0 & 0 & \bullet & 0 & 0 \\ 0 & 0 & 0 & 0 & \bullet & 0 \\ 0 & 0 & 0 & 0 & 0 & \bullet \end{pmatrix}$$

Rhombohedral

3
$$\begin{pmatrix} \bullet & \bullet & 0 & \bullet & \bullet & \odot \\ \odot & \bullet & 0 & \bullet & \bullet & \odot \\ \bullet & \bullet & \bullet & 0 & 0 & 0 \end{pmatrix}$$

32
$2 \mathbin{/\!/} Ox_1$
$$\begin{pmatrix} \bullet & \odot & 0 & \bullet & 0 & 0 \\ 0 & 0 & 0 & 0 & \bullet & \odot \\ 0 & 0 & 0 & 0 & 0 & 0 \end{pmatrix}$$

3m
$m \perp Ox_1$
$$\begin{pmatrix} 0 & 0 & 0 & 0 & \bullet & \odot \\ \odot & \bullet & 0 & \bullet & 0 & 0 \\ \bullet & \bullet & \bullet & 0 & 0 & 0 \end{pmatrix}$$

3m
$m \perp Ox_2$
$$\begin{pmatrix} \bullet & \odot & 0 & 0 & \bullet & 0 \\ 0 & 0 & 0 & \bullet & 0 & \odot \\ \bullet & \bullet & \bullet & 0 & 0 & 0 \end{pmatrix}$$

Hexagonal

6
$$\begin{pmatrix} 0 & 0 & 0 & \bullet & \bullet & 0 \\ 0 & 0 & 0 & \bullet & \bullet & 0 \\ \bullet & \bullet & \bullet & 0 & 0 & 0 \end{pmatrix}$$

6mm
$$\begin{pmatrix} 0 & 0 & 0 & 0 & \bullet & 0 \\ 0 & 0 & 0 & \bullet & 0 & 0 \\ \bullet & \bullet & \bullet & 0 & 0 & 0 \end{pmatrix}$$

622
$$\begin{pmatrix} 0 & 0 & 0 & \bullet & 0 & 0 \\ 0 & 0 & 0 & 0 & \bullet & 0 \\ 0 & 0 & 0 & 0 & 0 & 0 \end{pmatrix}$$

$\bar{6}$
$$\begin{pmatrix} \bullet & \odot & 0 & 0 & 0 & \odot \\ \odot & \bullet & 0 & 0 & 0 & \odot \\ 0 & 0 & 0 & 0 & 0 & 0 \end{pmatrix}$$

$\bar{6}2m$
$m \perp Ox_1$
$$\begin{pmatrix} 0 & 0 & 0 & 0 & 0 & \odot \\ \odot & \bullet & 0 & 0 & 0 & 0 \\ 0 & 0 & 0 & 0 & 0 & 0 \end{pmatrix}$$

$\bar{6}2m$
$m \perp Ox_2$
$$\begin{pmatrix} \bullet & \odot & 0 & 0 & 0 & 0 \\ 0 & 0 & 0 & 0 & 0 & \odot \\ 0 & 0 & 0 & 0 & 0 & 0 \end{pmatrix}$$

Cubic

432
$$\begin{pmatrix} 0 & 0 & 0 & 0 & 0 & 0 \\ 0 & 0 & 0 & 0 & 0 & 0 \\ 0 & 0 & 0 & 0 & 0 & 0 \end{pmatrix}$$

$\dfrac{23}{\bar{4}3m}$
$$\begin{pmatrix} 0 & 0 & 0 & \bullet & 0 & 0 \\ 0 & 0 & 0 & 0 & \bullet & 0 \\ 0 & 0 & 0 & 0 & 0 & \bullet \end{pmatrix}$$

●—● Equal components   ●—○ Components with equal absolute values but opposite signs
⊙ Component equal to minus twice the component shown as a full black circle to which it is connected

Table 15.3: Numerical values of the piezoelectric coefficients for some materials, in units 10^{-12} C N^{-1} (from [4])

KDP ($\bar{4}2m$)	d_{14}	d_{36}		
	1.4	23.2		
LiNbO$_3$ (3m)	d_{15}	d_{22}	d_{31}	d_{33}
	74	20.77	0.86	16.23
SiO$_2$ (32)	d_{11}	d_{14}		
Quartz	$-$ 2.3	$-$ 0.67		
GaAs ($\bar{4}3m$)	d_{14}			
	$-$ 2.7 (25°C) 2.63 (196°C)			
InSb ($\bar{4}3m$)	d_{14}			
	$-$ 2.35 (25°C)			

15.3.4. Longitudinal piezoelectricity surface

The longitudinal piezoelectricity surface is obtained by plotting, on any direction with unit vector $\mathbf{s}$, a length equal to the modulus of d'_{111} in the orthonormal set of axes $(Ox'_1x'_2x'_3)$, where Ox'_1 is parallel to $\mathbf{s}$. To represent this three-dimensional surface, we perform sections by planes going through the origin. The sign of d'_{111} is then indicated as we will see with the example below.

Assume the components of the piezoelectric tensor are known in the orthonormal axis system $(Ox_1x_2x_3)$ associated to the crystal axes, and denoted in this set of axes as d_{ijk}.

We obtain d'_{111} from the d_{ijk}'s through the usual rule (Sect. 9.4.1):

$$d'_{111} = a_{1i}a_{1j}a_{1k}d_{ijk}$$

where $\{a_{ij}\}$ is the transformation matrix for going over from the set of axes $(Ox_1x_2x_3)$ to the set $(Ox'_1x'_2x'_3)$. a_{ij} is the component parallel to axis Ox_j of the unit vector parallel to Ox'_1, here denoted as $\mathbf{s}$. Calling s_i the components of $\mathbf{s}$ in the axis system $(Ox_1x_2x_3)$, we then have

$$d'_{111} = s_is_js_kd_{ijk}. \tag{15.25}$$

We can interpret Equation (15.25) as follows. Assume the experimental conditions are such that, when applying a uniaxial stress at constant temperature to a piezoelectric crystal, the electric field in the crystal is zero (Fig. 15.3). According to (15.19), we would obtain a polarization $P_i = d_{ijk}T_{jk}$. In a set of axes $(Ox'_1x'_2x'_3)$ such that axis Ox'_1 is parallel to the uniaxial stress, the latter

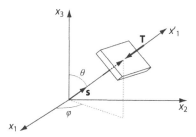

Figure 15.3: Schematic illustration of the meaning of longitudinal piezo-electricity, a measure of the polarization normal to the faces of a crystal to which a uniaxial stress is applied

has only one component, T'_{11}, and we obtain $P'_i = d'_{i11}T'_{11}$. We see that d'_{111} is the component of polarization parallel to the uniaxial stress. Thus, under these special experimental conditions, d'_{111} is the component of polarization normal to the face of the crystal to which a uniaxial stress is applied.

Application to a crystal with point group 4mm:

The non-zero components of the piezoelectric tensor are given in Table 15.2 and Equation (15.25) becomes

$$d'_{111} = s_3[(s_1^2 + s_2^2)(d_{15} + d_{31}) + s_3^2 d_{33}].$$

If we use the spherical coordinates of **s** $(r = 1, \theta, \varphi)$, we obtain

$$d'_{111} = \cos\theta[\sin^2\theta(d_{15} + d_{31}) + \cos^2\theta d_{33}].$$

This longitudinal piezoelectricity coefficient does not depend on azimuth φ. The longitudinal piezoelectricity surface is therefore a surface of revolution around the axis Ox_3, the 4-fold axis. Its section by a meridian plane is shown on Figure 15.4 for the case of barium titanate $BaTiO_3$ in its tetragonal phase (between 120°C and 5°C). We note that for $\theta > \pi/2$ the value of d'_{111} is negative. This is why the sign − is inserted in the center of the lobe corresponding to the half-space with negative x_3.

15.3.5. Other forms of piezoelectric coefficients

The above piezoelectric coefficients connect polarization to stress and relation (15.17b) shows that they are isothermal coefficients (at constant temperature). For some applications, it is more convenient to express polarization as a function of strain, be it through isothermal or adiabatic (constant entropy) coefficients. Let us focus on isothermal coefficients. The thermodynamic variables must then be temperature, electric field and strain. The thermodynamic function to consider is another Gibbs function G_2:

$$G_2 = U - \Theta S - E_i D_i,$$
$$dG_2 = -S d\Theta - D_i dE_i + T_{ij} dS_{ij}. \tag{15.26}$$

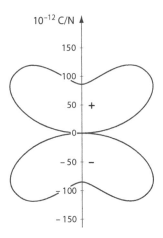

Figure 15.4: Intersection of the longitudinal piezoelectricity surface of a crystal of barium titanate (BaTiO$_3$), with point group 4mm, by a plane passing through the 4-fold axis, which is an axis of revolution for this surface

With the same approach as in Section 15.1.2 and Section 15.1.3, we obtain

$$\Delta S = (C/\Theta)\Delta\Theta + p_i E_i + \lambda_{kl} S_{kl}, \tag{15.27a}$$

$$D_j = p_j \Delta\Theta + \varepsilon_{ji} E_i + e_{jkl} S_{kl}, \tag{15.27b}$$

$$T_{mn} = -\lambda_{mn}\Delta\Theta - e_{imn} E_i + c_{mnkl} S_{kl}. \tag{15.27c}$$

Clearly the coefficients p_i are pyroelectric coefficients at constant temperature and strain, while in Equations (15.17) they were the constant temperature and zero stress coefficients. Their difference will be calculated in Section 15.4. We note that the coefficients e_{ijk} appear with opposite signs. The explanation is simple, since they are obtained through:

$$D_j = -\frac{\partial G_2}{\partial E_j} \text{ from (15.26) and } \frac{\partial D_j}{\partial S_{kl}} = -\frac{\partial^2 G_2}{\partial E_j \partial S_{kl}}.$$

We set:

$$e_{jkl} = -\frac{\partial^2 G_2}{\partial E_j \partial S_{kl}}. \tag{15.28}$$

On the other hand

$$T_{mn} = \frac{\partial G_2}{\partial S_{mn}} \quad \text{and} \quad \frac{\partial T_{mn}}{\partial E_i} = \frac{\partial^2 G_2}{\partial E_i \partial S_{mnl}} = -e_{imn}$$

from definition (15.28).

The opposite signs for the λ_{ij} have the same origin.

In piezoelectricity problems involving constant temperature, we then use the two equations:

$$D_j = \varepsilon_{ji} E_i + e_{jkl} S_{kl},$$
$$T_{mn} = -e_{imn} E_i + c_{mnkl} S_{kl}.$$

15.3.6. Applications

Materials used

Piezoelectric materials can be used under various forms:

- single crystal form: the best known is quartz, a natural crystal which is also manufactured, often as large single crystals with excellent quality (Fig. 1.3). Other important single crystal materials are berlinite ($AlPO_4$), gallium orthophosphate ($GaPO_4$), gallium arseniate ($GaAsO_4$), lithium niobate ($LiNbO_3$), lithium tantalate ($LiTaO_3$).

- ceramics: if the piezoelectric material is also ferroelectric, it can be used in the form of a polarized ceramic, *i.e.* of a compacted and sintered powder, the grains of which are reoriented by applying a strong electric field (see Sect. 15.2). Among the materials used in ceramic form are barium titanate ($BaTiO_3$), potassium niobate ($KNbO_3$), lead zirconate–titanate $Pb(Zr_x Ti_{1-x})O_3$, designated as PZT.

- thin crystal films

- polymers: some polymers such as polyvinylidene fluoride (PVDF) become piezoelectric if their molecules are oriented by a mechanical or electrical process.

Application areas

There are many areas of application, and the list below is not exhaustive.

- Converting an electrical wave into an acoustical wave, *i.e.* production of ultrasound, and conversely conversion of an acoustic wave into an electrical signal for ultrasound reception. These two effects make it possible to emit ultrasound waves and to receive those scattered by the medium. This is why they are used in medicine for echography and in sonars (the acronym for SOund Navigation And Ranging) to detect objects in water and determine their distance.[6]

- In the range of audible frequencies, production of microphones and loudspeakers, in particular for cellular telephones.

- Pressure sensors: pressure on the piezoelectric material generates a measurable electrical signal.

---

6. The sonar was invented by Paul Langevin and Constantin Chilowski in 1917 and used during World War I.

- Micro-manipulators: an applied voltage generates a strain which leads to micro-displacements.

- Piezoelectric resonators: elastic waves propagate in a crystal as modes which depend on its shape, dimensions and nature. The piezoelectric effect, which couples the mechanical and electric quantities, makes it possible to generate these vibration modes. This produces resonators which vibrate, when an electrical signal is applied, at a perfectly defined frequency, stable over long periods of time. Quartz watches thus use the resonance of a quartz tuning-fork shaped crystal to generate regular time pulses. The use of quartz crystals with special orientation, shape and dimensions make it possible to obtain resonators whose frequency practically does not vary with temperature. The resonance frequency of this type of oscillators is extremely sensitive to the mass of the crystal, yielding quartz balances capable of detecting extra masses on the order of an atomic monolayer.

- High voltage generation through application of a stress: this effect is used in gas lighters.

15.4. Principal and crossed effects under various conditions

In the preceding sections (except for Sect. 15.3.5), the various tensors representing principal or crossed effects connect a conjugate-type quantity to a thermodynamic coordinate-type quantity for constant values of the two conjugate quantities which do not enter the relation. For example tensor s_{mnkl} connects stress T_{kl} to strain S_{mn} at constant temperature and zero or constant electric field. Equation (15.17c) then yields:

$$S_{mn} = s_{mnkl}\big|_{E,\Theta}\, T_{kl}.$$

and $s_{mnkl}\big|_{E,\Theta}$ is the isothermal elastic coefficient. The same coefficient at constant entropy, $\Delta\mathcal{S} = 0$ and zero electric field is denoted as $s_{mnkl}\big|_{E,\mathcal{S}}$ and we now determine the relation connecting them. If $\Delta\mathcal{S} = 0$, the change in temperature is no more zero. It is given by Equation (15.17a), *i.e.*

$$0 = (C/\Theta)\Delta\Theta + \alpha_{kl} T_{kl} \quad \text{and} \quad \Delta\Theta = -(\Theta/C)\alpha_{kl} T_{kl}.$$

Equation (15.17c) becomes

$$S_{mn} = -\alpha_{mn}(\Theta/C)\alpha_{kl} T_{kl} + s_{mnkl} T_{kl} = (-\alpha_{mn}(\Theta/C)\alpha_{kl} + s_{mnkl}) T_{kl}$$

or
$$S_{mn} = s_{mnkl}\big|_{E,\mathcal{S}}\, T_{kl}$$

with
$$s_{mnkl}\big|_{E,\mathcal{S}} = s_{mnkl}\big|_{E,\Theta} - (\Theta/C)\alpha_{mn}\alpha_{kl}.$$

Table 15.4: Order of magnitude of the principal and crossed effects defined in this chapter, expressed in International System (SI) units

	$\Delta\Theta$	E_i	T_{kl}
ΔS	$(C/T)\ 10^4$	$\mathbf{p}\ 10^{-6}$	$[\alpha]\ 10^{-5}$
D_j	$\mathbf{p}\ 10^{-6}$	$\varepsilon_0[\varepsilon]\ 10^{-10}$	$[d]\ 10^{-12}$
S_{mn}	$[\alpha]\ 10^{-5}$	$[d]\ 10^{-12}$	$[s]\ 10^{-11}$

Table 15.4 allows us to calculate the order of magnitude of the difference between these isothermal and adiabatic coefficients, $viz.$ $10^{-4}.10^{-5}.10^{-5} = 10^{-14}$, while $s_{mnkl}|_{E,\Theta}$ is on the order of 10^{-11}, thus larger by three orders of magnitude. The difference is negligible.

This kind of calculation can be performed for all the other tensors, considered under various conditions. For example the difference between the zero electric field and the zero induction elastic constants, with both measurements performed at constant temperature, can be calculated. Generally, the relative difference between these two coefficients is much smaller than 1. This is not the case for the pyroelectric vector measured at zero stress, denoted by $p_i|_{T_{kl}=0}$, and at zero strain, $p_i|_{S_{kl}=0}$, the electric field being zero in either case. At zero stress and electric field, Equation (15.17b) can be written as: $P_j = p_j|_{T_{kl}=0}\,\Delta\Theta$. For a temperature variation $\Delta\Theta$ at zero strain, $S_{mn} = 0$, the stress T_{kl} is no more zero. It is given by Equation (15.17c):

$$0 = \alpha_{mn}\Delta\Theta + s_{mnkl}T_{kl}. \tag{15.29}$$

To obtain the value of T_{kl} from this equation, we use the contracted notation for T_{kl} and α_{mn} in order to use the elastic coefficients under the contracted form $s_{\mu\nu}$, and the matrix c inverse to matrix s.

Equation (15.29) can be written, for each value of β, as

$$s_{\beta\nu}T_\nu + \alpha_\beta\Delta\Theta = 0. \tag{15.30}$$

The matrix of compliance components $\{s_{\beta\nu}\}$ is the inverse of the matrix $\{c_{\lambda\mu}\}$ of stiffness components, which translates into relation

$$s_{\beta\nu}c_{\nu\mu} = \delta_{\beta\mu}. \tag{15.31}$$

We multiply Equation (15.30) by $c_{\mu\beta}$, then sum the six equations, with β varying over the range 1 to 6. We obtain:

$$c_{\mu\beta}s_{\beta\nu}T_\nu + c_{\mu\beta}\alpha_\beta\Delta\Theta = 0.$$

Using relation (15.31), the first term in this equation is equal to $\delta_{\mu\nu}T_\nu = T_\mu$, and we obtain

$$T_\mu = -c_{\mu\beta}\alpha_\beta\Delta\Theta$$

so that

$$T_{kl} = -c_{klmn}\alpha_{mn}\Delta\Theta$$

equivalent to (15.29).

Equation (15.17b) becomes

$$P_j = p_j|_{T_{kl}=0}\,\Delta\Theta - d_{jkl}c_{klmn}\alpha_{mn}\Delta\Theta$$
$$= p_j|_{S_{kl}=0}\,\Delta\Theta.$$

We obtain:

$$p_j|_{S_{kl}=0} = p_j|_{T_{kl}=0} - d_{jkl}c_{klmn}\alpha_{mn}.$$

The difference between the two pyroelectric effects is, according to Table 15.4, on the order of 10^{-6}, *i.e.* on the same order as the pyroelectric effect. This means that the pyroelectric vector at zero strain is difficult to measure, because it is extremely small. It is called primary pyroelectricity, while the term $d_{jkl}c_{klmn}\alpha_{mn}$ which is to be added to obtain the pyroelectric vector at constant stress is called secondary pyroelectricity. If the crystal is not submitted to stress during the change in temperature, it is free to deform through thermal expansion. This deformation produces, through piezoelectricity, a polarization which is secondary pyroelectricity.

15.5. Exercises

Exercise 15.1

Show that all the components of the piezoelectric tensor for group 432 are zero although this group is not centrosymmetric. Use the tensor form given in Table 15.2 for group 422, a subgroup of 432, and supplement it with the additional symmetry operation that leads to group 432.

Exercise 15.2

Consider a crystal with point group $\bar{4}2m$, assuming the form of the tensor is known for one of its subgroups, 222 or mm2. Choose the axis Ox_3 parallel to the $\bar{4}$ axis.

1. Determine the non-zero coefficients of the piezoelectric tensor in the two following cases:

    (a) Ox_1 is chosen parallel to one of the 2-fold axes,

    (b) Ox_1 is chosen perpendicular to one of the mirrors.

2. The compound $(NH_4)H_2PO_4$ (ammonium dihydrogen phosphate, ADP), with point group $\bar{4}2m$, has, in the axis system defined in question (1a), the piezoelectric coefficients:

    $d_{14} = d_{25} = 0.17 \times 10^{-11}$ C N^{-1} and $d_{36} = 5.17 \times 10^{-11}$ C N^{-1}.

    Determine its piezoelectric coefficients in the axis system defined in question (1b).

Exercise 15.3

Show that the longitudinal piezoelectricity surface for a crystal with point group 422 vanishes for all directions. (A similar result would be obtained for group 622). The form of the piezoelectric tensor for group 422 is given in Table 15.2.

Exercise 15.4

Quartz SiO_2 belong to group 32, for which the form of the piezoelectric tensor is given in Table 15.2 for the axis system where Ox_3 is parallel to the 3-fold axis and Ox_1 is parallel to a 2-fold axis.

1. A uniaxial stress T is applied to a quartz crystal along a direction in the plane (Ox_1, Ox_2). Call θ the angle between the direction of the stress and axis Ox_1.

    (a) Show that the obtained polarization $\mathbf{P}$ has a norm independent of θ.

    (b) How does the direction of $\mathbf{P}$ vary when θ varies?

    (c) What uniaxial stress should be applied to the crystal in order to obtain a polarization along Ox_2?

2. Calculate the longitudinal piezoelectric coefficient for a general direction. Represent the intersection of this surface with the plane (Ox_1, Ox_2). Deduce the overall look of the longitudinal piezoelectric surface.

Exercise 15.5

Zinc sulfide ZnS is a cubic crystal with point group $\bar{4}3m$.

1. A rectangle parallelepiped with sides parallel to the vectors $\mathbf{a}$, $\mathbf{b}$ and $\mathbf{c}$ of the unit cell is cut out of this crystal. What stress (or stresses) should be applied onto this parallelepiped to determine the value of d_{14}?

2. Calculate the longitudinal piezoelectric coefficient in a general direction defined by its direction cosines s_1, s_2, s_3. Express it as a function of the spherical coordinate angles θ and φ.

3. Calculate the values of φ and θ for which the longitudinal piezoelectric coefficient is maximum.

4. A cube of ZnS with sides parallel to directions [110], [$\bar{1}$10] and [001] is submitted to a uniform electric field $\mathbf{E}$, parallel to [001]. The value of this field is 5×10^5 V m^{-1}.

    (a) Show that the volume variation of the crystal under the effect of the applied electric field is zero.

    (b) Determine the relative variation in the lengths of the sides of the cube. The value of d_{14} is 2×10^{-11} C N^{-1}.

Exercise 15.6

1. Quartz, with point group 32, is piezoelectric. We try to predict, using Curie's principle, whether the application of a uniaxial stress along some simple crystallographic directions can lead to a polarization. Consider in turn the cases where:

   (a) the uniaxial stress is parallel to the 3-fold axis,

   (b) the uniaxial stress is parallel to a 2-fold axis,

   (c) the uniaxial stress is parallel to Ox_2, with Ox_1 being parallel to the 2-fold axis of (b),

   (d) the stress is a shear T_{12}.

2. Zinc oxide ZnO, with point group 6mm, is piezoelectric. We discuss the same questions as in (1), *viz.* whether a polarization can exist under the influence of a uniaxial stress. We choose an orthonormal axis system where Ox_3 is parallel to the 6-fold axis and Ox_1 is perpendicular to one of the mirrors. Consider the following cases:

   (a) the uniaxial stress is parallel to Ox_3,

   (b) the uniaxial stress is parallel to Ox_1,

   (c) the uniaxial stress is parallel to Ox_2.

   For each case, check the results by using the appropriate piezoelectric tensor, as given in Table 15.2.

Exercise 15.7

Consider a crystal with point group $\bar{4}2m$.

Choose an orthonormal axis system $(Ox_1x_2x_3)$ where Ox_1 and Ox_2 are parallel to the 2-fold axes. Using Curie's principle, determine whether a shear stress T_{12} allows a polarization to exist.

Same question if the shear stress rotates by 45° around axis Ox_3. Check the obtained results by using the piezoelectric tensor for group $\bar{4}2m$.

Complement 15C. Electrostriction and magnetostriction

Piezoelectricity is discussed in Section 15.3. It relates, through $S_{ij} = d_{ijk}E_k$, an electric field with the strain, linear with respect to the field, which it produces or, conversely, it relates the stress to the electrical polarization, again with a linear dependence. It is present only in crystals belonging to some point groups, and in particular lacking a center of symmetry.

Electrostriction is more general a property. It relates an electric field with the strain it produces, but now the strain is proportional to the square of the electric field in the case of an isotropic material, or bilinear in the field components for an anisotropic material: $S_{ij} = Q_{ijkl}E_kE_l$. This property, represented by a rank-4 tensor, is featured by all insulating materials (dielectrics), including liquids and crystals of any symmetry, but its intensity varies considerably.

Piezomagnetism, the magnetic analogue of piezoelectricity, was long believed to be impossible, until this was shown to be wrong in the late 1950's. It is however a weak effect, and is restricted to materials with special (magnetic) symmetry, such as the antiferromagnets MnF_2, CoF_2 and NiF_2.

Magnetostriction is the magnetic effect roughly corresponding to electrostriction. It too varies considerably among materials. Large effects are encountered in ferro- or ferrimagnets, *i.e.* materials which have, at the microscopic level, an ordered magnetic structure. The main contribution is due to the growth, under an applied magnetic field, of the domains[7] with magnetization direction near that of the field; at saturation, the strain is the spontaneous magnetostrictive strain, whereas the demagnetized state as well as the zero-field state involve an average over the domain magnetization directions. The most efficient magnetostrictive material is Terfenol D, a rare-earth–iron compound with formula[8] $Tb_xDy_{1-x}Fe_2$, with $x \sim 0.3$, in which strains of the order of 2×10^{-3} are encountered.

Both magnetostrictive and electrostrictive materials are important for applications, the more obvious ones being sensors and actuators. One of the advantages of the magnetic version is the fact that a magnetic field can be applied through external coils and magnetic circuits, and does not require voltage to be applied on the actuator or sensor itself.

---

7. See the footnote in Complement 8C.

8. The "nol" at the end of Terfenol has nothing to do with the international organic chemistry nomenclature, which links "ol" to alcohols: it just celebrates the (U.S.) Navy's Ordnance Laboratory, where research on magnetostrictive materials has long been very active toward the development of sonar for underwater sensing.

Further reading

E. du Trémolet de Lacheisserie, D. Gignoux and M. Schlenker (ed.), *Magnetism*, vol. 1: *Fundamentals* (Springer Verlag, New York and Heidelberg, 2005), chap. 12

E. du Trémolet de Lacheisserie, D. Gignoux and M. Schlenker (ed.), *Magnetism*, vol. 2: *Materials and Applications* (Springer Verlag, New York and Heidelberg, 2005), chap. 15

References

[1] L.D. Landau, E.M. Lifshitz and L.P. Pitaevskii, *Course of Theoretical Physics*, vol. 8, *Electrodynamics of Continuous Media*, 2nd edn. (Elsevier, Oxford, 1984)

[2] H.B. Callen, *Thermodynamics and an Introduction to Thermostatistics*, 2nd edn. (Wiley, New York, 1985)

[3] J.F. Nye, *Physical Properties of Crystals: Their Representation by Tensors and Matrices* (Oxford University Press, Oxford, 1985)

[4] *Landolt and Börnstein Tables*, vol. III.2 (Springer Verlag, Berlin, 1969)

Chapter 16

Light propagation in crystals

When an electromagnetic wave propagates in a crystal, the material becomes electrically polarized by the electric field of the wave since its electrical permittivity is different from that of vacuum. The propagation of electromagnetic waves is governed by Maxwell's equations. In the case of anisotropic materials, these yield the birefringence of crystals, which is thoroughly discussed.

16.1. Maxwell's equations

Assuming an uncharged, non-conducting material, Maxwell's equations have the form:

$$\text{div}\,\mathbf{D} = 0 \tag{16.1a}$$

$$\text{curl}\,\mathbf{E} = -\frac{\partial \mathbf{B}}{\partial t} \tag{16.1b}$$

$$\text{div}\,\mathbf{B} = 0 \tag{16.1c}$$

$$\text{curl}\,\mathbf{H} = \frac{\partial \mathbf{D}}{\partial t} \tag{16.1d}$$

E

where $\mathbf{D}$ is the electrical induction defined in (15.3) and again here in (16.3), the electric field, $\mathbf{B}$ the magnetic induction, $\mathbf{H}$ the magnetic field. Since we consider non-magnetic materials:

$$\mathbf{B} = \mu_0 \mathbf{H}$$

where μ_0 is the magnetic permeability of vacuum ($\mu_0 = 4\pi \times 10^{-7}$ in SI units), hence

$$\text{div}\,\mathbf{H} = 0 \tag{16.1e}$$

$$\text{curl}\,\mathbf{E} = -\mu_0 \frac{\partial \mathbf{H}}{\partial t}. \tag{16.1f}$$

Dielectric materials polarize under the influence of an oscillating electric field $\mathbf{E}$. The polarization $\mathbf{P}$ is related to the electric field by relation:

$$\mathbf{P} = \varepsilon_0 [\chi] \mathbf{E} \tag{16.2}$$

© Springer Science+Business Media Dordrecht 2014
C. Malgrange et al., *Symmetry and Physical Properties of Crystals*,
DOI 10.1007/978-94-017-8993-6_16

where $[\chi]$ is the electrical susceptibility tensor of the material, in the linear approximation which we adopt throughout this chapter. The materials for which this approximation is sufficient are linear materials. Induction $\mathbf{D}$ is then given by:

$$\mathbf{D} = \varepsilon_0\mathbf{E} + \mathbf{P} = \varepsilon_0\mathbf{E} + \varepsilon_0[\chi]\mathbf{E} = \varepsilon_0[\varepsilon]\mathbf{E}. \tag{16.3}$$

ε_0 is the permittivity of vacuum ($1/4\pi\varepsilon_0 = 9\times10^9$ in SI units) and $[\varepsilon] = [E]+[\chi]$ is the tensor of relative permittivity of the material, with $[E]$ the unit tensor.

This equation was already given in (15.3), in the discussion of permittivity in a static electric field. Here the electric field is rapidly variable, and the values of the coefficients of the relative permittivity tensor $[\varepsilon]$ can be very different from their value for a static field.

If the material is isotropic, tensor $[\varepsilon]$ reduces to a scalar, and the propagation equation for an electromagnetic wave is easily solved since condition (16.1a) then leads to div $\mathbf{E} = 0$. The solution for anisotropic materials is more complicated, and we restrict our search to sinusoidal solutions.

Also, we here consider only transparent materials, i.e. non absorbing materials, hence the components of tensor $[\varepsilon]$ are assumed to be real.

16.2. Light propagation in an isotropic material

We discuss propagation in either vacuum or an isotropic dielectric for which tensor $[\varepsilon]$ is a scalar, noted ε.

Using Maxwell's equations, we calculate:

$$\mathbf{curl\,curl\,E} = -\mu_0\frac{\partial}{\partial t}(\mathbf{curl\,H}) = -\mu_0\frac{\partial^2\mathbf{D}}{\partial t^2} = -\mu_0\varepsilon_0\varepsilon\frac{\partial^2\mathbf{E}}{\partial t^2}$$

and we then use the identity: $\mathbf{curl\,curl\,E} = \mathbf{grad}\,\mathrm{div}\,\mathbf{E} - \Delta\mathbf{E}$.

Since $\mathbf{D}$ and $\mathbf{E}$ are parallel (this will no more be the case in an anisotropic material), div $\mathbf{E} = 0$ from (16.1a), and we obtain:

$$\Delta\mathbf{E} - \varepsilon_0\mu_0\varepsilon\frac{\partial^2\mathbf{E}}{\partial t^2} = 0. \tag{16.4}$$

This equation has solutions of the form:

$$\mathbf{E} = \mathbf{E}_0 f\left(t \pm \frac{\mathbf{s}\cdot\mathbf{r}}{v}\right) \tag{16.5}$$

with $v = 1/\sqrt{\varepsilon\varepsilon_0\mu_0}$, i.e. plane waves propagating in the direction of unit vector $\mathbf{s}$ at velocity $\pm v$ (see (14.6) and Fig. 14.1).

Setting
$$t \pm \frac{\mathbf{s}\cdot\mathbf{r}}{v} = u,$$

$$\frac{\partial\mathbf{E}}{\partial x_i} = \pm\mathbf{E}_0\frac{s_i}{v}\frac{\partial f}{\partial u}, \qquad \frac{\partial^2\mathbf{E}}{\partial x_i^2} = \mathbf{E}_0\frac{s_i^2}{v^2}\frac{\partial^2 f}{\partial u^2}, \qquad \frac{\partial^2\mathbf{E}}{\partial t^2} = \mathbf{E}_0\frac{\partial^2 f}{\partial^2 u}$$

and Equation (16.4) is verified if $1/v^2 = \varepsilon\varepsilon_0\mu_0$.

In vacuum, $\varepsilon = 1$ and $v = c$. We find the well-known relation $\varepsilon_0\mu_0 c^2 = 1$. We deduce that, in a dielectric material, $v = c/\sqrt{\varepsilon}$.

By definition, the optical index, or refractive index, of a material is $n = c/v$, and the above relations indicate that $n^2 = \varepsilon$.

We would of course have obtained the same solution for the magnetic field, *i.e.* a plane wave propagating at velocity $v = c/\sqrt{\varepsilon}$.

16.3. Sinusoidal waves that are solutions of Maxwell's equations

Plane sinusoidal waves have the form:

$$\mathbf{E} = \mathbf{E}_0 \cos(\mathbf{k} \cdot \mathbf{r} - \omega t) \qquad \text{or} \qquad \mathbf{E} = \mathbf{E}_0 \, \Re\left[\exp i(\mathbf{k} \cdot \mathbf{r} - \omega t)\right] \qquad (16.6)$$

where $\mathbf{k} = k\mathbf{s}$ is the wave vector, ω is the angular frequency of the wave, and $\Re$ means "real part of".

We can write: $\cos(\mathbf{k} \cdot \mathbf{r} - \omega t) = \cos[\omega(t - k\mathbf{s} \cdot \mathbf{r}/\omega)]$. Comparing with Equation (16.5), we see that the propagation velocity v, or phase velocity, is such that:

$$v = \frac{\omega}{k}. \qquad (16.7)$$

Furthermore, $\mathbf{k} \cdot \mathbf{r} = k\mathbf{s} \cdot \mathbf{r} = k\ell$ (Fig. 14.1) if ℓ is the distance from origin of a wave plane, counted along an axis parallel to $\mathbf{s}$. The wavelength λ is such that, for a distance equal to λ, the argument of the cosine varies by 2π, hence $\lambda = 2\pi/k$.

Note: if we had chosen, as is customary in crystallography and in X-ray diffraction, to express waves in the form $\cos 2\pi(\mathbf{k} \cdot \mathbf{r} - \nu t)$ (Tab. 3.1), the norm of the wave vector would be equal to $1/\lambda$. The frequency ν is equal to $\omega/2\pi$.

The sinusoidal behavior of vectors $\mathbf{E}$ and $\mathbf{H}$ makes the differentiations of vectors $\mathbf{E}$ or $\mathbf{H}$ with respect to time and to the coordinates very easy:

$$\frac{\partial \mathbf{E}}{\partial t} = -i\,\omega\,\mathbf{E}, \qquad \operatorname{div}\mathbf{E} = i\,\mathbf{k} \cdot \mathbf{E} \qquad \text{and} \qquad \operatorname{curl}\mathbf{E} = i\,\mathbf{k} \times \mathbf{E}. \qquad (16.8)$$

The same relations are valid for vector $\mathbf{H}$.

Applying these relations to Equations (16.1a) and (16.1e), we obtain:

$$\mathbf{k} \cdot \mathbf{D} = 0 \qquad \text{and} \qquad \mathbf{k} \cdot \mathbf{H} = 0 \qquad (16.9)$$

so that vectors $\mathbf{D}$ and $\mathbf{H}$ are perpendicular to wave vector $\mathbf{k}$.

Furthermore, we deduce from Equations (16.1d) and (16.1f):

$$\mathbf{k} \times \mathbf{H} = -\omega\mathbf{D}, \qquad (16.10a)$$
$$\mathbf{k} \times \mathbf{E} = \mu_0\omega\mathbf{H}. \qquad (16.10b)$$

Vector $\mathbf{D}$ is perpendicular to $\mathbf{H}$ and to $\mathbf{k}$ from (16.10a), and $\mathbf{H}$ is perpendicular to $\mathbf{k}$ after (16.10b). We deduce that vectors $\mathbf{k}$, $\mathbf{D}$ and $\mathbf{H}$ form a direct trirectangular trihedron, whether the material is isotropic or not (Fig. 16.1a). Furthermore, $\mathbf{E}$ is perpendicular to $\mathbf{H}$ (16.10b), and therefore lies in the plane defined by $\mathbf{D}$ and $\mathbf{k}$. In an isotropic material, the trihedron defined by $\mathbf{k}$, $\mathbf{E}$ and $\mathbf{H}$ is also a direct trirectangular trihedron (Fig. 16.1b).

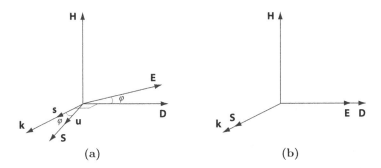

(a)   (b)

Figure 16.1: Relative orientations of vectors $\mathbf{H}$, $\mathbf{E}$, $\mathbf{D}$, $\mathbf{S}$ and $\mathbf{k}$. Vector $\mathbf{H}$ is perpendicular to the four other vectors. (a) Anisotropic material. (b) Isotropic material.

The direction of energy propagation, hence of the light rays, is that of Poynting's vector $\mathbf{S}$, the norm of which represents the energy that propagates across unit area during unit time, *i.e.* the intensity of the light beam propagating in the direction of $\mathbf{S}$:

$$\mathbf{S} = \mathbf{E} \times \mathbf{H}$$

so that $\mathbf{S}$ is perpendicular to $\mathbf{E}$ and $\mathbf{H}$. In an anisotropic material, the direction of energy propagation is thus usually different from that of the wave vector, and thus the propagation velocity of energy is different from the propagation velocity of the wave planes, which is the phase velocity defined in (16.7).

In an isotropic material, $\mathbf{S}$ is parallel to $\mathbf{k}$. It is then easy to calculate the norm of $\mathbf{S}$, because vectors $\mathbf{E}$, $\mathbf{H}$ and $\mathbf{S}$ form a trirectangular trihedron.

Using relation (16.10b), we find:

$$\|\mathbf{S}\| = \|\mathbf{E}\|\,\|\mathbf{H}\| = kE^2/\mu_0\omega = E^2/\mu_0 v = \varepsilon_0\varepsilon v E^2.$$

In an isotropic material, and in particular in air, the intensity of an electromagnetic wave is thus proportional to the square of the norm of the electric field.

16.4. Plane monochromatic wave in an anisotropic material

16.4.1. Basic equation

In an anisotropic material, the polarization $\mathbf{P}$ induced by an electric field $\mathbf{E}$ is, in general, no more parallel to $\mathbf{E}$. It is related to $\mathbf{E}$ by the electrical susceptibility tensor $[\chi]$:

$$\mathbf{P} = \varepsilon_0[\chi]\mathbf{E}$$

$$\text{and} \qquad \mathbf{D} = \varepsilon_0\mathbf{E} + \mathbf{P} = \varepsilon_0([E] + [\chi])\mathbf{E} = \varepsilon_0[\varepsilon]\mathbf{E}. \qquad (16.11)$$

$[\varepsilon]$ is the relative permittivity tensor of the crystal. This tensor is symmetric (Sect. 15.1.3).

Condition div $\mathbf{D} = 0$ does not lead to zero divergence for $\mathbf{E}$ any more, and the propagation equation obtained in (16.4) is no more valid. To simplify the problem, we seek, for the electromagnetic wave, a sinusoidal solution, so that the results of Section 16.3 can be used. We saw that the trihedron $\mathbf{D}$, $\mathbf{H}$, $\mathbf{k}$ is trirectangular. Vectors $\mathbf{E}$ and $\mathbf{D}$ are no more parallel, but enclose an angle φ. Since $\mathbf{E}$ is perpendicular to $\mathbf{H}$, it is in the plane defined by $\mathbf{D}$ and $\mathbf{k}$ (Fig. 16.1a). Poynting's vector is perpendicular to $\mathbf{E}$ and $\mathbf{H}$. It is therefore also in plane $(\mathbf{D}, \mathbf{k})$ and encloses with vector $\mathbf{k}$ the same angle φ.

Using Equations (16.1b) and (16.1d), we obtain:

$$\mathbf{curl\,curl\,E} = -\mu_0\frac{\partial^2\mathbf{D}}{\partial t^2}. \qquad (16.12)$$

Relations (16.8) make it possible to write Equation (16.12) in the form

$$-\mathbf{k} \times \mathbf{k} \times \mathbf{E} = \omega^2\mu_0\mathbf{D}. \qquad (16.13)$$

Furthermore, we know that any three vectors satisfy relation:

$$\mathbf{A} \times \mathbf{B} \times \mathbf{C} = (\mathbf{A} \cdot \mathbf{C})\mathbf{B} - (\mathbf{A} \cdot \mathbf{B})\mathbf{C} \qquad (16.14)$$

so that Equation (16.13) becomes:

$$-\mathbf{k}(\mathbf{k} \cdot \mathbf{E}) + k^2\mathbf{E} = \omega^2\mu_0\mathbf{D}.$$

Since the propagation velocity v of the wave (the phase velocity) is such that $v = \omega/k$ (Eq. (16.7)) and since the refractive index is defined by $n = c/v$, so that $n = ck/\omega$ and $\varepsilon_0\mu_0c^2 = 1$, we deduce that $k^2/(\omega^2\mu_0) = \varepsilon_0 n^2$. Setting, as before, $\mathbf{k} = k\mathbf{s}$, we obtain:

$$\mathbf{D} = \varepsilon_0 n^2[\mathbf{E} - \mathbf{s}(\mathbf{s} \cdot \mathbf{E})] \qquad (16.15a)$$

$$\text{and} \qquad [\varepsilon]\mathbf{E} = n^2[\mathbf{E} - \mathbf{s}(\mathbf{s} \cdot \mathbf{E})]. \qquad (16.15b)$$

Equation (16.15b) is actually a system of three equations:

$$[\varepsilon_{ij} + n^2 s_i s_j]E_j = n^2 E_i$$

which are linear and homogeneous equations for E_1, E_2 and E_3. For such a system to have solutions, its determinant must be zero. By expressing this condition, we obtain a relation giving the value of index n as a function of the propagation direction $\mathbf{s}$. Because calculating the determinant is tedious, we will use some calculational tricks which lead to the result more easily. We start from Equation (16.15a), and use the fact that $\mathbf{D} \cdot \mathbf{s} = 0$.

16.4.2. Birefringence

The second-rank, real and symmetric tensor $[\varepsilon]$ has three mutually orthogonal principal axes (Sect. 10.2.2). In the coordinate system of these principal axes, the tensor is diagonal:

$$\begin{pmatrix} \varepsilon_1 & & \\ & \varepsilon_2 & \\ & & \varepsilon_3 \end{pmatrix}$$

and, in the same frame, the components D_i and E_i of $\mathbf{D}$ and $\mathbf{E}$ respectively are related through:

$$D_1 = \varepsilon_0 \varepsilon_1 E_1, \qquad D_2 = \varepsilon_0 \varepsilon_2 E_2, \qquad D_3 = \varepsilon_0 \varepsilon_3 E_3. \qquad (16.16)$$

Equation (16.15a), when projected onto principal axis Ox_1, yields, using (16.16):

$$D_1 \left(1 - \frac{n^2}{\varepsilon_1}\right) = -s_1(\mathbf{s} \cdot \mathbf{E})$$

$$\text{or} \qquad D_1 = -\frac{\varepsilon_1 s_1}{\varepsilon_1 - n^2}(\mathbf{s} \cdot \mathbf{E}). \qquad (16.17a)$$

In the same way, we obtain:

$$D_2 = -\frac{\varepsilon_2 s_2}{\varepsilon_2 - n^2}(\mathbf{s} \cdot \mathbf{E}), \qquad (16.17b)$$

$$D_3 = -\frac{\varepsilon_3 s_3}{\varepsilon_3 - n^2}(\mathbf{s} \cdot \mathbf{E}). \qquad (16.17c)$$

Multiplying (16.17a) by s_1, (16.17b) by s_2 and (16.17c) by s_3, and adding them, we obtain for the left-hand side $D_i s_i = \mathbf{D} \cdot \mathbf{s} = 0$, hence

$$\sum_i \frac{s_i^2 \varepsilon_i}{\varepsilon_i - n^2} = 0. \qquad (16.18)$$

Expanding (16.18) yields:

$$s_1^2 \varepsilon_1(\varepsilon_2 - n^2)(\varepsilon_3 - n^2) + s_2^2 \varepsilon_2(\varepsilon_3 - n^2)(\varepsilon_1 - n^2)$$
$$+ s_3^2 \varepsilon_3(\varepsilon_1 - n^2)(\varepsilon_2 - n^2) = 0. \qquad (16.19)$$

This equation $f(n^2) = 0$ is a second-degree equation for n^2. It is easy to show that it has two solutions. We rank the values of ε_i in increasing order, for example $\varepsilon_1 < \varepsilon_2 < \varepsilon_3$. The value of $f(n^2 = \varepsilon_1)$ is positive, that of $f(n^2 = \varepsilon_2)$ is negative and that of $f(n^2 = \varepsilon_3)$ is positive. The continuous function $f(n^2)$ therefore goes to zero for two values of n^2, one between ε_1 and ε_2, the other between ε_2 and ε_3. One value of n^2 corresponds to a single value of n, because a negative index would be equivalent to a negative velocity, simply describing a wave propagating the other way. This is why only positive indices are considered.

For a given propagation direction, there are thus two values of the refractive index, hence two waves can propagate: they are called the eigenmodes. The material is said to be birefringent, or one says there is birefringence.

From now on, we set:

$$n_1 = \sqrt{\varepsilon_1}, \qquad n_2 = \sqrt{\varepsilon_2}, \qquad n_3 = \sqrt{\varepsilon_3} \tag{16.20}$$

and n_1, n_2 and n_3 are called the principal indices (meaning the indices associated to the principal values of tensor $[\varepsilon]$).

16.4.3. Index surface

The index surface is the locus of points M such that $\mathbf{OM} = n\,\mathbf{s}$, where n is the value of the index for a wave with wave-vector parallel to unit vector $\mathbf{s}$.[1] Since there are two values of n for one propagation direction $\mathbf{s}$, the index surface intersects any radius vector at two points.

Biaxial materials

We saw (Sect. 10.5) that symmetric rank-2 tensors, representing physical properties of triclinic, monoclinic or orthorhombic crystals, have three principal values which normally are different. Therefore, for such crystals, $\varepsilon_1 \neq \varepsilon_2 \neq \varepsilon_3$, and we obtain an index surface as shown on Figure 16.2a, where only one-eighth of the surface is represented because this surface is symmetrical with respect to planes $s_1 = 0$, $s_2 = 0$ and $s_3 = 0$.

If $s_1 = s_2 = 0$, and therefore $s_3 = 1$, Equation (16.19) leads to $n^2 = \varepsilon_1$ and $n^2 = \varepsilon_2$. The index surface intersects axis Ox_3 at $x_3 = \sqrt{\varepsilon_1} = n_1$ and $x_3 = \sqrt{\varepsilon_2} = n_2$. In the same way, it intersects axis Ox_1 at $x_1 = \sqrt{\varepsilon_2} = n_2$ and $x_1 = \sqrt{\varepsilon_3} = n_3$ and axis Ox_2 at $x_2 = \sqrt{\varepsilon_3} = n_3$ and $x_2 = \sqrt{\varepsilon_1} = n_1$.

It is easy to show that the intersections of the index surface by the planes $x_i = 0$ corresponding to $s_i = 0$ consist of a circle and an ellipse. These curves do not intersect in planes $x_1 = 0$ and $x_3 = 0$, but they do intersect in plane $x_2 = 0$. Because of the symmetry of the index surface, there are in this plane four points of intersection, deduced from one another by symmetries with respect

---

1. The index surface is also called the wave vector surface when the length along $\mathbf{s}$ is taken equal to kn instead of n, k being the modulus of the wave vector in vacuum.

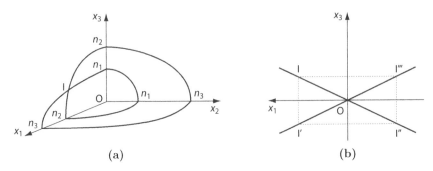

Figure 16.2: Index surface in the general case. (a) Intersection of the surface with the planes of the orthonormal trihedron formed by the principal axes of tensor $[\varepsilon]$. We assume $\varepsilon_1 < \varepsilon_2 < \varepsilon_3$ so that $n_1 < n_2 < n_3$. Due to its symmetry, it is sufficient to represent only one-eighth of the surface. (b) Relative orientation of the optical axes II'' and I'I''' in the plane perpendicular to Ox_2.

to planes $x_1 = 0$ and $x_3 = 0$ (Fig. 16.2b). There are thus two propagation direction, II'' and I'I''', for which the two indices coincide. For these two special directions, the crystal is no more birefringent. These two directions are called the optical axes of the crystal, whence the name biaxial crystals.

Uniaxial materials

For tetragonal, rhombohedral and hexagonal crystals, two of the principal values of tensor $[\varepsilon]$ are equal (Sect. 10.5.4). If we choose an axis system where Ox_3 is parallel to basis vector **c**, itself parallel to the rotation axis with order larger than 2 of the crystal, then $\varepsilon_1 = \varepsilon_2$. The value $n_o = n_1 = \sqrt{\varepsilon_1}$ is then called the ordinary index, and $n_e = n_3 = \sqrt{\varepsilon_3}$ is called the extraordinary index.

The index surface (16.19) then subdivides into two surfaces:

1. $\varepsilon_1 - n^2 = 0$, a sphere with radius $n = \sqrt{\varepsilon_1} = n_o$, called the ordinary sphere

2. $s_1^2 \varepsilon_1(\varepsilon_3 - n^2) + s_2^2 \varepsilon_1(\varepsilon_3 - n^2) + s_3^2 \varepsilon_3(\varepsilon_1 - n^2) = 0$, hence

$$\varepsilon_1(\varepsilon_3 - n^2)(s_1^2 + s_2^2) + \varepsilon_3(\varepsilon_1 - n^2)s_3^2 = 0 \text{ and}$$

$$\varepsilon_1\varepsilon_3 - n^2\varepsilon_1(s_1^2 + s_2^2) - n^2\varepsilon_3 s_3^2 = 0.$$

The index surface is defined as the locus of the ends of vectors **r** such that $\mathbf{r} = n\mathbf{s}$, hence $x_i = ns_i$. Hence the equation of the second surface is given by:

$$\varepsilon_1(x_1^2 + x_2^2) + \varepsilon_3 x_3^2 = \varepsilon_1\varepsilon_3,$$

$$\text{or} \qquad \frac{x_1^2 + x_2^2}{\varepsilon_3} + \frac{x_3^2}{\varepsilon_1} = 1,$$

$$\text{or again} \qquad \frac{x_1^2 + x_2^2}{n_e^2} + \frac{x_3^2}{n_o^2} = 1.$$

This is an ellipsoid of revolution around direction Ox_3, and tangent to the ordinary sphere along axis Ox_3 (Fig. 16.3). The two optical axes which were defined for biaxial materials now coincide. There is a single optical axis (whence the name uniaxial crystals), parallel to the n-fold rotation axis with order n larger than 2 (which itself is parallel to basis vector $\mathbf{c}$). If $n_e > n_o$, the crystal is said to be uniaxial positive (Fig. 16.3a), and if $n_e < n_o$ it is uniaxial negative (Fig. 16.3b).

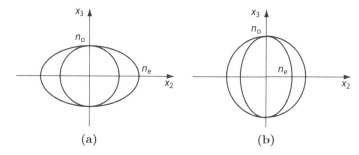

(a)   (b)

Figure 16.3: Index surface of a uniaxial crystal. Axis Ox_3 is an axis of revolution. (a) Uniaxial positive crystal. (b) Uniaxial negative crystal.

Cubic crystals

For cubic crystals, the three principal values of tensor $[\varepsilon]$ are equal, and these crystals are isotropic for electromagnetic wave propagation.

16.4.4. Index ellipsoid

We will see in the next section that determining the induction vectors $\mathbf{D}$ associated to a given propagation direction involves the tensor inverse to $[\varepsilon]$ and its associated quadric, called the index ellipsoid, which we now introduce.

The relative permittivity tensor $[\varepsilon]$ relates $\mathbf{E}$ to $\mathbf{D}/\varepsilon_0$ through:

$$\frac{\mathbf{D}}{\varepsilon_0} = [\varepsilon]\mathbf{E}. \qquad (16.21)$$

The inverse tensor, called the electrical impermeability tensor, relates $\mathbf{D}/\varepsilon_0$ to $\mathbf{E}$ through:

$$\mathbf{E} = [\eta]\frac{\mathbf{D}}{\varepsilon_0}. \qquad (16.22)$$

In the coordinate system of the principal axes of tensor $[\varepsilon]$, relation (16.21) takes the form:

$$\frac{D_1}{\varepsilon_0} = \varepsilon_1 E_1, \qquad \frac{D_2}{\varepsilon_0} = \varepsilon_2 E_2, \qquad \frac{D_3}{\varepsilon_0} = \varepsilon_3 E_3,$$

whence

$$E_1 = \frac{D_1}{\varepsilon_0}\left(\frac{1}{\varepsilon_1}\right) = \eta_1 \frac{D_1}{\varepsilon_0} \text{ if we set } \eta_1 = \frac{1}{\varepsilon_1}. \qquad (16.23)$$

In the same way, we obtain:

$$E_2 = \eta_2 \frac{D_2}{\varepsilon_0} \qquad \text{with } \eta_2 = \frac{1}{\varepsilon_2},$$

$$E_3 = \eta_3 \frac{D_3}{\varepsilon_0} \qquad \text{with } \eta_3 = \frac{1}{\varepsilon_3}.$$

Thus tensors $[\varepsilon]$ and $[\eta]$, inverse to each other, have the same principal axes, and their principal values are reciprocal to each other.

The equation of the representative quadric of tensor $[\eta]$, in the coordinate system of its principal axes, is:

$$\eta_1 x_1^2 + \eta_2 x_2^2 + \eta_3 x_3^2 = 1.$$

The values η_i are always positive since, from (16.23) and (16.20)

$$\eta_i = \frac{1}{\varepsilon_i} = \frac{1}{n_i^2}. \tag{16.24}$$

Hence the quadric is, in all cases, an ellipsoid, called the index ellipsoid. It can also be written as:

$$\frac{x_1^2}{n_1^2} + \frac{x_2^2}{n_2^2} + \frac{x_3^2}{n_3^2} = 1.$$

The semi-axes along the principal axis directions of this ellipsoid are n_1, n_2 and n_3.

For uniaxial crystals (tetragonal, rhombohedral and hexagonal), the index ellipsoid is of revolution around the axis of order larger than 2 which is always chosen as the Ox_3 axis. Earlier, we noted $n_1 = n_2 = n_o$ and $n_3 = n_e$. The index ellipsoid then has for its equation:

$$\frac{x_1^2 + x_2^2}{n_o^2} + \frac{x_3^2}{n_e^2} = 1. \tag{16.25}$$

16.4.5. Determining the induction vectors

We saw that the induction vector $\mathbf{D}$ is perpendicular to the wave vector, hence to unit vector $\mathbf{s}$. If we choose an orthonormal coordinate system $(Ox_1' x_2' x_3')$ such that Ox_3' is parallel to $\mathbf{s}$, vector $\mathbf{D}$ has only two non-zero components, D_1' and D_2'. They can be obtained by projecting Equation (16.15a), reproduced below, onto the plane $(Ox_1' x_2')$ perpendicular to $\mathbf{s}$.

$$\mathbf{D} = \varepsilon_0 \, n^2 \left[\mathbf{E} - \mathbf{s}(\mathbf{s} \cdot \mathbf{E})\right]. \tag{16.15a}$$

The second term in the right-hand side of this equation has zero projection, and we obtain:

$$D_i' = \varepsilon_0 \, n^2 \, E_i'.$$

Also, the projection of Equation (16.22) yields:

$$\varepsilon_0 E_i' = \eta_{ij}' D_j'$$

where the η'_{ij} are the components of the electrical impermeability tensor in the chosen axis system (which need not be that of the principal axes).

We obtain:

$$\eta'_{ij} D'_j = \frac{1}{n^2} D'_i, \qquad (16.26)$$

a system of two linear and homogeneous equations with two unknowns (D'_1 and D'_2), which is none other than the eigenvector equation of matrix $\{\eta'_{ij}\}$, a second-rank matrix whose eigenvalues are the values of $1/n^2$.

The representative surface of tensor $[\eta]$ is the index ellipsoid. In the subspace formed by Ox'_1 and Ox'_2, the representative surface of this same tensor is the intersection of the ellipsoid by the projection plane, hence an ellipse (Fig. 16.4). Therefore the solutions for $\mathbf{D}$ are vectors parallel to the axes of this ellipse. The eigenvalues are the reciprocals of the squares of the half-lengths of the axes of the ellipse (Sect. 10.2.3), hence the values of n are equal to the lengths of the semiaxes of the ellipse.

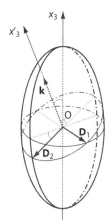

Figure 16.4: Determination of the induction vectors $\mathbf{D}_1$ and $\mathbf{D}_2$ of a wave with wave vector $\mathbf{k}$, propagating in a birefringent crystal. $\mathbf{D}_1$ and $\mathbf{D}_2$ are parallel to the axes of the ellipse along which the index ellipsoid intersects the plane normal to vector $\mathbf{k}$ and going through the center of the ellipsoid. The lengths of the semiaxes of the ellipse are the values of the respective indices.

Conclusion: To find the induction vectors of a wave propagating in a crystal with its wave vector parallel to unit vector $\mathbf{s}$, we consider the index ellipsoid of the crystal and its intersection by a plane going through the center of the ellipsoid, and perpendicular to vector $\mathbf{s}$. This intersection is an ellipse. The two waves which can propagate (the eigenmodes) have induction vectors parallel to the axes of this ellipse, and the corresponding indices are equal to the lengths of the semiaxes of the ellipse.

Using the index ellipsoid, we can retrieve the results obtained earlier about the existence of one or several optical axes. We demonstrate it for biaxial, then for uniaxial crystals.

Biaxial crystals

The lengths of the three semiaxes of the ellipsoid are different. We again rank them in order of increasing value, $n_1 < n_2 < n_3$. The intersections of the ellipsoid by a plane going through the origin are ellipses, but it can be shown that two of these intersections are circles. Consider a wave vector $\mathbf{k}$ in the plane (Ox_1, Ox_3), and enclosing angle β with Ox_3. We look for the ellipse along which the ellipsoid is intersected by the plane perpendicular to $\mathbf{k}$ going through O (Fig. 16.5). Its semiaxes are the semiaxis of the ellipsoid parallel to Ox_2, with length n_2, and the radius of the ellipse in plane (Ox_1, Ox_3) perpendicular to $\mathbf{k}$, with length n'. This length n' has a value between n_1 and n_3, and therefore there is an angle β for which this length is equal to n_2. The ellipse is then a circle, and wave propagation is isotropic. Let $\mathbf{k}_1$ be the wave vector thus defined. It is clear, from the symmetry with respect to plane (Ox_1, Ox_2), that the wave vector $\mathbf{k}_2$, with angle $\beta_2 = \pi - \beta$, also corresponds to a circular section of the ellipsoid. We retrieve the existence of two optical axes and their geometry in the plane perpendicular to Ox_2 (Fig. 16.2b).

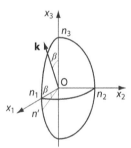

Figure 16.5: Intersection of the index ellipsoid by its planes of symmetry. A wave with wave vector $\mathbf{k}$ in plane (Ox_1, Ox_3) has as its indices n_2 and n' such that $n_1 < n' < n_3$.

Uniaxial crystals

The index ellipsoid is of revolution around the axis with order larger than 2. Therefore the section of this index ellipsoid by a plane perpendicular to this axis and going through the origin is a circle, and this axis is the optic axis. A general direction for wave vector $\mathbf{k}$ encloses angle θ with the optic axis. The plane normal to $\mathbf{k}$ and going through the origin intersects the ellipsoid along an ellipse for which one of the semiaxes is equal to n_o and the other to n, the value of which is between n_o and n_e (Fig. 16.6, where we chose, to make representation easier, the axis Ox_2 to be in the plane of the optic axis and wave vector $\mathbf{k}$).

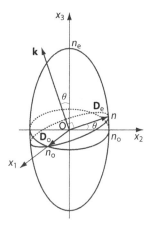

Figure 16.6: Determination of the induction vectors of a wave with wave vector **k** in the case of uniaxial crystals. Axis Ox_3 is parallel to the optical axis and Ox_1 perpendicular to the plane $(\mathbf{k}, Ox_3)$.

We conclude: for uniaxial crystals, the ordinary vibration $\mathbf{D}_o$ is perpendicular to the optic axis. The extraordinary vibration $\mathbf{D}_e$ is in the plane formed by wave vector $\mathbf{k}$ and the optic axis.

16.4.6. Direction of energy propagation

The aim of this section is to show that Poynting's vector is normal to the index surface.

We write relations (16.10a) and (16.10b) so that they explicitly show only vectors $\mathbf{E}$ and $\mathbf{H}$.

$$\mathbf{k} \times \mathbf{E} = \omega\mu_0\mathbf{H} \tag{16.27}$$

$$\mathbf{k} \times \mathbf{H} = -\omega\varepsilon_0[\varepsilon]\mathbf{E}. \tag{16.28}$$

Consider a variation $d\mathbf{k}$ in the wave vector. If the variation is general, it entails not only variations $d\mathbf{E}$ and $d\mathbf{H}$, but also $d\omega$ ($d\omega$ is zero only if $d\mathbf{k}$ is tangent to the index surface). Differentiating (16.27) and (16.28), we obtain:

$$d\mathbf{k} \times \mathbf{E} + \mathbf{k} \times d\mathbf{E} = \omega\mu_0 d\mathbf{H} + \mu_0\mathbf{H}d\omega, \tag{16.29}$$

$$d\mathbf{k} \times \mathbf{H} + \mathbf{k} \times d\mathbf{H} = -\omega\varepsilon_0[\varepsilon]d\mathbf{E} - \varepsilon_0[\varepsilon]\mathbf{E}d\omega. \tag{16.30}$$

We dot-multiply (16.29) with $\mathbf{H}$ and (16.30) with $\mathbf{E}$ after replacing $\mathbf{E} \cdot [\varepsilon]d\mathbf{E}$ by $[\varepsilon]\mathbf{E} \cdot d\mathbf{E}$, since $E_i\varepsilon_{ij}dE_j = \varepsilon_{ji}E_idE_j$ because $\varepsilon_{ij} = \varepsilon_{ji}$.

$$(\mathbf{H}, d\mathbf{k}, \mathbf{E}) + (\mathbf{H}, \mathbf{k}, d\mathbf{E}) = \omega\mu_0\mathbf{H} \cdot d\mathbf{H} + \mu_0 H^2 d\omega$$

$$-(\mathbf{H}, d\mathbf{k}, \mathbf{E}) + (\mathbf{E}, \mathbf{k}, d\mathbf{H}) = -\omega\varepsilon_0[\varepsilon]\mathbf{E} \cdot d\mathbf{E} - \mathbf{E} \cdot \mathbf{D}d\omega.$$

Subtracting both sides of these two equations :

$$2d\mathbf{k} \cdot (\mathbf{E} \times \mathbf{H}) + d\mathbf{E} \cdot (\mathbf{H} \times \mathbf{k} - \omega\varepsilon_0[\varepsilon]\mathbf{E})$$

$$-d\mathbf{H} \cdot (\mathbf{E} \times \mathbf{k} + \omega\mu_0\mathbf{H}) = (\mu_0 H^2 + \mathbf{E} \cdot \mathbf{D})d\omega. \tag{16.31}$$

The second and third terms are zero because of (16.28) and (16.27) respectively, and Equation (16.31) becomes:

$$2d\mathbf{k} \cdot (\mathbf{E} \times \mathbf{H}) = (\mu_0 H^2 + \mathbf{E} \cdot \mathbf{D})d\omega = 2ud\omega$$

if we note u the electromagnetic energy per unit volume (Born and Wolf, 1980)

$$u = \frac{1}{2}(\mathbf{E} \cdot \mathbf{D}) + \frac{1}{2}(\mathbf{H} \cdot \mathbf{B}).$$

Since we assume the material to be non-magnetic, $\mathbf{B} = \mu_0\mathbf{H}$. Using the relation that defines Poynting's vector, given in Section 16.3, $\mathbf{S} = \mathbf{E} \times \mathbf{H}$, we obtain

$$d\mathbf{k} \cdot \mathbf{S} = ud\omega.$$

Since $\mathbf{S} = u\mathbf{v}_S$ where $\mathbf{v}_S$ is the energy propagation velocity, we deduce:

$$\mathbf{v}_S \cdot d\mathbf{k} = d\omega.$$

The index surface is the constant ω surface in wave vector space. Therefore, if $d\mathbf{k}$ has any value in the plane tangent to the index surface, we have $d\omega = 0$ and $\mathbf{v}_S$ is perpendicular to $d\mathbf{k}$. Hence the ray direction, which is the direction of energy propagation $\mathbf{v}_S$, is perpendicular to the index surface. We note that $\mathbf{E}$, being perpendicular to $\mathbf{v}_S$, is tangent to the index surface.[2]

16.5. Refraction of a plane wave at the boundary between two materials

16.5.1. The wave-vectors follow the Snell-Descartes law

Boundary between two isotropic materials

When a wave which propagates in an isotropic first material (1) reaches the boundary with a second isotropic material (2), it splits into two waves: one, the refracted wave, propagates in material (2), the other one is reflected into material (1). Let $\mathbf{E}_1$, $\mathbf{k}_1$ be the electric field and the wave vector of the incident wave, $\mathbf{E}_2$ and $\mathbf{k}_2$ those of the refracted wave, $\mathbf{E}_3$ and $\mathbf{k}_3$ those of the reflected wave.

We have:
$$\mathbf{E}_1 = \mathbf{E}_{01} \exp i(\mathbf{k}_1 \cdot \mathbf{r} - \omega t),$$
$$\mathbf{E}_2 = \mathbf{E}_{02} \exp i(\mathbf{k}_2 \cdot \mathbf{r} - \omega t),$$
$$\mathbf{E}_3 = \mathbf{E}_{03} \exp i(\mathbf{k}_3 \cdot \mathbf{r} - \omega t).$$

The boundary condition on the electric field must be satisfied at all points on the separation surface and at all times, which implies that the space (and time)

---

2. The energy propagation velocity, *i.e.* the group velocity, can be written in the form $\mathbf{v}_S = \partial\omega/\partial\mathbf{k}$, a condensed notation for $v_{S1} = \partial\omega/\partial k_1$ etc. This is the generalization of the familiar expression for the group velocity in an isotropic material, $\partial\omega/\partial k$.

variations must be the same. We choose an orthonormal axis system $Oxyz$ so that the origin O is on the separation surface and Oz is perpendicular to this surface. At any point on the surface $z = 0$, we must therefore have:

$$k_{1x}x + k_{1y}y = k_{2x}x + k_{2y}y = k_{3x}x + k_{3y}y.$$

These equalities must hold whatever the values of x and y. This implies that

$$k_{1x} = k_{2x} = k_{3x} \qquad \text{and} \qquad k_{1y} = k_{2y} = k_{3y}.$$

Thus the tangential component of the wave vector is conserved in refraction and reflection.

Furthermore, in the isotropic materials we are considering in this section, the rays are parallel to the wave-vectors. This property leads to the Snell-Descartes laws, which have the following expression:

1. The refracted ray and the reflected ray are in the plane of incidence, defined as the plane containing the incident wave vector and the normal to the interface.

2. The reflection angle i_3 is equal to the incidence angle i_1. The refraction angle i_2 and the incidence angle i_1 are related by:

$$n_1 \sin i_1 = n_2 \sin i_2. \tag{16.32}$$

The derivation of these results is straightforward if we look at Figure 16.7a.

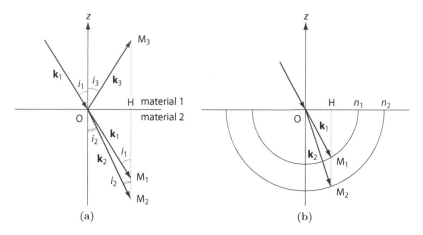

(a)   (b)

Figure 16.7: A wave with wave-vector $\mathbf{k}_1$ in an isotropic medium 1 gives rise to a reflected wave and a refracted wave. (a) Showing the continuity of the tangential component of the wave-vectors at the interface. (b) Construction of the refracted wave-vector $\mathbf{k}_2$ when medium 2 is isotropic.

The tangential component **OH** of the incident wave-vector $\mathbf{k}_1 = \mathbf{OM}_1$ is equal to the tangential components of the reflected wave-vector $\mathbf{k}_3 = \mathbf{OM}_3$ and of the refracted wave-vector $\mathbf{k}_2 = \mathbf{OM}_2$. Therefore, vectors $\mathbf{k}_3$ and $\mathbf{k}_2$ are in the plane of incidence. If n_1 and n_2 are the indices of materials 1 and 2 respectively,

$$k_1 = k_3 = k_0 n_1 \qquad \text{and} \qquad k_2 = k_0 n_2$$

where $k_0 = \omega/c$ is the norm of the wave-vector in vacuum.

Also

$$\text{OH} = k_1 \sin i_1 = k_3 \sin i_3 = k_2 \sin i_2. \tag{16.33}$$

Hence
$$i_1 = i_3,$$
$$n_1 \sin i_1 = n_2 \sin i_2.$$

The above properties make it easy to draw the refracted wave-vector $\mathbf{k}_2$ associated to a given incident wave-vector $\mathbf{k}_1$.

We draw, starting from point O located on the surface, the vector $\mathbf{k}_1 = \mathbf{OM}_1$, giving it length n_1 (which is equivalent to using k_0 as the unit). In medium 2, we draw a half-circle with radius n_2, centered at O. The normal to the interface going through M_1 intersects the half-circle with radius n_2 at M_2, the end of vector $\mathbf{k}_2 = \mathbf{OM}_2$.

In the following, we will focus solely on the refracted waves.

Interface between an isotropic material and an anisotropic material

The incident wave, with electric field **E**, gives rise to a reflected wave which we ignore, and to two refracted waves, with wave-vectors $\mathbf{k}_2$ and $\mathbf{k}_2'$, with electric fields $\mathbf{E}_2$ and $\mathbf{E}_2'$. This is double refraction.

$$\begin{aligned}
\mathbf{E}_1 &= \mathbf{E}_{01} \exp i(\mathbf{k}_1 \cdot \mathbf{r} - \omega t), \\
\mathbf{E}_2 &= \mathbf{E}_{02} \exp i(\mathbf{k}_2 \cdot \mathbf{r} - \omega t), \\
\mathbf{E}_2' &= \mathbf{E}_{02}' \exp i(\mathbf{k}_2' \cdot \mathbf{r} - \omega t).
\end{aligned}$$

The same argument as above shows that the tangential components of the wave-vectors are continuous at the interface between the two materials, and this leads to the Snell-Descartes laws *for the wave-vectors*. This is valid for the wave-vectors only, as, in a crystal, the rays are not necessarily parallel to the wave-vectors.

The refracted wave-vectors are determined using the same construction as above. As in Figure 16.8, we draw from a point O of the interface, chosen as origin, the incident wave-vector $\mathbf{k}_1 = \mathbf{OM}_1$, giving it length n_1, *i.e.* taking as the unit the wave-vector in vacuum k_0. We then draw the two sheets S and S' of the index surface, centered at O. The normal to the interface drawn from M_1 intersects the index surface at two points M_2 and M_2', the ends of wave-vectors $\mathbf{k}_2$ and $\mathbf{k}_2'$. We thus obtain:

$$n_1 \sin i_1 = n_2 \sin i_2 = n_2' \sin i_2'$$

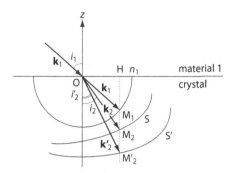

Figure 16.8: Construction of the wave-vectors of waves which can propagate in a birefringent crystal when the incident wave has wave-vector $\mathbf{k}_1$. S and S' are the two sheets of the index surface in the crystal.

where angles i_1, i_2 and i'_2 are the incidence angle and the refraction angles for the wave-vectors.

If material (1) were anisotropic, the same argument would apply to each of the wave-vectors in material (1), and each wave could give rise to two waves.

16.5.2. Application to uniaxial materials

The most frequent applications of crystal birefringence involve uniaxial crystals, for which the index surface consists of a sphere and an ellipsoid of revolution around the optic axis (parallel to the 3-, 4- or 6-fold axis of the crystal). The sphere and the ellipsoid are tangent to each other at their intersection with the optical axis (Fig. 16.3).

Several examples will be discussed as a function of the relative orientations of the plane of incidence and the optic axis of the crystal. We assume, for simplicity, that material 1 is isotropic, with index n_1, which makes it easy to draw the index surface related to the incident wave-vector, while retaining the full generality of the argument.

We take as the unit for the wave-vectors the value k_0 of the wave-vector in vacuum.

We recall that the plane of incidence is the plane containing the incident wave-vector and the normal to the interface. We define the principal section plane as the plane containing the optic axis and the normal to the entrance face of the crystal, *i.e.* the interface between the crystal and material 1, through which light enters the crystal.

(a) The optic axis is perpendicular to the plane of incidence

The index surface then intersects the plane of incidence along two concentric circles, with radii n_o and n_e (Fig. 16.9a). We obtain two different wave-vectors in the crystal. Since the rays are parallel to the normal to the index surface, the rays are, in this special case, parallel to the wavevectors. The directions of the corresponding induction vectors are obtained by recalling (see Sect. 16.4.5) that, in a uniaxial material, the ordinary vibration is perpendicular to the optic axis.

(b) The optic axis is parallel to the plane of incidence and to the entrance face

The plane of incidence intersects the index surface along a circle for the ordinary surface, and an ellipse for the extraordinary surface (Fig. 16.9b). There are two distinct wave-vectors in the crystal: $\mathbf{k}_2 = \mathbf{OM}_2 = \mathbf{k}_o$ for the ordinary ray, and $\mathbf{k}'_2 = \mathbf{OM}'_2 = \mathbf{k}_e$ for the extraordinary ray.

The ordinary ray is parallel to the ordinary wave-vector, while the extraordinary ray is parallel not to the extraordinary wave-vector but to the normal to the index surface at M'_2 (vector $\mathbf{u}$ on Fig. 16.9b).

(c) The optic axis is parallel to the plane of incidence and at any angle to the entrance face

This case is represented on Figure 16.9c. The ordinary ray is parallel to the ordinary wave-vector. The extraordinary ray is parallel to the normal at M'_2 to the index surface (vector $\mathbf{u}$ on the figure).

In case (a), the principal section plane is perpendicular to the plane of incidence, while it is parallel to it in cases (b) and (c). In the most general case, the plane of incidence and the principal section plane enclose any angle. This situation is more difficult to represent, because the extraordinary ray is no more in the plane of incidence.

16.6. Conclusion

We showed, in this chapter, that crystals which are anisotropic with respect to light propagation (*i.e.* all crystals except for cubic crystals) are birefringent: an incident plane wave, with given wave-vector, gives rise in the crystal to two waves propagating with two different indices and different wave-vectors. For each wave, the induction vector $\mathbf{D}$ is perpendicular to the wave-vector. At the interface between two materials, the tangential component of the wave-vector is conserved, hence a simple construction provides, for a given incident wave, the direction of the wave-vectors in the crystal. The corresponding rays are in general not parallel to the wave-vectors, but they are normal to the index surface.

Most applications use uniaxial birefringent crystals (tetragonal or rhombohedral or hexagonal). The index surface then consists of a sphere and an ellipsoid of revolution, tangent to each other at their intersection with the optical axis. The induction vector $\mathbf{D}_o$ of the ordinary wave is perpendicular to the optic axis and the induction vector $\mathbf{D}_e$ of the extraordinary wave is in the plane containing the extraordinary wave-vector and the optical axis.

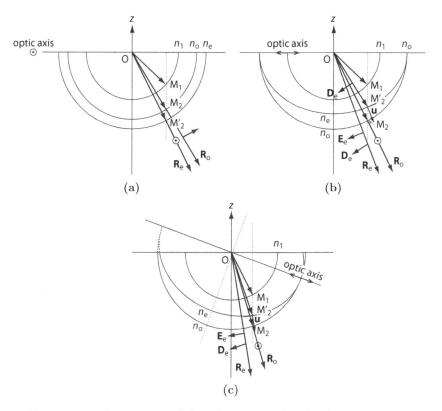

Figure 16.9: Construction of the refracted rays for a birefringent uniaxial crystal, using the index surface. The incident wave-vector is $k_1 = OM_1$ and the refracted wave-vectors are $k_o = k_2 = OM_2$ for the ordinary wave, and $k_e = k'_2 = OM'_2$ for the extraordinary wave. Vectors D_o and E_o are always parallel. (a) The plane of incidence is perpendicular to the optic axis. Vectors D_e and E_e are parallel. (b) The plane of incidence is parallel to the optic axis and the optic axis is in the plane of the surface. (c) The plane of incidence is parallel to the optic axis, which encloses a non-zero angle with the surface of the crystal. In cases (b) and (c), we note that the extraordinary ray R_e is parallel to the normal to the index surface at point M'_2, and that D_e is perpendicular to wave-vector k_e (while E_e is perpendicular to the ray R_e).

In this chapter, we assumed the permittivity of crystals to be real, which means that the crystals are not absorbing. This assumption of perfectly transparent crystals will be retained in all what follows. We make an exception for Figure 16.10, which shows a commonplace evidence of the anisotropic optical properties of zinc crystals, which are often large and conspicuous on zinc-plated metal surfaces, such as lamp poles or sign posts. The very fact that the crystals are visible with the naked eye is due to the dependence of the reflectivity on the crystal orientation of the surface.

Figure 16.10: A lamp pole surface, where the crystals comprising the zinc plating are conspicuous because of the anisotropic reflectivity of uniaxial zinc. The ribbon in the lower half is a centimeter scale.

The contents of this chapter make it possible to solve all problems dealing with light propagation in anisotropic crystals. We therefore thought it unnecessary to make the presentation more cumbersome by adding another approach, which uses the wave surface and the appropriate geometrical construction. This alternative presentation is given in Appendix 16A for the sake of those readers who are used to this approach for isotropic materials and would prefer to use it for anisotropic crystals too.

16.7. Exercises

Throughout the exercises for this chapter, the prisms are assumed to be right prisms having a right triangle with apical angle θ for their basis. For simplicity, they are called prisms with angle θ. The figures show the plane of incidence, which is always perpendicular to the lateral edges of the prism.

Exercise 16.1

Calculate the maximum value of the angle δ between the direction of the wave-vector and that of the associated ray in a uniaxial crystal with indices n_o and n_e. Find the numerical value for quartz, with $n_o = 1.544$ and $n_e = 1.553$.

Exercise 16.2

Consider a parallel-faced slab of calcite (a uniaxial negative crystal, meaning $n_e < n_o$), cut so that the optic axis is at 45° to the normal to the plate. The figure below shows the principal section plane of the plate, *i.e.* the plane containing the normal to the plate and the optic axis. The plate is illuminated by a beam of natural light, monochromatic and perpendicular to the plane of the plate.

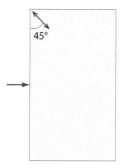

1. Draw the path of the light rays in the plate and after the plate. Indicate their polarization state.

2. Calculate the deviation Δ between the ordinary and extraordinary rays emerging from the plate, if the thickness e of the plate is 1 cm and the ordinary and extraordinary indices of calcite are respectively 1.6584 and 1.4865.

Exercise 16.3

A transparent uniaxial crystal is cut into a prism with apical angle $\theta = 30°$ (see figure below). The optic axis is parallel to the face containing AB and perpendicular to the lateral edge of the prism.

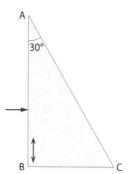

1. This prism is placed on a goniometer so that the entrance face for the beam is the face containing AB, and that this face is perpendicular to the incident beam. Draw the propagation directions of the beams in the prism

and at the exit of the prism. The deviation δ_o by the prism for the ordinary beam is 26.02°, and the deviation δ_e of the extraordinary beam is 18.01°. Find the values of n_o and n_e.

2. The prism is rotated around an axis parallel to its lateral side so that the entrance face is the one containing AC, and that it is normal to the incident beam. Draw the ray paths in the prism. Calculate the numerical values of the angular deviations produced by the prism for the ordinary and extraordinary rays.

Exercise 16.4

A Glan-Foucault polarizer consists of two identical prisms with angle θ, with their optic axes parallel to the lateral edges. The prisms are separated by a thin air layer (see the figure below, where the thickness of this layer is exaggerated).

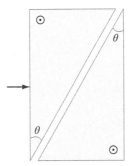

The prisms are made of calcite ($n_o = 1.658$ and $n_e = 1.486$). The device is illuminated by a monochromatic, unpolarized beam perpendicular to the entrance face. Calculate the range of angles θ such that there is only one beam at the exit of the device. Draw the ray paths in this case, and determine the polarization of the outgoing beam.

Exercise 16.5

A calcite prism ABC, with lateral sides parallel to the optic axis, is associated, as shown on the figure below, with a prism ACD made of glass with index n. The device is illuminated at normal incidence by a monochromatic beam.

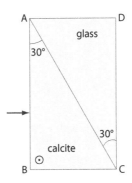

1. Draw the beam paths inside the two-prism system. Indicate the polarization state of each of the two waves at the exit of the two prisms.

2. Calculate the angle α between the two outgoing beams in the case of calcite, where $n_o = 1.6584$ and $n_e = 1.4865$ assuming the glass has index $n = 1.55$.

Exercise 16.6

A Rochon prism consists of two calcite prisms (ABC and ACD on the figure below), with angle 30°, placed at optical contact so that they form one composite parallel-faced slab. In prism ABC, the optic axis is perpendicular to the entrance face AB. In the second prism, the optic axis is parallel to the lateral edges. The device is illuminated at normal incidence by a monochromatic beam.

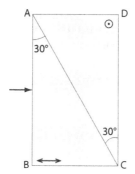

1. Draw the ray paths through the two-prism system. Indicate the polarization state for each of the two waves at the exit of the two prisms.

2. Calculate the angle α between the two outgoing beams in the case of calcite, for which $n_o = 1.6584$ and $n_e = 1.4865$.

Appendix 16A. Wave surface (or ray surface) and Huygens' construction

16A.1. Wave surface (or ray surface)

It is defined by considering a point source of electromagnetic wave at O. The wave surface is the locus of the points reached by the wave after a given time t, *i.e.* the equal-phase surface. It is thus the locus of points M such that $\mathbf{OM} = v_S\mathbf{u}$ where v_S is the propagation velocity of energy in the direction of unit vector $\mathbf{u}$, parallel to the Poynting vector $\mathbf{S}$.

We first search for the relation between the phase velocity v of the wave with wave-vector $\mathbf{k}$ in the crystal and the energy propagation velocity v_S for the same wave. A wave plane P (Fig. 16A.1) moves perpendicular to the wave-vector at velocity v, the phase velocity. After a second, it is at P′, a distance v from P. During the same time interval, energy propagates from A to B along the direction of the Poynting vector $\mathbf{S}$, with unit vector $\mathbf{u}$. Distance AB is therefore equal to v_S. Let φ be the angle between vectors $\mathbf{k}$ and $\mathbf{S}$. We see that

$$v_S = \frac{v}{\cos\varphi}. \tag{16A.1}$$

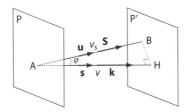

Figure 16A.1: Propagation of a wave plane from P to P′ over a time interval of one second. The wave-vector $\mathbf{k}$ is normal to P and P′.

We show that the wave surface can be easily deduced from the index surface because the equation describing the wave surface has a form similar to the equation yielding the index surface.

Consider Equation (16.10a) which we reproduce below, and cross-multiply both of its sides with $\mathbf{u}$, using for the left-hand side relation (16.14) for the double cross product:

$$\mathbf{k} \times \mathbf{H} = -\omega\mathbf{D} \tag{16.10a}$$

$$\mathbf{u} \times \mathbf{k} \times \mathbf{H} = \mathbf{k}(\mathbf{u} \cdot \mathbf{H}) - \mathbf{H}(\mathbf{u} \cdot \mathbf{k}) = -\mathbf{H}(\mathbf{u} \cdot \mathbf{k})$$

because $\mathbf{u}$ is perpendicular to $\mathbf{H}$. We deduce that:

$$\mathbf{u} \times \mathbf{D} = -\frac{1}{\omega}(\mathbf{u} \times \mathbf{k} \times \mathbf{H}) = \frac{1}{\omega}(\mathbf{u} \cdot \mathbf{k})\mathbf{H} = \frac{k}{\omega}(\mathbf{u} \cdot \mathbf{s})\mathbf{H} = \frac{1}{v}\cos\varphi\mathbf{H}$$

and, using (16A.1),

$$\mathbf{u} \times \mathbf{D} = \frac{1}{v_S}\mathbf{H} \quad \text{and} \quad \mathbf{H} = v_S(\mathbf{u} \times \mathbf{D}). \tag{16A.2}$$

The same process is applied to Equation (16.10b):

$$\mathbf{k} \times \mathbf{E} = \mu_0\omega\mathbf{H} \tag{16.10b}$$

$$\mathbf{u} \times \mathbf{k} \times \mathbf{E} = \mathbf{k}(\mathbf{u} \cdot \mathbf{E}) - \mathbf{E}(\mathbf{u} \cdot \mathbf{k}) = -k\mathbf{E}(\mathbf{u} \cdot \mathbf{s})$$

because $\mathbf{u}$ is perpendicular to $\mathbf{E}$.

$$\mathbf{u} \times \mathbf{H} = \frac{1}{\mu_0\omega}(\mathbf{u} \times \mathbf{k} \times \mathbf{E}) = -\frac{k}{\mu_0\omega}(\mathbf{u} \cdot \mathbf{s})\mathbf{E} = -\frac{\cos\varphi}{\mu_0 v}\mathbf{E} = -\frac{1}{\mu_0 v_S}\mathbf{E}$$

and
$$\mathbf{E} = -\mu_0 v_S(\mathbf{u} \times \mathbf{H}). \tag{16A.3}$$

Using (16A.3), then (16A.2), we obtain:

$$\mathbf{E} = -\mu_0 v_S^2(\mathbf{u} \times \mathbf{u} \times \mathbf{D}) = -\mu_0 v_S^2[\mathbf{u}(\mathbf{u} \cdot \mathbf{D}) - \mathbf{D}]$$

so that
$$\mathbf{E} = \mu_0 v_S^2[\mathbf{D} - \mathbf{u}(\mathbf{u} \cdot \mathbf{D})] \tag{16A.4}$$

which can be compared to Equation (16.15a), which we repeat:

$$\mathbf{D} = \varepsilon_0 n^2[\mathbf{E} - \mathbf{s}(\mathbf{s} \cdot \mathbf{E})]. \tag{16A.5}$$

These two equations have the same form. This is even more conspicuous if we rewrite them in order to show $\mathbf{E}$ and $\mathbf{D}/\varepsilon_0$ instead of $\mathbf{E}$ and $\mathbf{D}$:

$$\mathbf{E} = \frac{v_S^2}{c^2}\left[\frac{\mathbf{D}}{\varepsilon_0} - \mathbf{u}\left(\mathbf{u} \cdot \frac{\mathbf{D}}{\varepsilon_0}\right)\right] \tag{16A.6}$$

and
$$\frac{\mathbf{D}}{\varepsilon_0} = n^2[\mathbf{E} - \mathbf{s}(\mathbf{s} \cdot \mathbf{E})]. \tag{16A.7}$$

Equation (16A.7) relates $\mathbf{D}/\varepsilon_0$ to $\mathbf{E}$ by introducing the index n for the wave with wave-vector $\mathbf{k}$ parallel to unit vector $\mathbf{s}$ and perpendicular to $\mathbf{D}$. The surface $n = f(\mathbf{s})$ is the index surface. On the other hand, $\mathbf{D}/\varepsilon_0$ and $\mathbf{E}$ are related by the permittivity tensor $[\varepsilon]$.

Equation (16A.6) relates $\mathbf{E}$ to $\mathbf{D}/\varepsilon_0$ by introducing the propagation velocity of energy v_S/c (i.e. this velocity expressed in units of c, the velocity of light in vacuum) in the direction of Poynting's vector $\mathbf{S}$, parallel to unit vector $\mathbf{u}$ and perpendicular to $\mathbf{E}$. The surface $v_S/c = f(\mathbf{u})$ is the wave surface. On the other hand $\mathbf{E}$ and $\mathbf{D}/\varepsilon_0$ are related through the impermeability tensor $[\eta]$.

Thus Equation (16A.6) deduces from (16A.7), which was used throughout this chapter, by replacing $\mathbf{D}/\varepsilon_0$ by $\mathbf{E}$, $\mathbf{E}$ by $\mathbf{D}/\varepsilon_0$, $\mathbf{s}$ by $\mathbf{u}$ and n by v_S/c. Although they do not appear directly in the equations, it is clear that tensor $[\varepsilon]$ is replaced by $[\eta]$.

This correspondence makes it possible to deduce the equation of the wave surface from that of the index surface, given in (16.18) and expanded in (16.19):

$$\sum_i \frac{u_i^2 \eta_i}{\eta_i - \frac{v_S^2}{c^2}} = 0$$

and

$$u_1^2 \eta_1 \left(\eta_2 - \frac{v_S^2}{c^2} \right) \left(\eta_3 - \frac{v_S^2}{c^2} \right) + u_2^2 \eta_2 \left(\eta_3 - \frac{v_S^2}{c^2} \right) \left(\eta_1 - \frac{v_S^2}{c^2} \right)$$

$$+ u_3^2 \eta_3 \left(\eta_1 - \frac{v_S^2}{c^2} \right) \left(\eta_2 - \frac{v_S^2}{c^2} \right) = 0.$$

Since $\eta_i = 1/\varepsilon_i$, the wave surface has the same characteristics as the index surface provided n_i is replaced by $1/n_i$, if the quantity represented is v_S/c (or provided n_i is replaced by c/n_i if the quantity represented is v_S).

We thus obtain, for uniaxial crystals, a wave surface consisting of two sheets: a sphere with radius $1/n_o$ (the ordinary wave surface), and an ellipsoid (the extraordinary wave surface), with equation:

$$\frac{x_1^2 + x_2^2}{\eta_3} + \frac{x_3^2}{\eta_1} = 1$$

or

$$\frac{x_1^2 + x_2^2}{\frac{1}{n_e^2}} + \frac{x_3^2}{\frac{1}{n_o^2}} = 1$$

if the axis of order larger than 2 (the optic axis) is chosen as Ox_3.

This is an ellipsoid of revolution around Ox_3, tangent to the ordinary sphere (Fig. 16A.2). We showed that the normal to the index surface is parallel to Poynting's vector with unit vector $\mathbf{u}$. In the same way, the normal to the wave surface is parallel to unit vector $\mathbf{s}$, hence to the wave-vector $\mathbf{k}$, and the induction vector $\mathbf{D}$ is then tangent to the wave surface. Figure 16A.2 shows the wave surface for a uniaxial material, positive in (a), negative in (b).

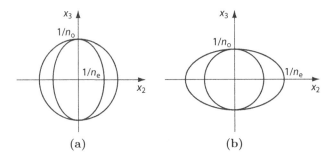

Figure 16A.2: Wave surfaces for uniaxial materials when the represented quantity is v_S/c: (a) Uniaxial positive; (b) Uniaxial negative

16A.2. Huygens' construction

It rests on the Huygens principle, which can be expressed on the basis of envelope waves. Consider a wave surface Σ_1 in a given material. One second later, this wave surface has become surface Σ_2, which is the envelope of the wave surfaces emitted by each point of Σ_1.

This result applies easily to the transmission of a plane wave from one isotropic material with index n to an anisotropic material. We recall that a wave plane is a wave surface for a point emitter located at infinity. Let $\mathbf{AO}$ be the incident ray direction in the isotropic material, and P the wave plane. Let Δ be the line along which the incident wave plane intersects the interface, one second after it went through O. The distance from Δ to plane P is equal to c/n if the isotropic material has index n. The wave plane in the crystal one second after it passed through O goes through line Δ and is tangent to the wave surface emitted at O (Fig. 16A.3). We thus obtain two wave planes, tangent respectively at M and M' to the two sheets Σ and Σ' of the wave surface. The rays propagate in the crystal along $\mathbf{OM}$ and $\mathbf{OM'}$. In a completely general case, points M and M' are not in the plane of incidence represented on the figure.

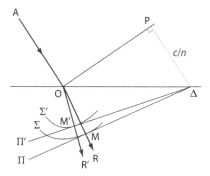

Figure 16A.3: Use of the wave surface to represent the transmission of a plane wave from an isotropic material into an anisotropic material (Huygens' construction). We draw the wave planes Π and Π' going through line Δ and tangent to the two sheets of the surface. Let M and M' be the contact points. The rays are given by directions $\mathbf{OM}$ and $\mathbf{OM'}$ whereas the wave-vectors are perpendicular to the wave planes.

Figure 16A.4 shows the application of this method in two special cases of uniaxial materials. In (a), the optic axis is parallel to the surface and to the plane of incidence. In (b), it has a general direction but the incident ray is perpendicular to the surface.

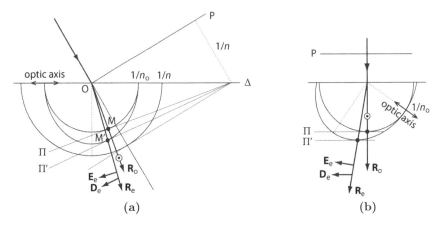

(a)   (b)

Figure 16A.4: Use of Huygens' construction to represent the transmission of a plane wave from an isotropic material with index n to an anisotropic uniaxial material (indices n_o and n_e) in the case when the optic axis is in the plane of incidence. The speed unit is here c as in Figure 16A.2. R_o and R_e are the ray directions for ordinary and extraordinary rays, respectively. (a) The optic axis is in the plane of the interface. (b) The optic axis is not in the plane of the interface, and the incident ray is perpendicular to the interface. Line Δ is then at infinity.

Chapter 17

Polarization of light
by crystals

In this chapter, we show how crystal birefringence can be used to change the polarization of an electromagnetic wave and thus obtain light in a well-defined polarization state. Various devices using this birefringence are described: linear polarizers and phase shifting plates.

17.1. Polarization state of an electromagnetic wave

We first define the polarization state for a plane and monochromatic electromagnetic wave propagating in vacuum. We consider the electric field vector rather than the magnetic field vector because the interaction of this wave with a material, assumed here to be non magnetic (Sect. 16.1), occurs mainly through the electric field. The electric field of the plane wave in vacuum is perpendicular to the wave-vector. Therefore, if we choose a coordinate system where Oz is parallel to the propagation direction, the electric field has only two components, E_x and E_y:

$$E_x = E_1 \cos(kz - \omega t + \alpha) \tag{17.1}$$
$$E_y = E_2 \cos(kz - \omega t + \beta) \tag{17.2}$$

The values of α and β depend on the origins of z and t chosen. Through an appropriate change in either of these origins, we can always obtain:

$$E_x = E_1 \cos(kz - \omega t) \tag{17.3}$$
$$E_y = E_2 \cos(kz - \omega t + \varphi) \qquad \text{where} \qquad \varphi = \beta - \alpha. \tag{17.4}$$

Because the wave is plane, it is sufficient to represent the electric field **E** along the propagation direction, Oz. Various cases will be investigated as a function of the value of φ.

© Springer Science+Business Media Dordrecht 2014
C. Malgrange et al., *Symmetry and Physical Properties of Crystals*,
DOI 10.1007/978-94-017-8993-6_17

17.1.1. Linearly polarized wave

Assume $\varphi = 0$ or π. Vector $\mathbf{E}$ remains parallel to itself, and encloses with axis Ox an angle θ such that $\tan\theta = E_2/E_1$ if $\varphi = 0$, and $\tan\theta = -E_2/E_1$ if $\varphi = \pi$ (Fig. 17.1). The polarization is called linear or rectilinear. At a given point on axis Oz, the end of vector $\mathbf{E}$ oscillates sinusoidally around O.

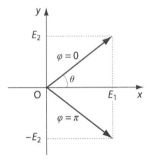

Figure 17.1: Linear polarizations obtained when the phase shift φ between the two perpendicular components is equal to 0 or to π

17.1.2. Circularly polarized wave

Assume $\varphi = \pm\pi/2$ and $E_1 = E_2$:

$$E_x = E_1 \cos(kz - \omega t) = E_1 \cos\psi$$
$$E_y = \mp E_1 \sin(kz - \omega t) = \mp E_1 \sin\psi$$

if we set $\psi = kz - \omega t$.

The projection m onto a plane perpendicular to Oz of the end M of the electric field vector describes a circle with radius E_1 since $E_x^2 + E_y^2 = E_1^2$.

At a given time t, ψ varies linearly with z and the end of the electric field vector describes a helix with axis Oz. If we consider the upper sign (corresponding to $\varphi = \pi/2$), the helix is left-handed and we will say that the polarization is left-circular. If $\varphi = -\pi/2$, the helix is right-handed[1] and the polarization is right-circular (Fig. 17.2). We note that a right-handed helix is transformed into a left-handed helix by a mirror.

At a given point of the propagation axis Oz (z fixed), the end of $\mathbf{E}$ describes, in the course of time, a circle. Consider, for simplicity, the situation at $z = 0$.

---

1. The screws used in mechanical applications are, except for highly exceptional cases, right-handed helices. The same applies to cork-screws. These right-handed screws and cork-screws penetrate into the material if they are rotated clockwise ("to the right") as seen by an operator facing the material.

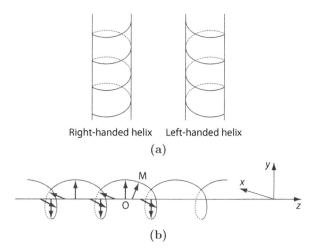

Right-handed helix   Left-handed helix

(a)

(b)

Figure 17.2: (a) Representation of a right-handed helix and a left-handed helix for which the envelope is materialized by the generators of a cylinder; (b) representation of the variation, along the propagation direction Oz, of the electric field vector of a right-circular wave at a given time. The locus of the ends of these vectors is a right-handed helix.

Right-circular polarization ($\varphi = -\pi/2$ and the lower sign for E_y) yields:

$$E_x = E_1 \cos \omega t$$
$$E_y = -E_1 \sin \omega t.$$

Thus, an observer located at a point with $z > 0$ (downstream along the propagation direction, hence looking at the source) sees the polarization vector **OM** rotating clockwise (Fig. 17.3a). Conversely, for left circular polarization, the same observer sees the polarization vector **OM** rotating counterclockwise (Fig. 17.3b).

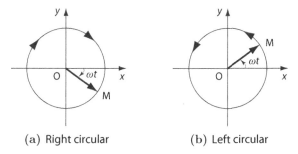

(a) Right circular   (b) Left circular

Figure 17.3: Representation as a function of time of the end M of the electric field vector at a given point for a right (a) and a left (b) circular wave. Axis Oz points in the propagation direction.

17.1.3. Elliptically polarized wave

Consider the most general case, where E_1, E_2 and φ have any values. We obtain the curve described by m, the projection of the end M of the electric field vector on the plane xOy, if we eliminate $(kz - \omega t)$ between Equations (17.3) and (17.4). We obtain:

$$\frac{E_x^2}{E_1^2} + \frac{E_y^2}{E_2^2} - \frac{2E_xE_y\cos\varphi}{E_1E_2} = \sin^2\varphi. \tag{17.5}$$

This is an ellipse, the axes of which are Ox and Oy if $\cos\varphi = 0$, hence for $\varphi = \pm\pi/2$. For other values of φ, it is easily shown that the axes of the ellipse enclose with axes Ox and Oy respectively the angle θ such that:

$$\tan 2\theta = \frac{2E_1E_2\cos\varphi}{E_1^2 - E_2^2}.$$

If $E_1 = E_2$, θ equals $\pm\pi/4$ for all values of φ. Figure 17.4 shows the trajectory described by the end of the electric field vector at a given point as a function of time, for various values of φ.

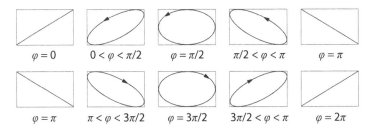

Figure 17.4: Curves described by the end of the electric field vector of a plane wave, at a given point, as a function of time, for various values of the phase shift φ between the two orthogonal components. The propagation direction points toward the observer.

Note: if we consider an electromagnetic wave in a birefringent crystal, the vector perpendicular to the wave-vector which characterizes the polarization state is the induction vector, and all the above results apply to the components D_x and D_y of induction.

17.1.4. Natural light

Classical light sources are made up of atoms oriented in all directions, with each of them emitting polarized light. The waves emitted by the various atoms are mutually incoherent. This is called "natural light" or "unpolarized light". This light is represented as the sum of two incoherent waves, each of them linearly polarized, with their polarizations in any two perpendicular directions. Usually, light is not completely unpolarized: it is partly polarized.

The polarization state is then conveniently described by the Stokes parameters which are defined in Section 17.5.

Here we consider totally polarized light, defined by its two orthogonal components, and sometimes totally unpolarized natural light.

17.2. Jones notation

Equations (17.3) and (17.4) can also be written as:

$$E_x = E_1 \Re \left[\exp i(kz - \omega t)\right] = \Re \left[E_1 \exp i(kz - \omega t)\right]$$
$$E_y = E_2 \Re \left[\exp i(kz - \omega t + \varphi)\right] = \Re \left[E_2 \exp i\varphi \exp i(kz - \omega t)\right]$$

where $\Re$ means "real part of".

The type of polarization depends on the values of E_1, E_2 and φ. The polarization of the wave is thus perfectly determined by the complex components:

$$E_x = E_1$$
$$E_y = E_2 \exp i\varphi$$

where φ is the phase shift between E_x and E_y.

In what follows, we will use this complex notation, called the Jones notation, invented by R. Clark Jones in 1941. The Jones vector of the previous field is then:

$$\begin{pmatrix} E_1 \\ E_2 \exp i\varphi \end{pmatrix}.$$

The amplitude of the electric field is equal to the modulus of this complex vector. If needed, the exact form of the wave is obtained by multiplying these complex components by $\exp i(kz - \omega t)$ and by taking their real parts.

A few examples of Jones vectors are given below, with a and b real numbers:

– Linear polarization: $\begin{pmatrix} a \\ b \end{pmatrix}.$

  The angle θ between the electric field and axis Ox is such that $\tan \theta = b/a$.

– Right circular polarization: $\begin{pmatrix} a \\ -ia \end{pmatrix}.$

– Left circular polarization: $\begin{pmatrix} a \\ ia \end{pmatrix}.$

  These Jones vectors for right circular and left circular polarization would have to be interchanged if the waves were described in the form $\exp i(\omega t - kz)$. In contrast, the rule given by Figure 17.3 to determine whether the polarization is right-handed or left-handed remains valid whatever the choice made for describing the wave.

– General polarization: $\begin{pmatrix} a \\ b \exp i\varphi \end{pmatrix}$.

17.3. Linear polarizers

A linear polarizer changes any incident light into a linearly polarized light, by acting differently upon the two orthogonal directions of linear polarization. One category of linear polarizers consists of materials which absorb much more efficiently one polarization direction than the perpendicular direction. This is, for example, the case of Polaroid® sheets, consisting of long polymer molecules, oriented by the application of a uniaxial stress and maintained in this orientation by enclosing the polymer in a more rigid plastic foil coating. These special molecules absorb the polarization which is parallel to their long side, and transmit the polarization which is perpendicular.

The other category splits the incident light into two beams with mutually orthogonal linear polarizations, retaining only one of the beams. One method is based on reflection at Brewster's angle. If light with any polarization state falls under this incidence angle i_B on a material with index n, glass for example, the component polarized parallel to the plane of incidence is not reflected. The other component, polarized perpendicular to the plane of incidence, is reflected, but only partly because part of the incident intensity is transmitted. This Brewster angle is such that $\tan i_B = n$.

The other, more efficient method, uses the birefringence properties of crystals. There are two types of such birefringent devices. The first type uses total reflection of one of the two beams (ordinary (o) or extraordinary (e)) on the exit face of a prism, and thus allows only one of the two linearly polarized beams to go through. In order for the outgoing beam not to be deviated, two prisms are mounted next to each other, separated by a thin air layer as in the Glan-Foucault prism (Fig. 17.5a and Exercise 16.4) or by an appropriate cement as in the Nicol prism, the oldest device, invented by William Nicol in 1828 (Fig. 17.5b), and described below. The other type of polarizing prism allows both beams to go through, but they diverge at the exit, thus yielding two distinct, linearly polarized beams (Fig. 17.5c and 17.5d, and Exercise 16.6).

The Nicol prism (sometimes called nicol) is an assembly of two calcite prisms (calcite is one of the crystal forms of calcium carbonate $CaCO_3$, also called Iceland spar because the first crystals were found in Iceland), cut in a special way. Calcite has point group $\bar{3}m$. It is therefore uniaxial, and the optic axis is the 3-fold axis. Calcite crystals feature cleavage planes deduced from one another by the 3-fold symmetry around the optic axis. An Iceland spar crystal is cleaved along these three planes, forming a parallelepiped in which two sides (AB and AD) are equal and the third (AA') is approximately three times longer than the two first sides (Fig. 17.6a). These relationships were calculated so

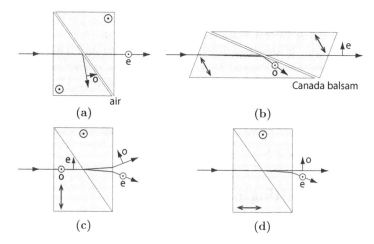

Figure 17.5: Schematic drawing of polarizing devices: (a) Glan-Foucault prism; (b) Nicol prism; (c) Wollaston prism; (d) Rochon prism. These prisms are made of calcite, a uniaxial negative crystal.

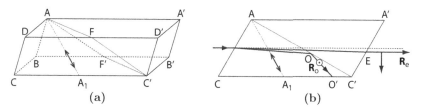

Figure 17.6: Nicol prism: (a) overall view; (b) principal section plane $AA'C'C$ (perpendicular to the entrance face $ABCD$ and containing the optical axis AA_1). Only the extraordinary ray emerges from the exit surface. It is slightly shifted with respect to the incident beam, the direction of which is shown by a dotted line.

that the diagonal plane $AA'C'C$ be perpendicular to face $ABCD$ used as the entrance face, and that it contain the optic axis AA_1. Plane $AA'C'C$ is thus the principal section plane.

The crystal is sawed along plane $AFC'F'$, perpendicular to the principal section plane. The two halves are then cemented together using a thin layer of Canada balsam, for which the index N is 1.55, a value between the ordinary index of calcite $n_o = 1.6584$ and its extraordinary index $n_e = 1.4864$ (Fig. 17.6b).

The incident beam is directed parallel to the large side AA' of the prism, so that the incidence plane coincides with the principal section plane. The rays are drawn as in Figure 16.9c. The ordinary ray can be shown by calculation to reach AC′ with incidence angle 75.7°, a value larger than the limiting angle,[2] 69.2°, corresponding to transition from the ordinary index of calcite to that of Canada balsam. It therefore undergoes total reflection. The index n for the extraordinary wave is equal to 1.53 (Exercise 16.5) for this orientation of the wave-vector. The extraordinary ray goes through the Canada balsam, then through the second prism, and emerges parallel to the incident beam and slightly shifted. The extraordinary ray is polarized parallel to the principal section plane. It is easy to rotate this polarization, just by rotating the prism around the incident beam.

17.4. Phase-shifting plates

Birefringent plates are used to introduce a phase shift between two perpendicular components of an electromagnetic wave. Consider a uniaxial crystal plate, with parallel faces, cut so that the optic axis is in the plane of the plate (Fig. 17.7).

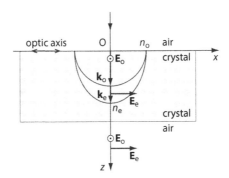

Figure 17.7: Propagation of the ordinary and extraordinary rays through a phase-shifting plate, a uniaxial plate with the optic axis parallel to its faces

Consider an incident beam perpendicular to the plate. The propagation geometry is then very simple. The ordinary and extraordinary wave-vectors are normal to the plate surface. The same applies for the directions of the ordinary and extraordinary rays in the crystal, and also in air at the exit of the plate. The index for the extraordinary ray is equal to n_e. For this simple geometry where the ordinary and extraordinary rays are parallel to the corresponding wave-vectors, the electric field vector $\mathbf{E}$ and induction vector $\mathbf{D}$ are parallel, and the polarization vector can be defined by either of these vectors: we choose $\mathbf{E}$. The polarization of the ordinary ray is perpendicular to the optic axis and that of the extraordinary ray is parallel to it. These two orthogonal

---

2. We recall that the limiting angle i_l for incidence of a wave on the surface separating a first material, with index n_1, from a second material, with index $n_2 < n_1$, corresponds to refraction angle 90° ($\sin i_l = n_2/n_1$). If the incidence angle is $\geq i_l$, there is no refracted beam: the wave is totally reflected.

directions are called the neutral lines of the plate. If we choose Oz parallel to the propagation direction and Ox parallel to the optic axis, the neutral lines of the plate are axes Ox and Oy. Consider an incident plane wave, the electric field of which on the entrance face of the crystal ($z = 0$) has complex components (Sect. 17.2):

$$\begin{pmatrix} E_1 \exp -i\omega t \\ E_2 \exp i\varphi \exp -i\omega t \end{pmatrix}.$$

In the plate, this wave gives rise to the ordinary wave, with polarization parallel to Oy, with index n_o, with the norm of its wave-vector $k\,n_o$, and to the extraordinary wave, with polarization parallel to Ox, with index n_e, and with the norm of its wave-vector $k\,n_e$. We recall that $k = \omega/c$ is the modulus of the wave-vector in vacuum. At a distance z from the entrance face, the wave has changed into:

$$E_x = E_1 \exp i(k\,n_e z - \omega t)$$
$$E_y = E_2 \exp i\varphi \exp i(k\,n_o z - \omega t).$$

At the exit of the plate with thickness e, the components E_y and E_x have an additional phase shift ψ such that $\psi = k(n_o - n_e)e$, which can also be written:

$$\psi = \frac{2\pi}{\lambda}(n_o - n_e)e.$$

At the exit, the wave is given in Jones notation by:

$$\begin{pmatrix} E_1 \\ E_2 \exp i\varphi \exp i\psi \end{pmatrix}.$$

The sign of ψ depends on the relative values of n_o and n_e, or more precisely on the indices for each of the neutral lines. We thus distinguish the fast neutral line, parallel to the electric field which propagates with the larger velocity and therefore the smaller index, and conversely the slow neutral line (smaller velocity and larger index). If Oy is the fast line, then the phase shift ψ is negative.

17.4.1. Half-wave plates

These are plates where the thickness is such that the phase shift ψ is equal to $(2p + 1)\pi$, with p an integer.

Assume the wave incident on the plate is linearly polarized, with components a and b along two orthogonal axes Ox and Oy, respectively parallel to the neutral lines of the plate.

State before the plate: $\begin{pmatrix} a \\ b \end{pmatrix}$.

State after the plate: $\begin{pmatrix} a \\ b \exp i\pi \end{pmatrix} = \begin{pmatrix} a \\ -b \end{pmatrix}$.

The vibration which, at the entrance to the plate, enclosed an angle θ with axis Ox such that $\tan\theta = b/a$ now encloses an angle $-\theta$ (Fig. 17.1).

A half-wave plate transforms a rectilinear vibration into another rectilinear vibration, symmetric of the first one with respect to its neutral lines.

17.4.2. Quarter-wave plates

These are plates whose thickness is such that the phase shift is equal to $(p + 1/2)\pi$ with p an integer, $i.e.$ $\pm\pi/2$ modulo 2π. These plates make it possible to change linear polarization into circular polarization. Consider linearly polarized light (natural light after passage through a linear polarizer) such that its polarization direction is at $45°$ to the neutral lines of the plates, chosen as the Ox and Oy axes.

The state before the plate is: $\begin{pmatrix} a \\ a \end{pmatrix}$

and after the plate: $\begin{pmatrix} a \\ a\exp -i\pi/2 \end{pmatrix} = \begin{pmatrix} a \\ -ia \end{pmatrix}$,

if the neutral line parallel to Oy is the fast line. After the plate, the polarization is right circular (Fig. 17.8).

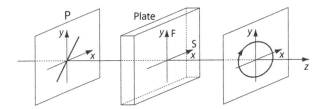

Figure 17.8: Schematic representation of a device providing right circular light starting from any incident polarization (generally an unpolarized beam). P is a linear polarizer. The plate is a quarter-wave plate. The fast F and slow S neutral lines are at an angle of $45°$ to the linear polarization transmitted by P.

A left circular polarization would be obtained:

- either by rotating the polarizer by $90°$

- or by leaving the polarizer in the same orientation and by rotating the plate by $90°$, which exchanges the slow and fast axes of the plate and now introduces a phase shift of $\pi/2$.

Quarter-wave plates are also used to analyse circular polarization. Consider a right-handed circularly polarized wave and let it go through a quarter-wave plate and then through a linear polarizer (called here analyzer) with a polarization direction at $45°$ to its neutral lines (Fig. 17.9). We choose axes Ox and Oy respectively parallel to the fast and slow neutral lines of the quarter-wave plate.

The polarization direction of the linear analyzer makes an angle of 45° with the Ox axis. The Jones vector for the right-circularly polarized wave is:

$$\begin{pmatrix} a \\ -ia \end{pmatrix}$$

and the intensity of the wave is $2a^2$. After the quarter wave plate which induces a phase-shift of $\pi/2$ between Oy and Ox components, the wave becomes

$$\begin{pmatrix} a \\ a \end{pmatrix}.$$

It is linearly polarized with the polarization parallel to the polarization direction of the linear analyzer. The wave is completely transmitted by the polarizer and its intensity is $2a^2$. If the incident wave had been a left-handed circularly polarized wave and if it was passed through the same device (quarter-wave plate and linear polarizer at 45° to the neutral lines), the transmitted intensity would have been zero. Such a device analyzes the intensity of the right-handed circular polarization. In contrast, the whole left-handed circularly polarized wave would have been transmitted through the same kind of device but with the direction of polarization of the linear polarizer at an angle of 135° with the Ox axis.

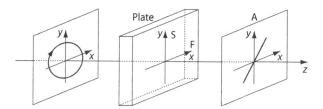

Figure 17.9: Schematic representation of a device which totally transmits right-handed circular polarized radiation and stops left-handed circular polarization

17.5. Partially polarized waves and Stokes parameters

The above sections dealt with perfectly monochromatic waves with perfectly defined polarization state, *i.e.* totally polarized waves. For these waves, the end of the electric field vector, at a given point, describes a perfectly defined curve, an ellipse in the general case. This ellipse reduces to a circle for circularly polarized waves, and to a straight line segment for linearly polarized waves. We mentioned that, in contrast, classical light sources emit totally unpolarized waves (Sect. 17.1.4). The end of the electric field vector moves randomly in the plane perpendicular to the propagation direction, and has no preferred orientation in this plane.

These two cases are the extreme situations of a totally polarized wave and a totally unpolarized wave. In many cases, we deal with a partially polarized wave, consisting of the superposition of monochromatic waves with various polarization states, obtained for instance through reflection or scattering, because these mechanisms affect polarization in a way which depends on the incidence and scattering angles.

We then define the polarization state of a wave propagating along direction Oz using the elements of the coherence matrix J:

$$J = \begin{pmatrix} J_{xx} & J_{xy} \\ J_{yx} & J_{yy} \end{pmatrix} = \begin{pmatrix} < E_x E_x^* > & < E_x E_y^* > \\ < E_y E_x^* > & < E_y E_y^* > \end{pmatrix}$$

where the angle brackets indicate that we are dealing with the average value of the quantity over all the waves in the beam.

The polarization state of the wave is now defined, using these matrix elements, through the Stokes parameters s_0, s_1, s_2 and s_3 [1] such that:

$$s_0 = J_{xx} + J_{yy} = < E_x E_x^* > + < E_y E_y^* >$$
$$s_1 = J_{xx} - J_{yy} = < E_x E_x^* > - < E_y E_y^* >$$
$$s_2 = J_{xy} + J_{yx} = 2 \Re < E_x E_y^* >$$
$$s_3 = -i(J_{xy} - J_{yx}) = 2 \Im < E_x E_y^* >$$

where $\Re$ stands for "real part of" and $\Im$ for "imaginary part of". It is clear that s_0 represents the intensity of the wave, and that s_1 represents the difference between the intensies transmitted by polarizers with polarization directions respectively parallel to Ox and Oy.

We now show that s_2 represents the difference between the intensities transmitted by linear polarizers with polarization directions at 45° and 135° to Ox respectively. For a wave with Jones vector

$$\begin{pmatrix} E_x \\ E_y \end{pmatrix},$$

the amplitudes transmitted by these polarizers are respectively:

$E_{(45°)} = (\sqrt{2}/2) (E_x + E_y)$ for the polarizer at 45° and

$E_{(135°)} = (\sqrt{2}/2) (E_x - E_y)$ for the polarizer at 135°.

The intensities are respectively:

$I(45°) = \frac{1}{2} (E_x + E_y)(E_x^* + E_y^*) = \frac{1}{2} (E_x E_x^* + E_y E_y^* + E_x E_y^* + E_x^* E_y)$

$I(135°) = \frac{1}{2} (E_x - E_y)(E_x^* - E_y^*) = \frac{1}{2} (E_x E_x^* + E_y E_y^* - E_x E_y^* - E_x^* E_y)$

$I(45°) - I(135°) = E_x E_y^* + E_x^* E_y = 2 \Re(E_x E_y^*)$

and for a partially polarized wave:

$I(45°) - I(135°) = 2 \Re < E_x E_y^* > = s_2$.

Let us now show that s_3 represents the difference between the intensities of the waves with right-circular and with left-circular polarization. We saw (Sect. 17.4.2) that the intensity of waves with right (respectively left) circular polarization can be measured using a device consisting of a quarter-wave plate and a linear polarizer with polarization direction at an angle of $45°$ (respectively $135°$) to the fast neutral line of the quarter-wave plate chosen as Ox axis. The amplitudes of the electric fields transmitted by these devices are noted $E(\pi/2, 45°)$ and $E(\pi/2, 135°)$ respectively. Similarly the transmitted intensities are noted $I(\pi/2, 45°)$ and $I(\pi/2, 135°)$ respectively. One gets:

$$E(\pi/2, 45°) = (\sqrt{2}/2)\,(E_x + iE_y) \text{ and}$$

$$E(\pi/2, 135°) = (\sqrt{2}/2)\,(E_x - iE_y)$$

where E_x and E_y are the components of the wave parallel to the directions of the neutral lines of the quarter-wave plate. Hence the transmitted intensities are:

$$I(\pi/2, 45°) = \tfrac{1}{2}\,(E_x + iE_y)(E_x^* - iE_y^*),$$

$$I(\pi/2, 135°) = \tfrac{1}{2}\,(E_x - iE_y)(E_x^* + iE_y^*).$$

$$I(\pi/2, 45°) - I(\pi/2, 135°) = i\,(E_y E_x^* - E_y^* E_x) = 2\,\Im(E_x E_y^*)$$

and for a partially polarized wave:

$$I(\pi/2, 45°) - I(\pi/2, 135°) = 2\,\Im < E_x E_y^* > = s_3.$$

The Poincaré coefficients P_1, P_2 and P_3 are defined as the Stokes parameters normalized by the intensity of the beam, so that:

$$P_1 = \frac{s_1}{s_0}, \quad P_2 = \frac{s_2}{s_0} \quad \text{and} \quad P_3 = \frac{s_3}{s_0}.$$

Thus P_1 is the linear polarization rate for axis directions Ox and Oy, P_2 that for directions at $45°$ and at $135°$ to axis Ox, and P_3 the circular polarization rate (the relative difference between the intensities of right- and left-circularly polarized waves).

A completely polarized wave is described by its Jones vector

$$\begin{pmatrix} a \\ b\exp i\varphi \end{pmatrix}$$

where a and b are real numbers. Then its Stokes parameters are:

$$s_0 = a^2 + b^2, \quad s_1 = a^2 - b^2, \quad s_2 = 2\,ab\,\cos\varphi \quad \text{and} \quad s_3 = 2\,ab\,\sin\varphi.$$

Thus, in this case, the four Stokes parameters are not independent, since

$$s_0^2 = s_1^2 + s_2^2 + s_3^2.$$

For a partially polarized wave, this relation has no reason to be valid. It can be shown that in all cases $s_0^2 \geq s_1^2 + s_2^2 + s_3^2$, the equal sign applying to a totally polarized wave [1]. It can be shown also that the Stokes parameters of a mixture of independent waves are sums of the respective Stokes parameters of the separate waves.

The Stokes parameters of a totally unpolarized wave are such that $s_1 = s_2 = s_3 = 0$.

Considering a wave with given Stokes parameters and denoting by a single symbol $\mathbf{s}$ the four Stokes parameters s_0, s_1, s_2 and s_3, one can write:

$$\mathbf{s} = \mathbf{s}^{(1)} + \mathbf{s}^{(2)}$$

with $\qquad \mathbf{s}^{(1)} = s_0 - \sqrt{s_1^2 + s_2^2 + s_3^2}, \, 0, \, 0, \, 0$

and $\qquad \mathbf{s}^{(2)} = \sqrt{s_1^2 + s_2^2 + s_3^2}, \, s_1, \, s_2, \, s_3$

$\mathbf{s}^{(1)}$ represents the unpolarized part of the wave and $\mathbf{s}^{(2)}$ the polarized part. Hence, the degree of polarization τ of the original wave, which is equal to the ratio of the intensity of the polarized part to the total intensity, is given by:

$$\tau = \frac{I_{\text{pol}}}{I_{\text{tot}}} = \frac{\sqrt{s_1^2 + s_2^2 + s_3^2}}{s_0}.$$

Exercice 17.7 deals with an illuminating, albeit theoretical, partially polarized wave situation.

17.6. Exercises

Exercise 17.1

Consider a plate of quartz (a uniaxial positive crystal, $n_e > n_o$), with parallel faces, cut so that the optic axis is parallel to the plate. A monochromatic parallel beam, with wavelength $\lambda = 590$ nm in vacuum, is incident on the plate, perpendicular to the plate.

1. Determine the neutral lines of the plate. Choose an orthonormal axis system with Oz parallel to the incident beam and Ox and Oy parallel to the neutral lines of the plate.

2. What phase shift does the plate introduce between the components parallel to Ox and Oy of an initially linear polarization? Determine the thickness e of the plate required to obtain a quarter-wave plate, then a half-wave plate. We have $n_o = 1.544$ and $n_e = 1.553$.

3. The incident beam is linearly polarized and its polarization is at $45°$ to the neutral lines of the plate. Show that, if the plate is a quarter-wave plate, the beam after the plate is circularly polarized. Is this a right- or a left-circular polarization? If it is right-circular (or left-circular), how should the device be changed to make it left-circular (or right-circular)?

4. What do we obtain if we place across the path of this circularly polarized beam another quarter-wave plate, identical to the first one, with its neutral lines parallel to those of the first one?

5.   What is the effect of a quarter-wave plate on natural light?

Exercise 17.2

Consider a quartz plate (quartz is a uniaxial positive crystal, $n_e > n_o$), cut parallel to the optic axis, and set between two crossed polarizers, *i.e.* polarizers with perpendicular polarization directions. A monochromatic parallel beam (with wavelength λ in vacuum) is incident perpendicular to the plate.

1.   The polarization direction of the beam incident on the plate is at angle α to the fast neutral line of the quartz plate. Calculate, as a function of the phase shift φ introduced by the plate, the intensity I of the beam after the analyzer.

2.   Choose $\alpha = 45°$. Determine the smallest thickness of the plate that makes it possible to obtain:

   i)   zero intensity,

   ii)   maximal intensity.

3. α is still equal to $45°$. The initial beam is now white (*i.e.* polychromatic), and we are interested in wavelengths in the visible range, $0.4\ \mu$m $< \lambda < 0.8\ \mu$m. What wavelengths are extinguished after the analyzer if $e = 0.2$ mm? Assume that $n_e - n_o$ varies little with wavelength and is equal to 0.01.

Exercise 17.3

A parallel, monochromatic, linearly polarized beam is normally incident on a birefringent plate, the neutral lines of which are oriented at $45°$ to the rectilinear vibration. The phase shift introduced by the plate between the extraordinary vibration ($// Oy$) and the ordinary vibration ($// Ox$) is φ.

1.   Show that, after the plate, the vibration is elliptic, the axes of the ellipse being respectively parallel and perpendicular to the direction of the rectilinear vibration.

2.   Calculate the axis ratio of the ellipse.

Exercise 17.4

Using two linear polarizers with known polarization directions, how can the neutral line directions of a birefringent plate be determined?

Exercise 17.5

Consider the Nicol prism described in Section 17.3. Figure 17.6b shows its principal section plane, coinciding with the plane of incidence. The angle $(\mathbf{AC}, \mathbf{AA_1})$ between the optic axis and the entrance face is equal to $45°$, angle $(\mathbf{CA}, \mathbf{CC'})$ is $71°$ and angle $(\mathbf{C'A}, \mathbf{C'C})$ is $22°$. The ordinary and extraordinary indices of calcite are $n_o = 1.6584$, $n_e = 1.4864$, and the index of Canada balsam is $N = 1.55$.

1. Calculate the angle of incidence of the ordinary ray on AC'. Show that this ray is totally reflected on the Canada balsam.

2. Determine the direction of the wave-vector, the index and the polarization of the extraordinary ray. Show that it goes through the Canada balsam.

3. What is the direction of the extraordinary ray in the prism? Show that the outgoing ray is very slightly shifted with respect to the incident beam.

Exercise 17.6

In view of a special optical application (generation of a wave with angular frequency 2ω starting from a wave with angular frequency ω, a process known as Second Harmonic Generation, SHG, see Complement 19C), we need to find a direction of the wave-vector $\mathbf{k}$ in a birefringent crystal such that $n(2\omega) = n(\omega)$. Because the index of a material varies with frequency, the solution can be found, in a uniaxial material, by associating the ordinary wave to one frequency and the extraordinary wave to the other one.

For a crystal of KH_2PO_4 (Potassium dihydrogen phosphate, KDP), which is uniaxial, find the angle θ the wave-vector $\mathbf{k}$ must enclose with the optic axis so that $n_e(2\omega) = n_o(\omega)$, taking into account that $n(2\omega) > n(\omega)$ and $n_e < n_o$. Useful values are:
for $\lambda_1 = 694.3$ nm (corresponding to ω): $n_o = 1.506$ and $n_e = 1.466$,
for $\lambda_2 = \lambda_1/2$: $n_o = 1.534$ and $n_e = 1.487$.

Exercise 17.7

1. We consider a monochromatic wave with Jones vector

$$\begin{pmatrix} a \\ ib \end{pmatrix}$$

in an axis system Oxy, where a and b are positive real numbers. What is, at a given point, the trajectory of the end of the electric field vector of this wave? What is its circular polarization rate ?

2. Same questions for a wave with Jones vector

$$\begin{pmatrix} b \\ ia \end{pmatrix}$$

3. We consider the wave formed by the superposition of the two above waves. Determine the Stokes parameters for this wave and its degree of polarization.

References

[1] M. Born and E. Wolf, *Principles of Optics*, 7th edn. (Cambridge University Press, Cambridge, 1999)

Chapter 18

Rotatory power and optical activity

Going through some crystals rotates the linear polarization of incident radiation. This phenomenon, known as rotatory power or circular birefringence, is described by a difference in refractive index for right- and left-hand circular polarization. It is interpreted through the gyrotropy tensor, related to the space dependence of the permittivity tensor. The investigation of this tensor makes it possible to determine what point groups a crystal must belong to in order to feature this property.

18.1. Definition of the rotatory power of a material

Rotatory power was discovered by François Arago in 1811 and Jean-Baptiste Biot in 1812. The first theoretical interpretation was given by Augustin Fresnel in 1824.

A material features rotatory power if it causes the plane of polarization of linearly polarized light[1] to rotate. This property is distinctly different from the previously discussed case of linear, or rectilinear, birefringence. It can exist in optically isotropic crystals (cubic crystals) and also in liquids or solutions. We will not deal with the rotatory power of liquids and solutions because they are outside the scope of this book, which focuses on crystalline solids. Also, the differences in refractive index involved in this phenomenon are much smaller than those corresponding to birefringence. Therefore, clearly evidencing rotatory power in anisotropic materials requires sending the beam parallel to the optic axis in uniaxial crystals (or to one of the two optic axes in biaxial crystals) in order to eliminate the effect of birefringence.

---

1. The plane of polarization is the plane containing the wave-vector and the polarization vector, which is perpendicular to it. In a birefringent crystal, the polarization vector is the induction vector **D**, while in air, where the electric field and induction are parallel, the description is based on the electric field.

© Springer Science+Business Media Dordrecht 2014
C. Malgrange et al., *Symmetry and Physical Properties of Crystals*,
DOI 10.1007/978-94-017-8993-6_18

Consider a material featuring rotatory power, for example a quartz plate in which the faces are perpendicular to the 3-fold axis. We illuminate this plate with a linearly polarized beam perpendicular to the plate. After going through the plate, the polarization has rotated by an angle α. This angle turns out experimentally to be proportional to the thickness e of the crystal. The rotatory power ρ is defined by:

$$\rho = \frac{\alpha}{e}. \tag{18.1}$$

If we exchange the entrance and exit faces of the plate, the polarization rotation angle α does not change sign.

The sign of the rotation angle depends on the crystal, which may be either dextrogyre (rotation to the right) or levogyre (rotation to the left). This requires defining what is meant by rotation to the right or to the left. A crystal is said to be dextrogyre if, for an observer looking at the source through the crystal, the plane of polarization of the light rotates to the right, *i.e.* clockwise (Fig. 18.1).

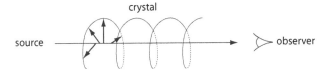

Figure 18.1: Rotation of the induction vector to the right for an observer looking at the source

The end of the polarization vector then describes, during propagation in the dextrogyre crystal, a left-handed helix (see Fig. 17.2). It was agreed to give positive values to α for dextrogyre materials. To indicate the order of magnitude of the rotatory polarization phenomenon, Figure 18.2 shows the value of the rotatory power ρ for quartz as a function of the wavelength of light in the visible range. We see that ρ is of the order of a few tens of degrees per millimeter of thickness traversed.

18.2. Fresnel's interpretation

Fresnel interpreted this phenomenon by showing that, if a material has different refractive indices for right- and left-hand circular polarizations, then this material shows rotatory power.

A linearly polarized electromagnetic wave can be split into two waves with equal amplitudes, one right-circular, the other left-circular. A wave propagating along Oz and polarized along Ox can be split, using the Jones notation (Sect. 17.2), in the following way:

$$\begin{pmatrix} 1 \\ 0 \end{pmatrix} = \frac{1}{2} \begin{pmatrix} 1 \\ i \end{pmatrix} + \frac{1}{2} \begin{pmatrix} 1 \\ -i \end{pmatrix}.$$

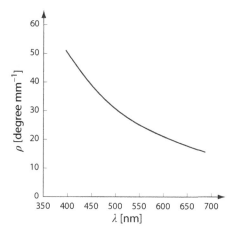

Figure 18.2: Rotatory power of quartz as a function of wavelength

The first term represents a left-circular vibration, and the second term a right-circular vibration (Sect. 17.2), when using waves with the form $\Re\left[\exp i(kz - \omega t)\right]$.

If the index of the material for the right-circular vibration, n_r, is different from the index, n_l, for the left-circular polarization, the initial wave, with wave-vector in air $k_0 = 2\pi/\lambda$, becomes after propagating over a distance z in the crystal:

$$
\frac{1}{2}\begin{pmatrix} 1 \\ i \end{pmatrix} \exp(ik_0 n_l z) + \frac{1}{2}\begin{pmatrix} 1 \\ -i \end{pmatrix} \exp(ik_0 n_r z)
$$
$$
= \frac{1}{2}\exp i\left(\frac{k_0(n_l + n_r)z}{2}\right)
$$
$$
\times \left[\begin{pmatrix} 1 \\ i \end{pmatrix}\exp i\left(\frac{k_0(n_l - n_r)z}{2}\right) + \begin{pmatrix} 1 \\ -i \end{pmatrix}\exp i\left(\frac{k_0(n_r - n_l)z}{2}\right)\right].
$$

We set:
$$
\frac{k_0(n_l + n_r)z}{2} = \psi \qquad \text{and} \qquad \frac{k_0(n_r - n_l)z}{2} = \alpha'.
$$

We obtain a vibration:

$$
\frac{1}{2}\exp i\psi \begin{pmatrix} \exp -i\alpha' + \exp i\alpha' \\ i\exp -i\alpha' - i\exp i\alpha' \end{pmatrix} = \exp i\psi \begin{pmatrix} \cos \alpha' \\ \sin \alpha' \end{pmatrix}
$$

i.e. a linearly polarized vibration, with polarization at an angle α' with respect to the axis Ox which is parallel to the initial polarization. The plane of polarization of the vibration has rotated by an angle α', proportional to the thickness z of crystal traversed, and such that:

$$
\alpha' = \frac{k_0(n_r - n_l)z}{2} = \frac{\pi(n_r - n_l)z}{\lambda}.
$$

Here, α' positive corresponds to the trigonometric sense, hence to a rotation to the left for an observer with the propagation vector of the incoming light pointing toward her, *i.e.* for an observer looking at the source: thus the crystal would be levogyre. Taking into account the convention on the sign of rotation defined in the preceding section, $\alpha' = -\alpha$, and the rotatory power $\rho = \alpha/z$ is equal to:

$$\rho = \frac{\pi(n_l - n_r)}{\lambda}. \tag{18.2}$$

The material is dextrogyre if $n_l > n_r$, hence if the propagation velocity v_r of right-circular polarized waves is larger than the propagation velocity v_l of the left-circular polarized waves, and conversely for a levogyre crystal. Dextrogyre quartz has a rotatory power equal to $18.8°$ per mm for wavelength 632.8 nm (red helium-neon laser). Therefore $n_l - n_r = 6.6 \times 10^{-5}$, a very small value compared to the birefringence of most crystals, which is on the order of 10^{-3}. For quartz, $n_e = 1.5533$, $n_o = 1.5442$ and $n_e - n_o = 9 \times 10^{-3}$.

18.3. Interpretation of rotatory power through the influence of the local environment

18.3.1. Effect of spatial dispersion

Until now, we assumed that the electrical polarization at a point depends only on the electric field at this point. Actually polarization can also depend on the local environment. This is described by a spatial dispersion term. The relation between $\mathbf{E}$ and $\mathbf{D}$ was, till now, described (see Eq. (16.22)) as:

$$E_m = \eta_{mn} \frac{D_n}{\varepsilon_0}. \tag{18.3}$$

To account for spatial dispersion, we must add to Equation (18.3) a term proportional to the space derivatives of induction:

$$E_m = \frac{1}{\varepsilon_0}\left(\eta_{mn} D_n + \gamma_{mnp} \frac{\partial D_n}{\partial x_p}\right). \tag{18.4}$$

We seek a solution in the form of a plane sinusoidal wave:

$$\mathbf{D} = \mathbf{D}_0 \exp i(\mathbf{k} \cdot \mathbf{r} - \omega t).$$

We then have:

$$\frac{\partial D_n}{\partial x_p} = ik_p D_n = iks_p D_n$$

where the s_i are the components of the unit vector parallel to $\mathbf{k}$ $(\mathbf{k} = k\mathbf{s})$.

To simplify notations, we set $\beta_{mnp} = k\gamma_{mnp}$, and Equation (18.4) becomes:

$$E_m = \frac{1}{\varepsilon_0}(\eta_{mn} D_n + i\beta_{mnp} s_p D_n). \tag{18.5}$$

It can be shown [1] that tensor β, called the gyration tensor, is antisymmetric with respect to the first two subscripts:

$$\beta_{mnp} = -\beta_{nmp}.$$

Among the 27 components β_{mnp}, 9 (those of the form β_{iil}) are therefore zero, and the remaining 18 (of the form β_{ijl} with $i \neq j$) are pairwise opposite. There are thus 9 independent components. We define two-subscript components G_{ml} through:

$$G_{ml} = \frac{1}{2}\delta_{mij}\beta_{ijl} \tag{18.6}$$

where δ_{mij} is the Levi-Civita symbol, or permutation tensor, which was defined in Section 10.6.2 and which is shown to be an axial rank-3 tensor in Section 18A.2. We see for example that:

$$G_{11} = \frac{1}{2}(\delta_{123}\beta_{231} + \delta_{132}\beta_{321}) = \frac{1}{2}(\beta_{231} - \beta_{321}) \text{ or } G_{11} = \beta_{231} = -\beta_{321}.$$

In the same way,

$$G_{22} = \beta_{312} = -\beta_{132}, \qquad G_{12} = \beta_{232} = -\beta_{322}, \qquad \text{etc.}$$

It is easily shown (Sect. 18A.3) that performing a change in frame on the δ_{mij}'s and the β_{ijl}'s transforms the G_{ml}'s like the components of an axial rank-2 tensor. The axial rank-2 tensor $[G]$ is called the gyrotropy tensor.[2] We will see that rotatory power is described by the symmetrical part of $[G]$ alone. Therefore all crystals for which $[G]$ is non zero will be called optically active, whether the symmetrical part is zero or not. Those for which the symmetrical part of $[G]$ is non zero (Sect. 18.5) will be said to feature rotatory power.

We will also show (Sect. 18A.4) that relation (18.6) leads to:

$$\beta_{ijl} = \delta_{ijn}G_{nl}$$

and Equation (18.5) becomes:

$$E_m = \frac{1}{\varepsilon_0}(\eta_{mn}D_n + i\delta_{mnj}G_{jp}s_pD_n). \tag{18.7}$$

We define axial vector $\overset{\smile}{\mathbf{g}}$ through its components $g_j = G_{jp}s_p$ and:

$$\overset{\smile}{\mathbf{g}} = [G]\mathbf{s}. \tag{18.8}$$

Equation (18.7) becomes:

$$E_m = \frac{1}{\varepsilon_0}(\eta_{mn}D_n + i\delta_{mnj}g_jD_n).$$

---

2. Some authors call $[G]$ the gyration tensor. Here, we call it the gyrotropy tensor, in order to clearly distinguish it from $[\beta]$, the gyration tensor (Eq. (18.6)).

The second term in the right-hand side is the m-th component of the cross product of vector $\mathbf{D}$ with vector $\overset{\smile}{\mathbf{g}}$, and we can write:

$$\mathbf{E} = \frac{1}{\varepsilon_0}([\eta]\mathbf{D} + i\mathbf{D} \times \overset{\smile}{\mathbf{g}}). \tag{18.9}$$

Optical activity introduces a purely imaginary term into the impermeability tensor.

We assumed, as in the preceding chapters (see Sect. 16.1), that the crystals are perfectly transparent, *i.e.* that absorption is zero. If it were not zero, absorption would correspond to a non-zero imaginary part in tensor $[\eta]$. The imaginary part which appears here in the relation between $\mathbf{D}$ and $\mathbf{E}$ has nothing to do with absorption. It introduces between $\mathbf{E}$ and $\mathbf{D}$ a phase-shift term which, as we will show further on, induces a difference between the indices of the material for the right- and left-circular polarizations which correspond to the eigenmodes. If absorption were to be taken into account, we would have to consider tensor $[\beta]$ as complex, and to deal with its imaginary part. A complete treatment taking into account absorption can be found in the books by Agranovich and Ginzburg [2] and Barron [3].

18.3.2. Wave propagation in optically active crystals

We saw (Eq. (16.15a)) that the propagation of a sinusoidal electromagnetic wave is described by the basic equation, reproduced here:

$$\mathbf{D} = \varepsilon_0 n^2[\mathbf{E} - \mathbf{s}(\mathbf{s} \cdot \mathbf{E})] \tag{18.10}$$

where $\mathbf{s}$ is a unit vector parallel to the wave-vector.

In Chapter 16, to obtain the propagation behavior of waves in a crystal, we associated to (18.10) relation $\mathbf{E} = [\eta]\mathbf{D}/\varepsilon_0$, which we must now replace with Equation (18.9).

Since $\mathbf{D}$ is perpendicular to $\mathbf{s}$ (Eq. (16.10a), a result which is easily retrieved by writing the dot product of $\mathbf{D}$ and $\mathbf{s}$ in (18.10)), the search for vectors $\mathbf{D}$ that are solutions of (18.9) and (18.10) is simplified by projecting these equations onto a plane perpendicular to $\mathbf{s}$. We choose Ox_3 to be parallel to $\mathbf{s}$, and we simplify the form of tensor $[\eta]$ by choosing for Ox_1 and Ox_2 the directions of the axes of the ellipse along which the index ellipsoid intersects the plane perpendicular to $\mathbf{s}$. In this coordinate system, tensor $[\eta]$ can be written as:

$$\begin{pmatrix} \eta_1 & 0 & \eta_{13} \\ 0 & \eta_2 & \eta_{23} \\ \eta_{13} & \eta_{23} & \eta_{33} \end{pmatrix}.$$

To simplify the analysis of the general case in what follows, we henceforth choose axis Ox_1 such that $\eta_1 > \eta_2$.

Projecting (18.10) onto Ox_1 and Ox_2 provides:

$$D_1 = \varepsilon_0 n^2 E_1 \tag{18.11a}$$

$$D_2 = \varepsilon_0 n^2 E_2. \tag{18.11b}$$

Projecting (18.9) yields:

$$E_1 = \eta_1 \frac{D_1}{\varepsilon_0} + ig_3 \frac{D_2}{\varepsilon_0} \qquad (18.12a)$$

$$E_2 = \eta_2 \frac{D_2}{\varepsilon_0} - ig_3 \frac{D_1}{\varepsilon_0} \qquad (18.12b)$$

because
$$\mathbf{D} \times \overset{\smile}{\mathbf{g}} = \begin{pmatrix} D_1 \\ D_2 \\ 0 \end{pmatrix} \times \begin{pmatrix} g_1 \\ g_2 \\ g_3 \end{pmatrix} = \begin{pmatrix} D_2 g_3 \\ -D_1 g_3 \\ D_1 g_2 - D_2 g_1 \end{pmatrix}.$$

Substituting in system (18.12) the values of E_1 and E_2 by their values as determined from system (18.11), we obtain:

$$\frac{1}{n^2} \frac{D_1}{\varepsilon_0} = \eta_1 \frac{D_1}{\varepsilon_0} + ig_3 \frac{D_2}{\varepsilon_0}$$

$$\frac{1}{n^2} \frac{D_2}{\varepsilon_0} = \eta_2 \frac{D_2}{\varepsilon_0} - ig_3 \frac{D_1}{\varepsilon_0}.$$

This system can be written as:

$$\frac{1}{n^2} \frac{\mathbf{D}}{\epsilon_0} = M \frac{\mathbf{D}}{\epsilon_0}$$

with

$$M = \begin{pmatrix} \eta_1 & ig_3 \\ -ig_3 & \eta_2 \end{pmatrix}.$$

The solutions are the eigenvectors of matrix M. The eigenvalues λ of this matrix are the values of $1/n^2$. They are obtained by solving equation:

$$(\eta_1 - \lambda)(\eta_2 - \lambda) - g_3^2 = 0. \qquad (18.13)$$

Only the component g_3 of $\overset{\smile}{\mathbf{g}}$ is used. The definition (18.8) of $\overset{\smile}{\mathbf{g}}$ makes it possible to write: $g_j = G_{jp} s_p$. Since, in the coordinate system we chose, $\mathbf{s} = (0, 0, 1)$, we deduce that $g_3 = G_{33}$. Here the axis system, which we note R, is associated to the direction of the wave-vector. The gyrotropy tensor is in general defined by its components G'_{ij} in an orthonormal axis system R' associated to the crystal lattice. Let $\{a_{ij}\}$ be the transformation matrix for going over from the orthonormal axis system R' associated to the crystal lattice to the orthonormal axis system R associated to the wave-vector direction. We obtain:

$$G_{33} = \Delta a_{3i} a_{3j} G'_{ij} = \Delta s'_i s'_j G'_{ij}$$

where s'_1, s'_2 and s'_3 are the components of vector $\mathbf{s}$ in the frame R' associated to the crystal lattice. The determinant Δ of the transformation matrix is equal to 1 because frames R and R' are both chosen to be direct. Finally:

$$g_3 = G_{33} = G'_{ij} s'_i s'_j. \qquad (18.14)$$

A special case: propagation along an optic axis

This is the most frequent case in experiments, because then birefringence has no effect, and optical activity is evidenced on its own. If **s** is parallel to an optic axis, then:

$$\eta_1 = \eta_2 = \frac{1}{n_o^2} \tag{18.15}$$

where n_o is the index for propagation along the optic axis. For a uniaxial crystal, n_o is the ordinary index.

Equation (18.13) has solutions $\lambda = \eta_1 \pm g$ where, to simplify notations, we note $g = g_3$.

The eigenvectors are then:

$\begin{pmatrix} 1 \\ -i \end{pmatrix}$ for $\lambda = \dfrac{1}{n^2} = \eta_1 + g$, which is a right-circular polarized wave;

$\begin{pmatrix} 1 \\ i \end{pmatrix}$ for $\lambda = \dfrac{1}{n^2} = \eta_1 - g$, which is a left-circular polarized wave (Sect. 17.2).

We retrieve Fresnel's assumption of two waves, right- and left-circular polarized, propagating with different indices, n_r and n_l.

Using (18.15), we obtain:

$$\frac{1}{n_r^2} = \frac{1}{n_o^2} + g \quad \text{and} \quad \frac{1}{n_l^2} = \frac{1}{n_o^2} - g$$

so that
$$\Delta \left(\frac{1}{n^2} \right) = -\frac{2\Delta n}{n^3} = -2g$$

where $\Delta n = n_l - n_r$. We obtain $\Delta n = n_l - n_r = g n_o^3$ and

$$\rho = \frac{\pi}{\lambda} g n_o^3 \tag{18.16}$$

with
$$g = G_{ij} s_i s_j, \tag{18.17}$$

where G_{ij}, s_i, s_j are components in the orthonormal axis system associated to the crystal.

General case

The eigenvalue Equation (18.13) has solution:

$$\lambda_\pm = \frac{(\eta_1 + \eta_2)}{2} \pm \frac{1}{2} [(\eta_1 - \eta_2)^2 + 4g^2]^{\frac{1}{2}}.$$

We recall that Ox_1 was chosen such that $\eta_1 > \eta_2$.

We define τ such that:

$$\frac{[(\eta_1 - \eta_2)^2 + 4g^2]^{\frac{1}{2}} - [\eta_1 - \eta_2]}{2} = \tau g,$$

and we note that $|\tau| \leq 1$ and $\tau g > 0$.

We obtain: $\qquad \lambda_+ = \eta_1 + \tau g \qquad$ and $\qquad \lambda_- = \eta_2 - \tau g.$

The associated eigenvectors are respectively:

$$\begin{pmatrix} 1 \\ -i\tau \end{pmatrix} \quad \text{and} \quad \begin{pmatrix} \tau \\ i \end{pmatrix},$$

i.e. elliptical vibrations, the axes of which are the directions of the inductions which would propagate in the material if optical activity were neglected.

We check that, for propagation along the optic axis, where $\eta_1 = \eta_2$, we obtain $\tau = 1$ if $g > 0$ and $\tau = -1$ if $g < 0$, and thus right- and left-circular polarizations. Further, if $g > 0$, $n_r < n_l$ (since $\tau = 1$) and ρ is positive while, if $g < 0$, $n_r > n_l$ and ρ is negative, in agreement with (18.16).

As soon as the propagation direction deviates by a few degrees from the direction of the optic axis, $\eta_1 - \eta_2$ becomes large compared to g. We then find $\tau \simeq g/(\eta_1 - \eta_2)$ so that $\tau \ll 1$. The waves have weak ellipticity. The major axes of these vibrations are parallel to the vibrations which would propagate if optical activity were neglected.

18.4. Effect of crystal symmetry on the gyrotropy tensor

The axial tensor $[G]$ has no reason to be symmetric. However, (18.17) shows that only the symmetrical part of $[G]$ is involved in rotatory power (18.16). The components of $[G]$ with different subscripts combine in (18.17) as, for example, $(G_{12}+G_{21})s_1 s_2$ etc. However, it is interesting to begin by considering the whole tensor which represents the total optical activity, keeping in mind that only the symmetrical part of $[G]$ is of interest for the rotatory power.

18.4.1. Centrosymmetric groups

The axial rank-2 tensor $[G]$ derives from the polar rank-3 tensor $[\beta]$. It is therefore clear that it is identically zero for all centrosymmetric crystal groups.

This result can be retrieved directly using axial tensor $[G]$. The components G'_{ij} of an axial rank-2 tensor in an orthonormal coordinate system $(Ox'_1x'_2x'_3)$ deduced from another orthonormal coordinate system $(Ox_1x_2x_3)$ through the transformation matrix $\{a_{ij}\}$ are expressed as a function of the components G_{ij} in the coordinate system $(Ox_1x_2x_3)$ through relation:

$$G'_{ij} = \Delta a_{im} a_{jn} G_{mn}, \qquad (18.18)$$

where Δ is the determinant of matrix $\{a_{ij}\}$ (see Eq. (18A.1)). If the crystal is centrosymmetric, a change from referential $(Ox_1x_2x_3)$ to the referential symmetric with respect to origin O must also leave the components of G unchanged, and $G'_{ij} = G_{ij}$. But in this transformation $a_{ij} = -\delta_{ij}$ and $\Delta = -1$. Hence:

$$G'_{ij} = -G_{ij}.$$

Finally

$$G'_{ij} = G_{ij} = -G_{ij} \quad \text{and} \quad G_{ij} = 0.$$

18.4.2. Non-centrosymmetric groups

We now examine some point groups using the direct inspection method.

Group 2

We choose Ox_3 to be parallel to the 2-fold axis. A rotation by π of the coordinate system $(Ox_1x_2x_3)$ around the 2-fold axis changes it into system $(Ox'_1x'_2x'_3)$. The components of a vector transform according to

$$x'_1 = -x_1, \quad x'_2 = -x_2 \quad \text{and} \quad x'_3 = x_3.$$

For a rotation, $\Delta = 1$, hence $[G]$ has form:

$$\begin{pmatrix} \bullet & \bullet & 0 \\ & \bullet & 0 \\ & & \bullet \end{pmatrix}$$

Group m

We choose axis Ox_3 perpendicular to the mirror. The operation of symmetry with respect to the mirror, when performed on the axis system, changes the components of a vector according to

$$x'_1 = x_1, \quad x'_2 = x_2 \quad \text{and} \quad x'_3 = -x_3.$$

For mirror symmetry, $\Delta = -1$, hence $[G]$ has form:

$$\begin{pmatrix} 0 & 0 & \bullet \\ & 0 & \bullet \\ & & 0 \end{pmatrix}$$

Using the preceding result, we immediately see that, if we add to this mirror a 2-fold axis perpendicular to the mirror, hence parallel to Ox_3, all terms in this tensor are zero. This is no surprise since group $2/m$ is centrosymmetric.

We note that the existence of a mirror implies that the trace of the matrix representing $[G]$ is zero.

Group 4

We choose axis Ox_3 parallel to the 4-fold axis. A rotation by $\pi/2$ of the coordinate system $(Ox_1 x_2 x_3)$ around the 4-fold axis changes it into system $(Ox_1' x_2' x_3')$, and the components of a vector transform according to

$$x_1' = x_2, \quad x_2' = -x_1 \quad \text{and} \quad x_3' = x_3.$$

For a rotation, $\Delta = 1$ and $[G]$ has form

$$\begin{pmatrix} \bullet & \bullet & 0 \\ \circ & \bullet & 0 \\ 0 & 0 & \bullet \end{pmatrix}$$

where the empty circle connected to the full circle means that the two connected coefficients have equal absolute values and opposite signs.

We notice that here $G_{12} = -G_{21}$, so that $[G]$ has an antisymmetric part. Rotatory power depends on the symmetric part of $[G]$, so that the tensor representing the rotatory power has form:

$$\begin{pmatrix} \bullet & 0 & 0 \\ 0 & \bullet & 0 \\ 0 & 0 & \bullet \end{pmatrix}$$

Group 422

Group 422 deduces from group 4 by addition of a two-fold axis perpendicular to the 4-fold axis. One chooses the Ox_3 axis parallel to the 4-fold axis and the Ox_1 and Ox_2 axes parallel to the two-fold axes. A rotation of the coordinate system $(Ox_1 x_2 x_3)$ by π around the 2-fold axis parallel to Ox_1 changes it into system $(Ox_1' x_2' x_3')$, and the components of a vector transform according to

$$x_1' = x_1, \quad x_2' = -x_2 \quad \text{and} \quad x_3' = -x_3.$$

$\Delta = 1$ and $G_{12}' = -G_{12} = G_{12} = 0$.

Tensor $[G]$ is now symmetric and identical to the symmetric part of $[G]$ for point group 4.

Group 4mm

Group 4mm deduces from group 4 by addition of a mirror parallel to the 4-fold axis. One chooses the Ox_3 axis parallel to the 4-fold axis and the Ox_1 and Ox_2 axes perpendicular to the mirrors. A mirror symmetry normal to Ox_1 transforms the coordinate system $(Ox_1 x_2 x_3)$ into system $(Ox_1' x_2' x_3')$, and the components of a vector transform according to

$$x_1' = -x_1, \quad x_2' = x_2 \quad \text{and} \quad x_3' = x_3.$$

For a mirror symmetry, $\Delta = -1$ and $G_{11} = G_{22} = G_{33} = 0$ whereas G_{12} and G_{21} are not zero. The $[G]$ tensor is now antisymmetric :

$$\begin{pmatrix} 0 & \bullet & 0 \\ \circ & 0 & 0 \\ 0 & 0 & 0 \end{pmatrix}$$

so that crystals belonging to class 4mm do not feature rotatory power.

Table 18.1 gives the form of the gyrotropy tensor $[G]$ for 18 out of the 21 non-centrosymmetric crystallographic groups. For the three other groups ($\bar{4}$3m, $\bar{6}$2m and $\bar{6}$), all components are zero. For three groups (4mm, 3m and 6mm), tensor $[G]$ is antisymmetric: thus they do not feature rotatory power, but they are optically active because of this antisymmetric $[G]$ tensor. We showed (Sect. 10.6) that a rank-2 antisymmetric polar tensor is equivalent to an axial vector. In the same way, it can be shown that a rank-2 antisymmetric axial tensor is equivalent to a polar vector. The gyrotropy tensor of crystal classes 4mm, 3m and 6mm thus has only a vectorial part. It is interesting to note that all the crystal classes for which the $[G]$ tensor involves a vectorial part (1, 2, m, mm2, 4, 4mm, 3, 3m, 6, 6mm) are just the pyroelectric classes. This vectorial part of the gyrotropy tensor can be evidenced using quite delicate reflectivity experiments.

Table 18.2 gives the form of the symmetric part of the gyrotropy tensor or rotatory power tensor.

18.5. Rotatory power and chirality

Table 18.3 gives examples of crystals featuring rotatory power, and the value of their rotatory power ρ (in absolute value) for a given wavelength. Wavelength 632.8 nm is that of the helium-neon laser.

The crystals mentioned in this table belong to one of the crystal groups called enantiomorphic or chiral groups. These groups, featuring neither a mirror nor a center of symmetry, and, in a more general way, no roto-inversion, are groups 1, 2, 222, 3, 32, 4, 422, 6, 622, 23, 432. An object is said to be chiral if it is not superimposable with its image in a mirror. This is the case for a hand (in Greek: $\chi\varepsilon\iota\rho$, hence the term chiral). The mirror image of a right hand is a left hand. In the same way, a helix is a chiral object: the mirror image of a right-handed helix is a left-handed helix, and conversely. Crystals belonging to these enantiomorphic groups can exist under two forms, called enantiomorphic, one dextrogyre, the other levogyre. Their structures deduce from each other through a mirror symmetry operation. For those whose space group has definite chirality because it includes a screw axis giving equivalent atom positions located on a right- or left-handed helix (3_1 and 3_2 for example),

Table 18.1: Form of the gyrotropy tensor $[G]$

Triclinic	1	$\begin{pmatrix} \bullet & \bullet & \bullet \\ \bullet & \bullet & \bullet \\ \bullet & \bullet & \bullet \end{pmatrix}$			

Monoclinic

2 $2 // Ox_3$ $\begin{pmatrix} \bullet & \bullet & 0 \\ \bullet & \bullet & 0 \\ 0 & 0 & \bullet \end{pmatrix}$ $2 // Ox_2$ $\begin{pmatrix} \bullet & 0 & \bullet \\ 0 & \bullet & 0 \\ \bullet & 0 & \bullet \end{pmatrix}$

m $m \perp Ox_3$ $\begin{pmatrix} 0 & 0 & \bullet \\ 0 & 0 & \bullet \\ \bullet & \bullet & 0 \end{pmatrix}$ $m \perp Ox_2$ $\begin{pmatrix} 0 & \bullet & 0 \\ \bullet & 0 & \bullet \\ 0 & \bullet & 0 \end{pmatrix}$

Orthorhombic

222 $\begin{pmatrix} \bullet & 0 & 0 \\ 0 & \bullet & 0 \\ 0 & 0 & \bullet \end{pmatrix}$ $mm2$ $\begin{pmatrix} 0 & \bullet & 0 \\ \bullet & 0 & 0 \\ 0 & 0 & 0 \end{pmatrix}$

Tetragonal

4 $\begin{pmatrix} \bullet & \bullet & 0 \\ \circ & \bullet & 0 \\ 0 & 0 & \bullet \end{pmatrix}$ 422 $\begin{pmatrix} \bullet & 0 & 0 \\ 0 & \bullet & 0 \\ 0 & 0 & \bullet \end{pmatrix}$ $4mm$ $\begin{pmatrix} 0 & \bullet & 0 \\ \circ & 0 & 0 \\ 0 & 0 & 0 \end{pmatrix}$

$\bar{4}$ $\begin{pmatrix} \bullet & \bullet & 0 \\ \bullet & \circ & 0 \\ 0 & 0 & 0 \end{pmatrix}$ $\bar{4}2m$ $2 // Ox_1$ $\begin{pmatrix} \bullet & 0 & 0 \\ 0 & \circ & 0 \\ 0 & 0 & 0 \end{pmatrix}$

Rhombohedral and Hexagonal

$3, 6$ $\begin{pmatrix} \bullet & \bullet & 0 \\ \circ & \bullet & 0 \\ 0 & 0 & \bullet \end{pmatrix}$ $\begin{matrix}32, 622 \\ 2 // Ox_1\end{matrix}$ $\begin{pmatrix} \bullet & 0 & 0 \\ 0 & \bullet & 0 \\ 0 & 0 & \bullet \end{pmatrix}$ $\begin{matrix}3m, 6mm \\ m \perp Ox_1\end{matrix}$ $\begin{pmatrix} 0 & \bullet & 0 \\ \circ & 0 & 0 \\ 0 & 0 & 0 \end{pmatrix}$

Cubic

$23, 432$ $\begin{pmatrix} \bullet & 0 & 0 \\ 0 & \bullet & 0 \\ 0 & 0 & \bullet \end{pmatrix}$

●—● Equal components   ●—○ Components with equal absolute values but opposite signs

the space groups of crystals of these two enantiomorphic forms are different. For example, right-handed quartz has space group $P3_121$ and left-handed quartz $P3_221$. The connection between the chirality of the crystal structure and the sign of the rotatory power has long been discussed, and it is no trivial matter. Rotatory power is due to the most polarizable atoms. If these atoms are not in equivalent crystal sites, they can actually form pseudo-helices with chirality opposite to the chirality of the crystal axis, as was shown by A.M. Glazer and K. Stadnicka [4]. The same authors showed that, in crystals where the screw axes have no definite chirality, such as 6_3 or 2_1 (Sect. 7.2.1), the most polarizable atoms form pseudo-helices with given chirality for a given enantiomorph.

It can be noticed that the trace of the matrix representing tensor $[G]$ in enantiomorphic classes is non zero whereas for the other crystals featuring rotatory power the trace is zero. This is not surprising since we already noticed that, for crystal classes with a mirror symmetry the trace of $[G]$ is zero (see group m in Sect. 18.4.2).

Table 18.2: Form of the symmetrical part of the gyrotropy tensor $[G]$ for the various point groups. Only the coefficients on the diagonal and in the part above it are represented.

Triclinic

1 $\quad$ $\begin{pmatrix} \bullet & \bullet & \bullet \\ & \bullet & \bullet \\ & & \bullet \end{pmatrix}$

Monoclinic

2 $\quad 2\ /\!/\ Ox_3$ $\begin{pmatrix} \bullet & \bullet & 0 \\ & \bullet & 0 \\ & & \bullet \end{pmatrix}$ $\quad 2\ /\!/\ Ox_2$ $\begin{pmatrix} \bullet & 0 & \bullet \\ & \bullet & 0 \\ & & \bullet \end{pmatrix}$

m $\quad m \perp Ox_3$ $\begin{pmatrix} 0 & 0 & \bullet \\ & 0 & \bullet \\ & & 0 \end{pmatrix}$ $\quad m \perp Ox_2$ $\begin{pmatrix} 0 & \bullet & 0 \\ & 0 & \bullet \\ & & 0 \end{pmatrix}$

Orthorhombic

222 $\begin{pmatrix} \bullet & 0 & 0 \\ & \bullet & 0 \\ & & \bullet \end{pmatrix}$ $\quad mm2$ $\begin{pmatrix} 0 & \bullet & 0 \\ & 0 & 0 \\ & & 0 \end{pmatrix}$

Tetragonal

$4, 422$ $\begin{pmatrix} \bullet\!\!\diagdown\!\!\bullet & 0 & 0 \\ & \bullet & 0 \\ & & \bullet \end{pmatrix}$ $\quad \bar{4}$ $\begin{pmatrix} \bullet\!\!\diagdown\!\!\circ & \bullet & 0 \\ & \circ & 0 \\ & & 0 \end{pmatrix}$ $\quad \begin{matrix}\bar{4}2m \\ 2\ /\!/\ Ox_1\end{matrix}$ $\begin{pmatrix} \bullet\!\!\diagdown\!\!\circ & 0 & 0 \\ & \circ & 0 \\ & & 0 \end{pmatrix}$

Rhombohedral and Hexagonal

$3, 32, 6, 622$ $\begin{pmatrix} \bullet\!\!\diagdown\!\!\bullet & 0 & 0 \\ & \bullet & 0 \\ & & \bullet \end{pmatrix}$

Cubic

$23, 432$ $\begin{pmatrix} \bullet\!\!\diagdown\!\!\bullet & 0 & 0 \\ & \bullet\!\!\diagdown\!\!\bullet & 0 \\ & & \bullet \end{pmatrix}$

$\bullet\!\!-\!\!\bullet$ Equal components $\qquad \bullet\!\!-\!\!\circ$ Components with equal absolute values but opposite signs

Table 18.3: Values of the rotatory power ρ of some crystals

Material	Point group	Space group	Wavelength (nm)	ρ (degree . mm^{-1})
Quartz (SiO$_2$)	32	P3$_1$21	643.8	18
			274.0	122.1
Berlinite (AlPO$_4$)	32	P3$_1$21	632.8	14.6
Cinnabar (HgS)	32	P3$_1$21	632.8	320
NaBrO$_3$	23	P2$_1$3	632.8	1.65
NaClO$_3$	23	P2$_1$3	632.8	2.44
α−LiIO$_3$	23	P6$_3$2	632.8	86.7

While rotatory power is possible for all crystals belonging to an enantiomorphic group, the converse property is not true: belonging to an enantiomorphic group is not a necessary condition for a crystal to feature rotatory power. This is shown on Table 18.2, where non-enantiomorphic groups m, mm2, $\bar{4}$ and $\bar{4}$2m are seen to feature rotatory power. However, the rotatory power of these crystals is often difficult to measure, because they do not necessarily feature rotatory power along the optical axis, and it is then necessary to measure a rotatory power combined with birefringence. The sign of the rotatory power in these crystals depends on the direction of the incident beam, and the crystal is dextrogyre or levogyre depending on this direction. Take the example of a crystal with point group mm2, for which tensor $[G]$ has only one non-zero component, G_{12}.[3] From (18.16) and (18.17), its rotatory power is proportional to $G_{12}s_1s_2$. If the optic axes for this crystal are in plane (Ox_1x_2), the sign of the rotatory power is positive along one of the axes, and negative along the other one, since the product s_1s_2 changes sign from one optic axis to the other. This is the case for sodium nitrite $NaNO_2$ in its ferroelectric phase (T < 163°C), where rotatory power was measured for wavelength 632.8 nm (Fig. 18.3). Along one of the optic axes, the rotatory power is 15.7°/mm, while it has the same value but opposite sign along the other optic axis.

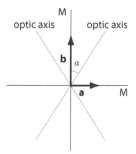

Figure 18.3: Plane (**a**, **b**), and therefore (Ox_1, Ox_2) for a crystal of sodium nitrite ($NaNO_2$). The angle α between the optic axes and one of the two mirrors M is equal to 33°.

These results are interpreted by taking into account the positions of the ions in the unit cell: they form a left-handed helix along an axis approximately parallel to one of the optic axes, and form a right-handed helix along an axis approximately parallel to the other optic axis. The two optic axes deduce from each other through a mirror operation which changes the right-handed helix into a left-handed helix and vice-versa. In contrast, if the optic axes are in the plane perpendicular to Ox_1 or to Ox_2 (a frequent case), the rotatory power along the optic axes is zero (s_1 or s_2 being equal to zero), and the rotatory power becomes very difficult to measure.

---

3. We recall that, for an orthorhombic crystal with point group mm2, the basis vector **c** (and thus axis Ox_3) is chosen parallel to the 2-fold axis.

18.6. Absorption and conclusion

We neglected absorption, which is extremely weak in transparent crystals at the wavelengths in the visible range considered here. We thus neglected the imaginary part of the index. The real and imaginary parts of the index are not independent. They are connected through integral relations (involving integration over the whole frequency range) known as the Kramers-Kronig relations (see [5]). As a consequence, a difference in index for two different polarizations leads to a notable difference in absorption near an absorption edge: this is dichroism. The birefringence discussed in Chapter 16 corresponds to a difference in index for two orthogonal linear polarizations. There is therefore a difference in absorption for these two polarizations, hence linear dichroism. The rotatory power is due to a difference in index for the right-circular and left-circular polarizations, which leads to a difference in absorption for these two polarizations, hence circular dichroism.

In the X-ray range, where crystals are not transparent, optical activity was experimentally evidenced recently, through dichroism measurements for natural rotatory power [6], and through reflectivity measurements for the vectorial part of $[G]$ [7].

We showed that, among the 21 non-centrosymmetric crystal classes, 18 are optically active. Out of these 18 classes, 15 feature rotatory power. Among these 15 classes, 11 are enantiomorphic, which means that crystals in these classes can exist under two different, mutually symmetric forms, one dextrogyre and the other levogyre. We have seen that the gyrotropic tensor of enantiomorphic classes has non-zero trace whereas, for the 4 non-enantiomorphic classes, the trace of $[G]$ is zero. For the 3 optically active classes which do not feature rotatory power, the axial tensor $[G]$ is antisymmetric and can be represented by a vector. This distinction between enantiomorphic crystals, non-enantiomorphic crystals which nevertheless feature rotatory power, and optically active crystals just on the basis of the vector part of $[G]$, was first introduced by Jerphagnon and Chemla in 1976 [8]. These authors split the symmetric part of $[G]$ into the sum of an isotropic tensor (proportional to the unit tensor), which they note $[A]$, and a tensor $[D]$. Thus the gyrotropy tensor $[G]$ is the sum of $[A]$, $[D]$ and the vector part $\mathbf{V}$ (equivalent to the antisymmetric part of $[G]$). For enantiomorphic crystals, $[A]$ is not zero, all crystals featuring rotatory power have $[A] + [D]$ non zero, and optically active crystals with no rotatory power have only $\mathbf{V}$ non zero.

Complement 18C. Magnetic optical rotation

The magneto-optical effects can be defined as the effects of a magnetic field on the propagation of an electro-magnetic wave, and can be described through the change in the high-frequency susceptibility tensor and refractive indices of a material due to the magnetic field. The Faraday magneto-optical effect, or magnetic gyrotropy, is the best-known member of this family. It concerns visible and near-visible light, but is now also an active field of research, using synchrotron radiation, in the X-ray spectral range.

The Faraday effect, discovered in 1845, consists in the rotation[4] of the plane of polarization of linearly polarized light under the influence of a magnetic field $\overset{\smile}{\mathbf{B}}$ or magnetization parallel to the propagation direction. It is thus reminiscent of natural rotatory power, discussed in this chapter. It too can be conveniently described through the difference in the refractive indices, hence the different phase velocities, associated to the right- and left-hand circular polarization states which are the eigenmodes. In materials with no spontaneous magnetization, such as dia- or paramagnets, the rotation α is proportional to the modulus of the magnetic field B and to the length of material d traversed by the light beam:

$$\alpha = k_V B d,$$

where k_V is the Verdet "constant", a material-dependent function of the temperature and wavelength, with values on the order of 10^1 rad $T^{-1}m^{-1}$. In materials with spontaneous magnetization, typically ferrimagnets which are fairly transparent in the visible and near infra-red range, the effect of the magnetic field is primarily to steer the distribution of local magnetization in the various domains (see note in Complement 8C), favoring those with direction close to the field, and tending toward saturation. The rotation is then

$$\alpha = \theta_F d,$$

with θ_F the rotation per unit length for magnetization parallel to the magnetic field. Values are on the order of 10^4 degree cm^{-1} for substituted iron garnets $(BiRGa)Fe_5O_{12}$ where R is a rare-earth element.

An essential feature that distinguishes magnetic and natural rotatory power is the behavior under reversal of the propagation direction of light. After two successive passes in opposite directions through a sample, the effect of spontaneous rotatory power cancels out. In the case of magneto-optical activity, in contrast, the rotation of the plane of polarization is doubled (Fig. 18C.1). This is usually designated as non-reciprocity, and is related to the non-invariance under time reversal (see Complement 8C) featured by a system with magnetization.

---

4. The rotation is actually complemented by the appearance of ellipticity.

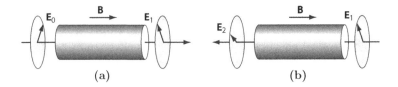

$$(a) \qquad\qquad\qquad\qquad (b)$$

Figure 18C.1: (a) Faraday rotation in an isotropic material (for example a liquid) submitted to magnetic field $\overset{\curvearrowright}{\mathbf{B}}$: the polarization of the light is rotated by the angle α between $\mathbf{E}_0$ and $\mathbf{E}_1$; it is here counterclockwise as seen by an observer looking at the light source. With just the magnetic field reversed, the rotation of the polarization becomes $-\alpha$, clockwise for the observer. (b) The propagation direction is reversed, but not the magnetic field: the rotation of the polarization is now clockwise for the observer (looking at the source, hence to the right), and therefore counterclockwise when looking along the propagation direction. Hence the overall rotation with respect to the laboratory frame, after two passes in opposite directions, is 2α.

The Faraday effect in ferrimagnets has applications such as magneto-optical isolators (preventing light from propagating in one direction), magnetic field sensors, or the observation of magnetic domains and of magnetic field maps.

Further reading

E. du Trémolet de Lacheisserie, D. Gignoux and M. Schlenker (ed.), *Magnetism*, vol. 1: *Fundamentals* (Springer Verlag, New York and Heidelberg, 2005), chap. 13

Appendix 18A. Axial tensors, or pseudo-tensors

18A.1. Definition of axial tensors, or pseudo-tensors

In Chapter 9, we defined a rank-n tensor as a mathematical object having, in a given orthonormal frame, 3^n components (if considering three-dimensional space) which, in a change of orthonormal frame, transform like the product of n components of vectors. This definition corresponds to polar tensors. Consider for example a rank-3 tensor, with components T_{ijk} in a given frame. After a change in frame characterized by transformation matrix A, with coefficients a_{ij}, the new components of this tensor become T'_{mnp} such that:

$$T'_{mnp} = a_{mi}a_{nj}a_{pk}T_{ijk}.$$

A polar vector is a rank-1 tensor.

We saw (Eq. (10.13)) that the components w_i of an axial vector, or pseudo-vector, transform in a change of orthonormal frame into w'_m such that:

$$w'_m = \Delta a_{mi}w_i$$

where Δ is the determinant of the transformation matrix, $+1$ for direct (or proper) transformations, and -1 for indirect (or improper) transformations.

More generally, pseudo-tensors (or axial tensors) of rank n are defined as mathematical objects having, in a given orthonormal frame, 3^n components which, in a change of frame, transform like the product of n components, multiplied by the determinant of the matrix transforming one frame into the other.

For example, a rank-3 axial tensor with components V_{ijk} in a given frame will, in another frame deduced from the previous one through the transformation matrix A, have components V'_{mnp} such that:

$$V'_{mnp} = \Delta a_{mi}a_{nj}a_{pk}V_{ijk}. \tag{18A.1}$$

18A.2. Levi-Civita tensor, or permutation tensor

In Section 10.6.2, we defined the Levi-Civita symbol, δ_{ijk}, such that, *whatever the coordinate system used,*

$\delta_{ijk} = 0$ if at least two of the subscripts i, j and k are equal,

$\delta_{ijk} = 1$ if the permutation ijk is direct (123, 231, 312),

$\delta_{ijk} = -1$ if it is indirect (132, 213, 321).

To show that this symbol is an axial tensor, we consider first the polar tensor of rank 3, ε_{ijk}, defined in a given system of axes, with basis vectors $\mathbf{e}_1$, $\mathbf{e}_2$ and $\mathbf{e}_3$ through:

$\varepsilon_{ijk} = 0$ if two of the subscripts i, j or k are equal,

$\varepsilon_{ijk} = 1$ if the circular permutation ijk is direct (123, 231, 312),

$\varepsilon_{ijk} = -1$ if it is indirect (132, 213, 321).

In another orthonormal axis system, with basis vectors e'_1, e'_2 and e'_3, with transformation matrix $\{a_{ij}\}$, these components become:

$$
\begin{aligned}
\varepsilon'_{mnp} &= a_{mi}a_{nj}a_{pk}\varepsilon_{ijk} \\
&= a_{m1}(a_{n2}a_{p3} - a_{n3}a_{p2}) + a_{m2}(a_{n3}a_{p1} - a_{n1}a_{p3}) + a_{m3}(a_{n1}a_{p2} - a_{n2}a_{p1}) \\
&= [a_{m1}(e'_n \times e'_p)_1 + a_{m2}(e'_n \times e'_p)_2 + a_{m3}(e'_n \times e'_p)_3] \\
&= e'_m \cdot (e'_n \times e'_p) = (e'_m, e'_n, e'_p).
\end{aligned}
$$

We thus see that:

$\varepsilon'_{mnp} = 0$ if two of the subscript m, n, p are equal,

$\varepsilon'_{mnp} = \Delta$ for $mnp = 123$, 231 and 312,

$\varepsilon'_{mnp} = -\Delta$ for $mnp = 132$, 213 and 321,

where Δ is the determinant of the transformation matrix.

This result shows that the Levi-Civita symbol, which retains the same value irrespective of coordinate changes, must therefore transform in a change of coordinate system in the following way:

$$
\delta'_{mnp} = \Delta a_{mi}a_{nj}a_{pk}\delta_{ijk}.
$$

It is therefore a rank-3 pseudo-tensor, or axial tensor, called the permutation tensor.

18A.3. The gyrotropy tensor $[G]$ is a rank-2 axial tensor

We show that the gyration tensor G_{ml} is a rank-2 axial tensor by investigating how it transforms in a coordinate system change with transformation matrix $\{a_{ij}\}$.

By definition (18.6),

$$
G_{ml} = \frac{1}{2}\delta_{mij}\beta_{ijl}
$$

where δ_{mij} is the permutation tensor, a rank-3 axial tensor, and β_{ijl} a polar rank-3 tensor. The new components G'_{uv} are such that:

$$
G'_{uv} = \frac{1}{2}\delta'_{uij}\beta'_{ijv} = \frac{1}{2}\Delta a_{um}a_{in}a_{jp}\delta_{mnp}a_{ik}a_{jl}a_{vw}\beta_{klw}.
$$

From Equation (9.10), $a_{in}a_{ik} = \delta_{nk}$ and $a_{jp}a_{jl} = \delta_{pl}$, hence we can write:

$$
G'_{uv} = \frac{1}{2}\Delta\delta_{nk}\delta_{pl}a_{um}a_{vw}\delta_{mnp}\beta_{klw} = \frac{1}{2}\Delta a_{um}a_{vw}\delta_{mnp}\beta_{npw} = \Delta a_{um}a_{vw}G_{mw}.
$$

This shows that $[G]$ is a rank-2 axial tensor.

18A.4. Relation between tensors $[G]$ and $[\beta]$

We calculate the product $\delta_{ijn}G_{nl}$ by using the definition (18.6) of G_{nl}.

$$\delta_{ijn}G_{nl} = \frac{1}{2}\delta_{ijn}\delta_{nmp}\beta_{mpl}.$$

For given i, j and n such that $\delta_{ijn} \neq 0$, the only non-zero products $\delta_{ijn}\delta_{nmp}$ are $\delta_{ijn}\delta_{nij}$ equal to 1, and $\delta_{ijn}\delta_{nji}$ equal to -1. Hence:

$$\delta_{ijn}G_{nl} = \frac{1}{2}(\beta_{ijl} - \beta_{jil}) = \beta_{ijl}.$$

We obtain the relation:

$$\beta_{ijl} = \delta_{ijn}G_{nl}.$$

References

[1] L.D. Landau, E.M. Lifshitz and L.P. Pitaevskii, *Course of Theoretical Physics*, vol. 8, *Electrodynamics of Continuous Media*, 2nd edn. (Elsevier, Oxford, 1984)

[2] V.M. Agranovich and V.L. Ginzburg, *Crystal Optics with Spatial Dispersion, and Excitons*, 2nd edn. (Springer-Verlag, 1984)

[3] L. Barron, *Molecular Light Scattering and Optical Activity*, 2nd edn. (Cambridge University Press, 2004)

[4] A.M. Glazer and K. Stadnicka, *J. Appl. Cryst.* **19**, 108–122 (1986)

[5] J.D. Jackson, *Classical Electrodynamics*, 3rd edn. (Wiley, 1998)

[6] J. Goulon *et al.*, X-ray natural circular dichroism in uniaxial gyrotropic single crystals of $LiIO_3$, *J. Chem. Phys.* **108**, 6394–6403 (1998)

[7] J. Goulon *et al.*, Vector part of optical activity probed with X-rays in hexagonal ZnO, *J. Phys. Condens. Matter* **19**, 156201–156219 (2007)

[8] J. Jerphagnon and D.S. Chemla, Optical activity in crystals, *J. Chem. Phys.* **65**, 1522–1529 (1976)

Chapter 19

Electro-optical and elasto-optical effects

This chapter introduces various effects through which the refractive index of a crystal (directly related to its permittivity tensor) can be altered through an external field, such as a static electric field or a strain. These changes to the index induced by an external, freely tailored field, have many applications. Some of these are introduced as the various effects are discussed.

19.1. Introduction

We saw that the optical properties of a crystalline material are directly related to the tensor of electrical impermeability $[\eta]$. This tensor is the inverse of the permittivity tensor $[\varepsilon]$, which is related to the electrical susceptibility tensor $[\chi]$ by relation:

$$[\varepsilon] = \varepsilon_0([E] + [\chi])$$

where $[E]$ is the identity tensor.

The electrical susceptibility $[\chi]$ describes the appearance of a polarization $\mathbf{P}$ (electric dipole moment per unit volume) under the action of the electric field $\mathbf{E}$ of the optical wave incident on the crystal.

In Chapter 16, which deals with crystal birefringence, we restricted the Maclaurin expansion of the polarization $\mathbf{P}$ of the material as a function of the electric field $\mathbf{E}$ with angular frequency ω, to the linear term:

$$\mathbf{P}(\omega) = \varepsilon_0[\chi]\mathbf{E}(\omega). \tag{19.1}$$

We then showed (Chap. 18) that optical activity is due to the influence of the local environment on the susceptibility, leading, in the expansion of $\mathbf{P}$ as a function of $\mathbf{E}$, to an additional term, proportional to the spatial derivatives of the components of the electric field:

$$\frac{P_i(\omega)}{\varepsilon_0} = \chi_{ij}E_j(\omega) + \chi_{ijk}\frac{\partial E_j(\omega)}{\partial x_k}. \tag{19.2}$$

© Springer Science+Business Media Dordrecht 2014
C. Malgrange et al., *Symmetry and Physical Properties of Crystals*,
DOI 10.1007/978-94-017-8993-6_19

The susceptibility can also be influenced by the application of a static electric field $\mathbf{E}^s$ which contributes to the expansion (19.2) an additional term of the form $\chi'_{ijk}E^s_j E_k(\omega)$:

$$\frac{P_i(\omega)}{\varepsilon_0} = \chi_{ij}E_j(\omega) + \chi_{ijk}\frac{\partial E_k(\omega)}{\partial x_j} + \chi'_{ijk}E^s_j E_k(\omega) + \dots \tag{19.3}$$

The expansion can be continued by adding a term depending on the square of the applied static electric field, and possibly other terms related to the influence on the susceptibility of other sorts of applied fields, such as a strain or a stress or a static magnetic field.

The existence of these additional terms in the expansion of the susceptibility leads to a dependence of the impermeability tensor $[\eta]$ on these external applied fields. This means that optical propagation is coupled, for example, with an applied electric field (electro-optical effects), a strain or stress (elasto-optical effects), or a magnetic field (magneto-optical effects). These couplings make it possible to alter optical propagation in crystals through these external fields, and they give rise to many applications.

We restrict discussion to a few examples which give insight on how to deal with these couplings. Section 19.2 introduces electro-optical effects, and Section 19.3 introduces elasto-optical effects, which are also called acousto-optical effects when the deformation results from the application of an acoustic wave.

19.2. Electro-optical effects

Applying a static electric field $\mathbf{E}^s$ to a crystal alters the dielectric impermeability tensor $[\eta]$. We can write:

$$\eta_{ij}(\mathbf{E}) - \eta_{ij}(\mathbf{E} = 0) = \Delta\eta_{ij} = r_{ijk}E_k + z_{ijkl}E_k E_l + \dots$$

To simplify notations, we replaced in this expression $\mathbf{E}^s$ by $\mathbf{E}$, because there is no ambiguity with the electric field $\mathbf{E}(\omega)$ of the optical wave, since $\mathbf{E}(\omega)$ is not involved in this expression for $[\eta]$.

The second-order term is much smaller than the first-order one. Therefore we investigate the linear term first. The quadratic term is evidenced when the linear term is zero. The coefficients r_{ijk} are the components of a rank-3 tensor since they relate rank-1 tensor $\mathbf{E}$ with rank-2 tensor $[\eta]$. In the same way, coefficients z_{ijkl} are the components of a rank-4 tensor.

19.2.1. Linear electro-optical effect, or Pockels effect

General properties

The changes induced on the tensor by a static electric field $\mathbf{E}$ are small, even for rather strong applied fields (10^4 V cm^{-1}). They are of interest only when they bring about a qualitative change to tensor $[\eta]$: for instance when a cubic, therefore optically isotropic, crystal becomes optically anisotropic (induced birefringence), or when a uniaxial crystal becomes biaxial. This change then has the advantage that it can be modulated by altering the amplitude of the electric field. The linear effect is described by:

$$\Delta\eta_{ij} = r_{ijk}E_k$$

where the r_{ijk} are the coefficients of the electro-optical Pockels rank-3 tensor, $[r]$.

Tensor $[\Delta\eta]$, the difference between two symmetric tensors, is symmetric, and the two-subscript notation involving i and j can be replaced by a single-subscript notation, as was done for the stress tensor $[T]$. We then obtain six components $\Delta\eta_\alpha$, where α ranges from 1 to 6, and tensor r_{ijk} can be written with two subscripts. We note that, in contrast to the piezoelectric tensor (Sect. 15.3.2), no doubling occurs when going over from the r_{ijk} to the $r_{\alpha k}$. We have $\Delta\eta_{23} = r_{23k}E_k$ and $\Delta\eta_{32} = r_{32k}E_k$, and the symmetry of $[\Delta\eta]$ leads to $r_{23k} = r_{32k}$.

We can write: $$\Delta\eta_4 = \Delta\eta_{23} = \Delta\eta_{32} = r_{4k}E_k$$

with $r_{4k} = r_{23k} = r_{32k}$, so that:

$$\Delta\eta_\alpha = r_{\alpha k}E_k.$$

In matrix notation, the vectors with components $\Delta\eta_\alpha$ and E_k are column vectors, and the components $r_{\alpha k}$ form a matrix with 6 rows and 3 columns, as for the inverse piezoelectric effect.

$$\begin{pmatrix} \Delta\eta_1 \\ \Delta\eta_2 \\ \Delta\eta_3 \\ \Delta\eta_4 \\ \Delta\eta_5 \\ \Delta\eta_6 \end{pmatrix} = \begin{pmatrix} r_{11} & r_{12} & r_{13} \\ r_{21} & r_{22} & r_{23} \\ r_{31} & r_{32} & r_{33} \\ r_{41} & r_{42} & r_{43} \\ r_{51} & r_{52} & r_{53} \\ r_{61} & r_{62} & r_{63} \end{pmatrix} \begin{pmatrix} E_1 \\ E_2 \\ E_3 \end{pmatrix}.$$

The effect of crystal symmetry on the number of independent components is the same as for the piezoelectric tensor. We therefore refer to Table 15.2, using the transposed matrices (Tab. 19.1a and 19.1b) and taking into account the fact that $r_{\alpha k} = r_{ijk}$ for $\alpha = 4, 5, 6$ whereas $d_{i\alpha} = 2d_{ijk}$. We recall that material rank-3 tensors are zero for all centrosymmetric crystals.

The order of magnitude of the non-zero components of tensor $[r]$ is 10^{-12} m V^{-1}, thus picometres per volt (pm V^{-1}).

Table 19.1a : Form of the matrices representing the electro-optical Pockels tensor (linear effect) for various point groups

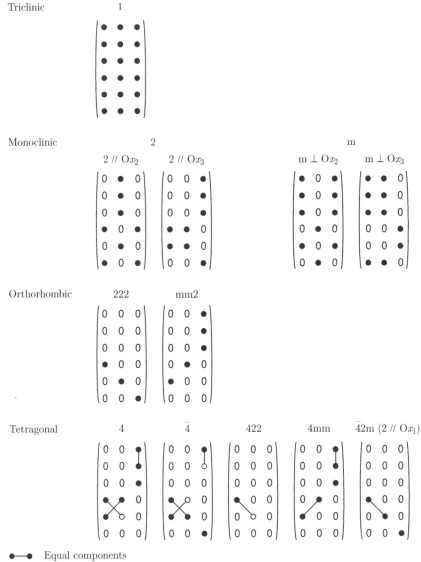

●——● Equal components
●——○ Components with equal absolute values but opposite signs

Table 19.1b : Form of the matrices representing the electro-optical Pockels tensor (linear effect) for various point groups (continued)

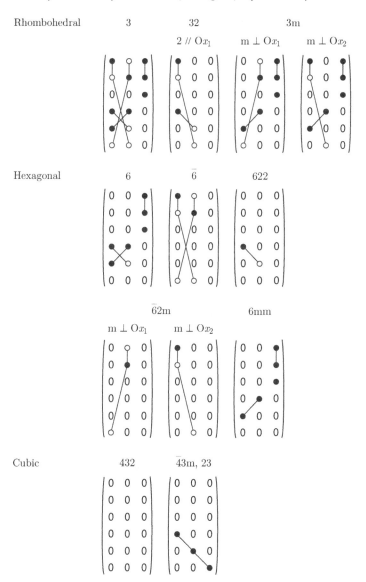

Rhombohedral   3   32   3m

                            $2 /\!/ Ox_1$ $m \perp Ox_1$ $m \perp Ox_2$

Hexagonal   6 $\bar{6}$ 622

                 $\bar{6}2m$ 6mm

                 $m \perp Ox_1$ $m \perp Ox_2$

Cubic   432 $\bar{4}3m$, 23

●——● Equal components
●——○ Components with equal absolute values but opposite signs

Example of a cubic crystal with point group 23

This corresponds to crystals of sodium bromate $NaBrO_3$ or of sodium chlorate $NaClO_3$.

Tensor $[r]$, expressed in the crystallographic axis system, has only three non-zero components which are equal to one another, $r_{41} = r_{52} = r_{63}$.

Consider a crystal plate of $NaBrO_3$ (or $NaClO_3$), with faces perpendicular to one of the 2-fold axes, chosen as Ox_3. We apply an electric field perpendicular to the plate (Fig. 19.1a).

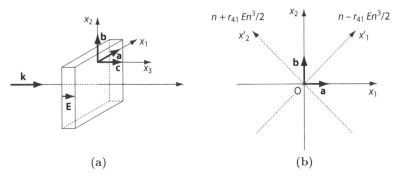

(a)   (b)

Figure 19.1: (a) Schematic representation of optical propagation in a crystal of $NaClO_3$ submitted to an electric field parallel to basis vector **c**; (b) directions of the neutral lines of the plate (dotted lines)

We have:

$$
\begin{pmatrix} \Delta\eta_1 \\ \Delta\eta_2 \\ \Delta\eta_3 \\ \Delta\eta_4 \\ \Delta\eta_5 \\ \Delta\eta_6 \end{pmatrix} = \begin{pmatrix} 0 & 0 & 0 \\ 0 & 0 & 0 \\ 0 & 0 & 0 \\ r_{41} & 0 & 0 \\ 0 & r_{41} & 0 \\ 0 & 0 & r_{41} \end{pmatrix} \begin{pmatrix} 0 \\ 0 \\ E \end{pmatrix} = \begin{pmatrix} 0 \\ 0 \\ 0 \\ 0 \\ 0 \\ r_{41}E \end{pmatrix}.
$$

$[\Delta\eta]$ has one non-zero component, and the electrical impermeability tensor $[\eta]$ becomes:

$$
\begin{pmatrix} \eta_0 & r_{41}E & 0 \\ r_{41}E & \eta_0 & 0 \\ 0 & 0 & \eta_0 \end{pmatrix}
$$

where η_0 is the value for zero electric field, $\eta_0 = 1/n^2$ (Eq. (16.24)), where n is the index of the crystal without electric field. The crystal is now birefringent. Axis Ox_3, parallel to one of the basis vectors of the cubic unit cell of the crystal (**c** on Fig. 19.1a) and to the applied electric field, is a principal axis of the tensor. In an axis system $(Ox_1'x_2'x_3)$ rotated by $45°$ around axis Ox_3, tensor $[\eta]$ becomes:

$$\begin{pmatrix} \dfrac{1}{n^2} + r_{41}E & 0 & 0 \\ 0 & \dfrac{1}{n^2} - r_{41}E & 0 \\ 0 & 0 & \dfrac{1}{n^2} \end{pmatrix}.$$

Axes Ox_1' and Ox_2' are principal axes. The principal values of tensor $[\eta]$ are the reciprocals of the squares of the indices n_i which define the index ellipsoid (16.24), hence:

$$\frac{1}{n_1^2} = \frac{1}{n^2} + r_{41}E \qquad \text{and} \qquad \frac{1}{n_2^2} = \frac{1}{n^2} - r_{41}E.$$

Suppose we send onto the crystal a plane electromagnetic wave with wave-vector $\mathbf{k}$ perpendicular to the plate, hence parallel to the electric field (Fig. 19.1a). The components of the electric field of the wave along Ox_1' and Ox_2' propagate with different indices n_1 and n_2 (Fig. 19.1b). The difference $\Delta n = n_2 - n_1$ is easily calculated, because the change induced by the electric field is very small. We obtain:

$$\Delta \left(\frac{1}{n^2} \right) = -2 \frac{\Delta n}{n^3} = -2 r_{41} E \tag{19.4}$$

whence we deduce:

$$\Delta n = n_2 - n_1 = n^3 r_{41} E.$$

The crystal plate has become birefringent, and its birefringence is proportional to the value of the applied electric field. Its neutral lines are parallel to Ox_1' and Ox_2', thus at $45°$ to vectors $\mathbf{a}$ and $\mathbf{b}$.

Example of a tetragonal crystal with point group $\bar{4}2m$: a crystal of KDP (potassium dihydrogen phosphate KH_2PO_4)

We consider a plate of KDP in which the faces are perpendicular to the optic axis ($\bar{4}$ axis parallel to $\mathbf{c}$), and an incident wave normal to the plate (Fig. 19.2a), thus with wave-vector parallel to the optic axis. When no electric field is applied, the plate is isotropic for the incident ray. The electro-optical tensor r_{ijk} for point group $\bar{4}2m$ has the form given below in Equation (19.5) if axis Ox_3 is chosen parallel to the $\bar{4}$ axis and axes Ox_1 and Ox_2 are parallel to the 2-fold axes. The latter are parallel to basis vectors $\mathbf{a}$ and $\mathbf{b}$ because the space group of KDP is $I\bar{4}2d$ (Sect. 7.4). Suppose we apply an electric field, with norm E, parallel to the optic axis, i.e. to the $\bar{4}$ axis. Tensor $\Delta \eta_\alpha$ is such that:

$$\begin{pmatrix} \Delta \eta_1 \\ \Delta \eta_2 \\ \Delta \eta_3 \\ \Delta \eta_4 \\ \Delta \eta_5 \\ \Delta \eta_6 \end{pmatrix} = \begin{pmatrix} 0 & 0 & 0 \\ 0 & 0 & 0 \\ 0 & 0 & 0 \\ r_{41} & 0 & 0 \\ 0 & r_{41} & 0 \\ 0 & 0 & r_{63} \end{pmatrix} \begin{pmatrix} 0 \\ 0 \\ E \end{pmatrix} = \begin{pmatrix} 0 \\ 0 \\ 0 \\ 0 \\ 0 \\ r_{63}E \end{pmatrix}. \tag{19.5}$$

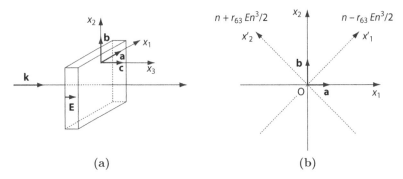

(a)   (b)

Figure 19.2: (a) Schematic representation of optical propagation in a crystal of KDP submitted to an electric field **E** parallel to basis vector **c**; (b) directions of the neutral lines of the plate (dotted lines)

We obtain $\Delta\eta_6 = r_{63}E$ and all the other terms are zero.

Tensor $[\eta]$ is now given by:

$$\begin{pmatrix} \dfrac{1}{n_o^2} & r_{63}E & 0 \\ r_{63}E & \dfrac{1}{n_o^2} & 0 \\ 0 & 0 & \dfrac{1}{n_e^2} \end{pmatrix}$$

where n_o and n_e are respectively the ordinary and extraordinary indices of the KDP crystal without applied electric field. Rotation of the axis system by 45° around the optic axis Ox_3 makes the tensor diagonal:

$$\begin{pmatrix} \dfrac{1}{n_o^2} + r_{63}E & 0 & 0 \\ 0 & \dfrac{1}{n_o^2} - r_{63}E & 0 \\ 0 & 0 & \dfrac{1}{n_e^2} \end{pmatrix} = \begin{pmatrix} \dfrac{1}{n_1^2} & 0 & 0 \\ 0 & \dfrac{1}{n_2^2} & 0 \\ 0 & 0 & \dfrac{1}{n_3^2} \end{pmatrix}$$

where n_1, n_2 and n_3 are the principal indices.

The index ellipsoid is no more an ellipsoid of revolution around Ox_3. Its principal axes are Ox_3 and two axes Ox_1' and Ox_2' at 45° to the axes Ox_1 and Ox_2. Ox_1' and Ox_2' are thus parallel to the mirrors in the crystal. We have:

$$\frac{1}{n_1^2} = \frac{1}{n_o^2} + r_{63}E \qquad \text{and} \qquad \frac{1}{n_2^2} = \frac{1}{n_o^2} - r_{63}E$$

whence we deduce as above, setting $\Delta n = n_2 - n_1$:

$$\Delta \left(\frac{1}{n^2} \right) = -2 \frac{\Delta n}{n_o^3} = -2r_{63}E,$$

$$\Delta n = n_2 - n_1 = n_o^3 r_{63} E. \tag{19.6}$$

The neutral lines of the plate are at $45°$ to axes **a** and **b** (Fig. 19.2b).

19.2.2. Applications of the linear electro-optical effect

Phase-shifting plates with variable delay

In the two special cases we discussed previously, applying an electric field normal to the plate changes isotropic propagation into anisotropic propagation. The neutral lines of the plate are at $45°$ to the vectors **a** and **b** parallel to the plane of the plate. The difference between the indices for waves with polarization parallel to these neutral lines is proportional to the amplitude E of the electric field and to an electro-optical coefficient r (r_{14} for NaBrO$_3$ and r_{63} for KDP):

$$\Delta n = n^3 r E.$$

If the beam traverses a thickness e of crystal, the components of the electric field of the wave, parallel to the neutral lines of the plate, feature, when leaving the plate, a phase difference φ:

$$\varphi = 2\pi \frac{(\Delta n)e}{\lambda} = 2\pi \frac{n^3 r E e}{\lambda} = 2\pi \frac{n^3 r V}{\lambda} \tag{19.7}$$

where V is the voltage applied between the two faces of the crystal.

The half-wave voltage V_π is defined as the voltage which must be applied to the crystal for the phase shift to be equal to π. It is obtained from Equation (19.6), yielding:

$$V_\pi = \frac{\lambda}{2n^3 r}. \tag{19.8}$$

The phase shift can then be written as:

$$\varphi = \frac{\pi V}{V_\pi}.$$

For a crystal of KDP, $r = r_{63} = 10.3$ pm V^{-1}, $n = 1.51$ for $\lambda = 545$ nm and $V_\pi = 7.8$ kV. For a crystal of NaClO$_3$, the half-wave voltage, for the same wavelength, is equal to 200 kV.

The same plates become quarter-wave plates ($\varphi = \pi/2$) if they are submitted to a voltage equal to half the half-wave voltage.

We note that the value of the phase shift does not depend on the crystal thickness, but on the voltage applied between its two faces. Thus, in principle, thin crystals can be used. However, since the voltages are rather large, the crystal must be thick enough to avoid electrical breakdown.

Longitudinal field intensity modulator

Consider a crystal of KDP cut into a rectangle parallelepiped with sides parallel to the basis vectors $\mathbf{a}$, $\mathbf{b}$ and $\mathbf{c}$ of the unit cell (Fig. 19.3). We choose an orthonormal coordinate system $(Ox_1x_2x_3)$ (the unit vectors on these axes are written as $\mathbf{i}$, $\mathbf{j}$ and $\mathbf{k}$) parallel to the basis vectors, and an axis system $(Ox_1'x_2'x_3)$ (with unit vectors $\mathbf{i}'$, $\mathbf{j}'$ and $\mathbf{k}$) rotated with respect to the first one by $45°$ around Ox_3.

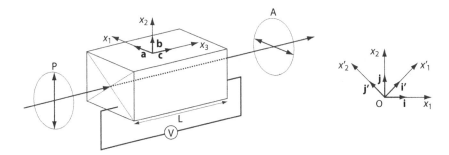

Figure 19.3: Intensity modulator with longitudinal field

We send onto this crystal a plane wave, with wave-vector $\mathbf{k}_o$ parallel to Ox_3, linearly polarized along Ox_2 thanks to a polarizer P. The crystal has length L along the propagation direction. Using transparent electrodes deposited on the entrance and exit surfaces of the crystal, we apply a voltage V, which creates in the crystal an electric field $E = V/L$ perpendicular to the plate. The electric field of the wave incident on the entrance face of the crystal has form $E_0(\exp -i\omega t)\mathbf{j}$, the origin of the axes being chosen on the entrance face of the crystal.

We saw (Sect. 19.2.1) that the neutral lines of the crystal are parallel to Ox_1' and Ox_2'. The incident polarization splits into two components parallel to the neutral lines $((\sqrt{2}/2)\,E_0\mathbf{i}'$ and $(\sqrt{2}/2)\,E_0\mathbf{j}')$, which propagate with different indices and thus shift their phase as propagation proceeds. At the crystal exit, the electric field of the wave can be written:

$$\mathbf{E} = E_0\frac{\sqrt{2}}{2}\exp i(k_0n_1L - \omega t)\,\mathbf{i}' + E_0\frac{\sqrt{2}}{2}\exp i(k_0n_2L - \omega t)\,\mathbf{j} \qquad (19.9)$$

with $\qquad n_1 = n_o - \dfrac{n_o^3 r_{63}E}{2}, \qquad n_2 = n_o + \dfrac{n_o^3 r_{63}E}{2} \qquad$ and $\qquad k_0 = \dfrac{2\pi}{\lambda}.$

We set $\qquad k_0(n_2 - n_1)L = 2\delta = \dfrac{2\pi}{\lambda}n_o^3 r_{63}EL = \dfrac{2\pi}{\lambda}n_o^3 r_{63}V$

so that $\qquad \delta = \dfrac{\pi}{\lambda}n_o^3 r_{63}V$

which only depends on the voltage V applied to the crystal.

Equation (19.8) can be written as:

$$\mathbf{E} = E_0 \frac{\sqrt{2}}{2} \exp i(k_0 n_1 L - \omega t)(\mathbf{i}' + \exp 2i\delta \mathbf{j}'), \qquad (19.10)$$

and, in Jones notation, the electric field has expression, in base $\{\mathbf{i}', \mathbf{j}'\}$:

$$\begin{pmatrix} E_0 \frac{\sqrt{2}}{2} \\ E_0 \frac{\sqrt{2}}{2} \exp 2i\delta \end{pmatrix}.$$

We see that, depending on the value 2δ of the phase shift between the two components, the wave can take on, at the crystal exit, any polarization (see Sect. 17.1.3). The crystal is a half-wave plate if the applied voltage V_π is such that $2\delta = \pi$, hence $V_\pi = \lambda/2n^3 r_{63}$, a value we already found in (19.7). We can thus write that $2\delta = \pi V/V_\pi$.

The electric field (19.9) can also be written, to within a phase factor:

$$\mathbf{E} = E_0 \frac{\sqrt{2}}{2} [E^{-i\delta} \mathbf{i}' + E^{i\delta} \mathbf{j}'].$$

A linear analyzer A, with polarization axis parallel to Ox_1, is placed after the crystal (Fig. 19.3). The crystal is thus set between crossed polarizers. The electric field after the analyzer is the sum of the projections onto Ox_1 of $\mathbf{E}$. The projection of vector $\mathbf{i}'$ is vector $(\sqrt{2}/2)\mathbf{i}$ and that of vector $\mathbf{j}'$ is $-(\sqrt{2}/2)\mathbf{i}$. We obtain an electric field $\mathbf{E}'$ such that

$$\mathbf{E}' = \frac{E_0}{2}[E^{-i\delta} - E^{i\delta}]\mathbf{i}$$
$$= -iE_0 \sin \delta \, \mathbf{i}.$$

The transmitted intensity I is proportional to the square of the amplitude of the electric field (Sect. 16.3), so that:

$$I = I_0 \sin^2 \delta = I_0 \sin^2 \left(\frac{\pi}{2} \frac{V}{V_\pi} \right)$$

where we note I_0 the intensity incident on the crystal.

If we apply a voltage equal to V_π, the intensity is maximum and equal to the intensity transmitted by the polarizer, whereas at zero voltage the intensity is zero (isotropic propagation, crystal placed between crossed polarizer and analyzer). This device can then be used as a beam switch, by changing from voltage V_π ($I = I_0$) to zero voltage ($I = 0$).

If a DC voltage equal to $V_\pi/2$ is sinusoidally modulated with amplitude much smaller than V_π, the intensity of the light beam is modulated quasi-sinusoidally. The slope of the function $I(V)$ is nearly constant in the vicinity of $V_\pi/2$ (Fig. 19.4), so that the relation between the voltage variation and the variation in transmitted intensity is quasi-linear. In this setup, the voltage V_π does not depend on the crystal length; it can therefore not be optimized. Also it is not easy to make transparent electrodes. The setup described next, with transverse field, is therefore preferred.

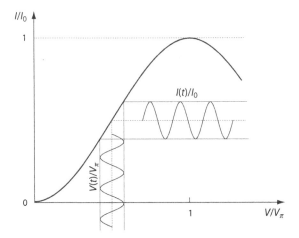

Figure 19.4: Analog modulation, using the experimental device schematically shown on Figure 19.3

Transverse field intensity modulator

The crystal used is again a rectangle parallelepiped of KDP, but cut so that its sides are parallel to Ox_3, Ox'_1 and Ox'_2, with the incident beam parallel to Ox'_1 (Fig. 19.5). It is placed between crossed polarizer and analyzer, the polarization direction of the analyzer being at an angle of $45°$ to the sides of the entrance face of the crystal. The electric field is parallel to Ox_3.

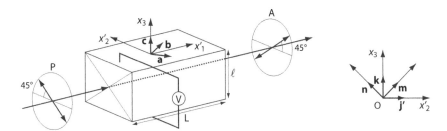

Figure 19.5: Intensity modulator with transverse field

The neutral lines of the plate are Ox_3 and Ox'_2, and only the index for the component parallel to Ox'_2 is modulated by the electric field. We note that, in the absence of the electric field, the crystal is birefringent for the chosen propagation direction. On the entrance face of the plate, the electric field of the light wave can be written as:

$$\mathbf{E} = E_0 \frac{\sqrt{2}}{2}(\mathbf{j}' + \mathbf{k}) \exp(-i\omega t)$$

taking the origin on the entrance face of the crystal.

After the plate, the electric field becomes:

$$\mathbf{E} = E_0 \frac{\sqrt{2}}{2} \left[\mathbf{j}' \exp ik_0 L \left(n_o + \frac{1}{2} n_o^3 r_{63} E \right) + \mathbf{k} \exp ik_0 L n_e \right] \exp(-i\omega t).$$

We set:

$$k_0 L (n_e - n_o - \frac{1}{2} n_o^3 r_{63} E) = 2\delta$$

and the electric field becomes:

$$\mathbf{E} = E_0 \frac{\sqrt{2}}{2} \exp ik_0 L \left(n_o + \frac{1}{2} n_o^3 r_{63} E \right) [\mathbf{j}' + \exp 2i\delta \ \mathbf{k}] \exp(-i\omega t)$$

and, in Jones notation in base $\{\mathbf{j}', \mathbf{k}\}$:

$$\begin{pmatrix} E_0 \frac{\sqrt{2}}{2} \\ E_0 \frac{\sqrt{2}}{2} \exp 2i\delta \end{pmatrix} \text{ or } \begin{pmatrix} E_0 \frac{\sqrt{2}}{2} \exp -i\delta \\ E_0 \frac{\sqrt{2}}{2} \exp i\delta \end{pmatrix}.$$

We can thus write the electric field, to within a phase factor:

$$\mathbf{E} = E_0 \frac{\sqrt{2}}{2} (\exp -i\delta \, \mathbf{j}' + \exp i\delta \, \mathbf{k}).$$

After the analyzer, the electric field $\mathbf{E}'$ is the projection of $\mathbf{E}$ on the unit vector $\mathbf{n}$ parallel to the analyzer (Fig. 19.5):

$$\mathbf{E}' = \frac{E_0}{2} [- \exp -i\delta + \exp i\delta] \mathbf{n} = i(E_0 \sin \delta) \mathbf{n}$$

and the intensity I of the light beam is:

$$I = I_0 \sin^2 \delta = I_0 \sin^2 \left[\frac{\pi L}{\lambda} \left(n_e - n_o - \frac{1}{2} n_o^3 r_{63} \frac{V}{\ell} \right) \right]$$

where ℓ is the thickness of the crystal in the direction of Ox_3 and V the applied voltage $(E = V/\ell)$. We see that the voltage V_π which changes the phase shift 2δ by π can be strongly decreased if the crystal specimen is long and thin, since:

$$V_\pi = \frac{\lambda}{n_o^3 r_{63}} \frac{\ell}{L}.$$

The disadvantage of this configuration is the strong birefringence, which induces a very large phase shift in the absence of an applied electric field. This birefringence is temperature-sensitive, and the operating point is unstable. To minimize this effect, two identical crystals, rotated by 90° around Ox_1', are placed one after the other, so as to cancel the total birefringence. The two crystals must then be submitted to opposite voltages, so that the electro-optical effects add (Exercise 19.1). The two crystals are placed in a thermally insulated box so that they are at the same temperature.

19.2.3. Quadratic electro-optical effect, or Kerr effect

General properties

This effect is much weaker than the linear effect. Therefore, it shows up appreciably only when the linear effect is zero, *i.e.* in centrosymmetric crystals and in liquids. It is defined by the following relation:

$$\eta_{ij}(\mathbf{E}) - \eta_{ij}(\mathbf{E} = \mathbf{0}) = \Delta\eta_{ij} = z_{ijkl}E_k E_l$$

where the z_{ijkl} are the components of the rank-4 electro-optical tensor, the Kerr tensor $[z]$.

$\Delta\eta_{ij}$ is symmetric and E_k and E_l can be interchanged. The matrix of coefficients z_{ijkl} can therefore be contracted. The transformation from z_{ijkl} to $z_{\alpha\beta}$ is performed by writing:

$$\begin{pmatrix} \Delta\eta_1 \\ \Delta\eta_2 \\ \Delta\eta_3 \\ \Delta\eta_4 \\ \Delta\eta_5 \\ \Delta\eta_6 \end{pmatrix} = \begin{pmatrix} z_{11} & z_{12} & z_{13} & z_{14} & z_{15} & z_{16} \\ z_{21} & z_{22} & z_{23} & z_{24} & z_{25} & z_{26} \\ z_{31} & z_{32} & z_{33} & z_{34} & z_{35} & z_{36} \\ z_{41} & z_{42} & z_{43} & z_{44} & z_{45} & z_{46} \\ z_{51} & z_{52} & z_{53} & z_{54} & z_{55} & z_{56} \\ z_{61} & z_{62} & z_{63} & z_{64} & z_{65} & z_{66} \end{pmatrix} \begin{pmatrix} E_1^2 \\ E_2^2 \\ E_3^2 \\ 2E_2 E_3 \\ 2E_3 E_1 \\ 2E_1 E_2 \end{pmatrix}.$$

Thus $z_{\alpha\beta} = z_{ijkl}$ whatever α and β, as in the case of the stiffness tensor. However, in contrast to the tensors of elasticity, the quadratic electro-optical tensor $[z]$ has no reason to be symmetric with respect to α and β, because no thermodynamic argument can be invoked. Thus, for a triclinic crystal (point groups 1 and $\bar{1}$), the 36 coefficients are independent. Tables 19.2a and 19.2b show the various non-zero terms of the tensor as a function of the crystal symmetry. The coefficients of this tensor are expressed in $m^2\, V^{-2}$ and are on the order of 10^{-18}.

Application to isotropic materials

Most applications of the Kerr electro-optical effect concern isotropic materials such as liquids. To know the form of the tensor, we start out from the tensor for the most symmetrical cubic materials (group m$\bar{3}$m) and then show, as was done for the stiffness coefficients (Sect. 13.6), that the three coefficients z_{11}, z_{12} and z_{44} are connected through relation:

$$z_{44} = \frac{1}{2}(z_{11} - z_{12}).$$

Suppose we apply an electric field $\mathbf{E}$. The form of tensor $z_{\alpha\beta}$ is independent of the choice of axes, since the material is isotropic. We can therefore choose the axis directions to suit ourselves, for example take Ox_1 parallel to the electric field.

Table 19.2a: Form of the matrices representing the coefficients $z_{\alpha\beta}$ (and $p_{\alpha\beta}$) for the various point groups

	P.G.	$z_{\alpha\beta}$
Triclinic	1 $\bar{1}$	$\begin{pmatrix} z_{11} & z_{12} & z_{13} & z_{14} & z_{15} & z_{16} \\ z_{21} & z_{22} & z_{23} & z_{24} & z_{25} & z_{26} \\ z_{31} & z_{32} & z_{33} & z_{34} & z_{35} & z_{36} \\ z_{41} & z_{42} & z_{43} & z_{44} & z_{45} & z_{46} \\ z_{51} & z_{52} & z_{53} & z_{54} & z_{55} & z_{56} \\ z_{61} & z_{62} & z_{63} & z_{64} & z_{65} & z_{66} \end{pmatrix}$
Monoclinic	2 m $\dfrac{2}{m}$ $2 \mathbin{/\!/} Ox_2$	$\begin{pmatrix} z_{11} & z_{12} & z_{13} & 0 & z_{15} & 0 \\ z_{21} & z_{22} & z_{23} & 0 & z_{25} & 0 \\ z_{31} & z_{32} & z_{33} & 0 & z_{35} & 0 \\ 0 & 0 & 0 & z_{44} & 0 & z_{46} \\ z_{51} & z_{52} & z_{53} & 0 & z_{55} & 0 \\ 0 & 0 & 0 & z_{64} & 0 & z_{66} \end{pmatrix}$
	2 m $\dfrac{2}{m}$ $2 \mathbin{/\!/} Ox_3$	$\begin{pmatrix} z_{11} & z_{12} & z_{13} & 0 & 0 & z_{16} \\ z_{21} & z_{22} & z_{23} & 0 & 0 & z_{26} \\ z_{31} & z_{32} & z_{33} & 0 & 0 & z_{36} \\ 0 & 0 & 0 & z_{44} & z_{45} & 0 \\ 0 & 0 & 0 & z_{54} & z_{55} & 0 \\ z_{61} & z_{62} & z_{63} & 0 & 0 & z_{66} \end{pmatrix}$
Orthorhombic	222 $mm2$ mmm	$\begin{pmatrix} z_{11} & z_{12} & z_{13} & 0 & 0 & 0 \\ z_{21} & z_{22} & z_{23} & 0 & 0 & 0 \\ z_{31} & z_{32} & z_{33} & 0 & 0 & 0 \\ 0 & 0 & 0 & z_{44} & 0 & 0 \\ 0 & 0 & 0 & 0 & z_{55} & 0 \\ 0 & 0 & 0 & 0 & 0 & z_{66} \end{pmatrix}$
Tetragonal	4 $\bar{4}$ $\dfrac{4}{m}$	$\begin{pmatrix} z_{11} & z_{12} & z_{13} & 0 & 0 & z_{16} \\ z_{12} & z_{11} & z_{13} & 0 & 0 & -z_{16} \\ z_{31} & z_{31} & z_{33} & 0 & 0 & 0 \\ 0 & 0 & 0 & z_{44} & z_{45} & 0 \\ 0 & 0 & 0 & -z_{45} & z_{44} & 0 \\ z_{61} & -z_{61} & 0 & 0 & 0 & z_{66} \end{pmatrix}$
	422 $4mm$ $\bar{4}2m$ $\dfrac{4}{m}mm$	$\begin{pmatrix} z_{11} & z_{12} & z_{13} & 0 & 0 & 0 \\ z_{12} & z_{11} & z_{13} & 0 & 0 & 0 \\ z_{31} & z_{31} & z_{33} & 0 & 0 & 0 \\ 0 & 0 & 0 & z_{44} & 0 & 0 \\ 0 & 0 & 0 & 0 & z_{44} & 0 \\ 0 & 0 & 0 & 0 & 0 & z_{66} \end{pmatrix}$

Table 19.2b: Form of the matrices representing the coefficients $z_{\alpha\beta}$ (and $p_{\alpha\beta}$) for the various point groups (continued); $z^* = \frac{1}{2}(z_{11} - z_{12})$

	P.G.	$z_{\alpha\beta}$
Rhomboedral	3 $\bar{3}$	$\begin{pmatrix} z_{11} & z_{12} & z_{13} & z_{14} & z_{15} & z_{16} \\ z_{12} & z_{11} & z_{13} & -z_{14} & -z_{15} & -z_{16} \\ z_{31} & z_{31} & z_{33} & 0 & 0 & 0 \\ z_{41} & -z_{41} & 0 & z_{44} & z_{45} & -z_{51} \\ z_{51} & -z_{51} & 0 & -z_{45} & z_{44} & z_{41} \\ -z_{16} & z_{16} & 0 & -z_{15} & z_{14} & z^* \end{pmatrix}$
	32 $3m$ $\bar{3}m$	$\begin{pmatrix} z_{11} & z_{12} & z_{13} & z_{14} & 0 & 0 \\ z_{12} & z_{11} & z_{13} & -z_{14} & 0 & 0 \\ z_{31} & z_{31} & z_{33} & 0 & 0 & 0 \\ z_{41} & -z_{41} & 0 & z_{44} & 0 & 0 \\ 0 & 0 & 0 & 0 & z_{44} & z_{41} \\ 0 & 0 & 0 & 0 & z_{14} & z^* \end{pmatrix}$
Hexagonal	6 $\bar{6}$ $\dfrac{6}{m}$	$\begin{pmatrix} z_{11} & z_{12} & z_{13} & 0 & 0 & z_{16} \\ z_{12} & z_{11} & z_{13} & 0 & 0 & -z_{16} \\ z_{31} & z_{31} & z_{33} & 0 & 0 & 0 \\ 0 & 0 & 0 & z_{44} & z_{45} & 0 \\ 0 & 0 & 0 & -z_{45} & z_{44} & 0 \\ -z_{16} & z_{16} & 0 & 0 & 0 & z^* \end{pmatrix}$
	622 $6mm$ $\bar{6}2m$ $\dfrac{6}{m}mm$	$\begin{pmatrix} z_{11} & z_{12} & z_{13} & 0 & 0 & 0 \\ z_{12} & z_{11} & z_{13} & 0 & 0 & 0 \\ z_{31} & z_{31} & z_{33} & 0 & 0 & 0 \\ 0 & 0 & 0 & z_{44} & 0 & 0 \\ 0 & 0 & 0 & 0 & z_{44} & 0 \\ 0 & 0 & 0 & 0 & 0 & z^* \end{pmatrix}$
Cubic	23 $m\bar{3}$	$\begin{pmatrix} z_{11} & z_{12} & z_{21} & 0 & 0 & 0 \\ z_{21} & z_{11} & z_{12} & 0 & 0 & 0 \\ z_{12} & z_{21} & z_{11} & 0 & 0 & 0 \\ 0 & 0 & 0 & z_{44} & 0 & 0 \\ 0 & 0 & 0 & 0 & z_{44} & 0 \\ 0 & 0 & 0 & 0 & 0 & z_{44} \end{pmatrix}$
	432 $\bar{4}3m$ $m\bar{3}m$	$\begin{pmatrix} z_{11} & z_{12} & z_{12} & 0 & 0 & 0 \\ z_{12} & z_{11} & z_{12} & 0 & 0 & 0 \\ z_{12} & z_{12} & z_{11} & 0 & 0 & 0 \\ 0 & 0 & 0 & z_{44} & 0 & 0 \\ 0 & 0 & 0 & 0 & z_{44} & 0 \\ 0 & 0 & 0 & 0 & 0 & z_{44} \end{pmatrix}$
Isotropic		$\begin{pmatrix} z_{11} & z_{12} & z_{12} & 0 & 0 & 0 \\ z_{12} & z_{11} & z_{12} & 0 & 0 & 0 \\ z_{12} & z_{12} & z_{11} & 0 & 0 & 0 \\ 0 & 0 & 0 & z^* & 0 & 0 \\ 0 & 0 & 0 & 0 & z^* & 0 \\ 0 & 0 & 0 & 0 & 0 & z^* \end{pmatrix}$

Tensor $[\Delta\eta]$ becomes:

$$
\begin{pmatrix} \Delta\eta_1 \\ \Delta\eta_2 \\ \Delta\eta_3 \\ \Delta\eta_4 \\ \Delta\eta_5 \\ \Delta\eta_6 \end{pmatrix} = \left(\begin{matrix} z_{11} & z_{12} & z_{12} & & & \\ z_{12} & z_{11} & z_{12} & & & \\ z_{12} & z_{12} & z_{11} & & & \\ & & & z_{44} & & \\ & & & & z_{44} & \\ & & & & & z_{44} \end{matrix} \right) \begin{pmatrix} E^2 \\ 0 \\ 0 \\ 0 \\ 0 \\ 0 \end{pmatrix} = \begin{pmatrix} z_{11}E^2 \\ z_{12}E^2 \\ z_{12}E^2 \\ 0 \\ 0 \\ 0 \end{pmatrix}
$$

and we obtain for tensor $[\eta]$:

$$
\begin{pmatrix} \dfrac{1}{n^2} + z_{11}E^2 & 0 & 0 \\ 0 & \dfrac{1}{n^2} + z_{12}E^2 & 0 \\ 0 & 0 & \dfrac{1}{n^2} + z_{12}E^2 \end{pmatrix}
$$

where n is the optical index of the liquid.

The material is no more isotropic: it has become uniaxial, with the optic axis parallel to the applied electric field, and with ordinary index n_o and extraordinary index n_e calculated as in (19.4) since the index change is very small:

$$
n_o = n - n^3 z_{12} \frac{E^2}{2} \quad \text{and} \quad n_e = n - n^3 z_{11} \frac{E^2}{2}.
$$

The index ellipsoid is of revolution around the direction of the electric field.

A Kerr cell contains a liquid to which an electric field $\mathbf{E}$ is applied normal to the propagation direction of the incident wave (Fig. 19.6a). Two waves propagate in the liquid, one with polarization parallel to $\mathbf{E}$ with index n_e, the other with index n_o (Fig. 19.6b). The difference Δn between the indices is given by:

$$
\Delta n = n_e - n_o = n^3 (z_{12} - z_{11}) \frac{E^2}{2} \tag{19.11}
$$

or

$$
\Delta n = -n^3 z_{44} E^2,
$$

which is often written as:

$$
n_e - n_o = K\lambda E^2 \tag{19.12}
$$

by setting

$$
K = \frac{n^3 (z_{12} - z_{11})}{2\lambda} = -\frac{n^3 z_{44}}{\lambda}. \tag{19.13}
$$

K is the Kerr constant.

If the material were not isotropic, but had point group 432 or m$\bar{3}$ or m$\bar{3}$m (cubic groups for which the linear electro-optical effect is zero), we would obtain the same result provided the electric field is applied parallel to one of the basis vector of the unit cell. The Kerr constant is equal to $n^3 (z_{12} - z_{11})/2\lambda$ (and is no more equal to $-n^3 z_{44}/\lambda$).

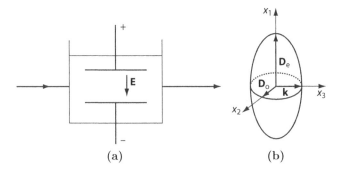

Figure 19.6: (a) Schematic representation of the principle of a Kerr cell; (b) construction of the polarization vectors using the index ellipsoid

19.3. Elasto-optical effects

19.3.1. Definition

Applying stress (tensor $[T]$) or a strain (tensor $[S]$) can lead to a variation $[\Delta\eta]$ of the electrical impermeability tensor $[\eta]$, expressed to first order by:

$$\Delta\eta_{ij} = p_{ijkl}S_{kl},$$
$$\Delta\eta_{ij} = \pi_{ijkl}T_{kl}.$$

The coefficients p_{ijkl} and π_{ijkl} are those of rank-4 tensors (called elasto-optical tensor and piezo-optical tensor respectively), since each of them connects a rank-2 tensor (the strain tensor $[S]$ or the stress tensor $[T]$) to another rank-2 tensor (tensor $[\Delta\eta]$).

Tensors $[S]$, $[T]$ and $[\Delta\eta]$ are symmetric. We can therefore contract the coefficients p_{ijkl} and π_{ijkl}, as was done previously for the elastic tensors and for the Kerr effect tensor.

The transformation rules for going from the S_{ij} to the S_α and from the T_{ij} to the T_α make it clear that:

$$p_{\alpha\beta} = p_{ijkl} \qquad \text{for all } \alpha \text{ and } \beta$$

and
$$\pi_{\alpha\beta} = \pi_{ijkl} \qquad \text{for } \alpha = 1 \text{ to } 6 \text{ and } \beta = 1 \text{ to } 3$$
$$\pi_{\alpha\beta} = 2\pi_{ijkl} \qquad \text{for } \alpha = 1 \text{ to } 6 \text{ and } \beta = 4 \text{ to } 6.$$

Matrices $p_{\alpha\beta}$ and $\pi_{\alpha\beta}$ have no reason to be symmetric with respect to α and β since no thermodynamic argument can be invoked. The effect of crystal symmetry on the number of independent coefficients in tensor $[p]$ is determined in the same way as for the Kerr effect tensor $[z]$. Table 19.2a and 19.2b are therefore valid for tensor $[p]$ too. The table of independent coefficients for tensor $[\pi]$ is slightly different (Tab. 19.3a and 19.3b) because contraction is not performed the same way.

Table 19.3a: Form of the matrices representing coefficients $\pi_{\alpha\beta}$ for the various point groups

P.G.	$\pi_{\alpha\beta}$
Triclinic 1 $\bar{1}$	$\begin{pmatrix} \pi_{11} & \pi_{12} & \pi_{13} & \pi_{14} & \pi_{15} & \pi_{16} \\ \pi_{21} & \pi_{22} & \pi_{23} & \pi_{24} & \pi_{25} & \pi_{26} \\ \pi_{31} & \pi_{32} & \pi_{33} & \pi_{34} & \pi_{35} & \pi_{36} \\ \pi_{41} & \pi_{42} & \pi_{43} & \pi_{44} & \pi_{45} & \pi_{46} \\ \pi_{51} & \pi_{52} & \pi_{53} & \pi_{54} & \pi_{55} & \pi_{56} \\ \pi_{61} & \pi_{62} & \pi_{63} & \pi_{64} & \pi_{65} & \pi_{66} \end{pmatrix}$
Monoclinic 2 m $\dfrac{2}{m}$ $2\,/\!/\,Ox_2$	$\begin{pmatrix} \pi_{11} & \pi_{12} & \pi_{13} & 0 & \pi_{15} & 0 \\ \pi_{21} & \pi_{22} & \pi_{23} & 0 & \pi_{25} & 0 \\ \pi_{31} & \pi_{32} & \pi_{33} & 0 & \pi_{35} & 0 \\ 0 & 0 & 0 & \pi_{44} & 0 & \pi_{46} \\ \pi_{51} & \pi_{52} & \pi_{53} & 0 & \pi_{55} & 0 \\ 0 & 0 & 0 & \pi_{64} & 0 & \pi_{66} \end{pmatrix}$
2 m $\dfrac{2}{m}$ $2\,/\!/\,Ox_3$	$\begin{pmatrix} \pi_{11} & \pi_{12} & \pi_{13} & 0 & 0 & \pi_{16} \\ \pi_{21} & \pi_{22} & \pi_{23} & 0 & 0 & \pi_{26} \\ \pi_{31} & \pi_{32} & \pi_{33} & 0 & 0 & \pi_{36} \\ 0 & 0 & 0 & \pi_{44} & \pi_{45} & 0 \\ 0 & 0 & 0 & \pi_{54} & \pi_{55} & 0 \\ \pi_{61} & \pi_{62} & \pi_{63} & 0 & 0 & \pi_{66} \end{pmatrix}$
Orthorhombic 222 $mm2$ mmm	$\begin{pmatrix} \pi_{11} & \pi_{12} & \pi_{13} & 0 & 0 & 0 \\ \pi_{21} & \pi_{22} & \pi_{23} & 0 & 0 & 0 \\ \pi_{31} & \pi_{32} & \pi_{33} & 0 & 0 & 0 \\ 0 & 0 & 0 & \pi_{44} & 0 & 0 \\ 0 & 0 & 0 & 0 & \pi_{55} & 0 \\ 0 & 0 & 0 & 0 & 0 & \pi_{66} \end{pmatrix}$
Tetragonal 4 $\bar{4}$ $\dfrac{4}{m}$	$\begin{pmatrix} \pi_{11} & \pi_{12} & \pi_{13} & 0 & 0 & \pi_{16} \\ \pi_{12} & \pi_{11} & \pi_{13} & 0 & 0 & -\pi_{16} \\ \pi_{31} & \pi_{31} & \pi_{33} & 0 & 0 & 0 \\ 0 & 0 & 0 & \pi_{44} & \pi_{45} & 0 \\ 0 & 0 & 0 & -\pi_{45} & \pi_{44} & 0 \\ \pi_{61} & -\pi_{61} & 0 & 0 & 0 & \pi_{66} \end{pmatrix}$
422 $4mm$ $\bar{4}2m$ $\dfrac{4}{m}mm$	$\begin{pmatrix} \pi_{11} & \pi_{12} & \pi_{13} & 0 & 0 & 0 \\ \pi_{12} & \pi_{11} & \pi_{13} & 0 & 0 & 0 \\ \pi_{31} & \pi_{31} & \pi_{33} & 0 & 0 & 0 \\ 0 & 0 & 0 & \pi_{44} & 0 & 0 \\ 0 & 0 & 0 & 0 & \pi_{44} & 0 \\ 0 & 0 & 0 & 0 & 0 & \pi_{66} \end{pmatrix}$

Table 19.3b: Form of the matrices representing coefficients $\pi_{\alpha\beta}$ for the various point groups (continued); $\pi^* = (\pi_{11} - \pi_{12})$

P.G.	$\pi_{\alpha\beta}$					
Rhombohedral 3 $\bar{3}$	π_{11}	π_{12}	π_{13}	π_{14}	π_{15}	$-2\pi_{61}$
	π_{12}	π_{11}	π_{13}	$-\pi_{14}$	$-\pi_{15}$	$2\pi_{61}$
	π_{31}	π_{31}	π_{33}	0	0	0
	π_{41}	$-\pi_{41}$	0	π_{44}	π_{45}	$-2\pi_{51}$
	π_{51}	$-\pi_{51}$	0	$-\pi_{45}$	π_{44}	$2\pi_{41}$
	π_{61}	$-\pi_{61}$	0	$-\pi_{15}$	π_{14}	π^*
32 3m $\bar{3}$m	π_{11}	π_{12}	π_{13}	π_{14}	0	0
	π_{12}	π_{11}	π_{13}	$-\pi_{14}$	0	0
	π_{31}	π_{31}	π_{33}	0	0	0
	π_{41}	$-\pi_{41}$	0	π_{44}	0	0
	0	0	0	0	π_{44}	$2\pi_{41}$
	0	0	0	0	π_{14}	π^*
Hexagonal 6 $\bar{6}$ $\dfrac{6}{m}$	π_{11}	π_{12}	π_{13}	0	0	$-2\pi_{61}$
	π_{12}	π_{11}	π_{13}	0	0	$2\pi_{61}$
	π_{31}	π_{31}	π_{33}	0	0	0
	0	0	0	π_{44}	π_{45}	0
	0	0	0	$-\pi_{45}$	π_{44}	0
	π_{61}	$-\pi_{61}$	0	0	0	π^*
622 6mm $\bar{6}$2m $\dfrac{6}{m}$mm	π_{11}	π_{12}	π_{13}	0	0	0
	π_{12}	π_{11}	π_{13}	0	0	0
	π_{31}	π_{31}	π_{33}	0	0	0
	0	0	0	π_{44}	0	0
	0	0	0	0	π_{44}	0
	0	0	0	0	0	π^*
Cubic 23 m$\bar{3}$	π_{11}	π_{12}	π_{13}	0	0	0
	π_{13}	π_{11}	π_{12}	0	0	0
	π_{12}	π_{13}	π_{11}	0	0	0
	0	0	0	π_{44}	0	0
	0	0	0	0	π_{44}	0
	0	0	0	0	0	π_{44}
432 $\bar{4}$3m m$\bar{3}$m	π_{11}	π_{12}	π_{12}	0	0	0
	π_{12}	π_{11}	π_{12}	0	0	0
	π_{12}	π_{12}	π_{11}	0	0	0
	0	0	0	π_{44}	0	0
	0	0	0	0	π_{44}	0
	0	0	0	0	0	π_{44}
Isotropic	π_{11}	π_{12}	π_{12}	0	0	0
	π_{12}	π_{11}	π_{12}	0	0	0
	π_{12}	π_{12}	π_{11}	0	0	0
	0	0	0	π^*	0	0
	0	0	0	0	π^*	0
	0	0	0	0	0	π^*

Tensors $[p]$ and $[\pi]$ are mutually related, since tensors $[S]$ and $[T]$ are: $[S] = [s][T]$ and $[T] = [c][S]$, hence:

$$p_{ijkl} = \pi_{ijmn} c_{mnkl} \qquad \text{and} \qquad \pi_{ijkl} = p_{ijmn} s_{mnkl}.$$

The coefficients p_{ijkl} are dimensionless since $[S]$ and $[\Delta\eta]$ are, and their maximum value is of the order of 1 (and usually closer to 0.1). The coefficients π_{ijkl} are dimensionally reciprocal stresses, and are of the order of 10^{-12} m^2 N^{-1}.

The variation in refractive index, under the effect of stress, is used to analogically evidence stresses in mechanical parts, normally metallic, through the use of models made of transparent and isotropic plastics such as acrylics (Plexiglas®, Lucite®, Perspex®, Altuglas®).

19.3.2. Application to the acousto-optical effects

An acoustic wave is a strain wave. In photo-elastic crystals, *i.e.* those featuring elasto-optical effects, the acoustic wave gives rise to variations in index which it may be of interest to create at will.

Consider for example a crystal of lead molybdate, PbMoO$_4$, with point group $4/m$. It is uniaxial, with optical axis parallel to the 4-fold axis. Suppose we excite in this crystal a longitudinal acoustic wave with wave-vector $\mathbf{K}$ parallel to the 4-fold axis. This wave creates a displacement $\mathbf{u}$ parallel to the 4-fold axis. We choose an orthonormal coordinate system with Ox_3 parallel to this 4-fold axis. The longitudinal displacement then has only one component u_3:

$$u_3 = a \sin(Kx_3 - \Omega t)$$

where Ω is the angular frequency of the acoustic wave.

This displacement induces a strain tensor which has only one non-zero term:

$$S_3 = \frac{\partial u_3}{\partial x_3} = aK \cos(Kx_3 - \Omega t) = S_0 \cos(Kx_3 - \Omega t),$$

where we set, to simplify the expression, $S_0 = aK$.

The strain induces a variation in the impermeability tensor given by:

$$
\begin{pmatrix} \Delta\eta_1 \\ \Delta\eta_2 \\ \Delta\eta_3 \\ \Delta\eta_4 \\ \Delta\eta_5 \\ \Delta\eta_6 \end{pmatrix}
=
\begin{pmatrix}
p_{11} & p_{12} & p_{13} & & & p_{16} \\
p_{12} & p_{11} & p_{13} & & & -p_{16} \\
p_{31} & p_{31} & p_{33} & & & \\
 & & & p_{44} & p_{45} & \\
 & & & -p_{45} & p_{44} & \\
p_{61} & -p_{61} & & & & p_{66}
\end{pmatrix}
\begin{pmatrix} 0 \\ 0 \\ S_3 \\ 0 \\ 0 \\ 0 \end{pmatrix}
=
\begin{pmatrix} p_{13}S_3 \\ p_{13}S_3 \\ p_{33}S_3 \\ 0 \\ 0 \\ 0 \end{pmatrix}.
$$

The crystal, when no acoustic wave is excited, has principal indices $n_1 = n_2 = n_o$ and $n_3 = n_e$, where n_o and n_e are the ordinary and extraordinary indices respectively. With the acoustic wave, tensor $[\eta]$ becomes:

$$\begin{pmatrix} \dfrac{1}{n_o^2} + p_{13}S_3 & 0 & 0 \\ 0 & \dfrac{1}{n_o^2} + p_{13}S_3 & 0 \\ 0 & 0 & \dfrac{1}{n_e^2} + p_{33}S_3. \end{pmatrix}.$$

The crystal remains uniaxial, with its optic axis parallel to Ox_3, but its ordinary and extraordinary indices have changed to become n'_o and n'_e respectively.

We obtain:

$$\frac{1}{n_o'^2} = \frac{1}{n_o^2} + p_{13}S_3 \qquad \text{and} \qquad \frac{1}{n_e'^2} = \frac{1}{n_e^2} + p_{33}S_3.$$

Thus
$$n'_o = n_o - \frac{1}{2}n_o^3 p_{13}S_0 \cos(Kx_3 - \Omega t),$$
$$n'_e = n_e - \frac{1}{2}n_e^3 p_{33}S_0 \cos(Kx_3 - \Omega t).$$

The acoustic wave creates, within the crystal, a modulation of the ordinary and extraordinary indices, periodic along Ox_3, *i.e.* a grating with period $\Lambda = 2\pi/K$ (Fig. 19.7a). The propagation velocity of this index modulation is that of the acoustic wave $(V = \Omega/K)$, on the order of a few $\mathrm{km\,s^{-1}}$, which is extremely small compared to the propagation velocity of light. During the time it takes light to propagate through the crystal, the grating can therefore be considered as motionless.

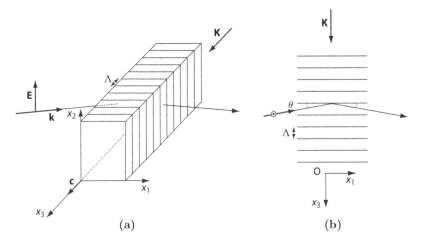

Figure 19.7: (a) Schematic representation of diffraction of a wave by a crystal of PbMoO$_4$ excited by an acoustic wave with wave-vector **K**; (b) Bragg diffraction on the acoustic grating

Suppose we send onto this crystal an optical wave with wave-vector $\mathbf{k}$ belonging to plane (Ox_1, Ox_3), linearly polarized along Ox_2. This wave propagates in the crystal with the index n'_o, modulated by the acoustic wave. It is diffracted by this refractive index grating, in the same way as an X-ray wave is diffracted by a crystal. The diffraction condition for a wave incident on the crystal at an angle θ with the lattice planes can be expressed, as for Bragg "reflection" (diffraction) (see Sect. 3.2.3), by:

$$2\Lambda \sin \theta = \lambda/n_o. \tag{19.14}$$

The refractive index does not show up in the expression for X-ray diffraction because it is then practically equal to 1.

Therefore, the beam is, after diffraction, deviated by an angle 2θ when the acoustic wave is excited (Fig. 19.7b). The device is an optical deflector. The value of theta is on the order of a degree, as shown by the calculation below.

For a crystal of PbMoO$_4$ and an optical wavelength of 500 nm in air, the ordinary index is 2.3. If the acoustic wave has frequency 500 MHz, its propagation velocity in the crystal is 3750 m s^{-1}. Its wavelength is 7.5 μm. The diffraction condition becomes $\sin \theta = 0.0145$ and $\theta = 0.83°$, so that the deviation of the beam is 1.67°.

We assumed, as a first approximation, that the index grating created by the acoustic wave is fixed. Actually, its motion leads, via the Doppler effect, to a (very small) variation $\Delta\omega$ of the diffracted wave frequency. This variation is given by:

$$\frac{\Delta\omega}{\omega} = \frac{2V_{/\!/}}{(c/n)}$$

where $V_{/\!/}$ is the projection of the acoustic wave velocity V onto the propagation direction of the optical wave. If the acoustic wave propagates toward the optical wave (as on Fig. 19.7b), $\Delta\omega$ is positive, while it is negative in the opposite case. Here $V_{/\!/} = V \sin \theta$, and

$$\frac{\Delta\omega}{\omega} = \frac{2V \sin \theta}{(c/n)}.$$

Taking into account (19.13) and $\omega = 2\pi c/\lambda$,

$$\Delta\omega = \frac{2\pi V}{\Lambda} = \Omega.$$

The diffracted wave has an angular frequency ω' slightly different from ω.

$$\omega' = \omega \pm \Omega.$$

The $+$ sign corresponds to the case of Figure 19.7, while the $-$ sign corresponds to propagation of the acoustic wave in the opposite direction.

This result can be interpreted in terms of particles. An electromagnetic wave with angular frequency ω and wave-vector $\mathbf{k}$ is a flow of photons with energy $\hbar\omega$ and momentum $\hbar\mathbf{k}$. The acoustic wave with angular frequency Ω and wave-vector $\mathbf{K}$ is a set of phonons with energy $\hbar\Omega$ and momentum $\hbar\mathbf{K}$. Diffraction of light by the acoustic wave can be viewed as a collision between particles. In the case of Figure 19.7, an incident photon $(\omega, \mathbf{k})$ and a phonon $(\Omega, \mathbf{K})$ disappear while a new photon $(\omega', \mathbf{k}')$ is created. Conservation of energy and of momentum imply:

$$\omega' = \omega + \Omega$$
$$\mathbf{k}' = \mathbf{k} + \mathbf{K}.$$

If the acoustic wave were propagating the other way, an incident photon $(\omega, \mathbf{k})$ would disappear and a phonon $(\Omega, \mathbf{K})$ and a diffracted photon $(\omega', \mathbf{k}')$ would be created, with

$$\omega' = \omega - \Omega$$
$$\mathbf{k}' = \mathbf{k} - \mathbf{K}.$$

19.4. Exercises

Electro-optical effect

Exercise 19.1.

A block of LiNbO$_3$ (point group 3m) has the shape of a rectangle parallelepiped. Its entrance face is parallel to one of the mirrors and one of the sides of this face is parallel to the 3-fold axis (coinciding with the optic axis). We choose an orthonormal axis set such that Ox_3 is parallel to the optic axis and Ox_1 is perpendicular to the entrance face. An electric field, parallel to Ox_3 and pointing toward the positive Ox_3 direction, is applied via transparent electrodes deposited on the faces perpendicular to Ox_3 (see figure below).

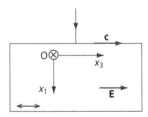

1. Show that the crystal remains uniaxial.

2. A beam of light, perpendicular to the plate and polarized linearly, is sent onto this crystal. Calculate the phase shift φ between the components parallel to Ox_2 and Ox_3 of the electric field of the wave, the thickness of crystal traversed being e.

3. After the first crystal there is a second crystal, identical to the first one
    (including for the position of the electrodes). It is rotated, with respect
    to the first one, by 90° around the propagation direction. It is submitted
    to an electric field with the same norm as that applied to the first crystal,
    but with inverse polarity. Calculate the phase shift φ' which will add
    to the phase shift φ between the same components of the incident wave.
    Calculate the thickness e of each crystal so that the total phase shift is
    equal to π.

    Numerically: $E = 500$ V cm^{-1}; $\lambda = 633$ nm and $n_o = 2.200$; $n_e = 2.286$;
    $r_{13} = 9.6 \times 10^{-12}$ m V^{-1}; $r_{33} = 30.9 \times 10^{-12}$ m V^{-1}.

Exercise 19.2

CdTe is a cubic crystal with point group $\bar{4}$3m. In a system of orthonormal axes
parallel to the basis vectors **a**, **b** and **c**, the electro-optical tensor has three non-
zero coefficients r_{41}, r_{52} and r_{63}. They are equal: $r_{41} = r_{52} = r_{63} = r$. For
CdTe, $r = 4.5 \times 10^{-12}$ m V^{-1}. The index n of CdTe for $\lambda = 1\,\mu$m is 2.84. An
electric field **E** is applied in the [111] direction.

1. Show that, under this electric field, the crystal becomes uniaxial. Deter-
    mine the direction of the optic axis. Calculate the ordinary and extraor-
    dinary indices for $E = 2 \times 10^5$ V m^{-1}.

2. A crystal of CdTe is cut into a rectangle parallelepiped. One of the sides
    of the entrance face of the crystal is parallel to [111]. An electric field is
    applied parallel to this [111] direction. A beam, with wavelength $\lambda = 1\,\mu$m,
    is perpendicular to the entrance face and traverses this block over a length
    $L = 8$ cm. What is the field required to obtain a half-wave plate?

Exercise 19.3

A crystal with point group $\bar{4}$2m is optically uniaxial. It is submitted to an
electric field parallel to the $\bar{4}$ axis. Using the Curie principle, determine whether
the crystal remains uniaxial.

Consider the same question if the crystal has point group 4mm and if the
electric field is applied parallel to the 4-fold axis.

Check the results using the linear electro-optical tensor for each of these point
groups.

Elasto-optical effect

Exercise 19.4

A crystal of NaCl, with point group m$\bar{3}$m, is submitted to a uniaxial stress T
along direction [1$\bar{1}$0].

1. Show that the crystal becomes biaxial. With $T = 2 \times 10^8$ N m^{-2},
    $n_0 = 1.54423$ and with $\pi_{11} = 0.25$, $\pi_{12} = 1.46$, $\pi_{44} = -0.85$ (in units

10^{-12} m^2 N^{-1}), calculate the values of the principal indices for this stressed crystal.

2. The entrance face of the crystal is a (111) surface, and it is illuminated in normal incidence. What are the waves which propagate in the crystal, and what are the corresponding indices?

Exercise 19.5

For point groups 432, $\bar{4}$3m, m$\bar{3}$m, show that the variation in refractive index produced by a hydrostatic pressure p is given by:

$$\Delta n = \frac{n^3}{6}(\pi_{11} + 2\pi_{12})p.$$

Show that Δn can also be written in the form $\Delta n = (n^3/6)\chi(p_{11} + 2p_{12})p$, where the p_{mn} are the components of the elasto-optical tensor, and χ the bulk compressibility of the crystal.

Acousto-optical effect

Exercise 19.6

Germanium is transparent in the infra-red range, and has point group m$\bar{3}$m. A transverse acoustic wave, with frequency $f = 500$ MHz, is excited in a crystal of germanium. The wave, with amplitude A, is polarized along [010] and propagates along direction [001] (see Sect. 14.3.1). Its propagation velocity is given by $v_t = \sqrt{c_{44}/\rho}$ and its value is 3580 m s^{-1} in germanium.

1. What strain does this wave produce?

2. Determine the effect on the propagation of a light wave induced by this strain, by calculating

(a) the induced index variations,

(b) the period of the index grating thus produced.

Complement 19C. Frequency doubling, or Second Harmonic Generation (SHG)

In this chapter, we investigated the influence of a static electric field on the polarization of a crystal by an electromagnetic wave. This is a special case of a non-linear effect connecting the polarization of a crystal to the electromagnetic field(s) that propagate in it. Equation (19.3) can be written, neglecting the spatial dispersion term:

$$\frac{P_i(\omega)}{\varepsilon_0} = \chi_{ij} E_j(\omega) + \chi_{ijk} E_j^s E_k(\omega) + \cdots$$

This is a special case of a more general relation:

$$\frac{P_i(\omega_3)}{\varepsilon_0} = \chi_{ij} E_j(\omega_3) + \chi_{ijk} E_j(\omega_1) E_k(\omega_2) + \cdots$$

which can be interpreted as follows. If two electromagnetic waves, with angular frequencies respectively ω_1 and ω_2, propagate in a crystal, they entail distortions of the electric charge distribution in the crystal, which can oscillate with angular frequency ω_3, a combination of the angular frequencies ω_1 and ω_2 ($\omega_3 = \omega_1 \pm \omega_2$), and thus radiate with the same angular frequency ω_3.

The Pockels effect (Sect. 19.2.1) corresponds to $\omega_1 = 0$, $\omega_2 = \omega$ and $\omega_3 = \omega_1 + \omega_2 = \omega$.

Second harmonic generation corresponds to $\omega_1 = \omega_2 = \omega$ and $\omega_3 = \omega_1 + \omega_2 = 2\omega$. This phenomenon was evidenced by Franken, Hill, Peters and Weinreich[1] in 1961. They focused the beam from a ruby laser (wavelength 694.3 nm) on a quartz crystal, and analyzed the outgoing beam with a prism. They observed, next to the beam with wavelength 694.3 nm, a beam with very low intensity, with wavelength half (347.1 nm) that of the incident beam, hence with twice the angular frequency. The second harmonic yield was then 10^{-8}. A schematic representation of the experiment is shown on Figure 19C.1. Because χ_{ijk} is a third-rank tensor property, second harmonic generation is possible only in non-centrosymmetric crystals.

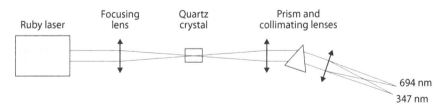

Figure 19C.1: Schematic representation of the setup for second harmonic generation

1. P.A. Franken *et al.*, Generation of optical harmonics, *Phys. Rev. Lett.* **7** (4), 118 (1961).

Using a material more efficient than quartz, a laser with higher intensity, and the phase match property discussed below makes it possible to obtain much better yields.

The investigation of the form of the electric fields shows that the ratio of the powers delivered at the angular frequencies 2ω and ω respectively is proportional to the beam power, to the angular frequency ω, and to the function

$$\frac{\sin^2 \frac{\Delta k L}{2}}{\left(\frac{\Delta k L}{2}\right)^2} \tag{19C.1}$$

where L is the length of crystal traversed and $\Delta k = k(2\omega) - 2k(\omega)$.

It is clear that the above function is maximal for $\Delta k = 0$ so that $k(2\omega) - 2k(\omega) = 0$, which means

$$n(2\omega)\frac{2\omega}{c} = 2n(\omega)\frac{\omega}{c},$$

so that

$$n(2\omega) = n(\omega). \tag{19C.2}$$

In general, the ordinary and extraordinary indices are increasing functions of ω and relation (19C.2) cannot be satisfied if both beams are either ordinary or extraordinary. However, Equation (19C.2) can be satisfied by using the ordinary wave at one frequency and the extraordinary wave at the other frequency. For a uniaxial negative crystal (such as KH_2PO_4, also called KDP), $n_o(\omega)$ must have a value between $n_e(2\omega)$ and $n_o(2\omega)$. There are then directions of the wave-vector for which the ordinary index at angular frequency ω is equal to the extraordinary index at angular frequency 2ω. These directions form a cone of revolution around the optic axis (Exercise 17.6). They are said to produce phase matching.

Second harmonic generation (SHG) is widely used, in particular to generate green light at wavelength 532 nm starting with a Nd:YAG laser which emits in the infra-red at 1064 nm.

Further reading

G. New, *Introduction to Non-Linear Optics* (Cambridge University Press, 2011)

A. Yariv, Y. Pochi, *Optical Waves in Crystals* (Wiley, 2003)

Chapter 20

Solutions to the exercises

20.1. Exercises for Chapter 2

Exercise 2.1

The figure below shows the plane perpendicular to the mirrors going through a given point M. O is the intersection of zz' with this plane. M' is the transform of M through m_1 and M'' the transform of M' through m_2.

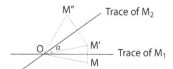

Figure 20.1

1. We see that $(\mathbf{OM}, \mathbf{OM}'') = 2\alpha$ and $m_2 m_1 = \text{Rot}_{zz'}(2\alpha)$.

2. By left-multiplying through m_2, we obtain $m_1 = m_2 \text{Rot}_{zz'}(2\alpha)$ since $(m_2)^2 = E$ with E being identity. In the same way, by right-multiplying by m_1 and using $(m_1)^2 = E$, we obtain $m_2 = \text{Rot}_{zz'}(2\alpha)m_1$.

3. The existence of an n-fold axis zz' leads to $\text{Rot}_{zz'}(2\pi/n)$ being a symmetry operation. m_1 is also assumed to be a symmetry operation. So is the product $\text{Rot}_{zz'}(2\pi/n)\, m_1 = m_2$. Therefore there is a mirror M_2 enclosing the angle π/n with M_1, etc.

4. If A is the rotation by $\pi/2$ around the axis of the pyramid and m_1, m_2, m_3, m_4 the operations of symmetry with respect to each of the 4 mirrors, the 8 symmetry operations are A, A^2, A^3, $A^4 = E$, m_1, m_2, m_3, m_4.

5. The symmetry group is not commutative. For example $m_1 m_2 = A$ and $m_2 m_1 = A^3$ if m_1 and m_2 are the mirror operations with respect to two mirrors deduced from each other through a 45° rotation around zz'.

© Springer Science+Business Media Dordrecht 2014
C. Malgrange et al., *Symmetry and Physical Properties of Crystals*,
DOI 10.1007/978-94-017-8993-6_20

Exercise 2.2

$C_2H_2Br_2$ molecule: a 2-fold axis going through the line that connects the
C atoms; 2 orthogonal mirrors going through the 2-fold axis, one containing
the H atoms, the other the Br atoms.

C_3H_4 molecule: the figure below shows the projection, with the ordinates
indicated, of the molecule onto a plane perpendicular to the carbon chain and
going through the central C atom. It shows that there is a $\bar{4}$ axis going through
the line that connects the carbons, with the center of symmetry of the rotoinversions coinciding with the central C atom. There are 2 orthogonal mirrors,
with traces M_1 and M_2, and two 2-fold axes in the plane of the figure and at
$45°$ to the mirrors.

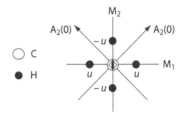

Figure 20.2

Exercise 2.3

The 4-fold axes are Ox, Oy and Oz, perpendicular to the faces and going
through the center O of the cube (Fig. 20.3). The 3-fold axes are the major
diagonals of the cube (DF, AG, CE and BH). The 2-fold axes connect the
middles of two sides in two opposite faces. Consider, for example, the axis
going through M, the middle of BC, and N the middle of EH. MN is parallel to
EB. But EB is perpendicular to AF and to AD. Thus it is perpendicular to the
plane ADGF. A rotation by π around MN thus transforms the cube apices A,
D, G and F into G, F, A and D respectively. It is also clear that this rotation
transforms C, B, E and H into B, C, H and E respectively. Thus the cube is
transformed into itself. MN is a 2-fold axis.

20.2. Exercises for Chapter 3

Exercise 3.1

1. The equation of the plane has form $hx + ky + lz = n$ with n an integer.
    It goes through the nodes $(1, 0, 0)$, $(0, 1, 0)$ and $(0, 0, 3)$, so that $h = n$,
    $k = n$, $3l = n$. The required plane is (331).

2. The (hkl) plane containing $[211]$ and $[120]$ satisfies equations $2h+k+l = 0$
    and $h + 2k = 0$, so that $h = -2k$ and $l = 3k$. The plane is $(\bar{2}13)$. It
    contains $[302]$.

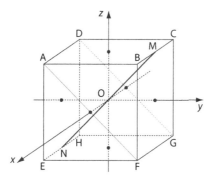

Figure 20.3

3. The required row $[uvw]$ satisfies equations $3u + v + w = 0$ and $u - v = 0$.
    Thus $u = v$ and $4u = -w$. The row is $[11\bar{4}]$.

Exercise 3.2

1. The required angle α is the angle between the normals, respectively $\mathbf{n}_1$
    and $\mathbf{n}_2$, to the planes (101) and $(10\bar{1})$. $\mathbf{n}_1 = [101]^*$ and $\mathbf{n}_2 = [10\bar{1}]^*$.

    $\mathbf{n}_1 \cdot \mathbf{n}_2 = \|\mathbf{n}_1\|\|\mathbf{n}_2\| \cos \alpha$ with $\mathbf{n}_1 = \mathbf{a}^* + \mathbf{c}^*$ and $\mathbf{n}_2 = \mathbf{a}^* - \mathbf{c}^*$.

    Since $\mathbf{a}^*$ and $\mathbf{c}^*$ are perpendicular:

    $$\cos \alpha = \frac{a^{*2} - c^{*2}}{a^{*2} + c^{*2}} = \frac{c^2 - a^2}{c^2 + a^2} = -0.413 \quad \text{and} \quad \alpha = 114.4°.$$

    Another method can be used in this simple case where the planes (101)
    and $(10\bar{1})$ are parallel to $\mathbf{b}$ and thus perpendicular to the plane of the
    figure, $(\mathbf{c}, \mathbf{a})$. We see immediately that $\tan \gamma = c/a$ so that $\gamma = 32.8°$.
    The angle $2\gamma = 65.6°$ is the supplement of α.

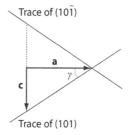

Figure 20.4

2. In this system $\mathbf{a}^*$ and $\mathbf{b}^*$ are respectively parallel to $\mathbf{a}$ and $\mathbf{b}$. Furthermore,
    $\mathbf{a}$ and $\mathbf{b}$ have the same norm, and the same applies to $\mathbf{a}^*$ and $\mathbf{b}^*$. Therefore
    $h\mathbf{a} + k\mathbf{b}$ is parallel to $h\mathbf{a}^* + k\mathbf{b}^*$, hence perpendicular to $(hk0)$. In contrast,
    $\mathbf{a}^*$ and $\mathbf{c}^*$ are respectively parallel to $\mathbf{a}$ and $\mathbf{c}$ but the ratios of their moduli
    are different since $\|\mathbf{a}^*\| = 1/a$ and $\|\mathbf{c}^*\| = 1/c$.

3. $d_{321} = \dfrac{1}{\|3\mathbf{a}^* + 2\mathbf{b}^* + \mathbf{c}^*\|}$. Thus $d_{321} = 1.169$ Å.

Exercise 3.3

1. (a) The equation of line D is $hu + kv = n$ an integer, u and v being the coordinates of the nodes on D. $2h = n$, $3k = n$, so that $n = 6$ and $h = 3$, $k = 2$. The equation is $3u + 2v = 6$.

    (b) 5 lines numbered from 1 to 5.

    (c) The rows have equation $3u + 2v = n$. For the row going through node $(1, 0)$, $n = 3$. For the row going through $(0, 1)$, $n = 2$ (Fig. 20.5).

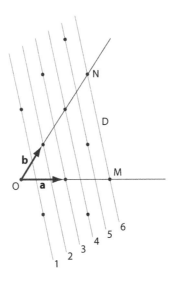

Figure 20.5

2. (a) The row that goes through M and N is $[2\bar{3}0]$. Plane (hkl) is parallel to $\mathbf{c}$, hence $l = 0$. It contains $[2\bar{3}0]$, hence $2h - 3k = 0$. The plane is (320). This result can be retrieved using the above figure. The lattice plane closest to origin intersects plane $(\mathbf{a}, \mathbf{b})$ along the line numbered 1, which intersects $\mathbf{a}$ at $\mathbf{a}/3$ and $\mathbf{b}$ at $\mathbf{b}/2$.

    (b) Plane of the form (32ℓ). **MP** parallel to $[20\bar{1}]$, hence $\ell = 6$, so that the family is (326).

Exercise 3.4

1. Figure 20.6a below, where $\mathbf{c}$ and $\mathbf{c}^*$ are taken to point upwards.
$$\|\mathbf{a}^*\| = \|\mathbf{b}^*\| = 2/a\sqrt{3}; \qquad \|\mathbf{c}^*\| = 1/c.$$

2. Set $\mathbf{V}^* = [110]^* = \mathbf{a}^* + \mathbf{b}^*$.
$$(\mathbf{V}^*)^2 = (\mathbf{a}^* + \mathbf{b}^*)(\mathbf{a}^* + \mathbf{b}^*) = \mathbf{a}^{*2} + \mathbf{b}^{*2} + 2\mathbf{a}^* \cdot \mathbf{b}^* = 3a^{*2}$$

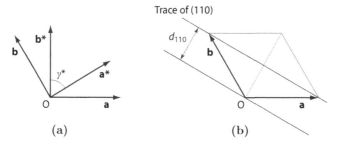

Figure 20.6

since $\mathbf{a}^* \cdot \mathbf{b}^* = (a^*)^2 \cos \gamma^* = (a^*)^2/2$.

$$\|\mathbf{V}^*\| = \sqrt{3}a^* = \frac{2}{a}.$$

$$d_{110} = \frac{1}{\|[110]^*\|} = \frac{a}{2}.$$

This result can, in this special case, be retrieved geometrically using Figure 20.6b.

3. Let α be the angle between the planes (120) and (100). It is also the angle between the vector $[120]^*$, in other words $\mathbf{a}^* + 2\mathbf{b}^*$, and $[100]^* = \mathbf{a}^*$. The calculation is:

$$(\mathbf{a}^* + 2\mathbf{b}^*) \cdot \mathbf{a}^* = \|\mathbf{a}^* + 2\mathbf{b}^*\| \, \|\mathbf{a}^*\| \cos \alpha.$$

This yields $\cos \alpha = 2/\sqrt{7}$ and $\alpha = 40.9°$.

Exercise 3.5

1. $\text{Area OMN} = \left\|\dfrac{\mathbf{a}}{h} \times \dfrac{\mathbf{b}}{k}\right\|; \quad \text{area OMQ} = \left\|\dfrac{\mathbf{a}}{h} \times \dfrac{\mathbf{d}}{i}\right\|; \quad \text{area OQN} = \left\|\dfrac{\mathbf{d}}{i} \times \dfrac{\mathbf{b}}{k}\right\|.$

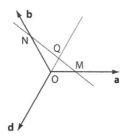

Figure 20.7

2. Area OMN = area OMQ + area OQN. Since the cross products have the same orientation, we can write:

$$\frac{\mathbf{a}}{h} \times \frac{\mathbf{b}}{k} = \frac{\mathbf{a}}{h} \times \frac{\mathbf{d}}{i} + \frac{\mathbf{d}}{i} \times \frac{\mathbf{b}}{k} \quad \text{and} \quad \mathbf{d} = -\mathbf{a} - \mathbf{b}.$$

Therefore: $\dfrac{\mathbf{a} \times \mathbf{b}}{hk} = -\dfrac{\mathbf{a} \times \mathbf{b}}{hi} - \dfrac{\mathbf{a} \times \mathbf{b}}{ik}$ and $\dfrac{1}{hk} = -\left(\dfrac{1}{hi} + \dfrac{1}{ik}\right)$

hence $h + k + i = 0$.

3. $(12\bar{3}1)$ plane; rotation $2\pi/3$ $(\bar{3}121)$; rotation $4\pi/3$ $(2\bar{3}11)$; symmetry with respect to $(\mathbf{a}, \mathbf{c})$ $(1\bar{3}21)$; symmetry with respect to $(\mathbf{b}, \mathbf{c})$ $(\bar{3}211)$.

Exercise 3.6

Unit cell EFGHKLMN. If there were nodes A, B, C and D at the centers of the faces EHNK, EFLK, FGML and HGMN respectively, then **AB** would be a lattice vector. But $\mathbf{AB} = \mathbf{NL}/2$, and there should be a node at the center of face KLMN, so that finally the lattice would be face-centered.

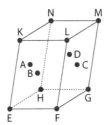

Figure 20.8

Exercise 3.7

CsCl has a cubic P lattice; the motif is: a pair of neighboring Cs^+ and Cl^- ions. NaCl has an F-centered cubic lattice; motif: one Na^+ and a neighboring Cl^-. ZnS has an F-centered cubic lattice; the motif is one Zn^{2+} and one neighboring S^{2-}. Diamond has an F-centered cubic lattice with the motif: C at $(0, 0, 0)$ and C at $(1/4, 1/4, 1/4)$. CaF_2 has an F-centered cubic lattice, with motif: Ca $(0, 0, 0)$ and two F atoms, at $(1/4, 1/4, 1/4)$ and $-(1/4, 1/4, 1/4)$ respectively for example. TiO_2 has a tetragonal P lattice, with a motif consisting of two TiO_2 groups, one around the Ti at $(0, 0, 0)$ and the other around the Ti at $(1/2, 1/2, 1/2)$. $CaTiO_3$ has a P cubic lattice; the motif is Ca at $(0, 0, 0)$, Ti at $(1/2, 1/2, 1/2)$ and 3 O at the centers of the three faces of the cubic unit cell, *viz.* at $(0, 1/2, 1/2)$, $(1/2, 0, 1/2)$, $(1/2, 1/2, 0)$.

It is clear that an atom in the motif can always be replaced by an identical atom displaced by a lattice vector. For example, in the motif indicated for CaF_2, the F atom at $(-1/4, -1/4, -1/4)$ could be replaced by an F atom at $(1/4, -1/4, 1/4)$ deduced through the lattice mode translation $1/2(\mathbf{a} + \mathbf{c})$, etc.

Exercise 3.8

1. For clarity, the atoms of one layer at height 0 are represented by full circles, connected by full lines forming hexagons. Those of the following layer are represented by empty circles connected by dashed lines which also form hexagons. They are at height $1/2$ (Fig. 20.9).

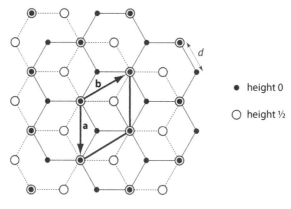

Figure 20.9

- ● height 0
- ○ height ½

2. The cell is hexagonal, with a value of c equal to twice the distance between layers, *viz.* $c = 6.70$ Å. The distance d between two neighboring atoms in a layer is such that $d = (2/3)a(\sqrt{3}/2) = a/\sqrt{3}$ and $a = \sqrt{3}d = 2.46$ Å. There are four atoms per unit cell.

Exercise 3.9

1. Consider a prism. The two Gd atoms in the bases each is counted for $1/2$. The 12 Ni atoms at the apices of the prism each count for $1/6$, thus in total 2 atoms. The atoms at the centers of the 6 faces each count for $1/2$, thus in total 3. The formula is $GdNi_5$.

2. The lattice is hexagonal and the motif, comprising a $GdNi_5$ group, is included in the shaded area on Figure 20.10.

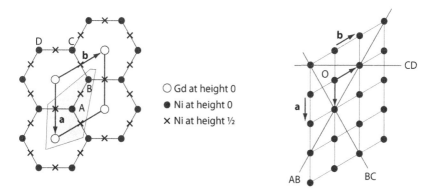

- ○ Gd at height 0
- ● Ni at height 0
- ✕ Ni at height ½

Figure 20.10

3. The face parallel to AB is a (110) plane, in hexagonal lattice notation $(11\bar{2}0)$. The face parallel to BC is deduced from the face parallel to AB by a rotation through $-2\pi/3$ (or $4\pi/3$), and the correct permutation of

indices (k becomes h, h becomes i) leads to plane ($1\bar{2}10$). In the same
way, the face parallel to CD, which is deduced from the face parallel to
AB by a rotation through $2\pi/3$, is a ($\bar{2}110$) plane. These results can be
checked on the reduced hexagonal lattice in Figure 20.10: the plane of the
family closest to origin, intersecting vector $\mathbf{a}$ at a/h and vector $\mathbf{b}$ at b/k,
has been represented for each face.

4. The angle between the ($1\bar{1}0$) and (110) planes is equal to $90°$.

Exercise 3.10

Projection of the nodes of a face-centered tetragonal lattice onto a plane perpendicular to $\mathbf{c}$.

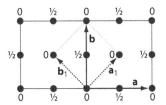

Figure 20.11

The unit cell defined by the basis vectors $\mathbf{a}_1$, $\mathbf{b}_1$, $\mathbf{c}$, half the size of the F($\mathbf{a},\mathbf{b}$)
unit cell, is centered tetragonal. Note: the same projection applies for an
F-centered cubic lattice, but then the centered lattice would have a $\mathbf{c}$ vector
with length $\sqrt{2}$ times larger than a_1 and b_1. It would no more be cubic.

Exercise 3.11

Projection of the nodes of a face-centered monoclinic lattice onto a plane perpendicular to $\mathbf{c}$, here chosen parallel to the 2-fold axis.

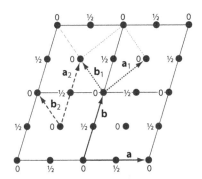

Figure 20.12

The basis vectors $\mathbf{a}_1$ and $\mathbf{b}_1$ define a monoclinic centered unit cell and the vectors $\mathbf{a}_2$ and $\mathbf{b}_2 = \mathbf{b}_1$ a base-centered monoclinic unit cell $(\mathbf{a}_2, \mathbf{c})$, with B lattice.

Exercise 3.12

Line zz', containing the major diagonal K′F of the face-centered cubic unit cell, is a 3-fold axis (Fig. 20.13a). A rotation by $2\pi/3$ around this axis changes L′ into G′, G′ into K and K into L′. The middle of L′G′, R′, is transformed into the middle of G′K, S′, and in the same way S′ is changed into Q′. Line zz' therefore intersects the plane P′ containing the equilateral triangle L′G′K at its center of gravity I′, which is also the center of gravity of the equilateral triangle R′S′Q′. In the same way, zz' intersects the plane P containing the equilateral triangle F′GL at its center of gravity I. Figure 20.13b is a projection of the planes P and P′ along zz': the projections of the nodes in plane P′ are represented by full circles and those of plane P by empty circles. The respective positions of the projections of triangles F′GL and L′G′K, whose centers of gravity I and I′ project at the same point, are easily deduced from the fact that lines F′G and KL′, GL and L′G′, and LF′ and KG′ are pairwise parallel. Figure 20.13b is identical to Figure 3.14c if it is complemented by the lattice translations. Since the lattice is face-centered cubic and not rhombohedral, the distance between the (111) planes is perfectly defined and equal to $a\sqrt{3}/3$ if a is the length of the basis vectors.

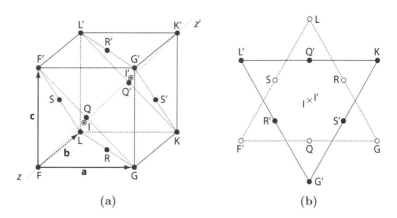

(a)   (b)

Figure 20.13

Exercises 3.13 and 3.14

They should be addressed in the same way as used in the course to determine the reciprocal lattice of a face-centered cubic lattice.

For the reciprocal lattice of a body-centered cubic lattice, we obtain a primitive cell with basis vectors

$$\mathbf{A}^* = \mathbf{b}^* + \mathbf{c}^*, \qquad \mathbf{B}^* = \mathbf{c}^* + \mathbf{a}^*, \qquad \mathbf{C}^* = \mathbf{a}^* + \mathbf{b}^*,$$

and thus a face-centered cubic lattice with basis vectors $2\mathbf{a}^*$, $2\mathbf{b}^*$ and $2\mathbf{c}^*$.

The reciprocal lattice of an A (base-centered) lattice has a primitive cell with basis vectors

$$\mathbf{A}^* = \mathbf{a}^*, \qquad \mathbf{B}^* = (\mathbf{b}^* - \mathbf{c}^*), \qquad \mathbf{C}^* = (\mathbf{b}^* + \mathbf{c}^*).$$

The lattice is an A-type base-centered lattice, with basis vectors $\mathbf{a}^*$, $2\mathbf{b}^*$ and $2\mathbf{c}^*$.

20.3. Exercises for Chapter 4

Exercise 4.1

Figure in plane Π :

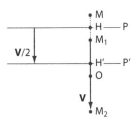

Figure 20.14

O is the middle of $\mathbf{M_1M_2} = \mathbf{V}$. $\mathbf{MM_2} = \mathbf{MM_1} + \mathbf{M_1M_2} = 2(\mathbf{MH} + \mathbf{M_1O}) = 2(\mathbf{MH} + \mathbf{V}/2) = 2\mathbf{MH'}$ if $\mathbf{HH'} = \mathbf{M_1O} = \mathbf{V}/2$.

Exercise 4.2

Consider a general point M. The figure is drawn in the plane going through M and perpendicular to zz'. The traces of P_1 and P_2 are represented in this plane.

1. An elementary geometrical argument shows that M_2 is the transform of M through a symmetry with respect to O (Fig. 20.15a).

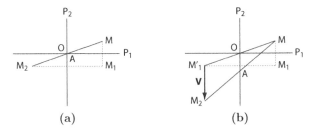

Figure 20.15

2. M_2 is the transform of M through the product operation (Fig. 20.15b). Line MM_2 intersects the trace of P_2 at A. Triangles MOA and $MM_1'M_2$

are similar. Since O is the middle of MM_1', we deduce that A is the middle of MM_2 and that $\mathbf{OA} = \mathbf{M_1'M_2}/2 = \mathbf{V}/2$, independently of the choice of point M in the plane. M_2 deduces from M through a rotation by π around an axis deduced from zz' through a translation by $\mathbf{V}/2$.

20.4. Exercises for Chapter 5

Exercise 5.1

1. In plane Π, shown on Figure 20.16b, the projection H_2 of M_2 deduces from the projection H of M by a rotation through 2θ around zz'. Also $\overline{H_2M_2} = -\overline{H_1M_1} = \overline{HM}$, hence M_2 deduces from M by a rotation through 2θ around zz'.

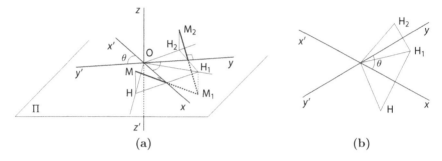

Figure 20.16

2. $A'A = R_\perp(2\theta)$. By right-multiplying this equation with A and noting that $A^2 = E$ (where E is the identity operation), we find: $A' = R_\perp(2\theta)A$. In the same way, left-multiplying by A' and knowing that $A'^2 = E$, we obtain $A = A' R_\perp(2\theta)$.

    If an object has an n-fold axis zz', then $R_{zz'}(2\pi/n)$ is a symmetry operation for this object. If it additionally has a 2-fold axis perpendicular to zz', the operation of rotation by π around this axis xx', noted A, is also a symmetry operation. So is the product $R_{zz'}(2\pi/n)A$. We just showed that $R_{zz'}(2\pi/n) A = A'$ where A' is a rotation by π around an axis perpendicular to zz' and enclosing with axis xx' an angle π/n. Repeating the argument, the object is shown to have n 2-fold axes, perpendicular to zz', and enclosing angles of π/n with each other.

Exercise 5.2

The stereographic projection of a general direction is shown on Figure 20.17a by the cross marked 1. Its transforms through rotations by $2\pi/3$ and $4\pi/3$ around the axis perpendicular to the projection plane are the crosses marked 2 and 3 respectively. The transforms of directions 1, 2 and 3 through the 2-fold axis whose stereographic projection consists of points A and B are the circles marked 4, 5 and 6 respectively. This figure shows the existence of two 2-fold axes enclosing angles of $2\pi/3$ and $4\pi/3$ with AB (Fig. 20.17b).

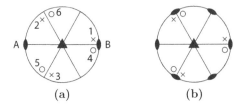

Figure 20.17

Exercise 5.3

1. We choose an orthonormal coordinate system $(Oxyz)$ such that Oy and Oz coincide with the 4-fold axes (the origin O is therefore located at their intersection). We define the operators for rotation by an angle θ around, say an axis Oy as $R_{Oy}(\theta)$, which means that, if $\mathbf{V'}$ is the vector obtained through rotation of a vector $\mathbf{V}$ by an angle θ around an axis Oy, then $\mathbf{V'} = R_{Oy}(\theta)\,\mathbf{V}$. The matrices representing the operators for rotations by $\pm\pi/2$ around Oy or Oz, in the axis system we defined, are given by:

$$R_{Oz}(\pi/2) = \begin{pmatrix} 0 & -1 & 0 \\ 1 & 0 & 0 \\ 0 & 0 & 1 \end{pmatrix} \qquad R_{Oz}(-\pi/2) = \begin{pmatrix} 0 & 1 & 0 \\ -1 & 0 & 0 \\ 0 & 0 & 1 \end{pmatrix}$$

$$R_{Oy}(\pi/2) = \begin{pmatrix} 0 & 0 & 1 \\ 0 & 1 & 0 \\ -1 & 0 & 0 \end{pmatrix} \qquad R_{Oy}(-\pi/2) = \begin{pmatrix} 0 & 0 & -1 \\ 0 & 1 & 0 \\ 1 & 0 & 0 \end{pmatrix}.$$

We deduce that $R_{Oy}(\pi/2)R_{Oz}(\pi/2) = \begin{pmatrix} 0 & 0 & 1 \\ 1 & 0 & 0 \\ 0 & 1 & 0 \end{pmatrix}$.

The composition, or product operation, of a rotation by $\pi/2$ around Oz with a rotation by $\pi/2$ around Oy is an operation which transforms basis vector $\mathbf{i}$ into $\mathbf{j}$, $\mathbf{j}$ into $\mathbf{k}$ and $\mathbf{k}$ into $\mathbf{i}$, $i.e.$ a rotation by $2\pi/3$ around the axis parallel to direction [111] in a cube with sides parallel to Ox, Oy and Oz.

The same result can be obtained by considering a cube ABCDEFGH (Fig. 20.18).

The rotation by $\pi/2$ around Oz transforms:

A, B, C, D, E, F, G, H into
B, C, D, A, F, G, H, E respectively.

The rotation by $\pi/2$ around Oy transforms:

F, G, C, B, E, H, D, A into
G, C, B, F, H, D, A, E respectively.

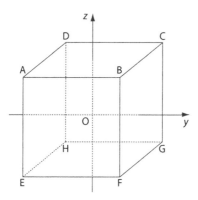

Figure 20.18

We thus see that, through rotations around Oz, A becomes B and B, through rotation around Oy, becomes F. Thus the composition (product) of the two operations finally transforms:

A, B, C, D, E, F, G, H into
F, B, A, E, G, C, D, H respectively.

Points B and H remain unaltered, and A becomes F, F becomes C and C becomes A. In the same way, E becomes G, G becomes D, D becomes E. The result is a rotation by $2\pi/3$ around the axis $[111]$ of the cube.

In the same way, we can show using either method that:

$$R_{Oy}(\pi/2)\,R_{Oz}(-\pi/2) = R_{[1\bar{1}1]}(-2\pi/3),$$
$$R_{Oy}(-\pi/2)\,R_{Oz}(\pi/2) = R_{[11\bar{1}]}(-2\pi/3),$$
$$R_{Oy}(-\pi/2)\,R_{Oz}(-\pi/2) = R_{[\bar{1}11]}(2\pi/3).$$

Thus the directions $[1\bar{1}1]$, $[11\bar{1}]$, $[\bar{1}11]$ are also 3-fold axes.

We show below the matrices representing rotations by $2\pi/3$ around the four cube diagonals:

$$R_{[111]}\,(2\pi/3) \qquad R_{[1\bar{1}1]}\,(2\pi/3) \qquad R_{[11\bar{1}]}\,(2\pi/3) \qquad R_{[\bar{1}11]}\,(2\pi/3)$$

$$\begin{pmatrix} 0 & 0 & 1 \\ 1 & 0 & 0 \\ 0 & 1 & 0 \end{pmatrix} \quad \begin{pmatrix} 0 & -1 & 0 \\ 0 & 0 & -1 \\ 1 & 0 & 0 \end{pmatrix} \quad \begin{pmatrix} 0 & 1 & 0 \\ 0 & 0 & -1 \\ -1 & 0 & 0 \end{pmatrix} \quad \begin{pmatrix} 0 & -1 & 0 \\ 0 & 0 & 1 \\ -1 & 0 & 0 \end{pmatrix}$$

We recall that the operation inverse to the rotation by $2\pi/3$ around one of these axes is the rotation by $-2\pi/3$ around the same axis, and that the associated matrix is the transpose of the matrix of the $2\pi/3$ rotation.

2. Performing the product of the operation consisting of a $\pm 2\pi/3$ rotation around one of the major diagonals of the cube by the operation consisting

of a $\pm 2\pi/3$ rotation around another cube diagonal, we obtain a rotation by π around Ox or Oy or Oz. Example:

$$R_{[1\bar{1}1]}\,(2\pi/3)R_{[111]}\,(2\pi/3) = \begin{pmatrix} 0 & -1 & 0 \\ 0 & 0 & -1 \\ 1 & 0 & 0 \end{pmatrix}\begin{pmatrix} 0 & 0 & 1 \\ 1 & 0 & 0 \\ 0 & 1 & 0 \end{pmatrix}$$

$$= \begin{pmatrix} -1 & 0 & 0 \\ 0 & -1 & 0 \\ 0 & 0 & 1 \end{pmatrix}$$

i.e. a rotation by π around Oz.

Exercise 5.4

Let K be the point in the stereographic projection which is located at the intersection of arc bFb′ and line aa′. This point K is the stereographic projection of the direction parallel to row [011] and pointing to the north, *i.e.* direction OP on Figure 20.19b. This figure shows the meridian plane containing direction [011]. We obtain: $OK = R\tan\alpha = R\tan(\theta/2) = R\tan(\pi/8) = R(\sqrt{2}-1)$. Therefore $aK = R\sqrt{2}$. Since $ab' = R\sqrt{2}$, the circle that goes through b′, K and b has for its center a.

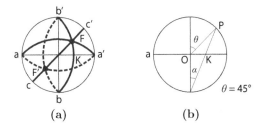

Figure 20.19

Exercise 5.5

Crystals with a hexagonal lattice can have either a hexagonal or a rhombohedral point group. The groups are as follows:
Holohedral: 6/mmm, order 24;
Hemihedral: $\bar{6}$2m, 6mm, 622, 6/m and $\bar{3}$m with order 12;
Tetartohedral: $\bar{6}$, 6, 3m, 32, $\bar{3}$ with order 6;
Ogdohedral: 3, with order 3.

The reader can check that the only cases of ogdohedral crystals are those with point group 3 and with a hexagonal lattice.

Exercise 5.6

1. Form $\{31\bar{4}0\}$

    (a) Group 6mm: to find the faces equivalent to face $(hkil)$ through the symmetry operations of group 6mm, it is sufficient to apply the various symmetry elements of the group, *viz.* rotations by $k\pi/3$ (k being an integer between 1 and 6), and then a mirror operation on the various faces obtained, hence 6 new faces. This gives a total of 12 planes if the normal to the planes is not a special direction.

The indices are obtained by remembering that plane $hkil$ intersects axes Ox, Oy and Ov, parallel respectively to **a**, **b** and **d**, at a/h, b/k and d/i. Rotations by $2\pi/3$ and $4\pi/3$ thus produce permutations among h, k and i and conserve l. Rotations by an odd number of times $\pi/3$ lead to the form $(\bar{h}\bar{k}\bar{i}l)$ and the planes deduced through permutations among $\bar{h}$, $\bar{k}$ and $\bar{i}$.

$(hkil)$ becomes $(ihkl)$ through a rotation by $2\pi/3$. It becomes $(kihl)$ through a rotation by $4\pi/3$. It becomes $(\bar{k}\bar{i}hl)$ through a rotation by $\pi/3$, $(\bar{h}\bar{k}\bar{i}l)$ through a rotation by π and $(\bar{i}\bar{h}kl)$ through a rotation by $5\pi/3$.

The operation of a mirror parallel to **c** and **a** conserves h and l and exchanges k and i.

Form $\{31\bar{4}0\}$ comprises 12 planes: planes $(31\bar{4}0)$, $(\bar{4}310)$, $(1\bar{4}30)$, $(\bar{1}\bar{4}30)$, $(\bar{3}\bar{1}40)$, $(4\bar{3}\bar{1}0)$ and their transforms through one of the mirrors, $(3\bar{4}10)$, $(\bar{4}130)$, $(13\bar{4}0)$, $(\bar{1}3\bar{4}0)$, $(\bar{3}4\bar{1}0)$, $(4\bar{1}\bar{3}0)$.

(b) Group $\bar{6}2$m: the planes equivalent to $(31\bar{4}0)$ are obtained by applying first the symmetry operations associated to the $\bar{6}$ axis, *i.e.* rotations by $2\pi/3$ and $4\pi/3$ around Oz, and a mirror perpendicular to Oz. The mirror symmetry transforms $(hkil)$ into $(hki\bar{l})$. Here $l = 0$ and plane $(31\bar{4}0)$ is a special plane because its normal is in the mirror plane. We must also consider the second generating element, which can be either one of the 2-fold axes or one of the mirrors. We choose a mirror, which we take to be parallel to **a**.

Form $\{31\bar{4}0\}$ comprises 6 planes: the planes $(31\bar{4}0)$, $(\bar{4}310)$, $(1\bar{4}30)$ and $(3\bar{4}10)$, $(\bar{4}130)$, $(13\bar{4}0)$.

2. Form $\{111\}$ for the cubic groups.

Form $\{111\}$ contains the set of faces perpendicular to all the directions equivalent to direction $[111]^*$. For a general direction, the number of equivalent directions is equal to the order n of the group. If the direction is located on a symmetry element to which are associated p symmetry operations, the number of equivalent directions is equal to n/p.

Direction $[111]^*$ is a three-fold axis in all the cubic groups. Furthermore, in those groups which also contain diagonal mirrors, *i.e.* groups $\bar{4}3$m and m$\bar{3}$m, it is also located on these mirrors, which divides by another factor 2 the number of equivalent directions. Note that, in this enumeration, it is sufficient to consider one of the 3 diagonal mirrors, because the two other mirrors are only the result of the existence of the 3-fold axis and of a mirror going through this axis.

- Group 23: form $\{111\}$ comprises plane (111) and planes $(1\bar{1}\bar{1})$, $(\bar{1}1\bar{1})$ and $(\bar{1}\bar{1}1)$, obtained by applying the 2-fold axes. It is distinct from form $\{\bar{1}\bar{1}\bar{1}\}$ because there is no center of symmetry. The latter form contains planes $(\bar{1}\bar{1}\bar{1})$, $(\bar{1}11)$, $(1\bar{1}1)$ and $(11\bar{1})$. We find 4 equivalent planes. This number is indeed equal to $n/p = 12/3 = 4$.

- Group $m\bar{3}$: the existence of the mirrors entails the existence of a center of symmetry. The forms $\{111\}$ and $\{\bar{1}\bar{1}\bar{1}\}$ are identical and comprise the eight planes (111), $(1\bar{1}\bar{1})$, $(\bar{1}1\bar{1})$, $(\bar{1}\bar{1}1)$ and $(\bar{1}\bar{1}\bar{1})$, $(\bar{1}11)$, $(1\bar{1}1)$, $(11\bar{1})$. We check that this number is equal to $n/p = 24/3 = 8$.

- Group 432: although this group does not feature a center of symmetry, the forms $\{111\}$ and $\{\bar{1}\bar{1}\bar{1}\}$ are identical because, for this special direction, the product of a rotation by π around Ox and a rotation by $\pi/2$ around Oy transforms direction $[111]^*$ into direction $[\bar{1}\bar{1}\bar{1}]^*$. The eight planes are equivalent $(n/p = 24/3 = 8)$.

- Group $\bar{4}3m$: same results as for 23. Forms $\{111\}$ and $\{\bar{1}\bar{1}\bar{1}\}$ are different. The order of the group is 24. Direction $[111]^*$ is a 3-fold axis and is located on a diagonal mirror, hence $24/6 = 4$ equivalent directions.

- Group $m\bar{3}m$: eight equivalent planes $(48/6)$, and the forms $\{111\}$ and $\{\bar{1}\bar{1}\bar{1}\}$ are equivalent because there is a center of symmetry.

20.5. Exercises for Chapter 7

Exercise 7.1

1. (a) Generating elements: $\bar{4}$ and 2-fold axes. The stereographic projection is performed onto a plane perpendicular to the $\bar{4}$-axis (Fig. 20.20a). Consider a general direction, with stereographic projection at 1 (cross), and its transforms 2 (circle), 3 (cross), and 4 (circle) through the symmetry operations associated to the $\bar{4}$-axis. The 2-fold axis has for its stereographic projection points A and B. The transforms of directions 1, 2, 3 and 4 through rotations by π around the 2-fold axis have stereographic projections 5, 6, 7 and 8 respectively. This shows the existence of the other 2-fold axis and of two mirrors at $45°$ to the 2-fold axes.

    Generating elements: $\bar{4}$-axis, and the mirror with the segment MM′ as its stereographic projection. Same approach: 1, 2, 3 and 4 are the stereographic projections of directions equivalent through the operations associated with the $\bar{4}$ axis. 5, 6, 7 and 8 are respectively the stereographic projections of the transforms of directions 1, 2, 3 and 4 through the mirror operation m. This shows the existence of the 2-fold axes located in the projection plane at $45°$ to the projections of the mirrors. The stereographic projection shows very clearly that, if an object has a $\bar{4}$ axis and a mirror containing this axis, then there are also 2-fold axes in a plane perpendicular to the $\bar{4}$ axis, going through the center of symmetry associated to this $\bar{4}$ axis, and enclosing an angle of $45°$ with the traces of the mirrors in this plane (see the C_3H_4 molecule of Exercise 2.2).

(b) The stereographic projections of the equivalent directions of group $4/mmm$, the point group of the tetragonal lattice, are given on Figure 20.20c, where the initial general direction is identical to that in Figures 20.20a and 20.20b. We see that Figures 20.20a and 20.20b represent subgroups of 20.20c. The basis vectors **a** and **b** of the lattice are parallel to the mirrors of group $4/mmm$, and can therefore be parallel either to the 2-fold axes or to the mirrors of group $\bar{4}2m$, leading to the space groups $P\bar{4}2m$ and $P\bar{4}m2$ respectively.

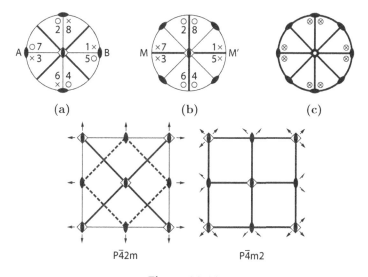

Figure 20.20

In order to determine all the symmetry elements present in the unit cell, we apply the rules for the product (composition) of symmetry operations (rotations and rotoinversion) with the lattice translations (Sect. 7.2.3), keeping in mind that a $\bar{4}$ axis contains a 2-fold axis.

The positions equivalent to the general position with coordinates x, y, z are:

$$P\bar{4}2m \quad x, y, z \quad y, \bar{x}, \bar{z} \quad \bar{x}, \bar{y}, z \quad \bar{y}, x, \bar{z}$$
$$\qquad\qquad x, \bar{y}, \bar{z} \quad y, x, z \quad \bar{x}, y, \bar{z} \quad \bar{y}, \bar{x}, z \; .$$
$$P\bar{4}m2 \quad x, y, z \quad y, \bar{x}, \bar{z} \quad \bar{x}, \bar{y}, z \quad \bar{y}, x, \bar{z}$$
$$\qquad\qquad \bar{x}, y, z \quad \bar{y}, \bar{x}, \bar{z} \quad x, \bar{y}, z \quad y, x, \bar{z}.$$

For both groups, the four positions in the first line are obtained by applying the $\bar{4}$ operations and their multiples. Those of the second line are obtained by applying to the first four positions the operation rotation by π around an axis parallel to **a** for P$\bar{4}$2m, and the operation mirror parallel to **a** for P$\bar{4}$m2.

2. Same approach as in (1). The point group 6/mmm of the hexagonal lattice contains 6 mirrors going through the 6-fold axis, which gives rise to two possibilities for placing the mirrors of point group 3m, a subgroup of the point group of the lattice.

If the lattice is rhombohedral, the point group $\bar{3}$m of the lattice only has 3 mirrors going through the 3-fold axis. They must coincide with the mirrors in the point group of the crystal.

Exercise 7.2

Group P4 contains a 4-fold axis. We choose the origin of the unit cell on this 4-fold axis, and apply the rules of Section 7.2.3 to obtain the product of the symmetry operations associated to the 4-fold axis with the lattice translations. The rotation by $\pi/2$ around the 4-fold axis at A, followed by a translation **a**, is equivalent to a rotation by $\pi/2$ around an axis going through K. There is a 4-fold axis at K. The operation rotation by π around the 4-fold axis going through A followed by a translation **a** is equivalent to a rotation by π around an axis going through L, the middle of $\mathbf{AB} = \mathbf{a}$. There is a 2-fold axis at L, etc.

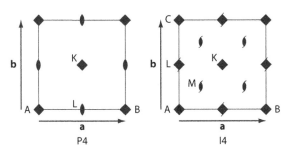

Figure 20.21

Group I4. Existence of a 4-fold axis and a translation $\frac{1}{2}(\mathbf{a} + \mathbf{b} + \mathbf{c})$. To perform the product of a rotation by $\pi/2$ around the 4-fold axis going through A with the translation, we decompose the translation into its component parallel to the axis, *viz.* $\mathbf{c}/2$, and its component perpendicular to the axis, $\frac{1}{2}(\mathbf{a} + \mathbf{b}) = \mathbf{AK}$. The result is a rotation by $\pi/2$ around an axis located at L, the middle of $\mathbf{AC} = \mathbf{b}$ (the point on the perpendicular bisector of AK at which AK subtends an angle $\pi/2$) followed by a translation by $\mathbf{c}/2$. Therefore there exists a 4_2 axis going through L the middle of AC.

The product (composition) of the rotation by π around the 4-fold axis going through A with the translation $\frac{1}{2}(\mathbf{a} + \mathbf{b} + \mathbf{c})$ is a rotation by π followed by a translation by $\mathbf{c}/2$ around an axis going through M the middle of AK, hence there exists a 2_1 axis going through M.

Exercise 7.3

The generating elements of this space group are the operations $(m_1, \mathbf{c}/2)$ and $(m_2, \mathbf{a}/2)$ (modulo the lattice translations). The common origin O of these two operations is located at a point on the intersection of these two mirrors. The product of these two symmetry operations with the same origin is performed by applying relation (4.8). We obtain:

$$(m_1, \mathbf{c}/2)(m_2, \mathbf{a}/2) = (m_1 m_2, m_1 \mathbf{a}/2 + \mathbf{c}/2) = (A, -\mathbf{a}/2 + \mathbf{c}/2),$$

where A is a rotation by π around the intersection line of the two mirrors. The translation part includes both a component $(\mathbf{c}/2)$ parallel to the axis, which is a 2_1 axis, and a component perpendicular to the axis, $(-\mathbf{a}/2)$. The latter can be canceled by displacing the axis: for a 2-fold axis, this implies translating it by $-\mathbf{a}/4$ (figure below). Thus there is a 2_1 axis at O' such that $\mathbf{OO'} = -\mathbf{a}/4$. The space group is written Pca2_1 (c-type mirror perpendicular to $\mathbf{a}$, a-type mirror perpendicular to $\mathbf{b}$, and 2_1 axis parallel to $\mathbf{c}$).

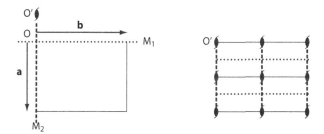

Figure 20.22

Exercise 7.4:

Np

1. Primitive lattice, since the atomic positions cannot be separated into various groups deduced from one another through one (or several) lattice mode translations, such as $\frac{1}{2}(\mathbf{a} + \mathbf{b})$ for a C lattice, etc.

2. The Np atom at A is located on a $\bar{4}$ axis, and we easily see two perpendicular pure mirrors going through A (Fig. 20.23a). The point group of this site is therefore $\bar{4}2m$. The Np atom at B is on a 4-fold axis and on two orthogonal pure mirrors. The site of atom B has point group 4mm. The point group of the crystal has subgroups $\bar{4}2m$ and 4mm. It is therefore 4/mmm.

3. There is therefore a mirror parallel to the plane of the figure. It is a glide mirror with translation $\frac{1}{2}(\mathbf{a} + \mathbf{b})$ at height 0 (n-type mirror). The diagonal mirrors going through A are glide mirrors with glide $\frac{1}{2}(\mathbf{a} + \mathbf{b})$ but those which go through B are pure mirrors. The space group is therefore P4/nmm (n° 129 in International Tables vol. A). Figure 20.23b shows the

mirrors present in the unit cell, and the rotation and rotoinversion axes perpendicular to the plane of the figure.

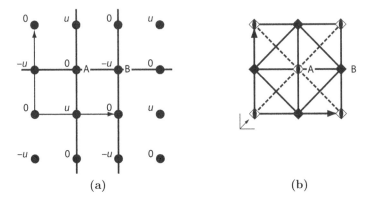

(a)   (b)

Figure 20.23

Exercise 7.5:

NiAs

1. Primitive lattice (same reason as in Exercise 7.4, question (1)). A 6_3 axis going through a Ni atom, 3 pure mirrors going through this 6_3 axis, and 3 c-type mirrors (glide $\mathbf{c}/2$) intersecting them. Space group P6_3mc (n° 186 in I.T. A) (the mirrors perpendicular to $\mathbf{a}$ and $\mathbf{b}$ are m-type mirrors).

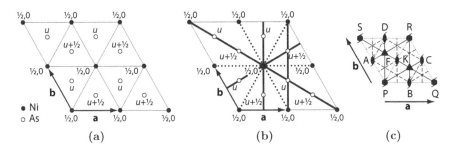

(a)   (b)   (c)

Figure 20.24

Figure (c) shows the symmetry axes and the mirrors present in the unit cell. We apply the rules of Section 7.2.3, taking into account the fact that a 6_3 axis contains a 3-fold axis and a 2_1 axis. The combination of the 3-fold axis at P and of the lattice translation $\mathbf{a}$ gives a 3-fold axis at point K, while the combination of axis 2_1 at P and of the lattice translation $\mathbf{a}$ gives a 2_1 axis at B. In order to find the effect of the combination of the pure mirror with trace SQ and of the lattice translation $\mathbf{a}$, we decompose $\mathbf{a}$ into its component parallel to the mirror, $(\mathbf{a} - \mathbf{b})/2$, and its component perpendicular to the mirror, $(\mathbf{a} + \mathbf{b})/2$. The result is the operation of a mirror with glide $(\mathbf{a} - \mathbf{b})/2$, the trace of which on the plane of the figure deduces from SQ through a translation by $(\mathbf{a} + \mathbf{b})/4$. We obtain the glide mirror with trace CD (dotted line, see Tab. 7.2). The combination of the c-mirror with trace PR and of the translation with vector $\mathbf{a}$ is equivalent to a mirror with trace BC (translation by half the component perpendicular to the mirror of vector $\mathbf{a}$) with glide equal to the component parallel to the mirror of $\mathbf{a}$, $viz.$ $(\mathbf{a} + \mathbf{b})/2$, which adds to $\mathbf{c}/2$ to yield a mirror with trace BC and glide $(\mathbf{a} + \mathbf{b} + \mathbf{c})/2$ (dotted line with long/short alternation).

The point symmetry of the Ni site is 6mm and that of the As atoms is 3m.

Exercise 7.6:

FeOCl

1. Primitive lattice (same reason as in Exercise 7.4, question (1)).

2.

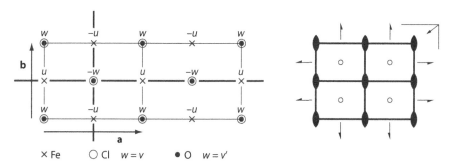

Figure 20.25

3. We immediately distinguish two pure mirrors perpendicular to the plane of the figure and perpendicular to each other. There is a glide mirror with glide vector $\frac{1}{2}(\mathbf{a} + \mathbf{b})$, parallel to the plane of the figure at height 0, and therefore a n mirror perpendicular to $\mathbf{c}$. The space group is Pmmn. The associated point group is mmm, a centrosymmetric group. There are therefore symmetry centers. The 2-fold axes parallel to $\mathbf{c}$ are pure axes while those parallel to $\mathbf{a}$ or $\mathbf{b}$ are 2_1 axes.

4. There are 8 positions (order of the group = 8) equivalent to the general position. Their coordinates are:

$$x, y, z \qquad x + \frac{1}{2}, y + \frac{1}{2}, -z$$
$$-x, y, z \qquad -x + \frac{1}{2}, y + \frac{1}{2}, -z$$
$$x, -y, z \qquad x + \frac{1}{2}, -y + \frac{1}{2}, -z$$
$$-x, -y, z \qquad -x + \frac{1}{2}, -y + \frac{1}{2}, -z.$$

20.6. Exercises for Chapter 8

Exercise 8.1

Figure 3.14b, reproduced below, shows a rhombohedral unit cell. In the close-packed structure ABCABC, points L, N, P and M are the centers of identical, mutually contacting atoms with radius ρ. Then LN $=$ LP $=$ PN $=2\rho$. The angle of the rhombohedron is then equal to 60°. A rhombohedron with angle 60° is the primitive cell for a face-centered cubic lattice.

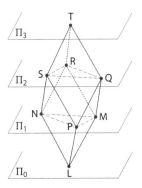

Figure 20.26

Exercise 8.2

1. Figure 20.27a shows the regular tetrahedron MNPL formed by the centers M, N and P of three mutually contacting atoms in a plane A and the center L of the atom in plane B which contacts them. H, the projection of L onto plane A, is the center of gravity (and orthocenter) of the equilateral triangle MNP, and MH $= (2/3)$MK. Also MK $= a(\sqrt{3}/2)$, hence MH $= a/\sqrt{3}$.

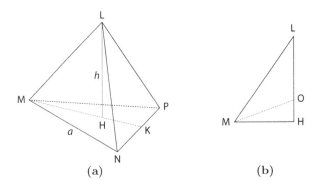

Figure 20.27

In the right triangle LHM, $h^2 = \text{ML}^2 - \text{MH}^2 = a^2(2/3)$. Hence $h = a\sqrt{2/3}$.

2. $c = 2h = 2a\sqrt{2/3}$ and $c/a = 2\sqrt{\dfrac{2}{3}} = 1.63$.

Exercise 8.3

1.   We use the figures and the results of Exercise 8.2. The center O of the
     tetrahedron with side a is located on the altitude LH, which is a 3-fold
     axis for the tetrahedron. We must have $OL = OM = d$ (Fig. 20.27b).

$$OH = h - d = a\sqrt{2/3} - d,$$
$$OM^2 = d^2 = MH^2 + OH^2 = a^2/3 + (a\sqrt{2/3} - d)^2 = d^2.$$

We deduce from the last equation: $a^2 - 2a\sqrt{2/3}d = 0$ and $d = \dfrac{a}{2}\sqrt{\dfrac{3}{2}}$.

Comparing d and h, we see that $d = 3h/4$.

The center of the regular tetrahedron is located on the height from a vertex,
at a distance from the vertex equal to $3/4$ of the height.

2. $d = \dfrac{a}{2}\sqrt{\dfrac{3}{2}} = r_1 + r_2$ and $r_2 = a/2$ hence $r_1 = a/2\left(\sqrt{\dfrac{3}{2}} - 1\right)$

and $\dfrac{r_1}{r_2} = \sqrt{\dfrac{3}{2}} - 1 = 0.225$.

3.   The positions of the Zn atoms are those of the centers of an ABABAB
     stacking. The calculated value of c/a is 1.636. The stacking of the centers
     is that of close packing, but here the atoms are not in contact. The value
     of parameter u is close to $0.375 = 3/8$.

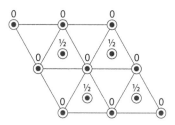

Figure 20.28

The heights indicated on the figure are those of the Zn atoms represented
by full circles. The S atoms (empty circles) project onto the Zn atoms
and are located a distance $3c/8$ above them. A Zn atom at height 0 and
the 3 neighboring Zn atoms at height $1/2$ form a regular tetrahedron with
height $c/2$. The S atom at height $3/8$ is a distance $3c/8$ from the vertex,
hence $3/4$ of the height of the tetrahedron. It is therefore at the center of
this tetrahedron. The wurtzite structure can thus be represented as an
ABABAB stacking, with S atoms in the tetrahedral sites. There is one
tetrahedral site per Zn atom. In the blende structure, the Zn atoms form

an f.c.c. stacking, *i.e.* an ABCABC stacking. The S atoms are in every other tetrahedral site, since there are twice as many tetrahedral sites as Zn atoms in the f.c.c. structure. We thus understand more clearly why ZnS can exist under these two forms, blende and wurtzite.

20.7. Exercises for Chapter 9

Exercise 9.1

1. $\{a_{ij}\}= \begin{pmatrix} -\frac{1}{2} & \frac{\sqrt{3}}{2} & 0 \\ -\frac{\sqrt{3}}{2} & -\frac{1}{2} & 0 \\ 0 & 0 & 1 \end{pmatrix}$

2.

$$T'_{11} = (1/4)\, T_{11} + (3/4)\, T_{22} - (\sqrt{3}/2)\, T_{12} = T_{11}$$

$$\text{hence } (3/4)\, (T_{22} - T_{11}) = (\sqrt{3}/2)\, T_{12}, \tag{20.1}$$

$$T'_{22} = (3/4)\, T_{11} + (1/4)\, T_{22} + (\sqrt{3}/2)\, T_{12} = T_{22}$$

$$\text{hence } (3/4)\, (T_{22} - T_{11}) = (\sqrt{3}/2)\, T_{12}, \tag{20.2}$$

$$T'_{12} = (\sqrt{3}/4)\, (T_{11} - T_{22}) - (1/2)\, T_{12} = T_{12}$$

$$\text{hence } (\sqrt{3}/4)\, (T_{22} - T_{11}) = (-3/2)\, T_{12}. \tag{20.3}$$

Relations (20.1) and (20.2) are identical. Relations (20.2) and (20.3) imply that $T_{12} = 0$ and $T_{11} = T_{22}$.

3.

$$T'_{13} = (-1/2)\, T_{13} + (\sqrt{3}/2)T_{23} = T_{13} \text{ hence } 3\, T_{13} = \sqrt{3}\, T_{23},$$

$$T'_{23} = (-\sqrt{3}/2)\, T_{13} - (1/2)\, T_{23} = T_{23} \text{ hence } 3\, T_{23} = -\sqrt{3}\, T_{13},$$

$$T'_{33} = T_{33}.$$

We deduce that $T_{13} = T_{23} = 0$.

Exercise 9.2

Since $[T]$ is a material tensor, each component must remain unchanged under a symmetry operation of the point group of the crystal, *i.e.* $T'_{ij} = T_{ij}$.

The operation of a mirror perpendicular to Ox_1 changes the x_i coordinates of a point M into coordinates x'_j such that:

$$x'_1 = -x_1 \qquad x'_2 = x_2 \qquad x'_3 = x_3.$$

Since the components T_{ij} transform like the product of coordinates $x_i x_j$, we obtain:

$$T'_{11} = T_{11}, \qquad\qquad T'_{22} = T_{22}, \qquad\qquad T'_{33} = T_{33},$$
$$T'_{12} = -T_{12} = T_{12} = 0, \quad T'_{13} = -T_{13} = T_{13} = 0, \quad T'_{23} = T_{23}.$$

We see that the components T_{ij} with the subscripts ij containing one and only one subscript equal to 1 are zero.

The operation of a mirror perpendicular to Ox_2 changes the coordinates x_i of a point M into coordinates x'_j such that:

$$x'_1 = x_1 \qquad x'_2 = -x_2 \qquad x'_3 = x_3,$$

and the components T_{ij} containing one and only one subscript equal to 2 are zero.

The existence of a third mirror entails that the components containing one and only one subscript equal to 3 are zero. But these components, T_{13} and T_{23}, are already zero because of the first two mirrors.

We thus note that the form of the tensor is the same for groups mm2 and mmm:

$$\begin{pmatrix} T_{11} & & \\ & T_{22} & \\ & & T_{33} \end{pmatrix}$$

in an orthonormal set of axes perpendicular to the mirrors.

20.8. Exercises for Chapter 10

Exercise 10.1

1. $\{T'_{ij}\} = \begin{pmatrix} 1 & 0 & -4 \\ 0 & 2 & 0 \\ -4 & 0 & 3 \end{pmatrix}$

2. $\{T''_{ij}\} = \begin{pmatrix} 1 & 0 & 4 \\ 0 & 2 & 0 \\ 4 & 0 & 3 \end{pmatrix}$

We could expect to find the tensor unchanged, since the basis vector **b**, perpendicular to **a** and **c**, is a 2-fold axis for the three monoclinic groups.

Exercise 10.2

1. (a) $\mathbf{j} = \begin{pmatrix} 7 \\ 0 \\ \sqrt{3} \end{pmatrix} E$ (b) $\mathbf{j} = \begin{pmatrix} 0 \\ 8 \\ 0 \end{pmatrix} E$ (c) $\mathbf{j} = \begin{pmatrix} \sqrt{3} \\ 0 \\ 5 \end{pmatrix} E$

(d) $\mathbf{j} = \begin{pmatrix} \frac{7+\sqrt{3}}{\sqrt{2}} \\ 0 \\ \frac{5+\sqrt{3}}{\sqrt{2}} \end{pmatrix} E$ (e) $\mathbf{E} = \begin{pmatrix} \frac{\sqrt{3}}{2} \\ 0 \\ \frac{1}{2} \end{pmatrix} E$; $\mathbf{j} = \begin{pmatrix} \frac{\sqrt{3}}{2} \\ 0 \\ \frac{1}{2} \end{pmatrix} 8E = 8\mathbf{E}$,

j is parallel to **E**, and this special direction for **E** is that of one of the principal axes of the tensor.

2.   (a) $j_n = 5E$.

(b) $j_n/E = 5$ is the value of the electrical conductivity of the material in direction $[001]$.

3.   The coordinate change matrix is given by:

$$\begin{pmatrix} \dfrac{1}{2} & 0 & -\dfrac{\sqrt{3}}{2} \\ 0 & 1 & 0 \\ \dfrac{\sqrt{3}}{2} & 0 & \dfrac{1}{2} \end{pmatrix}.$$

The form of the tensor is therefore:

$$[\sigma'] = \begin{pmatrix} 4 & & \\ & 8 & \\ & & 8 \end{pmatrix}$$

and this axis system is that of the principal axes of the tensor. This result is no surprise, since the initial form of the tensor shows that Ox_2 is a principal axis. Furthermore, we saw in question 1e) that the direction in plane (Ox_3, Ox_1) that encloses an angle of $60°$ with Ox_3 is a principal axis.

Exercise 10.3

1. $\{\sigma_{ij}\} = \begin{pmatrix} \sigma_1 & & \\ & \sigma_1 & \\ & & \sigma_3 \end{pmatrix}$ (see Sect. 10.5.4).

2. $\sigma_s = (s_1 \; s_2 \; s_3) \begin{pmatrix} \sigma_1 & & \\ & \sigma_1 & \\ & & \sigma_3 \end{pmatrix} \begin{pmatrix} s_1 \\ s_2 \\ s_3 \end{pmatrix} = (s_1 \; s_2 \; s_3) \begin{pmatrix} \sigma_1 s_1 \\ \sigma_1 s_2 \\ \sigma_3 s_3 \end{pmatrix}$

$= \sigma_1(s_1^2 + s_2^2) + \sigma_3 s_3^2.$

$$s_1 = \sin\theta \cos\varphi, \qquad s_2 = \sin\theta \sin\varphi, \qquad s_3 = \cos\theta.$$

$$\sigma_s = \sigma_1 \sin^2\theta + \sigma_3 \cos^2\theta.$$

σ_s is independent of φ. This is no surprise since the representative quadric for tensor $[\sigma]$ has revolution symmetry around the three-fold axis.

Exercise 10.4

1.   The symmetry operation associated to the presence of a $\bar{4}$ axis parallel to Ox_3 is the product of a rotation by $\pi/2$ around this axis by a symmetry with respect to origin O. This operation changes the coordinates x_i of a point M into coordinates x'_j such that:

$$x'_1 = -x_2 \qquad x'_2 = x_1 \qquad x'_3 = -x_3.$$

The components T'_{ij} are expressed as:

$$T'_{11} = T_{22} \qquad T'_{22} = T_{11} \qquad T'_{33} = T_{33},$$
$$T'_{12} = -T_{21} = -T_{12} \text{ (symmetric tensor)},$$
$$T'_{13} = T_{23} \qquad T'_{23} = -T_{13}.$$

Because $[T]$ is a material tensor, each component must remain unchanged under an operation of the point group of the crystal: $T'_{ij} = T_{ij}$.

We deduce:

$$T_{11} = T_{22} \qquad T_{12} = 0 \qquad T_{13} = T_{23} = 0.$$

The form of tensor $[T]$ is: $\begin{pmatrix} T_{11} & & \\ & T_{11} & \\ & & T_{33} \end{pmatrix}.$

2. For group $\bar{4}$2m, we must add another mirror perpendicular to axis Ox_1. The symmetry operation with respect to the mirror changes the coordinates x_i of a point M into coordinates x'_j such that:

$$x'_1 = -x_1, \qquad x'_2 = x_2, \qquad x'_3 = x_3.$$

This operation leaves the form of the tensor unchanged.

20.9. Exercises for Chapter 11

Exercise 11.1

1. In a coordinate system $(Ox'_1 x'_2 x'_3)$ where Ox'_1 is parallel to $\mathbf{n}$, tensor $[T]$ has the form:
$$\begin{pmatrix} T & 0 & 0 \\ 0 & 0 & 0 \\ 0 & 0 & 0 \end{pmatrix}.$$

In the coordinate system $(Ox_1 x_2 x_3)$, the components T_{ij} of the stress tensor are given by $T_{ij} = a_{ik} a_{jl} T'_{kl} = a_{i1} a_{j1} T$ since the only non-zero term T'_{ij} is $T'_{11} = T$. The only components which play a role are the components a_{i1} of the matrix for going over from system $(Ox'_1 x'_2 x'_3)$ to system $(Ox_1 x_2 x_3)$, i.e. the components n_1, n_2, n_3 of vector $\mathbf{n}$.

We obtain $T_{ij} = n_i n_j T$, i.e. the matrix:

$$\begin{pmatrix} n_1^2 & n_1 n_2 & n_1 n_3 \\ n_1 n_2 & n_2^2 & n_2 n_3 \\ n_1 n_3 & n_2 n_3 & n_3^2 \end{pmatrix} T.$$

Exercise 11.2

1. $\mathbf{T_n} = [T]\mathbf{n}$. Here only $T_{33} \neq 0$ and $(\mathbf{T_n})_3 = n_3 T$ and $(\mathbf{T_n})_2 = (\mathbf{T_n})_1 = 0$.

$$T_\nu = \mathbf{T_n} \cdot \mathbf{n} = n_3^2 T.$$

The shear component is the projection of $\mathbf{T_n}$ onto the surface. In the plane defined by $\mathbf{T_n}$ and $\mathbf{n}$ (shown in the figure):

$$\tau^2 = \|\mathbf{T_n}\|^2 - T_\nu^2 = n_3^2 T^2 - n_3^4 T^2 = n_3^2 T^2 (1 - n_3^2),$$

$$\tau = T n_3 \sqrt{1 - n_3^2}.$$

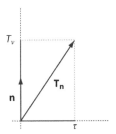

Figure 20.29

2. $\dfrac{\partial \tau}{\partial n_3} = \dfrac{1 - 2n_3^2}{\sqrt{1 - n_3^2}} T$, which vanishes for $n_3 = \pm \sqrt{2}/2$.

   $n_3 = \cos \theta$ where θ is the angle between $\mathbf{n}$ and Ox_3. The shear stress is maximal for the orientations of the surface for which the normal is on a cone with axis Ox_3 and apical angle $45°$. Its value is then $\tau = T/2$.

Exercise 11.3

The coordinate change matrix for the axis system is given by:

$$\begin{pmatrix} \frac{\sqrt{2}}{2} & \frac{\sqrt{2}}{2} & 0 \\ -\frac{\sqrt{2}}{2} & \frac{\sqrt{2}}{2} & 0 \\ 0 & 0 & 1 \end{pmatrix}.$$

In the new referential, the components T'_{ij} form the matrix:

$$\begin{pmatrix} T & 0 & 0 \\ 0 & -T & 0 \\ 0 & 0 & 0 \end{pmatrix}.$$

Exercise 11.4

1. Direction Ox_3 is a principal axis, with principal value 7.6. In the subspace (Ox_1, Ox_2), we solve: $\det (\{T_{ij}\} - \lambda E) = 0$. We obtain: $\lambda_1 = -2$, $\lambda_2 = 5.5$.

   For $\lambda_1 = -2$, the principal direction is defined by the relation $x_1 = -1.33 \, x_2$ between its components, so that the direction encloses an angle $-36.9°$ with axis Ox_1.

For $\lambda_2 = 5.5$, the principal direction is perpendicular to the above direction.

Thus, the principal directions deduce from directions Ox_1 and Ox_2 through a rotation by $-36.9°$ around axis Ox_3.

2. The position of the center C of the Mohr circle is given by: $x_C = (2.8 + 0.7)/2 = 1.75$ and angle $2\theta'$ is defined by: $\tan 2\theta' = 3.6/(2.8 - 1.75)$, so that $2\theta' = 73.74°$.

Thus the principal directions deduce from directions Ox_1 and Ox_2 by a rotation through $\theta = -\theta' = -36.9°$ around axis Ox_3.

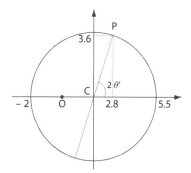

Figure 20.30

3. Let $\mathbf{n}$ be the vector $(\sqrt{3}/2, 1/2, 0)$. We have $(\mathbf{T_n})_i = T_{ij}n_j$, so that:

$$\begin{pmatrix} (\mathbf{T_n})_1 \\ (\mathbf{T_n})_2 \\ (\mathbf{T_n})_3 \end{pmatrix} = 10^7 \begin{pmatrix} 0.7 & 3.6 & 0 \\ 3.6 & 2.8 & 0 \\ 0 & 0 & 7.6 \end{pmatrix} \begin{pmatrix} \frac{\sqrt{3}}{2} \\ \frac{1}{2} \\ 0 \end{pmatrix} = 10^7 \begin{pmatrix} 2.4 \\ 4.5 \\ 0 \end{pmatrix}$$

and $\|\mathbf{T_n}\| = 5.1 \times 10^7$ N m^{-2}.

The value of the normal stress is:

$$T_\nu = \mathbf{T} \cdot \mathbf{n} = \begin{pmatrix} \frac{\sqrt{3}}{2} & \frac{1}{2} & 0 \end{pmatrix} \begin{pmatrix} 2.4 \\ 4.5 \\ 0 \end{pmatrix} 10^7$$

and $T_\nu = 4.3 \times 10^7$ N m^{-2}.

The shear stress has value: $\tau = \sqrt{\|\mathbf{T_n}\|^2 - T_\nu^2} = 2.7 \times 10^7$ N m^{-2}.

20.10. Exercises for Chapter 12

Exercise 12.1

1. $\{e_{ij}\} = \begin{pmatrix} 2 & -1 & 0 \\ 2 & -2 & 0 \\ 0 & 0 & 0 \end{pmatrix} 10^{-5}$; $\{S_{ij}\} = \begin{pmatrix} 2 & \frac{1}{2} & 0 \\ \frac{1}{2} & -2 & 0 \\ 0 & 0 & 0 \end{pmatrix} 10^{-5}$;

$$\{\Omega_{ij}\} = \begin{pmatrix} 0 & -\frac{3}{2} & 0 \\ \frac{3}{2} & 0 & 0 \\ 0 & 0 & 0 \end{pmatrix} 10^{-5}.$$

$[\Omega]$ represents a rotation around $\mathbf{n}$ $(0, 0, 1)$ through an angle $\frac{3}{2}10^{-5}$ rad (see Sect. 12.2.2).

2. $\mathbf{n} = \frac{1}{\sqrt{3}} \begin{pmatrix} 1 \\ 1 \\ 1 \end{pmatrix}$ and $\left. \dfrac{\Delta \ell}{\ell} \right|_{\mathbf{n}} = n_i\, n_j\, S_{ij} = \frac{1}{3}10^{-5}.$

Exercise 12.2

1. The displacement vectors $\Delta \mathbf{u} = \Delta \mathbf{r}' - \Delta \mathbf{r}$ associated to each side have the following values:

    For the side parallel to Ox_1: $\Delta \mathbf{u}_{x_1} = \begin{pmatrix} 0 \\ S \\ 0 \end{pmatrix}$.

    For the side parallel to Ox_2 : $\Delta \mathbf{u}_{x_2} = \begin{pmatrix} S \\ 0 \\ 0 \end{pmatrix}$.

    A square with unit side becomes a lozenge, the major diagonal of which is parallel to the diagonal of the undeformed square (figure below).

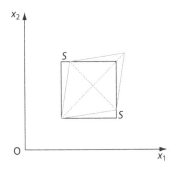

Figure 20.31

2. We want to find the eigenvalues and eigenvectors of the matrix $\{S_{ij}\}$ associated to the tensor:

$$\{S_{ij}\} = \begin{pmatrix} 0 & S & 0 \\ S & 0 & 0 \\ 0 & 0 & 0 \end{pmatrix}.$$

    The eigenvalues are found by solving $\det (\{S_{ij}\} - \lambda E) = 0$. We obtain three distinct principal values: $\lambda_1 = S$, $\lambda_2 = -S$ and $\lambda_3 = 0$. The eigenvector associated to λ_1 is parallel to the vector $(1, 1, 0)$, the eigenvector associated to λ_2 is parallel to vector $(-1, 1, 0)$. The eigenvector associated to λ_3 is along Ox_3.

The frame of the principal axes is rotated by $45°$ around Ox_3 with respect to the initial frame. In this axis system, the matrix representing the tensor is:

$$\begin{pmatrix} S & 0 & 0 \\ 0 & -S & 0 \\ 0 & 0 & 0 \end{pmatrix}.$$

Exercise 12.3

1. $\mathbf{n} = \begin{pmatrix} \cos\alpha \\ \sin\alpha \\ 0 \end{pmatrix}$, $\left.\dfrac{\Delta L}{L}\right|_{\mathbf{n}} = n_i n_j S_{ij} = \eta\cos^2\alpha - 0.3\eta\sin^2\alpha$

    so that $\left.\dfrac{\Delta L}{L}\right|_{\mathbf{n}} = \eta(1.3\cos^2\alpha - 0.3)$. For $\alpha < 61.3°$, $L' > L$.

2. After deformation, $\Delta\mathbf{u} = [S]\Delta\mathbf{r}$ and $\Delta\mathbf{r} = \begin{pmatrix} \cos\alpha \\ \sin\alpha \\ 0 \end{pmatrix} L$, so that

    $$\Delta u_1 = \eta L \cos\alpha \qquad \text{and} \qquad \Delta u_2 = -0.3\eta L \sin\alpha.$$

    We project this displacement vector $\Delta\mathbf{u}$ onto the direction perpendicular to the direction defined by $\begin{pmatrix} -\sin\alpha \\ \cos\alpha \\ 0 \end{pmatrix}$ (Fig. 20.32).

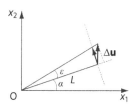

Figure 20.32

We obtain:

$$-\eta L\cos\alpha\sin\alpha - 0.3\eta L\cos\alpha\sin\alpha = -1.3\eta\cos\alpha\sin\alpha L$$

and the expression for angle ε is:

$$\varepsilon = -1.3\eta\cos\alpha\sin\alpha = -0.65\eta\sin 2\alpha.$$

Exercise 12.4

$$S_{ij} = \begin{pmatrix} 4 & -1 & 2 \\ -1 & 1 & 0 \\ 2 & 0 & 3 \end{pmatrix} 10^{-6}, \qquad \Omega_{ij} = \begin{pmatrix} 0 & 3 & -4 \\ -3 & 0 & 0 \\ 4 & 0 & 0 \end{pmatrix} 10^{-6}.$$

From the results of Section 12.2.2, the rotation axis is parallel to the unit vector $\mathbf{n}$, with direction cosines n_1, n_2, n_3 such that $n_1 \Delta\alpha = 0$, $n_2 \Delta\alpha = -4 \times 10^{-6}$ and $n_3 \Delta\alpha = -3 \times 10^{-6}$, where $\Delta\alpha$ is the rotation angle.

The rotation is performed around the direction of unit vector

$$\mathbf{n} = \begin{pmatrix} 0 \\ -\frac{4}{5} \\ -\frac{3}{5} \end{pmatrix}.$$

We can rewrite matrix $\{\Omega_{ij}\}$ in the form:

$$\begin{pmatrix} 0 & \frac{3}{5} & -\frac{4}{5} \\ -\frac{3}{5} & 0 & 0 \\ \frac{4}{5} & 0 & 0 \end{pmatrix} 5 \times 10^{-6}$$

and $\Delta\alpha$ has the value $\Delta\alpha = 5 \times 10^{-6}$ rad around $\mathbf{n}$.

Exercise 12.5

1. The form of the thermal expansion tensor is

$$\begin{pmatrix} \alpha_1 & & \\ & \alpha_1 & \\ & & \alpha_3 \end{pmatrix}$$

in an orthonormal set of axes $(Ox_1x_2x_3)$ where Ox_3 is parallel to the basis vector $\mathbf{c}$ of the hexagonal structure and Ox_1 is any direction in the plane $(\mathbf{a}, \mathbf{b})$.

$$\alpha_1 = \alpha_2 = \left(\frac{a_{100} - a_{20}}{a_{20}}\right)\frac{1}{\Delta T} = -1.5 \times 10^{-6} \,^\circ\mathrm{C}^{-1},$$

$$\alpha_3 = \left(\frac{c_{100} - c_{20}}{c_{20}}\right)\frac{1}{\Delta T} = 28 \times 10^{-6} \,^\circ\mathrm{C}^{-1}.$$

2. Let $\mathbf{s}$ be a unit vector with arbitrary direction. Its components in spherical coordinates are:

$$s_1 = \sin\theta\cos\varphi, \qquad s_2 = \sin\theta\sin\varphi, \qquad s_3 = \cos\theta.$$

The value of the thermal expansion in direction $\mathbf{s}$ is given by:

$$\alpha_s = (s_1, s_2, s_3)\begin{pmatrix} \alpha_1 & & \\ & \alpha_1 & \\ & & \alpha_3 \end{pmatrix}\begin{pmatrix} s_1 \\ s_2 \\ s_3 \end{pmatrix} = \alpha_1\sin^2\theta + \alpha_3\cos^2\theta.$$

$$\alpha_s = 0 \text{ if } \tan^2\theta_0 = -\frac{\alpha_3}{\alpha_1} \text{ or } \theta_0 = \pm\arctan\sqrt{\frac{\alpha_3}{|\alpha_1|}} + p\pi \ (p \text{ an integer}).$$

By definition, $0 < \theta_0 < \pi$ and $\theta_0 = 77°$ or $\theta_0 = 103°$, *i.e.* all directions on a cone with apex O and apical angle 77°.

The representative surface for this tensor comprises a one-sheet hyperboloid and a two-sheet hyperboloid. These two hyperboloids are tangent to the same cone with angle $2\theta_0$ and with the apex at the origin of the axes. In the directions of the cone which is the asymptote to the hyperboloids, the relative elongation is zero because the radius vector r tends toward infinity.

3. We consider the direction $\varphi = 0$, $\theta = \theta_0$ *i.e.* the unit vector $\mathbf{s}$ $(\sin \theta_0, 0, \cos \theta_0)$. The displacement vector $\Delta \mathbf{u}$ has components $(\alpha_1 \sin \theta_0, 0, \alpha_3 \cos \theta_0)$. It is easy to check that $\Delta \mathbf{u} \cdot \mathbf{s} = \alpha_1 \sin^2 \theta_0 + \alpha_3 \cos^2 \theta_0 = 0$.

The rotation angle ε we want to determine is equal to the norm of $\Delta \mathbf{u}$.

$$
\begin{aligned}
\|\Delta \mathbf{u}\| &= (\alpha_1^2 \sin^2 \theta_0 + \alpha_3^2 \cos^2 \theta_0)^{1/2} = |\alpha_1 \sin \theta_0| \left(1 + \tan^4 \theta_0 \frac{\cos^2 \theta_0}{\sin^2 \theta_0}\right)^{1/2} \\
&= |\alpha_1 \sin \theta_0| (1 + \tan^2 \theta_0)^{1/2} = |\alpha_1| |\tan \theta_0|
\end{aligned}
$$

Thus $\|\Delta \mathbf{u}\| = \sqrt{\alpha_3 |\alpha_1|}$ and ε is equal to 6.48×10^{-6} rad $°C^{-1} \simeq 1.3$ arcsec $°C^{-1}$.

Exercise 12.6

$\dfrac{\Delta a}{a} = \alpha_{11} \Delta T$ and $\dfrac{\Delta c}{c} = \alpha_{33} \Delta T$.

Let φ be the angle between the planes (100) and (101) (figure below).

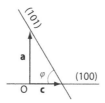

Figure 20.33

We have: $\tan \varphi = \dfrac{a}{c}$, or $\varphi = 57.21°$.

Hence:

$$
\frac{d(\tan \varphi)}{\tan \varphi} = \frac{\Delta \varphi}{\sin \varphi \cos \varphi} = \frac{2\Delta \varphi}{\sin 2\varphi} = \frac{\Delta a}{a} - \frac{\Delta c}{c} = (\alpha_{11} - \alpha_{33}) \Delta T.
$$

We get: $\alpha_{11} - \alpha_{33} = -2.1 \times 10^{-6}$ $°C^{-1}$.

On the other hand, the volume expansion is 23.4×10^{-6} so that: $2\alpha_{11} + \alpha_{33} = 23.4 \times 10^{-6}$ $°C^{-1}$.

Solving this system of equations, we obtain: $\alpha_{11} = 7.1 \times 10^{-6}$ °C^{-1} and $\alpha_{33} = 9.2 \times 10^{-6}$ °C^{-1}.

Exercise 12.7

1. Tetragonal phase between 0 and 120°C:

$$\Delta a = 0.011 \text{ Å and } \alpha_{11} = \alpha_{22} = \frac{1}{\Delta T}\frac{\Delta a}{a} = 2.3 \times 10^{-5},$$

$$\Delta c = -0.013 \text{ Å and } \alpha_{33} = \frac{1}{\Delta T}\frac{\Delta c}{c} = -2.7 \times 10^{-5}.$$

The thermal expansion tensor is:

$$\begin{pmatrix} 2.3 & & \\ & 2.3 & \\ & & -2.7 \end{pmatrix} 10^{-5} \text{ °C}^{-1}.$$

Let **n** be any direction with direction cosines n_1, n_2 and n_3. The value α_n of α in direction **n** is given by: $\alpha_n = n_i\alpha_{ij}n_j$, so that:

$$\alpha_n = \alpha_{11}(n_1^2 + n_2^2) + \alpha_{33}n_3^2$$

which vanishes for $-\dfrac{\alpha_{33}}{\alpha_{11}} = \dfrac{n_1^2 + n_2^2}{n_3^2}$.

If **n** is expressed in spherical coordinates (θ, φ), then $\tan^2\theta = -\alpha_{33}/\alpha_{11} = 1.171$ and the angle between the direction we want to determine and Ox_3 is $(\pm 47.3° + p\,180°)$, hence $47.3°$ or $132.7°$.

2. From Figure 20.34, we have:

$$\frac{b_o}{2} = b_m \cos\frac{\alpha}{2} \qquad \frac{c_o}{2} = b_m \sin\frac{\alpha}{2}.$$

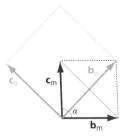

Figure 20.34

For the orthorhombic unit cell at $T = 0$°C, $\alpha = 90.20°$ and we obtain:

$$a_o = 3.988 \text{ Å}, \qquad b_o = 5.664 \text{ Å}, \qquad c_o = 5.684 \text{ Å}.$$

For the orthorhombic unit cell at $T = -100$°C, $\alpha = 90.40°$ and we obtain:

$$a_o = 3.978 \text{ Å}, \qquad b_o = 5.654 \text{ Å}, \qquad c_o = 5.694 \text{ Å}.$$

The thermal expansion tensor in the orthorhombic phase can be written, in the orthonormal axis system parallel to the basis vectors of the orthorhombic unit cell:

$$\begin{pmatrix} 2.5 & & \\ & 1.8 & \\ & & -1.8 \end{pmatrix} 10^{-5}\ {}^{\circ}\mathrm{C}^{-1}.$$

3. (a) Cubic $\leftrightarrow$ tetragonal transition at 120°C:

In an orthonormal axis system parallel to the basis vectors of the cubic unit cell, the strain tensor is:

$$\begin{pmatrix} -2 & & \\ & -2 & \\ & & 2.7 \end{pmatrix} 10^{-3}.$$

(b) Strain tensor describing the transition from the cubic unit cell at 120°C to the orthorhombic unit cell at 0°C:

The orthorhombic unit cell must be compared with the equivalent unit cell in the cubic phase, *i.e.* with a double unit cell with basis vectors $\mathbf{a}'$, $\mathbf{b}'$ and $\mathbf{c}'$. $\mathbf{a}'$ is parallel to $\mathbf{a}$ and $a' = 4.010$ Å, $\mathbf{b}'$ and $\mathbf{c}'$ are parallel respectively to the diagonals of the $(\mathbf{b}, \mathbf{c})$ faces, and such that $b' = c' = a\sqrt{2} = 5.671$ Å. The strain tensor is diagonal in an axis system parallel to the basis vectors of the orthorhombic unit cell, and:

$$S_{11} = \frac{a_o - a'}{a'} = -5.5 \times 10^{-3}, \quad S_{22} = \frac{b_o - b'}{b'} = -1.2 \times 10^{-3},$$

$$S_{33} = \frac{c_o - c'}{c'} = 2.2 \times 10^{-3}.$$

The strain tensor, when expressed in the axis system of the orthorhombic unit cell, is thus

$$\begin{pmatrix} -5.5 & & \\ & -1.2 & \\ & & 2.2 \end{pmatrix} 10^{-3}.$$

(c) Strain tensor describing the transition from the cubic unit cell to the rhombohedral unit cell:

When going over from the cubic unit cell to the rhombohedral unit cell, the [111] axis of the cube remains a 3-fold axis, and the mirrors containing this axis (the diagonal mirrors of the cube) remain mirrors for the rhombohedral unit cell. The basis vectors $\mathbf{a} = \mathbf{OA}$ and $\mathbf{c} = \mathbf{OC}$ of the cubic unit cell become vectors $\mathbf{OA}'$ and $\mathbf{OC}'$ such that:

$$\mathbf{OA}' \begin{cases} a'(1 - \frac{\varphi^2}{2}) \\ a'\varphi\frac{\sqrt{2}}{2} \\ a'\varphi\frac{\sqrt{2}}{2} \end{cases} \quad \text{and} \quad \mathbf{OC}' \begin{cases} a'\varphi\frac{\sqrt{2}}{2} \\ a'\varphi\frac{\sqrt{2}}{2} \\ a'(1 - \frac{\varphi^2}{2}) \end{cases}$$

where φ is the angle of rotation of $\mathbf{a}$, $\mathbf{b}$ and $\mathbf{c}$ in their respective diagonal mirror (Fig. 20.35).

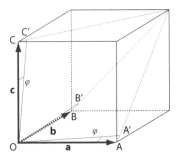

Figure 20.35

$$S_{11} = S_{22} = S_{33} = \frac{a'(1 - \varphi^2/2) - a}{a} \approx \frac{a' - a}{a} = -3.5 \times 10^{-3}.$$

$$S_{12} = S_{13} = \varphi\frac{\sqrt{2}}{2}.$$

Calculating angle φ as a function of α, we obtain $\mathbf{OA'} \cdot \mathbf{OC'} = a'^2 \cos\alpha = a'^2 \varphi\sqrt{2}$ and $\varphi = \frac{\cos\alpha}{\sqrt{2}}$.

Thus $S_{12} = S_{13} = \frac{\cos\alpha}{2} = 1.2 \times 10^{-3}$.

The component S_{23} is obtained in the same way.

The strain tensor describing the transition from the cubic unit cell to the orthorhombic unit cell, expressed in the axis system parallel to the basis vectors of the cubic unit cell, is:

$$\begin{pmatrix} -3.5 & 1.2 & 1.2 \\ 1.2 & -3.5 & 1.2 \\ 1.2 & 1.2 & -3.5 \end{pmatrix} 10^{-3}.$$

20.11. Exercises for Chapter 13

Exercise 13.1

By adding to the symmetry elements of group $\bar{4}2m$ a 3-fold axis parallel to the major diagonal of a cube with sides parallel to the basis vectors $\mathbf{a}$, $\mathbf{b}$ and $\mathbf{c}$, we obtain point group $\bar{4}3m$. Such a symmetry operation changes the coordinates x_i of a point M into coordinates x'_j such that:

$$x'_1 = x_2 \qquad x'_2 = x_3 \qquad x'_3 = x_1.$$

We obtain: $s_{11} = s_{22} = s_{33}, \qquad s_{44} = s_{55} = s_{66}, \qquad s_{12} = s_{13} = s_{23}.$

Exercise 13.2

1. We choose an orthonormal axis system $(Ox_1x_2x_3)$ parallel to the basis vectors **a**, **b** and **c** of the cubic unit cell. From the generalized Hooke's law, $S_\alpha = s_{\alpha\beta}T_\beta$ with $T_1 = T$ and $T_2 = T_3 = T_4 = T_5 = T_6 = 0$. The resulting strain tensor is:

$$\begin{pmatrix} s_{11} & & \\ & s_{12} & \\ & & s_{12} \end{pmatrix} T.$$

The relative elongation in direction [100] is $s_{11}T$.

2. In the axis system $(Ox_1'x_2'x_3)$, rotated through $45°$ around Ox_3 with respect to the initial frame, the stress tensor is such that $T_{11}' = T$ and that the other components are zero.

In the frame $(Ox_1x_2x_3)$ linked to the cube axes, this tensor becomes:

$$\begin{pmatrix} T/2 & T/2 & 0 \\ T/2 & T/2 & 0 \\ 0 & 0 & 0 \end{pmatrix}$$

and the strain tensor is thus:[1]

$$S_{ij} = \begin{pmatrix} s_{11} + s_{12} & \dfrac{s_{44}}{2} & \\ \dfrac{s_{44}}{2} & s_{11} + s_{12} & \\ & & 2s_{12} \end{pmatrix} \dfrac{T}{2}.$$

The relative elongation in direction [110] is obtained through relation (12.12):

$$\left.\frac{\Delta\ell}{\ell}\right|_{\mathbf{w}} = w_i w_j S_{ij}.$$

We obtain:

$$\left.\frac{\Delta\ell}{\ell}\right|_{[110]} = \frac{T}{4}[2(s_{11} + s_{12}) + s_{44}].$$

3. If the solid is isotropic, these two strains must be identical, so that:

$$\frac{1}{4}[2(s_{11} + s_{12}) + s_{44}] = s_{11} \quad \text{or} \quad s_{44} = 2(s_{11} - s_{12}).$$

Exercise 13.3

We directly apply the generalized Hooke's law (relation (13.1a)): $S_\alpha = s_{\alpha\beta}T_\beta$ with $T_1 = T_2 = T_3 = -p$ and $T_4 = T_5 = T_6 = 0$, using the elasticity tensor for group $4/mmm$ given in Table 13.5. We obtain :

$$-p \begin{pmatrix} s_{11} + s_{12} + s_{13} & & \\ & s_{11} + s_{12} + s_{13} & \\ & & 2s_{13} + s_{33} \end{pmatrix}.$$

---

1. We recall that $S_{12} = S_{21} = S_6/2$.

The numerical values are:

$$S_{11} = S_{22} = -0.61 \times 10^{-4}; \quad S_{33} = -0.68 \times 10^{-4}$$
$$\text{and} \quad \frac{\Delta V}{V} = S_{11} + S_{22} + S_{33} = -1.9 \times 10^{-4}.$$

Exercise 13.4

We consider a uniaxial stress T, parallel to a direction $\mathbf{w}$ with direction cosines w_1, w_2 and w_3. In an orthonormal axis system $(Ox_1'x_2'x_3')$ where Ox_1' is parallel to $\mathbf{w}$, the stress tensor has a single non-zero component, $T_{11}' = T$.

Let T_{ij} be the components of this tensor in the axis system $(Ox_1x_2x_3)$ parallel to the cube axes. We have $T_{mn} = a_{im}a_{jn}T_{ij}'$ where $\{a_{ij}\}$ is the transformation matrix for going over from system $(Ox_1x_2x_3)$ to system $(Ox_1'x_2'x_3')$.

Here $T_{mn} = a_{1m}a_{1n}T_{11}' = a_{1m}a_{1n}T$ with $a_{11} = w_1$, $a_{12} = w_2$ and $a_{13} = w_3$.

$$T_{11} = w_1^2 T, \qquad T_{22} = w_2^2 T, \qquad T_{33} = w_3^2 T,$$
$$T_{12} = w_1 w_2 T, \quad T_{13} = w_1 w_3 T, \quad T_{23} = w_2 w_3 T.$$

We showed (Sect. 12.4) that: $\dfrac{\Delta V}{V} = S_{11} + S_{22} + S_{33}$.

$$S_{11} = s_{11}T_{11} + s_{12}(T_{22} + T_{33}) = s_{11}w_1^2 T + s_{12}T(w_2^2 + w_3^2),$$
$$S_{22} = s_{11}T_{22} + s_{12}(T_{11} + T_{33}) = s_{11}w_2^2 T + s_{12}T(w_1^2 + w_3^2),$$
$$S_{33} = s_{11}T_{33} + s_{12}(T_{11} + T_{22}) = s_{11}w_3^2 T + s_{12}T(w_1^2 + w_2^2)$$

and $\dfrac{\Delta V}{V} = [s_{11}(w_1^2 + w_2^2 + w_3^2) + 2s_{12}(w_1^2 + w_2^2 + w_3^2)]T = (s_{11} + 2s_{12})\,T$
since $w_1^2 + w_2^2 + w_3^2 = 1$.

Exercise 13.5

It was shown in Section 13.7 that, in the cubic system, the reciprocal of Young's modulus in a given direction, defined by unit vector $\mathbf{w}$ with components w_1, w_2 and w_3, is given by:

$$\frac{1}{E_w} = s_{11} - 2\left(s_{11} - s_{12} - \frac{s_{44}}{2}\right)(w_2^2 w_3^2 + w_1^2 w_2^2 + w_1^2 w_3^2) = A.$$

E_w has its maximum value if A is minimum. If $s_{11} - s_{12} - s_{44}/2$ is positive, A is minimum if $B = w_2^2 w_3^2 + w_1^2 w_2^2 + w_1^2 w_3^2$ is maximum.

In spherical coordinates $(\theta,\ \varphi)$, the components of $\mathbf{w}$ are:

$$w_1 = \sin\theta\,\cos\varphi, \qquad w_2 = \sin\theta\,\sin\varphi, \qquad w_3 = \cos\theta$$

and $\quad B = \dfrac{\sin^2 2\theta}{4} + \sin^2 2\varphi\,\dfrac{\sin^4 \theta}{4}.$

B is the sum of two terms which are always positive, and only the second term depends on φ. Therefore B is maximum, whatever the value of θ is, when $\sin^2 2\varphi$ is maximum, hence for $2\varphi = \frac{\pi}{2} + p\pi$ $i.e.$ $\varphi = \frac{\pi}{4} + p\frac{\pi}{2}$.

For this value of φ, $$B = \frac{\sin^2 2\theta}{4} + \frac{\sin^4 \theta}{4}.$$

$$\frac{dB}{d\theta} = \cos 2\theta \sin 2\theta + \cos \theta \sin^3 \theta$$
$$= 2\cos^3 \theta \sin \theta - \sin^3 \theta \cos \theta$$
$$= \sin \theta \cos \theta (3\cos^2 \theta - 1).$$

$\frac{dB}{d\theta} = 0$ for $\theta = 0$, $\theta = \pi/2$ and for θ such that $\cos \theta = 1/\sqrt{3}$, thus $\theta = 54.7°$, which is the angle between direction $[111]$ and Ox_3 in the cubic system.

For $\theta = 0$: $B = 0$
For $\theta = \dfrac{\pi}{2}$: $B = 1/4$
Pour $\theta = 54.7°$: $B = 1/3$, the maximum value.

This shows that the value of Young's modulus for a cubic material is extremal in direction $[111]$.

Exercise 13.6

In an orthonormal coordinate system $(Ox'_1 x'_2 x'_3)$ where direction Ox'_1 is parallel to the $[110]$ direction of the cubic lattice, the stress tensor has only one non-zero component, $T'_{11} = T$.

In the orthonormal frame $(Ox_1 x_2 x_3)$ with axes parallel to the basis vectors of the cubic unit cell, this tensor becomes:

$$\begin{pmatrix} T/2 & T/2 & 0 \\ T/2 & T/2 & 0 \\ 0 & 0 & 0 \end{pmatrix}.$$

The resulting strain is given, in the axis system of the cube, by:

$$S_{ij} = \begin{pmatrix} s_{11} + s_{12} & \dfrac{s_{44}}{2} & \\ \dfrac{s_{44}}{2} & s_{11} + s_{12} & \\ & & 2s_{12} \end{pmatrix} \dfrac{T}{2}.$$

We recall that $S_6 = 2S_{12} = 2S_{21}$.

The relative elongation in a given direction is given by (12.12):

$$\left. \frac{\Delta\ell}{\ell} \right|_{\mathbf{w}} = w_i w_j S_{ij}.$$

Here, $\left. \dfrac{\Delta\ell}{\ell} \right|_{\mathbf{w}} = w_1^2 S_{11} + w_2^2 S_{22} + 2w_1 w_2 S_{12} + w_3^2 S_{33}$. We obtain:

$$\left. \frac{\Delta\ell}{\ell} \right|_{[110]} = \frac{T}{2}\left(s_{11} + s_{12} + \frac{s_{44}}{2} \right), \qquad \left. \frac{\Delta\ell}{\ell} \right|_{[001]} = s_{12}T,$$

$$\left. \frac{\Delta\ell}{\ell} \right|_{[1\bar{1}0]} = \frac{T}{2}\left(s_{11} + s_{12} - \frac{s_{44}}{2} \right).$$

Numerically:
$s_{12} = -6 \times 10^{-13}$ m^2 N^{-1}, $s_{11} = 9.6 \times 10^{-12}$ m^2 N^{-1}, $s_{44} = 78 \times 10^{-12}$ m^2 N^{-1}.

Exercise 13.7

1. The uniaxial stress applied to the crystal is parallel to direction $[111]$, with unit vector $\mathbf{w}$, with direction cosines $w_1 = w_2 = w_3 = 1/\sqrt{3}$.

    In a coordinate system where Ox_1' is parallel to $[111]$, the only non-zero component of the stress tensor is $T_{11}' = T$. In the orthonormal coordinate system with axes parallel to the basis vectors of the cubic unit cell, this tensor becomes (see the solution to Exercise 11.1):

$$\frac{1}{3} \begin{pmatrix} T & T & T \\ T & T & T \\ T & T & T \end{pmatrix}.$$

2. The resulting strains are:

$$S_1 = S_2 = S_3 = \frac{T}{3}(s_{11} + 2s_{12}), \qquad S_4 = S_5 = S_6 = \frac{T}{3}s_{44}.$$

3. The basis vectors $\mathbf{a}$, $\mathbf{b}$, $\mathbf{c}$ of the cubic unit cell become respectively, after deformation, the vectors $\mathbf{a}'$, $\mathbf{b}'$, $\mathbf{c}'$ such that

$$\mathbf{a}' = \begin{pmatrix} 1 + S_1 \\ S_6/2 \\ S_6/2 \end{pmatrix} a, \quad \mathbf{b}' = \begin{pmatrix} S_6/2 \\ 1 + S_1 \\ S_6/2 \end{pmatrix} a, \quad \mathbf{c}' = \begin{pmatrix} S_6/2 \\ S_6/2 \\ 1 + S_1 \end{pmatrix} a.$$

    We note that the components of these vectors can be deduced from one another by circular permutation. Therefore these vectors deduce from one another by rotation through $2\pi/3$ around direction $[111]$. The unit cell has become rhombohedral.

    The parameter a' of the rhombohedral unit cell is given by:

$$a' = a(1 + S_1) = a \left[1 + \frac{T}{3}(s_{11} + 2s_{12}) \right].$$

    $\mathbf{a}' \cdot \mathbf{b}' = \|\mathbf{a}'\| \|\mathbf{b}'\| \cos\gamma'$. Neglecting the second order terms, we obtain:

$$\cos\gamma' = S_6 = \cos(\pi/2 - \varepsilon) = \sin\varepsilon, \quad \text{so that} \quad \varepsilon = S_6 = \frac{T}{3}s_{44}.$$

Exercise 13.8

1. In an orthonormal coordinate system $(Ox_1x_2x_3)$ with axes parallel to the sides of the parallelepiped, the uniaxial stress has only one non-zero component, $T_{11} = -T$. The resulting strain tensor is:

$$-\begin{pmatrix} s_{11} & & \\ & s_{12} & \\ & & s_{12} \end{pmatrix} T.$$

As a result, $\dfrac{\Delta L}{L} = -s_{11}T = -1.49 \times 10^{-2}$ and $\Delta L = -74.5$ μm

$\dfrac{\Delta\ell}{\ell} = -s_{12}T = 0.63 \times 10^{-2}$ and $\Delta\ell = 6.3$ μm.

2. The orthonormal axis system $(Ox_1'x_2'x_3)$, with axes parallel to the sides of the rectangle parallelepiped, is deduced from the crystallographic axis system $(Ox_1x_2x_3)$ by a rotation through $+45°$ around Ox_3. In the axis system $(Ox_1'x_2'x_3)$, the tensor has only one non-zero component, $T_{11} = -T$. In the coordinate system $(Ox_1x_2x_3)$, this tensor becomes

$$\begin{pmatrix} -T/2 & -T/2 & \\ -T/2 & -T/2 & \\ & & 0 \end{pmatrix}.$$

The components of the corresponding strain tensor are:

$$S_1 = (s_{11} + s_{12})\left(-\frac{T}{2}\right) = S_2, \qquad S_3 = -s_{12}T,$$

$$S_4 = S_5 = 0 \qquad \text{and} \qquad S_6 = -\frac{T}{2}s_{44}.$$

In the coordinate system $(Ox_1'x_2'x_3)$, these components become:

$$S_1' = -\frac{T}{4}[2(s_{11} + s_{12}) + s_{44}] = -0.76 \times 10^{-2},$$

$$S_2' = -\frac{T}{4}[2(s_{11} + s_{12}) - s_{44}] = 0.63 \times 10^{-2},$$

$$S_3' = -s_{12}T = 0.975 \times 10^{-3}.$$

$$S_4' = S_5' = S_6' = 0.$$

Since the non-diagonal terms are zero, the rectangle parallelepiped remains a rectangle parallelepiped. The sides AB and BC have changed respectively by Δ(AB)$= -7.6$ μm and Δ(BC) $= 6.3$ μm. Length L has changed by $\Delta L = 0.49$ μm.

20.12. Exercises for Chapter 14

In all exercises, use is made of Table 14.1 to determine the Christoffel matrix, and of Tables 13.5a and 13.5b to know which components of the stiffness tensor $[c]$ are non-zero.

Exercise 14.1

1. (a) $\mathbf{k}$ is parallel to [001]. In the orthonormal coordinate system $(Ox_1x_2x_3)$ where Ox_3 is parallel to $\mathbf{c}$, $s_1 = s_2 = 0$ and $s_3 = 1$, so that Christoffel's matrix is given by

$$\begin{pmatrix} c_{55} & c_{45} & c_{35} \\ c_{45} & c_{44} & c_{34} \\ c_{35} & c_{34} & c_{33} \end{pmatrix}.$$

From Table 13.5a, $c_{35} = c_{34} = 0$ and the matrix becomes:

$$\begin{pmatrix} c_{55} & c_{45} & 0 \\ c_{45} & c_{44} & 0 \\ 0 & 0 & c_{33} \end{pmatrix}.$$

(b) The propagation velocities are obtained from the eigenvalues of this matrix. They are given by:

$$v_1 = \sqrt{\frac{\lambda_1}{\rho}} \quad \text{and} \quad v_2 = \sqrt{\frac{\lambda_2}{\rho}}$$

with $\lambda_{1,2} = \dfrac{c_{44} + c_{55} \pm \sqrt{(c_{44} - c_{55})^2 + 4c_{45}^2}}{2}$,

$$v_3 = \sqrt{\frac{c_{33}}{\rho}}.$$

The wave which propagates at velocity v_3 is polarized parallel to direction $[001]$: it is a longitudinal wave.

The waves which propagate with velocities v_1 and v_2 are transverse waves.

The polarization of the wave with velocity v_1 encloses with direction Ox_1, in plane (Ox_1, Ox_2), an angle θ_1 such that:

$$\theta_1 = \arctan\left(\frac{\lambda_1 - c_{55}}{c_{45}}\right) = \arctan\left[\frac{(c_{44} - c_{55}) - \sqrt{(c_{44} - c_{55})^2 + 4c_{45}^2}}{2c_{45}}\right].$$

The polarization of the wave with velocity v_2 encloses with direction Ox_1 the angle θ_2 such that:

$$\theta_2 = \arctan\left(\frac{\lambda_2 - c_{55}}{c_{45}}\right).$$

We check that $\theta_2 - \theta_1 = \pi/2$.

2. For the orthorhombic system, $c_{45} = 0$ (see Tab. 13.5a).

Christoffel's matrix takes the form:

$$\begin{pmatrix} c_{55} & 0 & 0 \\ 0 & c_{44} & 0 \\ 0 & 0 & c_{33} \end{pmatrix}.$$

Waves with polarization parallel to Ox_1 and Ox_2, hence transverse, propagate with the respective velocities v_1 and v_2, given by:

$$v_1 = \sqrt{\frac{c_{55}}{\rho}} \quad \text{and} \quad v_2 = \sqrt{\frac{c_{44}}{\rho}}.$$

The wave with polarization parallel to Ox_3, hence longitudinal, propagates with velocity v_3 given by:

$$v_3 = \sqrt{\frac{c_{33}}{\rho}}.$$

3. Christoffel's matrix is :

$$\begin{pmatrix} c_{55} & c_{45} & c_{35} \\ c_{45} & c_{44} & c_{34} \\ c_{35} & c_{34} & c_{33} \end{pmatrix}.$$

From Tables 13.5a and 13.5b, all point groups with an axis of order larger than 2 feature $c_{45} = c_{35} = c_{34} = 0$. Furthermore, $c_{44} = c_{55}$. Christoffel's matrix thus becomes:

$$\begin{pmatrix} c_{44} & & \\ & c_{44} & \\ & & c_{33} \end{pmatrix}.$$

There is a longitudinal wave, with propagation velocity v_ℓ given by:

$$v_\ell = \sqrt{\frac{c_{33}}{\rho}}.$$

The transverse waves have their polarization along any direction in the plane (Ox_1, Ox_2) and they propagate with velocity v_t given by:

$$v_t = \sqrt{\frac{c_{44}}{\rho}}.$$

4. For a general propagation direction in plane (Ox_1, Ox_2), we have: $s_1 \neq 0$, $s_2 \neq 0$ and $s_3 = 0$. Christoffel's matrix is:

$$\begin{pmatrix} \Gamma_{11} & \Gamma_{12} & \\ \Gamma_{12} & \Gamma_{22} & \\ & & \Gamma_{33} \end{pmatrix}$$

since $\Gamma_{13} = \Gamma_{23} = 0$ for a hexagonal crystal. Here:

$$\Gamma_{11} = c_{11}s_1^2 + c_{66}s_2^2,$$
$$\Gamma_{22} = c_{66}s_1^2 + c_{22}s_2^2,$$
$$\Gamma_{33} = c_{55}s_1^2 + c_{44}s_2^2 = c_{44}(s_1^2 + s_2^2) = c_{44},$$
$$\Gamma_{12} = (c_{12} + c_{66})s_1 s_2.$$

One wave with polarization parallel to the 6-fold axis can propagate. It is transverse, with propagation velocity given by:

$$v = \sqrt{\frac{c_{44}}{\rho}}.$$

Exercise 14.2

1. $\mathbf{k}$ is parallel to [100], hence $s_1 = 1$, $s_2 = s_3 = 0$, and Christoffel's matrix contains only 3 coefficients: $\Gamma_{11} = c_{11}$, $\Gamma_{22} = c_{44}$, $\Gamma_{33} = c_{44}$.

   One longitudinal wave can propagate, with velocity v_ℓ given by:

   $$v_\ell = \sqrt{\frac{c_{11}}{\rho}}.$$

   The transverse waves have any polarization direction in the plane (Ox_2, Ox_3), and their propagation velocity v_t is given by:

   $$v_t = \sqrt{\frac{c_{44}}{\rho}}.$$

2. $\mathbf{k}$ is parallel to [111], hence $s_1 = s_2 = s_3 = 1/\sqrt{3}$, and Christoffel's matrix takes the form:

   $$\frac{1}{3}\begin{pmatrix} c_{11} + 2c_{44} & c_{12} + c_{44} & c_{12} + c_{44} \\ c_{12} + c_{44} & c_{11} + 2c_{44} & c_{12} + c_{44} \\ c_{12} + c_{44} & c_{12} + c_{44} & c_{11} + 2c_{44} \end{pmatrix}.$$

   The propagation velocities are obtained from $\det(\{\Gamma_{ij}\} - \lambda E) = 0$.

   If we set $(c_{11} + 2c_{44})/3 = a$ and $(c_{12} + c_{44})/3 = b$:

   $$\det(\{\Gamma_{ij}\} - \lambda E) = (a - \lambda)[(a - \lambda)^2 - b^2] - b^2[a - \lambda - b] + b^2[b - (a - \lambda)].$$

   We must solve:

   $$(a - \lambda)(a - \lambda + b)(a - \lambda - b) - 2b^2(a - \lambda - b) = 0,$$
   $$(a - \lambda - b)[(a - \lambda)(a - \lambda + b) - 2b^2] = 0.$$

   We find $\lambda_1 = a - b$ and $\lambda^2 - \lambda(2a + b) + a^2 + ab - 2b^2 = 0$, so that $\lambda_2 = a - b = \lambda_1$ and $\lambda_3 = a + 2b$.

   The eigenvector associated to $\lambda_3 = a + 2b$ turns out to be parallel to [111], so that the wave is longitudinal with propagation velocity:

   $$v_\ell = \sqrt{\frac{c_{11} + 2c_{12} + 4c_{44}}{3\rho}}.$$

   The other two eigenvalues are equal and correspond to transverse waves with any polarization direction in the plane perpendicular to [111]. Their propagation velocity is given by:

   $$v_t = \sqrt{\frac{c_{11} - c_{12} + c_{44}}{3\rho}}.$$

3. Numerically:

The velocity of the transverse waves found in question (1) yields c_{44}:
$c_{44} = 4.40 \times 10^{11}$ N m^{-2}.

c_{11} and c_{12} are determined from the results of (2):

$$c_{11} + 2c_{12} = 3\rho v_\ell^2 - 4c_{44}$$
$$c_{11} - c_{12} = 3\rho v_t^2 - c_{44}.$$

We find: $c_{11} = 10.9 \times 10^{11}$ N m^{-2} and $c_{12} = -3.18 \times 10^{11}$ N m^{-2}.

Exercise 14.3

The wave-vector **k** is parallel to [110]: $s_1 = s_2 = \dfrac{1}{\sqrt{2}}$ and $s_3 = 0$.

Christoffel's matrix is: $\dfrac{1}{2} \begin{pmatrix} c_{11} + c_{66} & c_{12} + c_{66} & 0 \\ c_{12} + c_{66} & c_{11} + c_{66} & 0 \\ 0 & 0 & 2c_{44} \end{pmatrix}$.

$$\det(\{\Gamma_{ij}\} - \lambda E) = (c_{44} - \lambda)\left(\frac{c_{11} + c_{12} + 2c_{66}}{2} - \lambda\right)\left(\frac{c_{11} - c_{12}}{2} - \lambda\right).$$

The propagation velocities are obtained from $\det(\{\Gamma_{ij}\} - \lambda E) = 0$. Here c_{44} is an eigenvalue with eigenvector [001].

We find: $v_1 = \sqrt{\dfrac{c_{44}}{\rho}}$, polarization [001], transverse wave;

$v_2 = \sqrt{\dfrac{c_{11} - c_{12}}{2\rho}}$, polarization $[\bar{1}10]$, transverse wave;

$v_3 = \sqrt{\dfrac{c_{11} + c_{12} + 2c_{66}}{2\rho}}$, polarization [110], longitudinal wave.

20.13. Exercises for Chapter 15

Note: in this section, the coefficients of the piezoelectric tensors which vanish are represented by small dots.

Exercise 15.1

The piezoelectric tensor for group 422 has the following form in an orthonormal axes system $(Ox_1x_2x_3)$ where Ox_3 is parallel to the 4-fold axis and Ox_1 and Ox_2 are parallel to the 2-fold axes:

$$\begin{pmatrix} \cdot & \cdot & \cdot & d_{14} & \cdot & \cdot \\ \cdot & \cdot & \cdot & \cdot & -d_{14} & \cdot \\ \cdot & \cdot & \cdot & \cdot & \cdot & 0 \end{pmatrix}.$$

To determine the form of the tensor for group 432, we just have to add a 3-fold axis, parallel to the diagonal of the cube, because the 2-fold axes perpendicular to the diagonal planes result from the existence of the 4-fold and 3-fold axes. We then use the direct inspection method applied to a rotation by $2\pi/3$ around the major diagonal of the cube. This operation changes the coordinates x_i of a point M into coordinates x'_j such that:

$$x'_1 = x_2, \qquad x'_2 = x_3, \qquad x'_3 = x_1.$$

We obtain:

$$d'_{14} = 2d'_{132} = 2d_{213} = d_{25} = -d_{14} = d_{14}, \quad \text{whence} \quad d_{14} = 0.$$

Exercise 15.2

1. (a) We consider the stereographic projection of the equivalent directions for group 222 (Fig. 20.36a). The Ox_1 axis is parallel to one of the 2-fold axes. If we add the operation of symmetry with respect to a mirror enclosing an angle of $45°$ with Ox_1, we add the equivalent directions 5, 6, 7 and 8, which are the transforms of directions 1, 2, 3 and 4 respectively. We see that this leads to group $\bar{4}2m$.

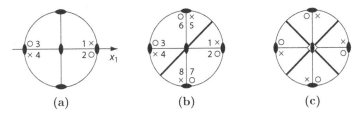

(a)   (b)   (c)

Figure 20.36

The form of the piezoelectric tensor for group 222 (in an orthonormal set of axes parallel to the 2-fold axes) is (Tab. 15.2):

$$\begin{pmatrix} . & . & . & d_{14} & . & . \\ . & . & . & . & d_{25} & . \\ . & . & . & . & . & d_{36} \end{pmatrix}.$$

The operation of symmetry with respect to the mirror changes the coordinates x_i of a point M into coordinates x'_j such that:

$$x'_1 = x_2, \qquad x'_2 = x_1, \qquad x'_3 = x_3.$$

This leads to $d'_{132} = d_{231} = d_{132}$ whence $d_{25} = d_{14}$.

The transformation of d_{312} yields nothing more. The form of the tensor is:

$$\begin{pmatrix} . & . & . & d_{14} & . & . \\ . & . & . & . & d_{14} & . \\ . & . & . & . & . & d_{36} \end{pmatrix}.$$

(b) In the same way, considering subgroup mm2, the addition of the rotation by π around an axis at $45°$ to the mirrors leads to group $\bar{4}2m$.

The form of the tensor for group mm2 (in an orthonormal system with axes perpendicular to the mirrors) is:

$$\begin{pmatrix} \cdot & \cdot & \cdot & \cdot & d_{15} & \cdot \\ \cdot & \cdot & \cdot & d_{24} & \cdot & \cdot \\ d_{31} & d_{32} & d_{33} & \cdot & \cdot & \cdot \end{pmatrix}.$$

A rotation by π around an axis at $45°$ to the mirrors changes coordinates x_i of a point M into coordinates x'_j such that:

$$x'_1 = x_2, \qquad x'_2 = x_1, \qquad x'_3 = -x_3.$$

Therefore:

$$\begin{aligned} d'_{131} &= -d_{232} = d_{131} && \text{whence} && d_{15} = -d_{24}; \\ d'_{311} &= -d_{322} = d_{311} && \text{whence} && d_{31} = -d_{32}; \\ d'_{333} &= -d_{333} = d_{333} && \text{whence} && d_{33} = 0. \end{aligned}$$

The form of the tensor is:

$$\begin{pmatrix} \cdot & \cdot & \cdot & \cdot & d_{15} & \cdot \\ \cdot & \cdot & \cdot & -d_{15} & \cdot & \cdot \\ d_{31} & -d_{31} & \cdot & \cdot & \cdot & \cdot \end{pmatrix}.$$

2. The tensor which we obtained in (1a), relative to an axis system $(Ox_1x_2x_3)$, must now be expressed in a new axis system $(Ox'_1x'_2x'_3)$, deduced from the first one through a rotation by $45°$ around axis Ox_3. Knowing the form of the tensor in the new axes reduces the task to the calculation of two of them: d'_{15} and d'_{31}.

We apply the rule $d'_{ijk} = a_{il}a_{jm}a_{kn}d_{lmn}$, where matrix $\{a_{ij}\}$ has the form:

$$\begin{pmatrix} \dfrac{\sqrt{2}}{2} & \dfrac{\sqrt{2}}{2} & 0 \\ -\dfrac{\sqrt{2}}{2} & \dfrac{\sqrt{2}}{2} & 0 \\ 0 & 0 & 1 \end{pmatrix}.$$

We obtain:

$$d'_{15} = 2d'_{131} = (d_{132} + d_{231}) = \frac{d_{14} + d_{25}}{2} = d_{14}$$

so that $d'_{15} = d_{14} = 0.17 \times 10^{-11} \text{ C N}^{-1}$.

$$d'_{31} = d'_{311} = \frac{d_{312} + d_{321}}{2} = \frac{d_{36}}{2}$$

whence $d'_{31} = \dfrac{d_{36}}{2} = 2.6 \times 10^{-11} \text{ C N}^{-1}$.

Exercise 15.3

For group 422 (as well as for group 622), the form of the piezoelectric tensor, in the orthonormal axis system $(Ox_1x_2x_3)$ where Ox_3 is parallel to the 4-fold axis (or the 6-fold axis) and Ox_1 and Ox_2 are parallel to the 2-fold axes, has the form:

$$\begin{pmatrix} . & . & . & d_{14} & . & . \\ . & . & . & . & -d_{14} & . \\ . & . & . & . & . & 0 \end{pmatrix}.$$

We saw (Sect. 15.3.4) that the longitudinal piezoelectric surface is defined by the set of points P such that the length of the vector **OP**, parallel to the direction of the unit vector **s** (s_1, s_2, s_3), is given by $s_i s_j s_k d_{ijk}$ (Eq. (15.25)).

Here, the only coefficients playing a part are $d_{14} = d_{123} + d_{132}$ and $d_{25} = d_{213} + d_{231} = -d_{14}$.

We find: $OP = s_1 s_2 s_3 (d_{14} + d_{25})$ and, since $d_{25} = -d_{14}$, we have $OP = 0$. The longitudinal piezoelectric surface vanishes for all directions.

Exercise 15.4

The form of the piezoelectric tensor, for group 32, in the orthonormal axis system $(Ox_1x_2x_3)$ defined in the text, is:

$$\begin{pmatrix} d_{11} & -d_{11} & . & d_{14} & . & . \\ . & . & . & . & -d_{14} & -2d_{11} \\ . & . & . & . & . & . \end{pmatrix}.$$

1. (a) In the orthonormal system $(Ox_1'x_2'x_3)$, rotated by angle θ with respect to $(Ox_1x_2x_3)$, the applied stress tensor has only one non-zero component, $T_{11}' = T$.

   In the frame $(Ox_1x_2x_3)$, this tensor becomes:

   $$\begin{pmatrix} \cos^2\theta & \cos\theta\sin\theta & 0 \\ \cos\theta\sin\theta & \sin^2\theta & 0 \\ 0 & 0 & 0 \end{pmatrix} T.$$

   We find for the components of polarization **P**:

   $$\mathbf{P} = \begin{pmatrix} d_{11}(\cos^2\theta - \sin^2\theta) \\ -d_{11}\sin 2\theta \\ 0 \end{pmatrix} T = \begin{pmatrix} d_{11}\cos 2\theta \\ -d_{11}\sin 2\theta \\ 0 \end{pmatrix} T.$$

   Its norm is: $\|\mathbf{P}\| = d_{11}T\sqrt{\cos^2 2\theta + \sin^2 2\theta} = d_{11}T$. It is independent of θ.

   (b) **P** encloses with direction Ox_1 the angle φ such that $\varphi = -2\theta$. Thus the polarization rotates twice as fast as the stress in the plane (Ox_1, Ox_2), and in the opposite direction.

(c) From question (1a), we see that $\mathbf{P}$ will be parallel to Ox_2 if $\cos 2\theta = 0$, hence when $\theta = \pi/4 + p\,\pi/2$ with p an integer. It is therefore sufficient to apply a uniaxial stress at $45°$ to the axes Ox_1 and Ox_2.

2. We saw that $d'_{111} = s_i s_j s_k d_{ijk}$ (Eq. (15.25)), so that here:

$$d'_{111} = s_1 d_{111}(s_1^2 - 3s_2^2).$$

Neither d_{14} nor $d_{25} = -d_{14}$ are involved, since the corresponding terms vanish.

In spherical coordinates (θ, φ), we obtain:

$$d'_{111} = d_{111}\sin^3\theta(\cos^3\varphi - 3\sin^2\varphi\cos\varphi) = d_{111}\sin^3\theta\cos 3\varphi.$$

The intersection of this surface with the plane (Ox_1, Ox_2) is obtained for $\theta = \pi/2$ and $d'_{111} = d_{111}\cos 3\varphi$ (Fig. 20.37, showing d'_{111}/d_{111}). The curve is periodic vs φ, with period $2\pi/3$, in agreement with the presence of the 3-fold axis.

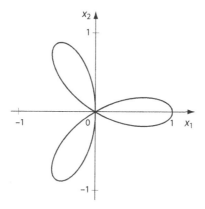

Figure 20.37

Exercise 15.5

1. The piezoelectric tensor for group $\bar{4}3m$ is given by:

$$\begin{pmatrix} . & . & . & d_{14} & . & . \\ . & . & . & . & d_{14} & . \\ . & . & . & . & . & d_{14} \end{pmatrix}.$$

Due to the form of this tensor, we must apply $T_4 = T_{23}$, or $T_5 = T_{13}$, or $T_6 = T_{12}$.

2. Using (15.25):

$$d'_{111} = s_i s_j s_k d_{ijk} = 3s_1 s_2 s_3 d_{14}$$

so that $d'_{111} = 3d_{14}\cos\theta\sin^2\theta\,\cos\varphi\sin\varphi = \dfrac{3}{2}d_{14}\cos\theta\sin^2\theta\sin 2\varphi$.

3. d'_{111} is maximal for $\varphi = \pi/4 + p\,\pi/2$ (p an integer) and for $\dfrac{\partial d'_{111}}{\partial \theta} = 0$.

$$\frac{\partial d'_{111}}{\partial \theta} = \frac{3}{2} d_{14} \sin 2\varphi \, \sin \theta (2\cos^2 \theta - \sin^2 \theta) = \frac{3}{2} d_{14} \sin 2\varphi \, \sin \theta (3\cos^2 \theta - 1)$$

which vanishes for $\sin \theta = 0$ so that $\theta = 0$ or π but, for these values, d'_{111} is zero, hence minimal.

d'_{111} is maximal for $3\cos^2 \theta - 1 = 0$, hence $\cos \theta = 1/\sqrt{3}$, or $\theta = 54.7°$. This value of θ, along with $\varphi = 45°$, are the spherical coordinates of direction [111] in the cube.

4.   In the orthonormal set of axes $(Ox_1x_2x_3)$ parallel to the basis vectors of the cubic unit cell, the only non-zero component of strain is S_6, with $S_6 = d_{14}E$. The strain tensor can be written as:

$$\begin{pmatrix} 0 & \dfrac{d_{14}E}{2} & 0 \\ \dfrac{d_{14}E}{2} & 0 & 0 \\ 0 & 0 & 0 \end{pmatrix}.$$

The relative volume variation is given by the trace of this tensor, hence $\dfrac{\Delta V}{V} = 0$.

To determine the new shape of the solid, we express the tensor in a coordinate system where Ox'_1 is parallel to [110], Ox'_2 is parallel to $[\bar{1}10]$ and Ox_3 is unchanged. The transformation matrix from the referential $(Ox_1x_2x_3)$ to this frame is:

$$\begin{pmatrix} \dfrac{\sqrt{2}}{2} & \dfrac{\sqrt{2}}{2} & 0 \\ -\dfrac{\sqrt{2}}{2} & \dfrac{\sqrt{2}}{2} & 0 \\ 0 & 0 & 1 \end{pmatrix}.$$

The components S'_{ij} of the strain tensor in the frame $(Ox'_1x'_2x'_3)$ are given by: $S'_{ij} = a_{ik}a_{jl}S_{kl}$. We obtain:

$$S'_{33} = S'_{31} = S'_{32} = S'_{12} = 0,$$

$$S'_{11} = S_{12} = \frac{d_{41}E}{2} = 5 \times 10^{-6},$$

$$S'_{22} = -S_{12} = -\frac{d_{41}E}{2} = -5 \times 10^{-6}.$$

Exercise 15.6

A uniaxial stress has symmetry group $\dfrac{\infty}{m} m$, with the ∞-order axis parallel to the stress-producing force.

The polarization (with symmetry ∞m), if it exists, must be at least as symmetric as the intersection of the crystal's point group and of $\dfrac{\infty}{m}$ m.

1.   (a) The intersection of group 32 and of $\dfrac{\infty}{m}$ m (with the ∞ order axis parallel to the 3-fold axis) is 32. No polarization can exist.

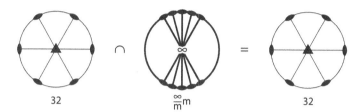

Figure 20.38

(b) The intersection is group 2, where the 2-fold axis is parallel to the uniaxial stress. This group is consistent with a polarization parallel to this 2-fold axis, and thus parallel to the applied stress.

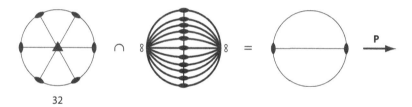

Figure 20.39

(c) The intersection is again group 2, with the 2-fold axis perpendicular to the uniaxial stress. This group is consistent with a polarization parallel to the 2-fold axis, hence perpendicular to the applied stress.

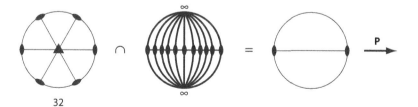

Figure 20.40

(d) A shear stress T_{12} has symmetry group mmm, where two of the mirrors are at an angle of 45° to the axes Ox_1 and Ox_2. The intersection of groups 32 and mmm is empty. A polarization can exist but, if it does, its direction cannot be predicted.

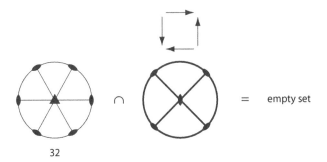

Figure 20.41

2. For ZnO, with point group 6mm:

(a) Uniaxial stress parallel to Ox_3: the intersection of the symmetry groups of the crystal and of the stress is group 6mm. There can exist a polarization parallel to Ox_3.

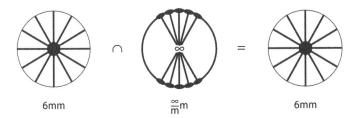

6mm $\frac{\infty}{m}m$ 6mm

Figure 20.42

(b) Uniaxial stress parallel to Ox_1: the intersection of the symmetry groups of the crystal and of the stress is group mm2. There can exist a polarization parallel to Ox_3.

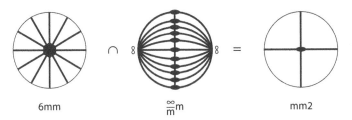

6mm $\frac{\infty}{m}m$ mm2

Figure 20.43

(c) The same result as in (b) is obtained for a uniaxial stress parallel to Ox_2.

Exercise 15.7

1. The intersection of group $\bar{4}2m$ and of the group, mmm, of the shear stress in the configuration of Figure 20.44a is group mm2. There can exist a polarization parallel to the 2-fold axis because mm2 is a subgroup of ∞m if the ∞-order axis is parallel to the 2-fold axis.

2. The intersection of group $\bar{4}2m$ with the group mmm of the shear stress in the configuration of Figure 20.44b leads to group 222, which is not a subgroup of ∞m. No polarization can exist.

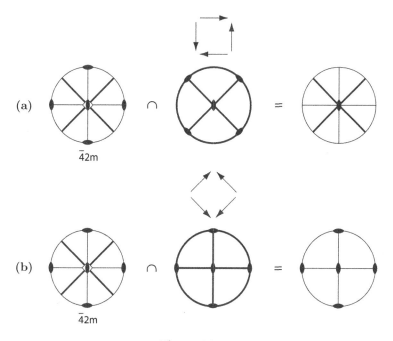

Figure 20.44

20.14. Exercises for Chapter 16

Note: for clarity, the difference between n_e and n_o is very much exaggerated in all figures.

Exercise 16.1

The light ray is parallel to the normal to the index surface. For the ordinary ray, $\delta = 0$. For the extraordinary ray, we draw the normal to the index surface at P_e, the intersection of the index surface and the wave vector starting at O. Let φ and θ be the angles of the normal to the index surface and the wave-vector, respectively, with axis Ox_2.

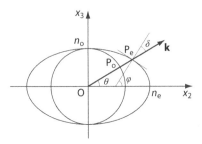

Figure 20.45

We have $\delta = \varphi - \theta$. The equation of the extraordinary index surface is given by:

$$\frac{x_2^2}{n_e^2} + \frac{x_3^2}{n_o^2} = 1.$$

The normal to this ellipsoid, at P_e (x_2, x_3), has components: $\left(\dfrac{2x_2}{n_e^2}, \dfrac{2x_3}{n_o^2} \right)$ and

$$\tan \varphi = \frac{x_3}{n_o^2} \frac{n_e^2}{x_2} = \frac{n_e^2}{n_o^2} \tan \theta = v \tan \theta \quad \text{if we set } v = \frac{n_e^2}{n_o^2},$$

$$\tan \delta = \tan(\varphi - \theta) = \frac{\tan \varphi - \tan \theta}{1 + \tan \varphi \tan \theta} = \frac{(v-1)\tan \theta}{1 + v \tan^2 \theta}.$$

The maximum value of δ is obtained for:

$$\frac{d(\tan \delta)}{d\theta} = \frac{(1 + \tan^2 \theta)(v-1)}{(1 + v \tan^2 \theta)^2} \left[1 - v \tan^2 \theta \right] = 0,$$

so that

i) $(v - 1) = 0 \rightarrow v = 1$. This is the special trivial case where $n_e = n_o$ and $\delta = 0$, the minimum value.

ii) $1 - v \tan^2 \theta = 0 \rightarrow \tan^2 \theta = \dfrac{1}{v} = \dfrac{n_o^2}{n_e^2} \rightarrow \tan \delta = \dfrac{n_e^2 - n_o^2}{2n_o n_e} \approx \dfrac{n_e - n_o}{n_e}$ if $n_e \approx n_o$.

Numerically: for quartz, $\delta = 0.33°$.

Since n_o and n_e are not very different, the angle θ for which δ is maximal is not very different from $45°$.

Exercise 16.2

The ordinary ray goes through the plate without being deviated. The extraordinary ray is parallel to the normal to the extraordinary index surface at P_e. The same approach as in the above exercise shows that the components of the normal to the index surface which are respectively parallel to Ox_2 and Ox_3 are $2x_2/n_e^2$ and $2x_3/n_o^2$. Let φ and θ be the angles respectively enclosed with axis Ox_3 by the normal to the index surface and the wave-vector.

$$\tan \varphi = \frac{x_2}{n_e^2} \frac{n_o^2}{x_3}.$$

Since, at P_e, $x_3 = x_2$, $\tan \varphi = \dfrac{n_o^2}{n_e^2}$.

Numerically: $\varphi = 51.2°$.

The deviation δ between the ordinary and extraordinary rays is $\delta = \varphi - \theta$ $= 6.2°$. In the plate, the extraordinary wave-vector is perpendicular to the plate (Descartes' law at the entrance surface). At the exit of the plate, it remains normal to the plate. The shift $\Delta = e \tan \delta = 0.11$ cm $= 1.1$ mm.

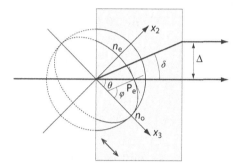

Figure 20.46

Exercise 16.3

1. At the entrance of the prism, the ordinary and extraordinary waves propagate with no deviation of either the wave-vector or the ray direction. The ordinary ray is polarized perpendicular to the optic axis, hence parallel to the edge of the prism. δ_o and δ_e are the deviations of the ordinary and extraordinary rays respectively.

   The incidence angle i on the exit surface is equal to $30°$ for both waves. Let r_o and r_e be the refraction angles for the ordinary and extraordinary waves respectively. In air, the directions of the wave-vector and of the ray coincide. The deviations δ_o and δ_e of the ordinary and extraordinary rays are $\delta_o = r_o - i$ and $\delta_e = r_e - i$.

   Hence: $r_o = \delta_o + i = 56.02°$ and $r_e = \delta_e + i = 48.01°$.

   Applying Descartes' laws yields $n_o = 1.6585$ and $n_e = 1.4865$.

2. At the prism entrance, the ordinary and extraordinary wave-vectors are not deviated (application of Descartes' laws to the wave-vectors). The same applies to the ordinary ray, but the extraordinary ray propagates along the normal at P_e to the index surface. The incidence angle of the ordinary and extraordinary wave-vectors on the exit surface is $i = 30°$,

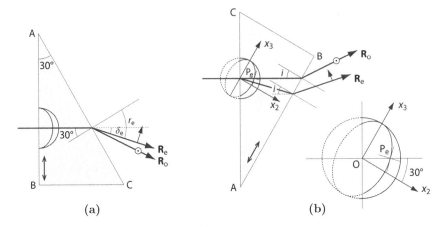

Figure 20.47

and the application of Descartes' laws (to the wave-vectors) provides the direction of the wave-vectors in air at the prism exit. This is also the direction of the rays.

For the ordinary ray, we find $r_o = 56.02°$, and the deviation, as above, is $\delta_o = 26.02°$.

For the extraordinary ray, we must determine the value n of the extraordinary index, equal to the radius vector of the index surface at P_e. The coordinates of P_e are $x_3 = n \sin \theta$ and $x_2 = n \cos \theta$ with $\theta = 30°$.

The following relation must hold:

$$\frac{n^2 \cos^2 \theta}{n_e^2} + \frac{n^2 \sin^2 \theta}{n_o^2} = 1 \quad \text{so that} \quad n^2 = 1/\left(\frac{\cos^2 \theta}{n_e^2} + \frac{\sin^2 \theta}{n_o^2}\right).$$

We find $n = 1.5245$.

Thus: $n \sin i = \sin r_e$ so that $r_e = 49.66°$ and $\delta_e = 19.66°$.

Exercise 16.4

The incident beam goes through the first prism undeviated. At the exit from this prism, the calcite-air interface can give rise to total reflection. For the ordinary ray, total reflection occurs for an angle of incidence on the interface equal to the angle θ of the prisms, such that $n_o \sin \theta > 1$, so that $\theta > 37.09°$. For the extraordinary ray, total reflection occurs for an angle θ such that $n_e \sin \theta > 1$, so that $\theta > 42.29°$.

Thus, for an angle θ such that $37.09° < \theta < 42.29°$, the ordinary ray is totally reflected and the extraordinary ray is transmitted by the prism, with its polarization in the plane defined by the optic axis and the wave-vector.

The polarization is perpendicular to the plane of the figure.

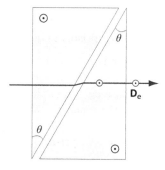

Figure 20.48

Exercise 16.5

1. In a uniaxial material, the electrical induction vector **D** of the ordinary
    wave is perpendicular to the optic axis, hence in this case in the plane of
    the figure. For the extraordinary wave, **D** is perpendicular to the plane of
    the figure.

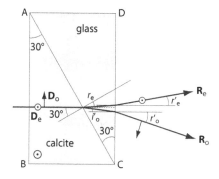

Figure 20.49

2. In the calcite prism, propagation occurs as in an isotropic material with
    an index n_o for the ordinary wave and an index n_e for the extraordinary
    wave. The angle of incidence for each of these waves at the calcite-glass
    interface is $i = 30°$. Applying Descartes' laws to this interface yields:

    i)   for the ordinary wave, $n_o \sin i = n \sin r_o$ where n is the index of the
         glass. We obtain: $r_o = 32.34°$.

    ii) for the extraordinary wave, $n_e \sin i = n \sin r_e$, and $r_e = 28.65°$.

    At the glass–air interface, the angles of incidence are respectively $i_o = 2.34°$
    and $i_e = 1.35°$. Descartes' laws yield, for the refraction angles: $r'_o = 3.63°$
    and $r'_e = 2.09°$, respectively. The deviation between the two outgoing
    beams is $\alpha = r'_o + r'_e = 5.72°$.

Exercise 16.6

The ordinary ray goes through both prisms undeviated.

In the first prism, the wave-vector is parallel to the optic axis, and the index is n_o irrespective of the polarization. In the second prism, the index surface intersects the plane of the figure along two circles, with radii n_o and n_e. The polarization of the ordinary ray is normal to the optic axis, and that of the extraordinary ray is perpendicular to the plane of the figure. For the latter polarization, the index is n_o in the first prism, and n_e in the second one. At the interface between the two prisms, the angle of incidence i is $30°$ and the refraction angle r_e is given by $n_o \sin i = n_e \sin r_e$. We obtain $r_e = 33.9°$. The wave-vector and the ray are parallel. The angle of incidence on the exit surface is $3.9°$ and the angle of refraction is equal to the angle α we are seeking.

$n_e \sin 3.9° = \sin \alpha$. Thus $\alpha = 5.8°$.

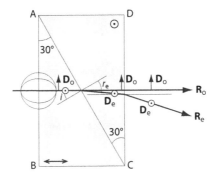

Figure 20.50

20.15. Exercises for Chapter 17

Exercise 17.1

1. For normal incidence, the wave-vectors and the rays for both the ordinary and the extraordinary waves go through the plate without being deviated. The polarization of the ordinary ray is perpendicular to the optic axis, and we choose this direction as the Oy axis. That of the extraordinary ray is parallel to the plane of the figure, and this will be the direction of axis Ox.

2. The phase shift φ between a wave polarized parallel to Oy and a wave polarized parallel to Ox is given by:

$$\varphi = \frac{2\pi(n_o - n_e)e}{\lambda}.$$

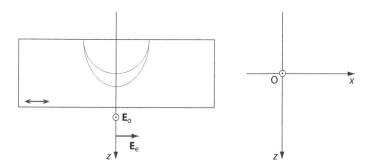

Figure 20.51

Since the crystal is positive, φ is negative for all values of e. The wave is a quarter-wave plate if $\varphi = -\frac{\pi}{2}(1 + 2p)$, with p a positive integer, so that:

$$e = \frac{\lambda}{4(n_e - n_o)}(1 + 2p).$$

Numerically, $e = 16.4 \ \mu\text{m} + p(32.8) \ \mu\text{m}$.

If $e = 16.4 \ \mu\text{m}$, $\varphi = -\pi/2$; if $e = 49.2 \ \mu\text{m}$, $\varphi = -3\pi/2$, which is equivalent to $+\pi/2$.

For a half-wave plate, $\varphi = -\pi(1 + 2p)$, so that:

$$e = \frac{\lambda}{2(n_e - n_o)}(1 + 2p).$$

Numerically, $e = 32.8 \ \mu\text{m} + p(65.6) \ \mu\text{m}$.

3. After the quarter-wave plate, the electric field is, in Jones notation;

$$E_x = \frac{\sqrt{2}}{2}E$$

$$E_y = \frac{\sqrt{2}}{2}E\exp(i\varphi) = -i\frac{\sqrt{2}}{2}E$$

for $\varphi = -\pi/2$ ($e = 16.4 \ \mu\text{m}$). After the plate, the wave is right-circular polarized.

A left-circular polarization would be obtained by rotating the plate through $\pi/2$ around the incident beam. The phase shift would then be $+\pi/2$. Another approach consists in turning the polarizer by $\pi/2$ around the incident beam, so that the polarization is in the second quadrant.

4. The second quarter-wave plate contributes an additional phase shift by $-\pi/2$, so that:

$$\frac{\sqrt{2}}{2}E\begin{pmatrix} 1 \\ -ie^{-i\frac{\pi}{2}} \end{pmatrix} = \frac{\sqrt{2}}{2}E\begin{pmatrix} 1 \\ -1 \end{pmatrix},$$

a wave with linear polarization parallel to the second bisector direction. This result is no surprise since the two quarter-wave plates together form a half-wave plate.

5. Natural light is unpolarized, and it can be described as the sum of two components with perpendicular linear polarizations, with a phase shift that varies randomly in time. A quarter-wave plate contributes an additional phase shift by $\pi/2$ to each of these phase shifts, which vary randomly as before. After the plate, light therefore remains unpolarized.

Exercise 17.2

1. The neutral lines are respectively parallel and perpendicular to the optic axis. In the case of quartz, the ordinary polarization (perpendicular to the optic axis) corresponds to the fast line, and the extraordinary polarization to the slow line (parallel to the optic axis). We choose axes Ox and Oy respectively parallel to the fast and slow neutral lines.

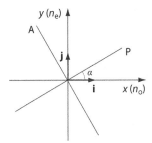

Figure 20.52

2. After the polarizer (Fig. 20.52), the electric field in Jones notation is:
$$\begin{pmatrix} E_0 \cos \alpha \\ E_0 \sin \alpha \end{pmatrix}.$$

After the plate, it becomes:
$$\begin{pmatrix} E_0 \cos \alpha \\ E_0 e^{i\varphi} \sin \alpha \end{pmatrix},$$

where $\varphi = \dfrac{2\pi}{\lambda}(n_e - n_o)e$ is the phase shift introduced by the plate. We can write:
$$\mathbf{E} = E_o \left(\cos \alpha \, \mathbf{i} + e^{i\varphi} \sin \alpha \, \mathbf{j} \right).$$

After the analyzer (angle α with axis Oy), the amplitude E of the electric field is given by:
$$E = E_o \cos \alpha \sin \alpha (e^{i\varphi} - 1).$$

The intensity I is proportional to:
$$|E|^2 = E \times E^* = E_o^2 \frac{\sin^2 2\alpha}{4} (e^{i\varphi} - 1)(e^{-i\varphi} - 1)$$

so that:

$$I = I_0 \sin^2 2\alpha \sin^2 \frac{\varphi}{2}.$$

3. If $\alpha = 45°$, then $I = I_0 \sin^2 \frac{\varphi}{2}$.

i) $I = 0$ corresponds to a phase shift $\varphi = 2p\pi$ (p an integer), *i.e.* either no plate, or a plate with thickness such that the phase shift is a multiple of 2π.

ii) I is maximum for $\varphi = \pi(1 + 2p)$, *i.e.* a half-wave plate:

$$e = \frac{\lambda}{2(n_e - n_0)}(1 + 2p) \quad \text{and} \quad e_{\min} = \frac{\lambda}{2(n_e - n_0)}.$$

4. $e = 0.2$ mm. The extinguished wavelengths are such that:

$$\varphi = 2p\pi = (n_e - n_0)e\frac{2\pi}{\lambda} \quad \text{so that} \quad \lambda = \frac{(n_e - n_0)e}{p}.$$

For $p < 3$ and $p > 5$, the wavelengths are not all in the visible range.

$p = 3$: $\lambda = 0.667\ \mu$m, $\quad p = 4$: $\lambda = 0.5\ \mu$m, $\quad p = 5$: $\lambda = 0.4\ \mu$m.

Exercise 17.3

1. The components of the electric field after the plate are:

$$E_x = E \cos \omega t$$
$$E_y = E \cos(\omega t + \varphi).$$

To determine the curve drawn out by the end of vector **E** as t varies, we eliminate time between the expressions of E_x and E_y. We have:

$$\cos \omega t = \frac{E_x}{E} \quad \text{and} \quad \sin \omega t = -\frac{E_y}{E \sin \varphi} + \frac{\cos \omega t \cos \varphi}{\sin \varphi} = \frac{E_x \cos \varphi - E_y}{E \sin \varphi}$$

so that:

$$\frac{E_x^2}{E^2} + \frac{(E_x \cos \varphi - E_y)^2}{E^2 \sin^2 \varphi} = 1$$

which is the equation of an ellipse. After the plate, the wave is therefore elliptically polarized.

To determine the directions of the axes of the ellipse, we express the equation of the ellipse in an axis system $(OXYZ)$ rotated by an angle θ with respect to the neutral lines of the plate, and we seek the value of θ such that, in this new frame, the coefficient of the crossed term $(E_X E_Y)$ be zero. The calculation yields $\theta = 45°$, whatever the value of φ. This result is valid only because the incident polarization is at $45°$ to the neutral lines of the plate.

The axes of the ellipse are at $45°$ to the neutral lines of the plate, hence respectively parallel and perpendicular to the direction of the incident vibration.

2. If $\theta = 45°$: $E_x = \dfrac{\sqrt{2}}{2}(E_X - E_Y)$ and $E_y = \dfrac{\sqrt{2}}{2}(E_X + E_Y)$, and the equation of the ellipse takes the form:

$$\frac{E_X^2}{2E^2\cos^2\varphi/2} + \frac{E_Y^2}{2E^2\sin^2\varphi/2} = 1.$$

The ratio $\dfrac{B}{A}$ of the ellipse axes has the value: $\dfrac{B}{A} = \tan\dfrac{\varphi}{2}$.

Exercise 17.4

If we rotate the plate, between crossed polarizer and analyzer, around the incident beam, which is perpendicular to the plate, extinction will occur when the neutral lines of the plate are respectively parallel to the polarization directions of the polarizer and the analyzer.

Exercise 17.5

1. The ordinary ray is refracted under an angle r with the normal to the entrance face such that $\sin i = n_o \sin r$, with $i = 19°$. We obtain $r = 11.3°$. The incident beam is thus deviated by 7.7°. Since the incident beam is at an angle of 22° to AC′, the ordinary wave-vector, which coincides with the ordinary ray, encloses with AC′ an angle of 14.3°, and the angle of incidence on AC′ is equal to 75.7°. The limiting angle i_ℓ for refraction of the ordinary ray on Canada balsam is given by $n_o \sin i_\ell = N = 1.55$, so that $i_\ell = 69.2°$. Thus the ordinary ray is totally reflected on the Canada balsam.

2. Figure 20.53 shows how to draw the extraordinary wave-vector $\mathbf{OM_2}$. On this figure, $\mathrm{OH} = \sin i = 0.326$. If $\mathbf{m}$ is a unit vector on the entrance face, $\mathbf{OM_2} \cdot \mathbf{m} = \mathrm{OH}$.

   $\mathbf{m} = (\sqrt{2}/2)(-\mathbf{i} + \mathbf{j})$ where $\mathbf{i}$ and $\mathbf{j}$ are unit vectors along Ox and Oy, respectively parallel to the axes of the index surface (Ox is parallel to the optic axis). Setting $\mathbf{OM_2} = x\mathbf{i} + y\mathbf{j}$, we obtain $\mathbf{OM_2} \cdot \mathbf{m} = \mathrm{OH} = (\sqrt{2}/2)(-x + y) = 0.326$ and $-x + y = 0.460$.

   On the other hand, M_2 is on the extraordinary index surface, the equation of which is:

   $$\frac{x^2}{n_o^2} + \frac{y^2}{n_e^2} = 1.$$

   Solving the two equations for x and y, we obtain $x = 0.828$ and $y = 1.288$. The extraordinary index n is equal to $(x^2 + y^2)^{1/2}$, so that $n = 1.53$, which is less than N. Thus the extraordinary ray propagates in the Canada balsam, and then in the second part of the prism practically without being shifted because the balsam layer is very thin. The extraordinary vibration is in the plane defined by the wave-vector and the optic axis, hence in the plane of Figure 17.6b. At the crystal exit, the beam is thus polarized in the plane defined by the incident ray and the optic axis.

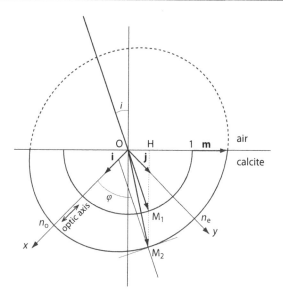

Figure 20.53

3. The direction of the extraordinary ray in the Nicol prism is that of the normal to the index surface at point M_2. The index surface has equation:

$$f(x,y) = \frac{x^2}{n_o^2} + \frac{y^2}{n_e^2} - 1 = 0$$

and its normal has components $\dfrac{\partial f}{\partial x}$ and $\dfrac{\partial f}{\partial y}$, hence $\dfrac{2x}{n_o^2}$ and $\dfrac{2y}{n_e^2}$. This normal is at an angle φ to Ox, such that:

$$\tan \varphi = \frac{y\, n_o^2}{x\, n_e^2} = 1.936 \qquad \text{and} \qquad \varphi = 62.7°.$$

The angle between the ray and the normal to the entrance face is equal to $17.7°$, a deviation of only $1.3°$ with respect to the incident ray.

Exercise 17.6

The condition being sought is $n_e(2\omega) = n_o(\omega)$. This is equivalent to determining the intersection of the ordinary sheet of the index surface for angular frequency ω with the extraordinary sheet for angular frequency 2ω.

The equation of the ordinary sheet for ω is: $x^2 + y^2 = n_o^2$.

The equation of the extraordinary sheet for 2ω is: $\dfrac{x^2}{n_o'^2} + \dfrac{y^2}{n_e'^2} = 1$.

Point P belongs to the ordinary sheet, so that: $x_P = n_o \cos\theta$ and $y_P = n_o \sin\theta$ for ω, and it must verify the equation of the extraordinary sheet for 2ω, so that:

$$\frac{n_o^2 \cos^2\theta}{n_o'^2} + \frac{n_o^2 \sin^2\theta}{n_e'^2} = 1.$$

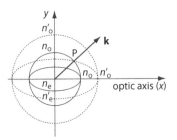

Figure 20.54

The required angle θ is thus defined by:

$$\sin^2 \theta = \frac{\dfrac{1}{n_o^2} - \dfrac{1}{n_o'^2}}{\dfrac{1}{n_e'^2} - \dfrac{1}{n_o'^2}}.$$

Numerically, we get: $\sin^2 \theta = 0.584$, so that $\theta = 49.8°$.

Exercise 17.7

1. We switch from the Jones vector to the real form of the wave by multiplying both components by $\exp -i\omega t$ and taking their real parts:

$$\begin{pmatrix} a \\ ib \end{pmatrix} \rightarrow \begin{pmatrix} a \exp -i\omega t \\ b \exp -i(\omega t - \pi/2) \end{pmatrix} \rightarrow \begin{pmatrix} a \cos \omega t \\ b \sin \omega t \end{pmatrix}.$$

The end of the electric field vector $\mathbf{E}$ describes in the counterclockwise direction an ellipse with major and minor axes Ox and Oy (Fig. 20.55-1).

We have:
$$s_3 = 2\,\Im(E_x E_y^*) = 2\,\Im(-iab) = -2ab,$$
$$s_0 = a^2 + b^2,$$
$$P_3 = -2ab/(a^2 + b^2),$$

so that the polarization fraction is negative, corresponding to more photons with left-circular polarization.

2. We obtain an ellipse with major and minor axes parallel to Ox and Oy, but the extreme values on the axes are interchanged. The polarization rate is the same as for wave $\begin{pmatrix} a \\ ib \end{pmatrix}$ (Fig. 20.55-2).

3. We calculate the Stokes parameters for the superposition of the two waves:

$$< E_x E_x^* >= (a^2 + b^2)/2,$$
$$< E_y E_y^* >= (b^2 + a^2)/2,$$
$$< E_x E_y^* >= (-iab - iab)/2 = -iab.$$
$$s_0 = a^2 + b^2, \; s_1 = 0, \; s_2 = 0 \text{ and } s_3 = -2ab.$$

The circular polarization rate is $s_3 = -2ab/(a^2 + b^2)$. The degree of polarization is no more equal to 1 as for the waves considered separately, and

$$\tau = \frac{\sqrt{s_1^2 + s_2^2 + s_3^2}}{s_0} = \frac{2ab}{(a^2 + b^2)}.$$

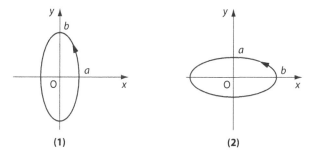

Figure 20.55

20.16. Exercises for Chapter 19

Exercise 19.1

1.
$$\begin{pmatrix} \Delta\eta_1 \\ \Delta\eta_2 \\ \Delta\eta_3 \\ \Delta\eta_4 \\ \Delta\eta_5 \\ \Delta\eta_6 \end{pmatrix} = \begin{pmatrix} & -r_{22} & r_{13} \\ & r_{22} & r_{13} \\ & & r_{33} \\ & r_{51} & \\ r_{51} & & \\ -2r_{22} & & \end{pmatrix} \begin{pmatrix} 0 \\ 0 \\ E \end{pmatrix} = \begin{pmatrix} r_{13}E \\ r_{13}E \\ r_{33}E \\ 0 \\ 0 \\ 0 \end{pmatrix}$$

and $\{\eta_{ij}\} = \begin{pmatrix} \eta_0 + \Delta\eta_1 & & \\ & \eta_0 + \Delta\eta_2 & \\ & & \eta_e + \Delta\eta_3 \end{pmatrix}$.

with $\Delta\eta_1 = \Delta\eta_2 = r_{13}E$ and $\Delta\eta_3 = r_{33}E$.

The material remains uniaxial, and the direction of the optic axis is retained. This result can be retrieved directly by applying Curie's principle. The intersection of the point group of the crystal, 3m, with the symmetry group of the electric field (a polar vector), ∞m where the ∞-order axis is parallel to the field, is here 3m and therefore the material remains uniaxial.

One calculates the new ordinary and extraordinary indices n_o' and n_e'.

$$\Delta\eta_i = \Delta\left(\frac{1}{n_i^2}\right) = -2\frac{\Delta n}{n_i^3} \qquad \text{and} \qquad \Delta n = -\frac{n_i^3 \Delta\eta_i}{2}$$

and

$$n_o' = n_o - \frac{n_o^3}{2} r_{13}E \qquad n_e' = n_e - \frac{n_e^3}{2} r_{33}E.$$

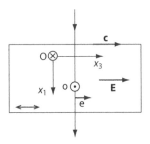

Figure 20.56

2. The ordinary wave is polarized perpendicular to the optic axis, hence in this case parallel to Ox_2. The extraordinary wave is polarized along Ox_3. The phase shift between the component of the incident wave parallel to Ox_2 and that parallel to Ox_3 is:

$$\varphi = \frac{2\pi E}{\lambda}(n'_o - n'_e) = \frac{2\pi E}{\lambda}[(n_o - n_e) - \frac{E}{2}(r_{13}n_o^3 - r_{33}n_e^3)].$$

3. We call $(Ox_1x'_2x'_3)$ the orthonormal coordinate system deduced from $(Ox_1x_2x_3)$ through a rotation by 90° around the incident beam, $i.e.$ the coordinate system linked to the crystallographic axes of the second crystal. In this frame, we obtain for the variations in the ordinary and extraordinary indices the same result as in (2), but with E changed into $-E$, so that:

$$n'_o = n_o + \frac{n_o^3}{2}r_{13}E \qquad n'_e = n_e + \frac{n_e^3}{2}r_{33}E.$$

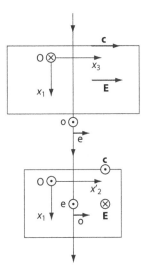

Figure 20.57

For the second crystal, the component of the wave parallel to Ox_2 is the extraordinary wave, and that parallel to Ox_3 is the ordinary wave. The phase shift φ' is:

$$\varphi' = \frac{2\pi e}{\lambda}(n'_e - n'_o) = \frac{2\pi e}{\lambda}[(n_e - n_o) - \frac{E}{2}(r_{13}n_o^3 - r_{33}n_e^3)].$$

The total phase shift after the two crystal is $\varphi + \varphi'$ and

$$\varphi + \varphi' = \frac{2\pi e}{\lambda}E(r_{33}n_e^3 - r_{13}n_o^3) = \frac{2\pi e}{\lambda}\Delta n.$$

where $\Delta n = E(r_{33}n_e^3 - r_{13}n_o^3)$.

The main part of the phase shift, that independent of E, has thus been eliminated, and the phase shift is proportional to E.

$$\varphi + \varphi' = \frac{2\pi e}{\lambda}E(r_{33}n_e^3 - r_{13}n_o^3) = \frac{2\pi e}{\lambda}\Delta n = \pi \ ; e = \frac{\lambda}{2\Delta n}.$$

Numerically: $\Delta n = 1.33 \times 10^{-5}$ and $e = 23.8$ mm.

Exercise 19.2

1.
$$\begin{pmatrix} \Delta\eta_1 \\ \Delta\eta_2 \\ \Delta\eta_3 \\ \Delta\eta_4 \\ \Delta\eta_5 \\ \Delta\eta_6 \end{pmatrix} = \frac{E}{\sqrt{3}}\begin{pmatrix} & & \\ r & & \\ & r & \\ & & r \end{pmatrix}\begin{pmatrix} 1 \\ 1 \\ 1 \end{pmatrix} = \begin{pmatrix} 0 \\ 0 \\ 0 \\ \alpha \\ \alpha \\ \alpha \end{pmatrix} \text{ with } \alpha = \frac{rE}{\sqrt{3}}$$

and $\{\eta_{ij}\} = \begin{pmatrix} \eta_0 & \alpha & \alpha \\ \alpha & \eta_0 & \alpha \\ \alpha & \alpha & \eta_0 \end{pmatrix}$.

The eigenvalues of $[\eta]$ are obtained by solving the equation: $\det(\{\eta_{ij}\} - \lambda E) = 0$. The same calculation was performed in Exercise 14.2, question (2). We obtain three eigenvalues: $\lambda_1 = \eta_0 + 2\alpha$ and $\lambda_2 = \lambda_3 = \eta_0 - \alpha$. The direction of the eigenvector associated with eigenvalue $\eta_0 + 2\alpha$ is [111]. All the vectors in the plane perpendicular to [111], hence (111), are eigenvectors for eigenvalue $\eta_0 - \alpha$. The optic axis is parallel to [111].

Since $\Delta\left(\frac{1}{n_i^2}\right) = -2\frac{\Delta n}{n_i^3} = \Delta\eta_i$, we obtain:

- for the ordinary index: $\Delta\left(\frac{1}{n^2}\right) = -\alpha$ and $n_o = n + n^3\frac{rE}{2\sqrt{3}}$,

- for the extraordinary index: $\Delta\left(\frac{1}{n^2}\right) = 2\alpha$ and $n_e = n - n^3\frac{rE}{\sqrt{3}}$.

Numerically: $n_o = 2.84 + 6 \times 10^{-6}$ and $n_e = 2.84 - 1.2 \times 10^{-5}$.

2. The geometry is the classic geometry of phase-shifting plates, *viz.* optical axis in the plane of the plate. The neutral lines are the direction of the

optic axis (extraordinary index) and the perpendicular direction in the plane of the plate (ordinary index).

$$n_e - n_o = -3n^3 \frac{rE}{2\sqrt{3}} = -n^3 \frac{rE\sqrt{3}}{2}.$$

The path difference between the two waves is given by: $\delta = (n_e - n_o)L$. For a half-wave plate, $\delta = \pm\lambda/2$, so that

$$-n^3 \frac{rE\sqrt{3}}{2} L = \frac{\lambda}{2} \quad \text{and} \quad E = \frac{\lambda}{\sqrt{3}n^3 rL}.$$

Numerically: $E = 7 \times 10^4$ V m^{-1}.

Exercise 19.3

The symmetry group of an electric field is ∞m, the ∞-order axis being parallel to the electric field. The intersection of the point group of the crystal with the electric field is mm2. The crystal has become biaxial. The axes of the ellipsoid representing the permittivity tensor are parallel to the mirrors and at 45° to the 2-fold axes.

The intersection of group 4mm and ∞m (∞-order axis parallel to the 4-fold axis) is 4mm. The crystal remains uniaxial.

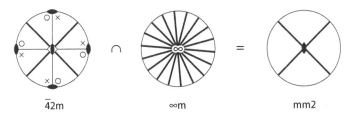

Figure 20.58

Exercise 19.4

1. In the orthonormal set of axes where Ox_3 is parallel to [001] and Ox_1 is parallel to [1$\bar{1}$0], the applied stress has expression:

$$\begin{pmatrix} T & 0 & 0 \\ 0 & 0 & 0 \\ 0 & 0 & 0 \end{pmatrix}.$$

In a coordinate system rotated by 45° around Ox_3 (the frame linked to the crystallographic axes), it becomes:

$$\begin{pmatrix} T/2 & -T/2 & 0 \\ -T/2 & T/2 & 0 \\ 0 & 0 & 0 \end{pmatrix}.$$

We obtain:

$$\begin{pmatrix} \Delta\eta_1 \\ \Delta\eta_2 \\ \Delta\eta_3 \\ \Delta\eta_4 \\ \Delta\eta_5 \\ \Delta\eta_6 \end{pmatrix} = \begin{pmatrix} \pi_{11} & \pi_{12} & \pi_{12} & 0 & 0 & 0 \\ \pi_{12} & \pi_{11} & \pi_{12} & 0 & 0 & 0 \\ \pi_{12} & \pi_{12} & \pi_{11} & 0 & 0 & 0 \\ 0 & 0 & 0 & \pi_{44} & 0 & 0 \\ 0 & 0 & 0 & 0 & \pi_{44} & 0 \\ 0 & 0 & 0 & 0 & 0 & \pi_{44} \end{pmatrix} \begin{pmatrix} T/2 \\ T/2 \\ 0 \\ 0 \\ 0 \\ -T/2 \end{pmatrix}$$

$$= \frac{1}{2} \begin{pmatrix} (\pi_{11}+\pi_{12})T \\ (\pi_{11}+\pi_{12})T \\ 2\pi_{12}T \\ 0 \\ 0 \\ -\pi_{44}T \end{pmatrix}.$$

Going over to the two-subscript notation for $[\Delta\eta]$, we obtain for $[\eta]$:

$$\begin{pmatrix} \eta_o + (\pi_{11}+\pi_{12})T/2 & -\pi_{44}T/2 & \\ -\pi_{44}T/2 & \eta_o + (\pi_{11}+\pi_{12})T/2 & \\ & & \eta_o + \pi_{12}T \end{pmatrix}.$$

In the initial axis system, we obtain for $[\eta]$:

$$\begin{pmatrix} \eta_o + (\pi_{11}+\pi_{12}+\pi_{44})T/2 & & \\ & \eta_o + (\pi_{11}+\pi_{12}-\pi_{44})T/2 & \\ & & \eta_o + \pi_{12}T \end{pmatrix}.$$

The eigenvector associated to eigenvalue $\eta_o + (\pi_{11} + \pi_{12} - \pi_{44})T/2$ is along Ox_2, *i.e.* [110], and that associated to the eigenvalue $\eta_o + (\pi_{11} + \pi_{12} + \pi_{44})T/2$ is parallel to Ox_1, *i.e.* to $[1\bar{1}0]$.

The values of the principal indices are:

$$n_1 = n_o - \frac{n_o^3}{4}(\pi_{11} + \pi_{12} + \pi_{44})T$$

$$n_2 = n_o - \frac{n_o^3}{4}(\pi_{11} + \pi_{12} - \pi_{44})T$$

$$n_3 = n_o - \frac{n_o^3}{2}\pi_{12}T.$$

Numerically: $n_1 = 1.54407$, $n_2 = 1.54375$ and $n_3 = 1.54396$.

2. In a cubic crystal, direction [111] is in the plane defined by directions [110] and [001], which are two of the principal axes of the index ellipsoid. The intersection of this ellipsoid with a plane perpendicular to [111] is an ellipse with one axis parallel to Ox_1. The index for the wave with electrical induction vector parallel to this direction is therefore n_1. The second axis of the ellipse is in plane (Ox_2, Ox_3). It is perpendicular to the direction [111] of the wave-vector. On the figure below, OP is the semiaxis we are seeking, and its length is equal to n.

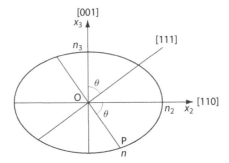

Figure 20.59

Let x_2 and x_3 be the coordinates of P such that: $x_2 = n\cos\theta$ and $x_3 = -n\sin\theta$. The equation of the ellipse going through P is:

$$\frac{x_2^2}{n_2^2} + \frac{x_3^2}{n_3^2} = 1.$$

We obtain:

$$n = \frac{1}{\sqrt{\left(\dfrac{\cos^2\theta}{n_2^2} + \dfrac{\sin^2\theta}{n_3^2}\right)}}$$

with $\theta = \arccos\dfrac{1}{\sqrt{3}} = 54.7°$, so that $n = 1.54402$.

Exercise 19.5

We apply relation $\Delta\eta_\alpha = \pi_{\alpha\beta}T_\beta$ with $T_1 = T_2 = T_3 = -p$ and $T_4 = T_5 = T_6 = 0$ by using the form of tensor $[\pi]$ for the cubic groups we are considering.

We obtain:

$$\begin{pmatrix} \Delta\eta_1 \\ \Delta\eta_2 \\ \Delta\eta_3 \\ \Delta\eta_4 \\ \Delta\eta_5 \\ \Delta\eta_6 \end{pmatrix} = \begin{pmatrix} \pi_{11} & \pi_{12} & \pi_{12} & 0 & 0 & 0 \\ \pi_{12} & \pi_{11} & \pi_{12} & 0 & 0 & 0 \\ \pi_{12} & \pi_{12} & \pi_{11} & 0 & 0 & 0 \\ 0 & 0 & 0 & \pi_{44} & 0 & 0 \\ 0 & 0 & 0 & 0 & \pi_{44} & 0 \\ 0 & 0 & 0 & 0 & 0 & \pi_{44} \end{pmatrix} \begin{pmatrix} -p \\ -p \\ -p \\ 0 \\ 0 \\ 0 \end{pmatrix}$$

$$= \begin{pmatrix} -(\pi_{11}+2\pi_{12})p \\ -(\pi_{11}+2\pi_{12})p \\ -(\pi_{11}+2\pi_{12})p \\ 0 \\ 0 \\ 0 \end{pmatrix},$$

so that $\{\Delta\eta_{ij}\} = \begin{pmatrix} -(\pi_{11}+2\pi_{12})p & & \\ & -(\pi_{11}+2\pi_{12})p & \\ & & -(\pi_{11}+2\pi_{12})p \end{pmatrix}.$

The crystal remains isotropic, but its index changes since:

$$\Delta\eta = \Delta\left(\frac{1}{n^2}\right) = \frac{-2\Delta n}{n^3} = -(\pi_{11} + 2\pi_{12})p.$$

Finally:

$$\Delta n = \frac{n^3}{2}(\pi_{11} + 2\pi_{12})p.$$

We now express the elasto-optical coefficients $\pi_{\alpha\beta}$ as a function of the coefficients $p_{\alpha\beta}$.

$\Delta\eta_\alpha = p_{\alpha\beta}S_\beta$ and, since $S_\beta = s_{\beta\gamma}T_\gamma$ we obtain: $\Delta\eta_\alpha = p_{\alpha\beta}s_{\beta\gamma}T_\gamma = \pi_{\alpha\gamma}T_\gamma$.
Thus $\pi_{\alpha\gamma} = p_{\alpha\beta}s_{\beta\gamma}$ and we obtain:

$$\pi_{11} = p_{11}s_{11} + p_{12}s_{21} + p_{13}s_{31} = p_{11}s_{11} + 2p_{12}s_{12},$$
$$\pi_{12} = p_{11}s_{12} + p_{12}s_{22} + p_{13}s_{32} = (p_{11} + p_{12})s_{12} + p_{12}s_{11},$$
$$\pi_{11} + 2\pi_{12} = (p_{11} + 2p_{12})(s_{11} + 2s_{12}) = \frac{\chi}{3}(p_{11} + 2p_{12})$$

since the bulk compressibility χ of a cubic crystal is equal to $3(s_{11} + 2s_{12})$ (Sect. 13.8.1). Finally:

$$\Delta n = \frac{n^3}{6}\chi(p_{11} + 2p_{12})p.$$

Exercise 19.6

1. The displacement **u** associated to the acoustic wave has, in the orthonormal system with axes parallel to the basis vectors of the cubic unit cell, components:

$$u_1 = 0, \qquad u_2 = A\cos(\Omega t - Kx_3), \qquad u_3 = 0$$

with $\Omega = 2\pi f$ if f is the frequency of the acoustic wave.

$$S_{ij} = \frac{1}{2}\left(\frac{\partial u_i}{\partial x_j} + \frac{\partial u_j}{\partial x_i}\right)$$

and, here, only $S_{23} = S_{32}$ are non-zero.

$$S_{23} = S_{32} = \frac{AK}{2}\sin(\Omega t - Kx_3) \text{ and } S_4 = 2S_{23} = AK\sin(\Omega t - Kx_3).$$

2. (a) We use relation $\Delta\eta_n = p_{nm}S_m$.

$$\begin{pmatrix} \Delta\eta_1 \\ \Delta\eta_2 \\ \Delta\eta_3 \\ \Delta\eta_4 \\ \Delta\eta_5 \\ \Delta\eta_6 \end{pmatrix} = \begin{pmatrix} p_{11} & p_{12} & p_{12} & 0 & 0 & 0 \\ p_{12} & p_{11} & p_{12} & 0 & 0 & 0 \\ p_{12} & p_{12} & p_{11} & 0 & 0 & 0 \\ 0 & 0 & 0 & p_{44} & 0 & 0 \\ 0 & 0 & 0 & 0 & p_{44} & 0 \\ 0 & 0 & 0 & 0 & 0 & p_{44} \end{pmatrix} \begin{pmatrix} 0 \\ 0 \\ 0 \\ S_4 \\ 0 \\ 0 \end{pmatrix} = \begin{pmatrix} 0 \\ 0 \\ 0 \\ p_{44}S_4 \\ 0 \\ 0 \end{pmatrix},$$

and $\{\eta_{ij}\} = \begin{pmatrix} \eta_o & & \\ & \eta_o & \Delta\eta_4 \\ & \Delta\eta_4 & \eta_o \end{pmatrix}$ avec $\Delta\eta_4 = p_{44}S_4$.

In a coordinate system rotated by $45°$ around Ox_1 with respect to the initial frame, this tensor is expressed as:

$$\{\eta_{ij}\} = \begin{pmatrix} \eta_o & & \\ & \eta_o + \Delta\eta_4 & \\ & & \eta_o - \Delta\eta_4 \end{pmatrix}.$$

The three indices are:

$$n_1 = n_o,$$

$$n_2' = n_o - \frac{n_o^3}{2}p_{44}AK\sin(\Omega t - Kx_3),$$

$$n_3' = n_o + \frac{n_o^3}{2}p_{44}AK\sin(\Omega t - Kx_3).$$

The system becomes biaxial, and the principal indices n_2' and n_3' vary periodically along direction Ox_3.

(b) Let D be the period of the refraction grating. We have $D = \dfrac{2\pi}{K}$ with $K = \dfrac{\Omega}{v_t} = \dfrac{2\pi f}{v_t}$, so that $D = \dfrac{v_t}{f}$.

Numerically, $D = 7.16 \ \mu$m.

General references

Crystallography and symmetry

Th. Hahn (ed.), *International Tables for Crystallography*, vol. A, *Space-group Symmetry*, 5th edn. (Kluwer Academic Publishers, Dordrecht, 2005)

Note: a paperback *Brief Teaching Edition*, useful to practice the use of these Tables (but not as a complete reference), is available at a much lower price. Theo Hahn (ed.), Kluwer Academic Publishers, Dordrecht, 2005.

H. Curien, *Les groupes en cristallographie*, in *Théorie des groupes en physique classique et quantique*, vol. 2, ed. by Th. Kahan (Dunod, Paris, 1971)

C. Giacovazzo (ed.), *Fundamentals of Crystallography* (International Union of Crystallography, Oxford University Press, 1992)

C. Hammond, *Introduction to Crystallography* (Oxford University Press, 1992)

D. Schwarzenbach and G. Chapuis, *Cristallographie* (Presses Polytechniques et Universitaires Romandes, Lausanne, 2008)

J. Sivardière, *Description de la symétrie. Des groupes de symétrie aux structures fractales* (EDP Sciences, Les Ulis, 2004)

J. Sivardière, *Symétrie et propriétés physiques. Du principe de Curie aux brisures de symétrie* (EDP Sciences, Les Ulis, 2004)

B.K. Vainshtein, *Fundamentals of Crystals*, vol. 1, 2, 3, 4 (Springer Verlag, Berlin, 1994)

M. Van Meerssche and J. Feneau-Dupont, *Introduction à la cristallographie et à la chimie structurale* (Peeters, Louvain-la-neuve, 1984)

Tensors and physical properties of crystals

A. Authier (ed.), *International Tables for Crystallography*, vol. D, *Physical Properties of Crystals* (Kluwer Academic Publishers, Dordrecht, 2003)

S. Bhagavantam, *Crystal Symmetry and Physical Properties* (Academic Press, London, 1966)

C. Malgrange et al., *Symmetry and Physical Properties of Crystals*, DOI 10.1007/978-94-017-8993-6

L.D. Landau, E.M. Lifshitz and L.P. Pitaevskii, *Course of Theoretical Physics*, vol. 8, *Electrodynamics of Continuous Media*, 2nd edn. (Elsevier, Oxford, 1984)

D.R. Lovett, *Tensor Properties of Crystals* (Adam Hilger, Bristol, 1990)

R.E. Newnham, *Properties of Materials. Anisotropy, Symmetry, Structure* (Oxford University Press, 2005)

J.F. Nye, *Physical Properties of Crystals: Their Representation by Tensors and Matrices* (Oxford University Press, Oxford, 1985)

D.E. Sands, *Vectors and Tensors in Crystallography* (Dover Publications, Mineola, New York, 1995)

L. Schwartz, *Les tenseurs* (Hermann, Paris, 1975)

Y. Sirotine and M. Chaskolskaia, *Fondements de la physique des cristaux* (Mir, Moscow, 1984)

W.A. Wooster, *Tensors and Group Theory for the Physical Properties of Crystals* (Clarendon Press, Oxford, 1973)

Continuum mechanics, elasticity and thermodynamics

D. Calecki, *Physique des milieux continus*, vol. 1, *Mécanique et thermodynamique* (Hermann, Paris, 2007)

H.B. Callen, *Thermodynamics and an Introduction to Thermostatistics*, 2nd edn. (Wiley, New York, 1985) (Do not use printings earlier than the sixth)

P. Germain and P. Muller, *Introduction à la mécanique des milieux continus* (Masson, Paris, 1995)

L.D. Landau and E.M. Lifshitz, *Course of Theoretical Physics*, vol. 7, *Theory of Elasticity*, 3rd edn. (Elsevier, Oxford, 1986)

J. Salençon, *Mécanique des milieux continus*, vol. 1: *Concepts généraux*, vol. 2: *Elasticité – Milieux curvilignes* (Ellipses, Paris, 1988)

I.S. Sokolnikoff, *Mathematical Theory of Elasticity* (Mc Graw Hill, New York, 1956)

Optics and crystal optics

M. Born and E. Wolf, *Principles of Optics*, 7th edn. (Cambridge University Press, Cambridge, 1999)

G.R. Fowles, *Introduction to Modern Optics* (Dover Publications, New York, 1989)

A. Yariv and P. Yeh, *Optical Waves in Crystals* (Wiley Interscience Publication, Wiley, New York, 1984)

Sundry topics

B.A. Auld, *Acoustic Fields and Waves in Solids*, 2nd edn. (R.E. Krieger Pub. Co, Malabar, Florida, 1990)

P.G. de Gennes and J. Prost, *The Physics of Liquid Crystals* (Oxford University Press, Oxford, 1993)

S.R. Elliot, *Physics of Amorphous Materials*, 2nd edn. (Longman Scientific and Technical, Harlow, Essex, 1990)

C. Janot, *Quasicrystals: a Primer* (Clarendon Press, Oxford University Press, 1992)

C. Janot and J.M. Dubois, *Les quasicristaux, matière à paradoxes* (EDP Sciences, Les Ulis, 1998)

T. Janssen, G. Chapuis and M. De Boissieu, *Aperiodic Crystals: from Modulated Phases to Quasicrystals* (Oxford University Press, Oxford, 2007)

I.C. Khoo, *Liquid Crystals*, 2nd edn. (Wiley, Hoboken, 2007)

M. Prutton, *Introduction to Surface Physics* (Clarendon Press, Oxford, 1995)

D. Royer and E. Dieulesaint, *Ondes élastiques dans les solides*, vol. 1: *Propagation libre et guidée*, vol. 2: *Génération, interaction acousto-optique, applications* (Masson, Paris, 1996)

D.P. Woodruff and T.A. Delchar, *Modern Techniques of Surface Science*, 2nd edn. (Cambridge University Press, 1994)

J. Zelenka, *Piezoelectric Resonators and their Applications* (Elsevier, Amsterdam, 1986)

Index

© Springer Science+Business Media Dordrecht 2014
C. Malgrange et al., *Symmetry and Physical Properties of Crystals*,
DOI 10.1007/978-94-017-8993-6

The authors

Cécile Malgrange is Emeritus Professor at Pierre & Marie Curie University, in Paris. After her French doctorate (Thèse d'Etat) obtained in 1969, she taught at the university of Paris and became a Professor at Paris Diderot University in Paris in 1978. She lectured on most fields of basic physics (mechanics, statistical physics, optics, quantum mechanics) and gave a course on crystallography and the physical properties of crystals. In 1983, she took part in the creation of a curriculum on physics and applications. Her research work was performed mostly in the Mineralogy and Crystallography Laboratory (now Institut de Minéralogie, de Physique des Matériaux et de Cosmochimie – IMPMC). It focused on X-ray optics based on Bragg diffraction, for which she is a world-renowned specialist (propagation in distorted crystals and development of X-ray optical elements, in particular phase plates, now in use on many synchrotron beam lines). She also worked on ferroelectric crystals and epilayers on semiconductors, using mainly X-ray diffraction techniques.

Christian Ricolleau is the leader of the Advanced Transmission Electron Microscopy and Nanostructures (Me^-ANS) group at the Materials and Quantum Phenomena laboratory (Paris Diderot University and CNRS). He is a physicist in materials science with much experience in electron microscopy (mainly in high resolution, diffraction and energy filtered TEM techniques). He has extensive knowledge in the growth of metallic nanoparticles on oxide substrates since it has been, in addition to the new development in advanced TEM techniques, his main field of research since 1999. After his PhD thesis on the wetting phenomenon of antiphase boundaries in binary alloys, he got the position of Assistant Professor in the Mineralogy Crystallography Lab in Paris Diderot University. In 2000, he participated in the creation of the Materials and Quantum Phenomena laboratory and, in 2004, he received the position of Professor at the Paris Diderot University. Since 2008, he has been the director of the Nanoalloys Research Network of the CNRS.

© Springer Science+Business Media Dordrecht 2014
C. Malgrange et al., *Symmetry and Physical Properties of Crystals*,
DOI 10.1007/978-94-017-8993-6

Michel Schlenker is Emeritus Professor at Grenoble-Alpes University – Grenoble Institute of Technology. He obtained his French doctorate (Thèse d'Etat) in physics in Grenoble, France, in 1970. Starting 1961, he taught, mostly basic physics, materials, crystallography and diffraction methods, at Grenoble University (now Université Joseph Fourier), then as a Professor of Physics at Grenoble Institute of Technology. His research was performed at the Massachusetts Institute of Technology (USA), IBM's San Jose (CA) Laboratory, Institut Laue Langevin (the European neutron source), and the European Synchrotron Radiation Facility (ESRF), with the Louis Néel magnetism laboratory of CNRS (now Institut Néel), Grenoble as his permanent base. His work centered on non-conventional, diffraction-based imaging approaches, using in particular neutrons and synchrotron radiation for investigations of defects and domains, specially in antiferromagnets. He was co-editor, then editor, of the Journal of Applied Crystallography, one of the journals of the International Union of Crystallography.

Printed in the United States
By Bookmasters